Colloquium of Advanced Materials (CAMS2016)

IOP Conference Series: Materials Science and Engineering
Volume 165

Johor, Malaysia
18 – 19 December 2016

Editors:

Al Emran Ismail
Abdul Latif Mohd Tobi
Waluyo Adi Siswanto

ISBN: 978-1-5108-3527-6
ISSN: 1757-8981

Printed from e-media with permission by:

Curran Associates, Inc.
57 Morehouse Lane
Red Hook, NY 12571

Some format issues inherent in the e-media version may also appear in this print version.

Printed by Curran Associates, Inc. (2017)

For permission requests, please contact the Institute of Physics
at the address below.

Institute of Physics
Dirac House, Temple Back
Bristol BS1 6BE UK

Phone: 44 1 17 929 7481
Fax: 44 1 17 920 0979

techtracking@iop.org

Additional copies of this publication are available from:

Curran Associates, Inc.
57 Morehouse Lane
Red Hook, NY 12571 USA
Phone: 845-758-0400
Fax: 845-758-2633
Email: curran@proceedings.com
Web: www.proceedings.com

Table of contents

Volume 165

Colloquium of Advanced Mechanics (CAMS2016)
18–19 December 2016, Johor, Malaysia

Accepted papers received: 16 December 2016
Published online: 12 January 2017

Preface

COLORED

SIMReg

NOVIA

SPEAR

ADCARe

IOP Conf. Series: Materials Science and Engineering **165** (2017) 011001 doi:10.1088/1757-899X/165/1/011001

Preface

CAMS2016 – Preface for Conference Proceedings

A.E Ismail, A. L Mohd Tobi and W. A. Siswanto

CAMS2016 is an internal colloquium on advance mechanics held between 18-19 December 2016 on the campus of Universiti Tun Hussein Onn Malaysia in Johor, Malaysia. The program included many topic related to mechanical and materials engineering as well as those in related field.

The meeting is conducted and managed by five research groups of SIMReG (Structural Integrity and Monitoring Research Group), COLORED (Crashworthiness and Collisions Research Group), ADCARe (Advance Dynamics Control and Automation Research Group), NOVIA (Noise And Vibration Analysis Research Group) and SPEAR (Sports Engineering Advancement Research Group). CAMS2016 is coordinated by the center of research MPROVE (Mechanical Failure Prevention and Reliability).

CAMS2016 has attracted 48 paper submissions and after the peer-review process, 36 papers have been accepted. The participants in the colloquium include researchers, students and lecturers within the campus. The contents of presentations hopefully will enhance the quality of research conducted and therefore will contribute to advancement of knowledge in the research..

This volume was edited by Abdul Latif Mohd Tobi, Al Emran Ismail, Waluyo Adi Siswanto, Saifulnizan Jamian, Nor Azali Azmir, Noormaziah Jafferi, Eliza Yusup, Kamarul Azhar Kamarudin, Ong Pauline, Zaleha Mohamad, Mohd Amran Madlan, and Noradila Abdul Latif.

IOP Conf. Series: Materials Science and Engineering **165** (2017) 011002 doi:10.1088/1757-899X/165/1/011002

LIST OF COMMITTEE: CAMS2016

1. Chairman CAMS2016

Dr. Al Emran Ismail
Crashworthiness and Collisions Research Group (COLORED)
Department of Engineering Mechanics,
Faculty of Mechanical and Manufacturing Engineering,
Universiti Tun Hussein Onn Malaysia (UTHM),
86400 Parit Raja, Batu Pahat,
Johor, MALAYSIA
Email: emran@uthm.edu.my

2. Vice Chairman CAMS2016
Publication Chair CAMS2016

Dr. Abdul Latif Mohd Tobi
Structural Integrity and Monitoring Research Group (SIMReG)
Department of Engineering Mechanics,
Faculty of Mechanical and Manufacturing Engineering,
Universiti Tun Hussein Onn Malaysia (UTHM),
86400 Parit Raja, Batu Pahat,
Johor, MALAYSIA
Email: abdlatif@uthm.edu.my

3. Website and Online Systems CAMS2016

Assoc. Prof. Dr. Waluyo Adi Siswanto
Sports Engineering Advancement Research (SPEAR)
Department of Engineering Mechanics,
Faculty of Mechanical and Manufacturing Engineering,
Universiti Tun Hussein Onn Malaysia (UTHM),
86400 Parit Raja, Batu Pahat,
Johor, MALAYSIA
Email: waluyo@uthm.edu.my

4. Payment and Registration CAMS2016

Dr. Saifulnizan Jamian
Crashworthiness and Collisions Research Group (COLORED)
Department of Engineering Mechanics,
Faculty of Mechanical and Manufacturing Engineering,
Universiti Tun Hussein Onn Malaysia (UTHM),
86400 Parit Raja, Batu Pahat,
Johor, MALAYSIA
Email: saifulnz@uthm.edu.my

IOP Conf. Series: Materials Science and Engineering **165** (2017) 011003 doi:10.1088/1757-899X/165/1/011003

EDITORIAL: CAMS2016

1. Dr. Abdul Latif Mohd Tobi
   Structural Integrity and Monitoring Research Group (SIMReG)
   Department of Engineering Mechanics,
   Faculty of Mechanical and Manufacturing Engineering,
   Universiti Tun Hussein Onn Malaysia (UTHM),
   86400 Parit Raja, Batu Pahat,
   Johor, MALAYSIA
   Email: abdlatif@uthm.edu.my

2. Dr. Al Emran Ismail
   Crashworthiness and Collisions Research Group (COLORED)
   Department of Engineering Mechanics,
   Faculty of Mechanical and Manufacturing Engineering,
   Universiti Tun Hussein Onn Malaysia (UTHM),
   86400 Parit Raja, Batu Pahat,
   Johor, MALAYSIA
   Email: emran@uthm.edu.my

3. Assoc. Prof. Dr. Waluyo Adi Siswanto
   Sports Engineering Advancement Research (SPEAR)
   Department of Engineering Mechanics,
   Faculty of Mechanical and Manufacturing Engineering,
   Universiti Tun Hussein Onn Malaysia (UTHM),
   86400 Parit Raja, Batu Pahat,
   Johor, MALAYSIA
   Email: waluyo@uthm.edu.my

4. Dr. Saifulnizan Jamian
   Crashworthiness and Collisions Research Group (COLORED)
   Department of Engineering Mechanics,
   Faculty of Mechanical and Manufacturing Engineering,
   Universiti Tun Hussein Onn Malaysia (UTHM),
   86400 Parit Raja, Batu Pahat,
   Johor, MALAYSIA
   Email: saifulnz@uthm.edu.my

5. Mr. Mohd Amran Hj. Madlan
   Noise and Vibration Analysis Research Group (NOVIA)
   Department of Engineering Mechanics,
   Faculty of Mechanical and Manufacturing Engineering,
   Universiti Tun Hussein Onn Malaysia (UTHM),
   86400 Parit Raja, Batu Pahat,
   Johor, MALAYSIA
   Email: amran@uthm.edu.my

6. Dr. Nor Azali Azmir
   Noise and Vibration Analysis Research Group (NOVIA)
   Department of Engineering Mechanics,
   Faculty of Mechanical and Manufacturing Engineering,
   Universiti Tun Hussein Onn Malaysia (UTHM),
   86400 Parit Raja, Batu Pahat,
   Johor, MALAYSIA
   Email: azali@uthm.edu.my

7. Dr. Noormaziah Jafferi
Advance Dynamics Control and Automation Research Group (ADCARe)
Department of Engineering Mechanics,
Faculty of Mechanical and Manufacturing Engineering,
Universiti Tun Hussein Onn Malaysia (UTHM),
86400 Parit Raja, Batu Pahat,
Johor, MALAYSIA
Email: normazia@uthm.edu.my

8. Dr. Eliza M Yusup
Sports Engineering Advancement Research (SPEAR)
Department of Engineering Mechanics,
Faculty of Mechanical and Manufacturing Engineering,
Universiti Tun Hussein Onn Malaysia (UTHM),
86400 Parit Raja, Batu Pahat,
Johor, MALAYSIA
Email: elizay@uthm.edu.my

9. Dr. Kamarul Azhar Kamarudin
Crashworthiness and Collisions Research Group (COLORED)
Department of Engineering Mechanics,
Faculty of Mechanical and Manufacturing Engineering,
Universiti Tun Hussein Onn Malaysia (UTHM),
86400 Parit Raja, Batu Pahat,
Johor, MALAYSIA
Email: kamarula@uthm.edu.my

10. Dr. Ong Pauline
Advance Dynamics Control and Automation Research Group (ADCARe)
Department of Engineering Mechanics,
Faculty of Mechanical and Manufacturing Engineering,
Universiti Tun Hussein Onn Malaysia (UTHM),
86400 Parit Raja, Batu Pahat,
Johor, MALAYSIA
Email: ongp@uthm.edu.my

11. Dr. Zaleha Mohamad
Structural Integrity and Monitoring Research Group (SIMReG)
Department of Engineering Mechanics,
Faculty of Mechanical and Manufacturing Engineering,
Universiti Tun Hussein Onn Malaysia (UTHM),
86400 Parit Raja, Batu Pahat,
Johor, MALAYSIA
Email: zaleha@uthm.edu.my

12. Dr. Noradila Abdul Latif
Structural Integrity and Monitoring Research Group (SIMReG)
Department of Engineering Mechanics,
Faculty of Mechanical and Manufacturing Engineering,
Universiti Tun Hussein Onn Malaysia (UTHM),
86400 Parit Raja, Batu Pahat,
Johor, MALAYSIA
Email: noradila@uthm.edu.my

Peer review statement

All papers published in this volume of *IOP Conference Series: Materials Science and Engineering* have been peer reviewed through processes administered by the proceedings Editors. Reviews were conducted by expert referees to the professional and scientific standards expected of a proceedings journal published by IOP Publishing.

Solutions of Stress Intensity Factors of Multiple Internal Axial Cracks in Hollow Cylinders

Al Emran Ismail

Faculty of Mechanical and Manufacturing Engineering, Faculty of Mechanical and Manufacturing Engineering, Unviersiti Tun Hussein Onn Malaysia, 86400 Johor, Malaysia.

emran@uthm.edu.my / al_emran@uthm.edu.my

Abstract. This paper solved numerically the stress intensity factors of multiple internal axial cracks in hollow cylinders under internal pressure. Semi-elliptical shaped cracks are modelled using ANSYS finite element program and the stress intensity factors are calculated using interaction domain integral which is based on the J-integral. There are two important parameters are considered such as crack aspect ratio, a/t is 0.2, 0.4 and 0.75 and crack spacing distance, d/c is 0.4, 0.7 and 1.1. For multiple internal axial cracks under internal pressure, stress intensity factors along the crack front decreased when crack aspect ratio, a/t is increased. It is also revealed that the role of crack spacing on the stress intensity factor is insignificant especially for the shallow cracks (a/t < 0.4). However, for the crack having a/t = 0.75, increasing the distance of d/c capable to increase the SIFs due to interaction effect.

1. Introduction

Mechanical reliability assessment considering fracture mechanics approach became an important tool in analyzing the role of flaws on the structural integrity especially on the fitness-for-service (FFS) procedures and life-extension programs of critically designed components [1]. Pipes are widely used to transport gas or liquid medium in the fastest way. Like other structures, pipe is also exposed to the external elements for examples harsh environment and other type of surface defects leading to form a premature cracks [2] and under certain circumstances, internal crack could be occurred frequently [3]. Most of the previous works focused on the single crack formed whether internally [4] or externally [5]. European flaw assessment procedure called SINTAP for an example [6] also recently covered the full scale pipe tests containing through wall and surface cracks under internal pressure and four pint bending moments. The goal is to predict the failure loads and the marginal safety.

Due to harsh environment, the formation of multiple cracks are unavoidable since surface corrosion capable to degrade such surface on the various locations. According to literature, there are tremendous amount of works dealing with the single crack [7-10]. Some of these works investigated the crack internally or externally especially on the surfaces of pipes or hollow cylinders. However, it is hard to find the behavior of multiple cracks occurred either on the external or internal surfaces of pipes under internal pressure.

Therefore, this paper presents a preliminary work to numerically solve the problems of multiple internal axial crack in pipes. There are three relative crack depth ratio, a/t is used such as 0.2, 0.4 and 0.75 with a normalized crack length, $\rho = 0.5$. It is also assumed that the pipe followed the rule of R/t =

0.5. In order to study the effect of distance between two crack, the ratio of d/c = 0.4, 0.7 and 1.1 is used. ANSYS finite element program through the use of ANSYS Parametric Design Language (APDL) is used to facilitate the process and the interaction domain integral method is used to calculate the stress intensity factor. Then, the roles of stress intensity factors on the crack geometries are investigated and discussed.

2. Methodology

Axial internal crack is numerically modelled using ANSYS finite element software. Figure 1 shows the axial internal crack in a hollow cylinder subjected to internal pressure where t is a thickness, d is a distance between two cracks, a is a minor, c is a major radiuses of semi-elliptical shaped cracks, P is a point along the crack front and x is its distance measured from the central point of cracks. Since the crack geometries are symmetrical, only a quarter finite element model is constructed as in Figure 2.

Special attention is given to the crack tip where singular iso-parametric element is used and the other region of cylinder is modelled using regular element. Stress intensity factor is determined for ten points along the crack front based on the domain integral method. In this work, the ratio of the inner radius of the cylinder to the thickness, R/t = 0.5 is selected, the normalized crack length, ρ = 0.5 (as in equation 1) is considered, relative crack depth, a/t is 0.1, 0.4 and 0.75 and the distance between the cracks, d/c are varied such as 0.4, 0.7 and 1.1. The stress intensity factor, K is then normalized according to equation (2). In order to analyze the role between two cracks, an interaction factor, ψ is introduced and it is defined as a ratio between the SIFs obtained from two and single cracks. It is determined according to equation (3).

$$\rho = \frac{c}{\sqrt{R_m t}} \tag{1}$$

$$F = \frac{K}{\sigma\sqrt{\pi a}} \tag{2}$$

$$\psi = \frac{F_{\text{multiple crack}}}{F_{\text{single crack}}} \tag{3}$$

where, $\sigma = \dfrac{PR_i}{t}$

Before the model is further used, it is very important to validate with the previous model (Kim et al., 2004). Since multiple internal axial cracks are not available then single crack of similar geometry is used as in Figure 3. For validation purpose, only the stress intensity factors at the deepest point along the crack are considered (x/h = 0.0). It is found that the present model is well similar with the existing model.

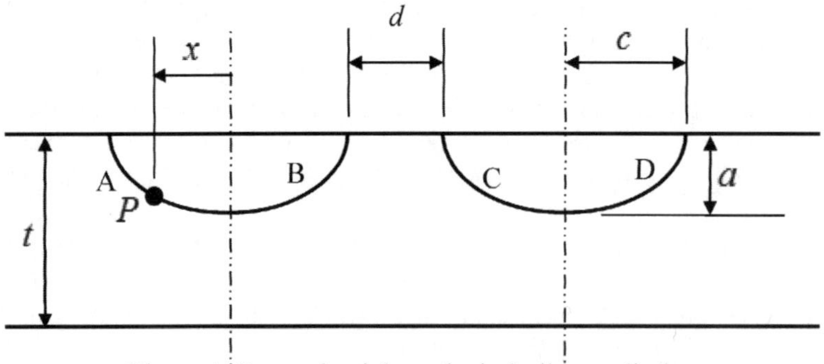

Figure 1. Internal axial cracks in hollow cylinder

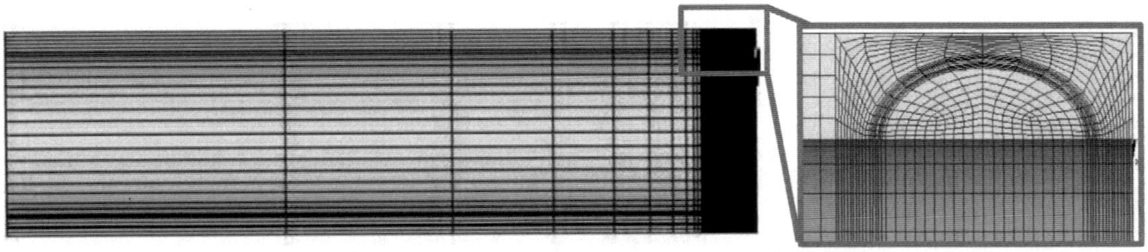

Figure 2 A quarter finite element model of multiple internal axial cracks in hollow cylinder

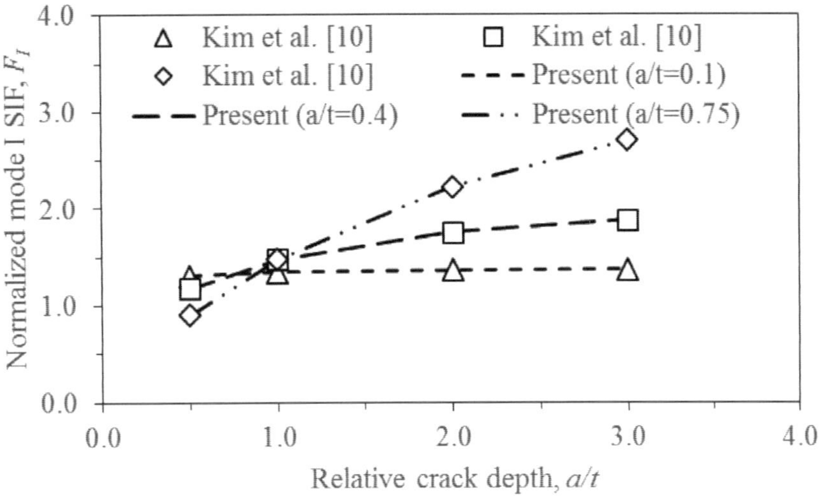

Figure 3 Finite element validations

3. Results and Discussion

In this work, there are two main crack fronts as in Figure 1. Then, it is divided into four sections A, B, C and D. According to the numerical analysis, crack fronts B and C interacted to each other while A and D remain intact. Therefore, for discussion crack fronts B and C are emphasized. On the other hand, since the crack is symmetrical either crack front B and C is presented. Figure 4 shows the effect of crack aspect ratios, *a/t* on the normalized mode I stress intensity factor (SIF) when different values of crack spacing, *d/c* is used. Figure 4 also reveals that the SIFs have an identical pattern where lower values of *a/t* produced higher SIFs and higher *a/t* produced lower SIFs. For the crack with $a/t = 0.2$, the SIFs are almost flattened along the crack front. This is due to since for $a/t = 0.2$, the crack front is almost flattened. This behavior of SIFs are almost similar with the results calculated by Kim et al. [10]. In the outer region of crack front, the SIFs showing some irregularities due to singularity problem since the SIFs are sharply increased.

Figure 5 shows the effect of crack distance, *d/c* on the SIFs for a certain value of *a/t*. It is indicated that the role of *d/c* only significant for the case of higher value of *a/t* for an example $a/t = 0.75$. Figure 5(a) shows the effect of *d/c* on the SIFs for $a/t = 0.2$. It is revealed that the SIFs along the crack fronts are almost identical meaning that varying *d/c* insignificantly affected the SIFs. This is due to for $a/t = 0.2$, the integrity of the tube is strong enough to sustain under the action of internal pressure and the effect of such defects are minimized. However, when deeper cracks are used ($a/t > 0.4$), the effect of *d/c* became obvious where the factor of *d/c* capable to decrease the SIFs as in Figures 5(b) and 5(c). This phenomenon is called the crack shielding since the formation of multiple cracks are not contributed to the increment of SIFs.

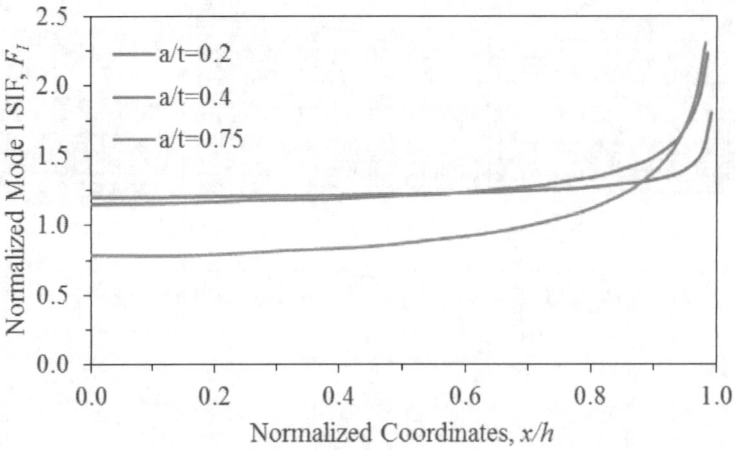

(a)

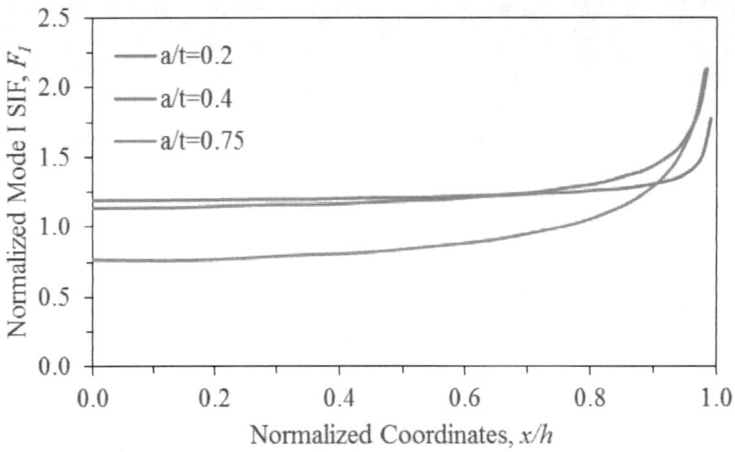

(b)

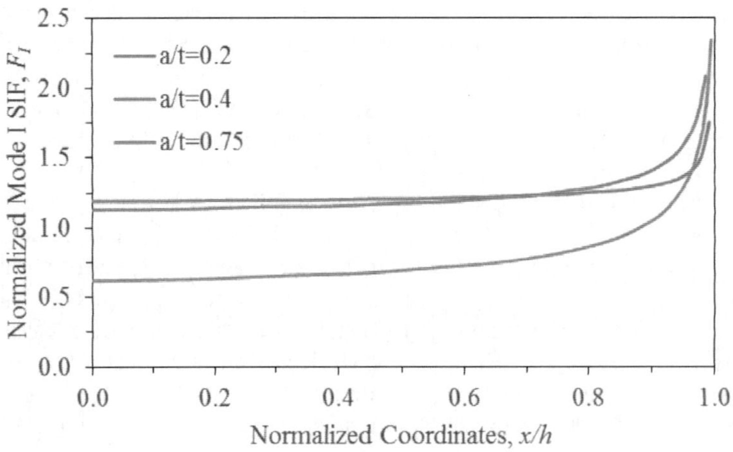

(c)

Figure 4 Stress intensity factors along the crack fronts when crack aspect ratio, *a/t* is varied for different crack spacing distance, *d/c* (a) 0.4, (b) 0.7 and (c) 1.1.

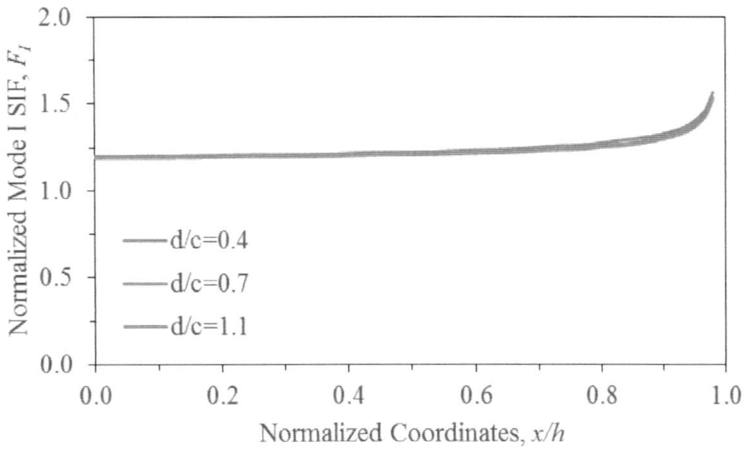

(a)

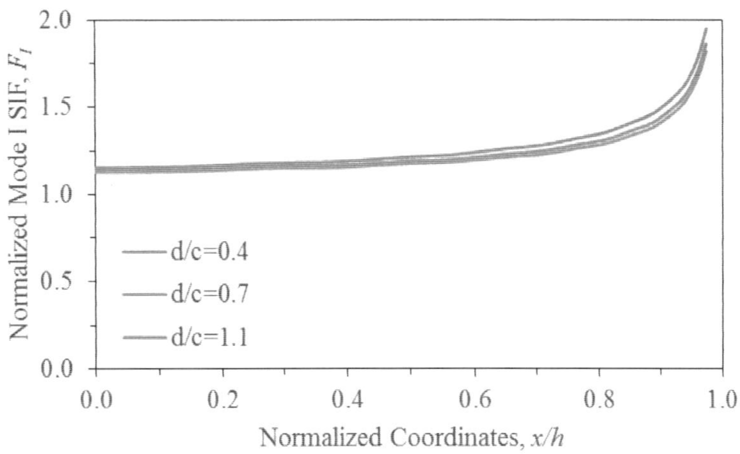

(b)

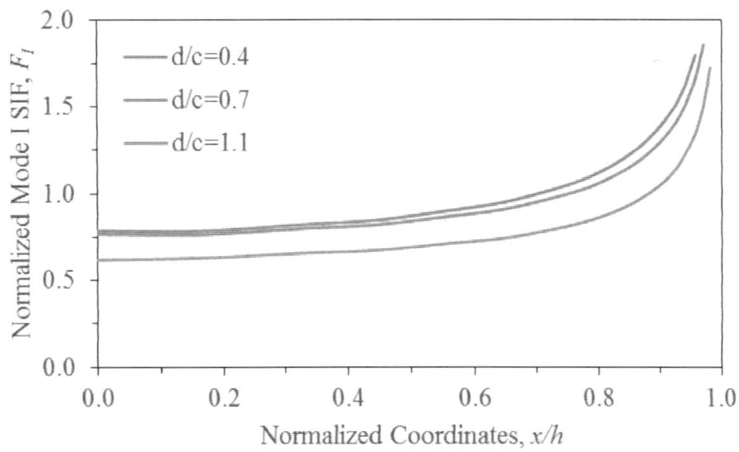

(c)

Figure 5 Stress intensity factors along the crack fronts when crack spacing, *d* is varied for different crack aspect ratio, *a/t* (a) 0.2, (b) 0.4 and (c) 0.75.

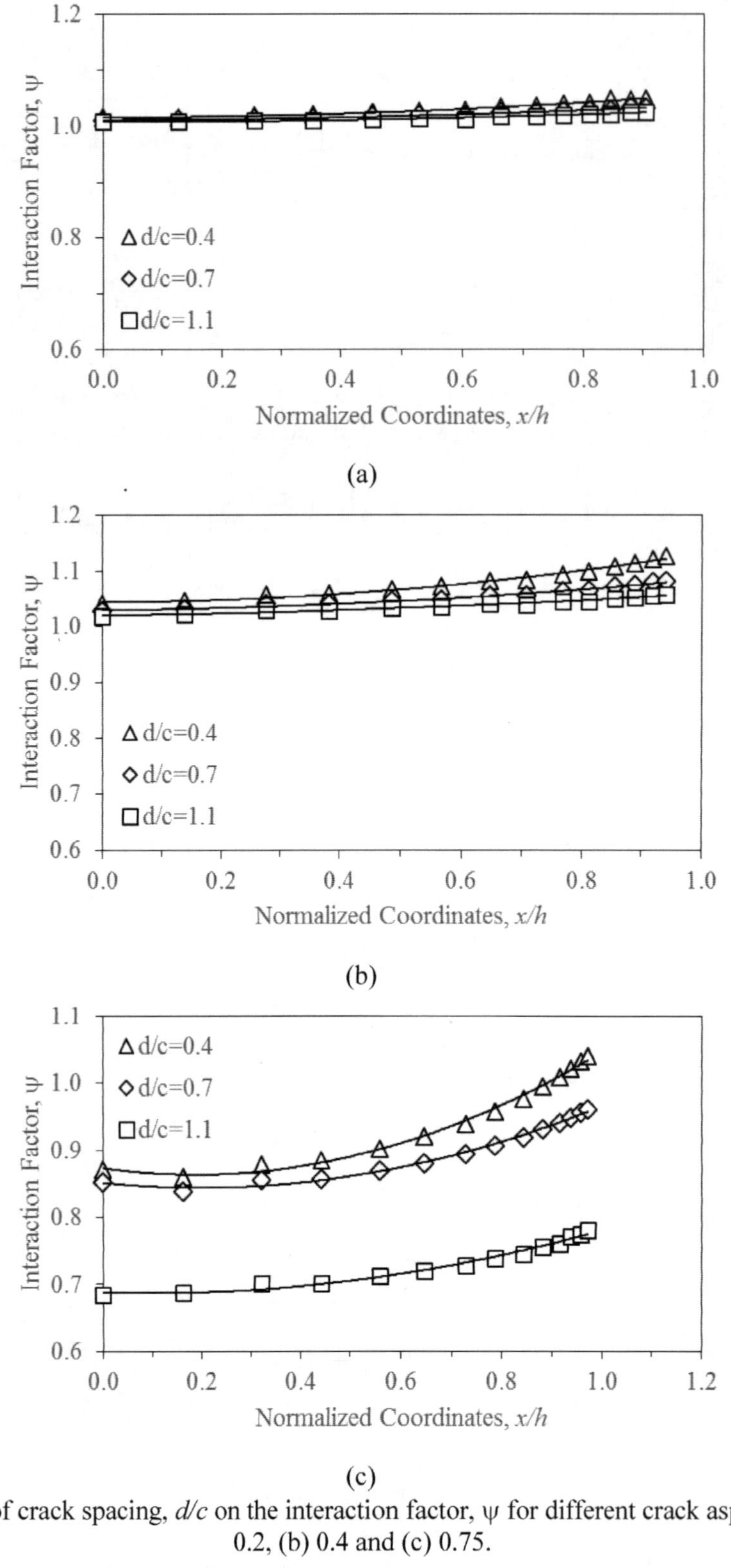

Figure 6 Effect of crack spacing, *d/c* on the interaction factor, ψ for different crack aspect ratio, *a/t* (a) 0.2, (b) 0.4 and (c) 0.75.

Figure 6 reveals a strong evidence the role of *d/c* on the SIFs obtained from multiple cracks. Figure 6(a) indicates the factor of crack interactions for *a/t* = 0.2 where different *d/c* is used. The factor of *d/c* is significant when higher *a/t* is implement as in Figures 6(b) and 6(c). According to Figure 6(c), the effect of *d/c* is tremendous especially on the deeper cracks. The existence of multiple internal cracks positioned axially are not contributed to increase the SIFs but it is capable to decrease which is indicated the shielding behavior.

4. Conclusion

In this work, ANSYS finite element program is used to model and analyze the roles of crack geometries (*a/t* = 0.2, 0.4 and 0.75) and crack inter-distance (*d/c* = 0.4, 0.7 and 1.1) on the SIFs along the crack front. According to the numerical analysis, it can be concluded that the parameter of *d/c* played an important role in lowering the SIFs especially when deeper crack is used (*a/t* = 0.75). If shallow crack is used (*a/t* = 0.2), the effect of *d/c* is insignificant where the SIFs along the crack fronts are almost identical. The decrement of SIFs for the case of cracks with *a/t* = 0.75 due to the phenomenon called a shielding effect.

Acknowledgement

Author acknowledged Universiti Tun Hussein Onn Malaysia and Ministry of Higher Education of Malaysia for financially sponsoring this work through Fundamental Research Grant Scheme (FRGS) 2016/1 vot. 1592.

References

[1] Mario SG Chiodo and Claudio R 2010 *Engineering Fracture Mechanics* 7 415-436.
[2] Shangtong Y, Li CQ and Yang Wei 2016 *Engineering Fracture Mechanics* 153 50-60.
[3] Lu YL, Zhang SJ, Huang XP and Huang J 2003 *International Journal of Pressure Vessel and Piping* 80 167-178.
[4] Ayatollaho MR and Khoramishad 2010 *International Journal of Pressure Vessel and Piping* 87 165-169.
[5] Brighenti R 2000 *Int. J. of Fatigue* 22 559-567.
[6] Zerbst U, Hamann R and Wohlschlegel A 2000 *International Journal of Pressure Vessel and Piping* 77 697-702.
[7] Schmitt W and Keim E 1979 *International Journal of Pressure Vessel and Piping* 7 106-118.
[8] Raju IS and Newmann Jr. JC 1982 *Journal of Pressure Vessel Technology* 104 293-298.
[9] Carpinteri A, Brighenti R and Spagnoli A 1998 *Nuclear Engineering and Design* 185 1-10.
[10] Kim YJ, Kim JS, Park YJ and Kim YJ 2004 *Engineering Fracture Mechanics* 71 925-944.
[11] Predan J, Mocilnik V and Gubeljak N 2013 *Engineering Fracture Mechanics* 105 152-168.
[12] Li CQ and Fu G and Yang W 2016 *Engineering Fracture Mechanics* 165 72-86.
[13] Ismail AE, Ariffin AK, Abdullah S and Ghazali MJ 2012 Meccanica 47 1141-1156.
[14] Ismail AE, Ariffin AK, Abdullah S and Ghazali MJ 2012 *Indian Journal of Engineering and Materials Science* 19 5-16.
[15] Ismail AE, Azlan MA, Mohd Tobi AL and Ahmad MH 2016 *IOP Conf. Series: Materials Science and Engineering* 152 012049.
[16] Ismail AE, Ariffin AK, Abdullah S and Ghazali MJ 2011 *International Journal of Automotive Technology* 12(2) 225-232.
[17] Ismail AE, Ariffin AK, Abdullah S, Ghazali MJ, Abdulrazzaq M and Daud R 2011 *Journal of Zhejing University – Science A (Applied Physics and Engineering)* 13(1) 1-8.
[18] Ismail AE 2013 *Applied Mechancis and Materials* 315 665-669.
[19] Ismail AE 2013 *Applied Mechancis and Materials* 315 655-659.

Residual stresses in shape memory alloy fiber reinforced aluminium matrix composite

Tang Tsz Loong,[1] Saifulnizan Jamian,[2] Al Emran Ismail,[2] Nik Hisyammudin Muhd Nur,[1] Yoshimi Watanabe[3]

[1]Faculty of Mechanical and Manufacturing Engineering, Universiti Tun Hussein Onn Malaysia, Johor, Malaysia

[2]Crashworthiness and Collisions Research Group, Faculty of Mechanical and Manufacturing Engineering, Universiti Tun Hussein Onn Malaysia, Johor, Malaysia

[3]Department of Mechanical Engineering, Nagoya Institute of Technology, Nagoya, Japan

saifulnz@uthm.edu.my

Abstract. Process-induced residual stress in shape memory alloy (SMA) fiber reinforced aluminum (Al) matrix composite was simulated by ANSYS APDL. The manufacturing process of the composite named as NiTi/Al is start with loading and unloading process of nickel titanium (NiTi) wire as SMA to generate a residual plastic strain. Then, this plastic deformed NiTi wire would be embedded into Al to become a composite. Lastly, the composite is heated form 289 K to 363 K and then cooled back to 300 K. Residual stress is generated in composite because of shape memory effect of NiTi and mismatch of thermal coefficient between NiTi wire and Al matrix of composite. ANSYS APDL has been used to simulate the distribution of residual stress and strain in this process. A sensitivity test has been done to determine the optimum number of nodes and elements used. Hence, the number of nodes and elements used are 15680 and 13680, respectively. Furthermore, the distribution of residual stress and strain of nickel fiber reinforced aluminium matrix composite (Ni/Al) and titanium fiber reinforced aluminium matrix composite (Ti/Al) under same simulation process also has been simulated by ANSYS APDL as comparison to NiTi/Al. The simulation results show that compressive residual stress is generated on Al matrix of Ni/Al, Ti/Al and NiTi/Al during heating and cooling process. Besides that, they also have similar trend of residual stress distribution but difference in term of value. For Ni/Al and Ti/Al, they are 0.4% difference on their maximum compressive residual stress at 363K. At same circumstance, NiTi/Al has higher residual stress value which is about 425% higher than Ni/Al and Ti/Al composite. This implies that shape memory effect of NiTi fiber reinforced in composite able to generated higher compressive residual stress in Al matrix, hence able to enhance tensile property of the composite.

1. Introduction

Shape memory alloys are a group of metallic alloys that can return to their original form in terms of shape and size after subjected to a memorization process between two transformation phases, which is temperature dependent and this transformation phenomenon is known as the shape memory effect [1].

The shape memory effect of shape memory alloy is used to design smart material. This type of smart material also known as SMA composite [2] and the design concept have been proposed by the Watanabe et. al [3], where the illustration of SMA composite design concept is show as Figure-1. The metal matrix composite fiber reinforced with SMA material will have better material characteristic and be safer because compressive stress is often seems as good stress for matrix composite because it can increase fatigue life of the component by reduce micro-crack of composite [2,4]. Therefore, knowledge and understanding of residual stress at matrix of SMA composite is very important because by knowing the distribution of residual stress at the design process would help the research to design better SMA composite and hence improve the performance of composite.

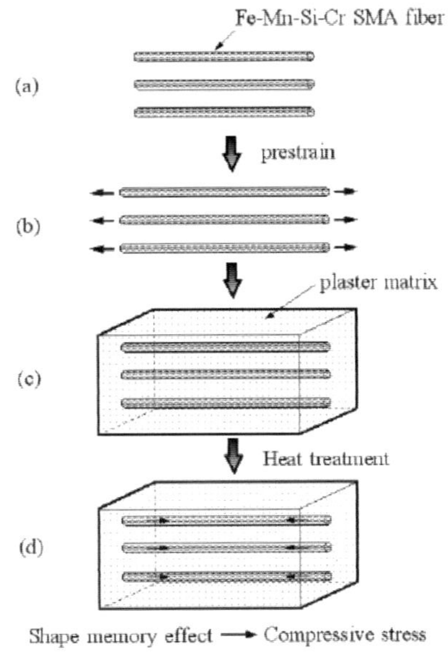

Figure-1: Design concept of SMA composite [3]

2. Methodology

This research is focus on simulation of three composite models. The composite models are NiTi/Al, Ni/Al and Ti/Al. The flow chart of the simulation and the geometry of the composite models is shown in Figure-2 and Figure-3, respectively. ANSYS APDL is used to simulate the residual stress of the composites. There are four types of material involves i.e. Al and three types of fiber, Ni, Ti and NiTi. The material property of them is shown as Table-1 and Table-2. The technique to determine the phase transformation temperatures in NiTi are briefly explained in Tan et al [5].

Quarter model is used in this simulation as shown in Figure-4. Quarter model is chosen because it can reduce great number of nodes and elements in finite element furthermore the model is symmetric to both loading and geometry condition. The boundary condition in this case is symmetric boundary condition and would get symmetric answer.

Table-1: Material properties

Parameter	**Symbol**	**Al** [6]	**Ni** [7]	**Ti** [8]	**Unit**
Young Modulus	E	70	202.5	103	GPa
Poisson Ratio	v	0.33	0.29	0.32	-
Thermal Expansion	α	23.5e-6	12.3e-6	8.6e-6	K^{-1} or $^oC^{-1}$
Yield strength	σ_y	160	185	155.5	MPa
Tensile strength	σ_s	170	434	215.5	MPa

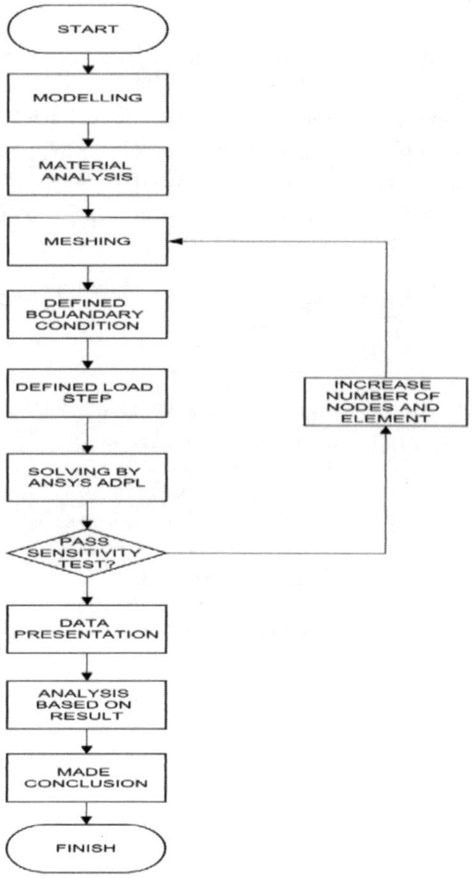

Figure-2: Simulation flowchart

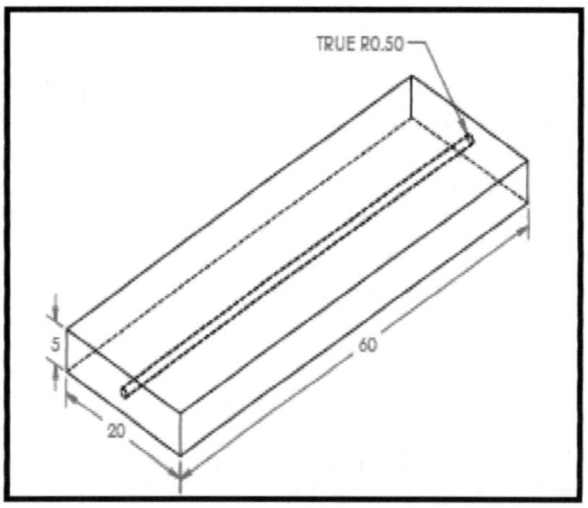

Figure-3: The model for simulation

Table-2: Nickel Titanium material property [9]

Parameter	Symbol	Value/Desciption	Unit
A) Elastic model			
Young modulus at austenite phase	E	70e3	MPa
Poisson ratio	v	0.3	-
B) SMA model			
Maximum transformation strain	$\bar{\varepsilon}_L$	0.07	MPa
Hardening parameter	h	500	MPa
Temperature scaling factor	β	8,3	MPa/K
Reference temperature	M_s	289	K
Hysteresis loop amplitude	R	120MPa	K
Lode dependency parameter	m	0	-
Young modulus at martensite phase	E_m	70e3	MPa

Then, the quarter model would subject to loading step follow the design method present in Figure-1 and the simulation step that follow the steps of Figure-2 has been modified and is shown as flow diagram in Figure-5. In order to simplified the model, the NiTi wire is embedded in Al matrix using similar technique in Jamian et al. [10, 11]. After solve all the load step, the data of residual stress would be able to read and plot out. However, before to start the simulation, sensitivity test shall conduct to determine the optimum number of nodes and elements

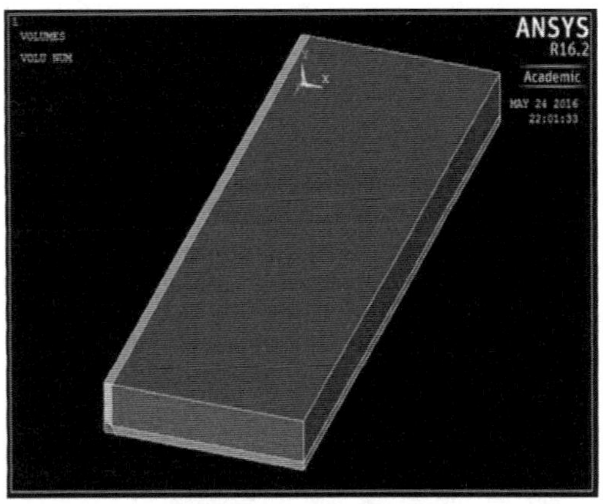

Figure-4: Quarter model

3. Results and Discussion

Before start doing simulation on NiTi/Al composite, modeling of NiTi wire has been done to determine the tensile loading required to generate maximum residual strain. The stress-strain relationship of NiTi is shown in Figure-6. Figure-6 indicated the maximum residual strain of NiTi wire is about 5.6 % and the maximum pressure required to generate this strain is about 190MPa.

Furthermore, as shown in Figure-6 the tensile loading required to induce phase transformation from multi variant oriented martensite to single variant oriented martensite is about 150 MPa. The residual strain of NiTi wire can be generated by subjected tensile loading 150MPa above but the residual strain is reach maximum when the tensile loading reached 190MPa. In this work, tensile loading 240MPa is chosen to generate maximum residual strain for NiTi wire before placed into Al matrix.

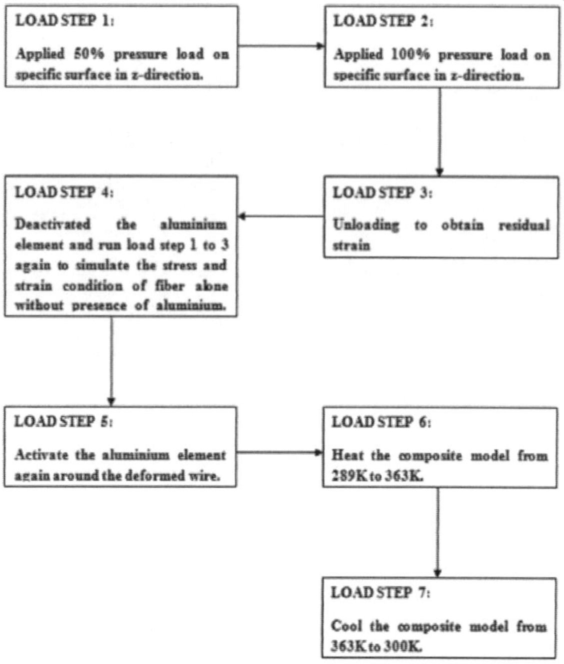

Figure-5: Load step involved in simulation

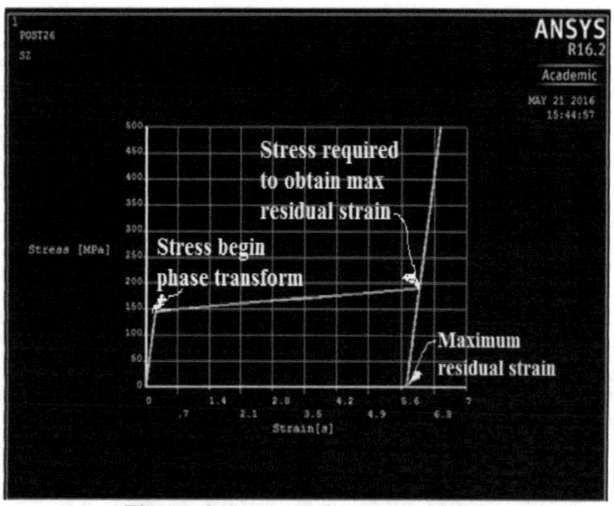

Figure-6: Stress-strain graph of NiTi

The maximum residual strain will be treated as maximum residual strain for simulation of NiTi model. Then, to determine the optimum number of elements a sensitivity test has been conducted. In sensitivity test, a testing node has been chosen to compare the result of stress in z-direction, Sz with difference value of elements and nodes. The result of sensitivity test has been shown as Table-3. The optimum number of element and node is 13680 and 15680, respectively. This number of number and nodes has been used to simulate in other two composite models.

Table-3: Result of sensitivity test

Element	Node	Stress in z-direction, Sz (MPa)	error
1890	2432	198.4	-
4455	5440	199.52	1.12
8385	9856	199.87	0.35
9675	11264	199.91	0.04
11520	13328	199.96	0.05
12960	14896	199.99	0.03
13680	15680	199.99	0

The residual stress distribution of NiTi/Al composite generated during manufacturing process has been simulated by ANSYS APDL. There are total 7 load steps involved in the simulation process and the time interval for each load step is 1s. Therefore, the total time taken of the simulation process is 7s. The time taken for each load step is set like this because the variable of the model is time invariant and easier to analyze during discussion. The result of stress in z-direction, Sz distribution for every 1s or load step is shown as Figure-7. The stress from origin to 1.802mm away in x-direction is showed in all graphs. Furthermore, the contour plot of each load steps has been shown in Figure-7(a) to (g) for the better inspection on the residual stress distribution.

Figure-7(a) indicates the stress distribution of NiTi/Al composite at time=1s. The tensile load applied on the NiTi wire is 120MPa which is the half of the tensile loading. Based on Figure-7(a), the stress is constant between range of 0mm to 0.36mm. The stress value is 120MPa which is equal to the tensile load applied on the NiTi wire. The stress value is decline slowly after 0.36mm and become zero around 0.78mm because there is modified Al matrix constraints the deformation of NiTi wire.

Figure-7(b) indicates the stress increased from 120MPa to 240MPa between range of 0mm to 0.36mm at time=2s. This stress value is same as the tensile loading applied on NiTi wire. The stress value is declined after 0.36mm and reach zero at 0.78mm.

After loading process is completed, unloading process is started at load step 3 and the residual stress distribution at time=3s is shown as Figure-7(c). By inspected on Figure-7(c), the residual stress is significantly small and can be treated as stress free after unloading. The stress is no exactly zero because numerical method is used to obtain the value while running the solver.

Initially, the NiTi wire is pulled alone without presence of Al matrix because ANSYS APDL is unable to add Al matrix element to the NiTI wire after run the solver. Therefore, the complete model of NiTi/Al composite is built first although the Al is does not exist at the beginning of the simulation process. Then, the Al element are deactivated at load step 4 and run the first three load step again to eliminate load effect of Al elements on the NiTi wire. After deactivation of Al matrix, the stress distribution is change but still significant small because the Al and NiTi wire is suppose in stress free condition as shown in Figure-7(d).

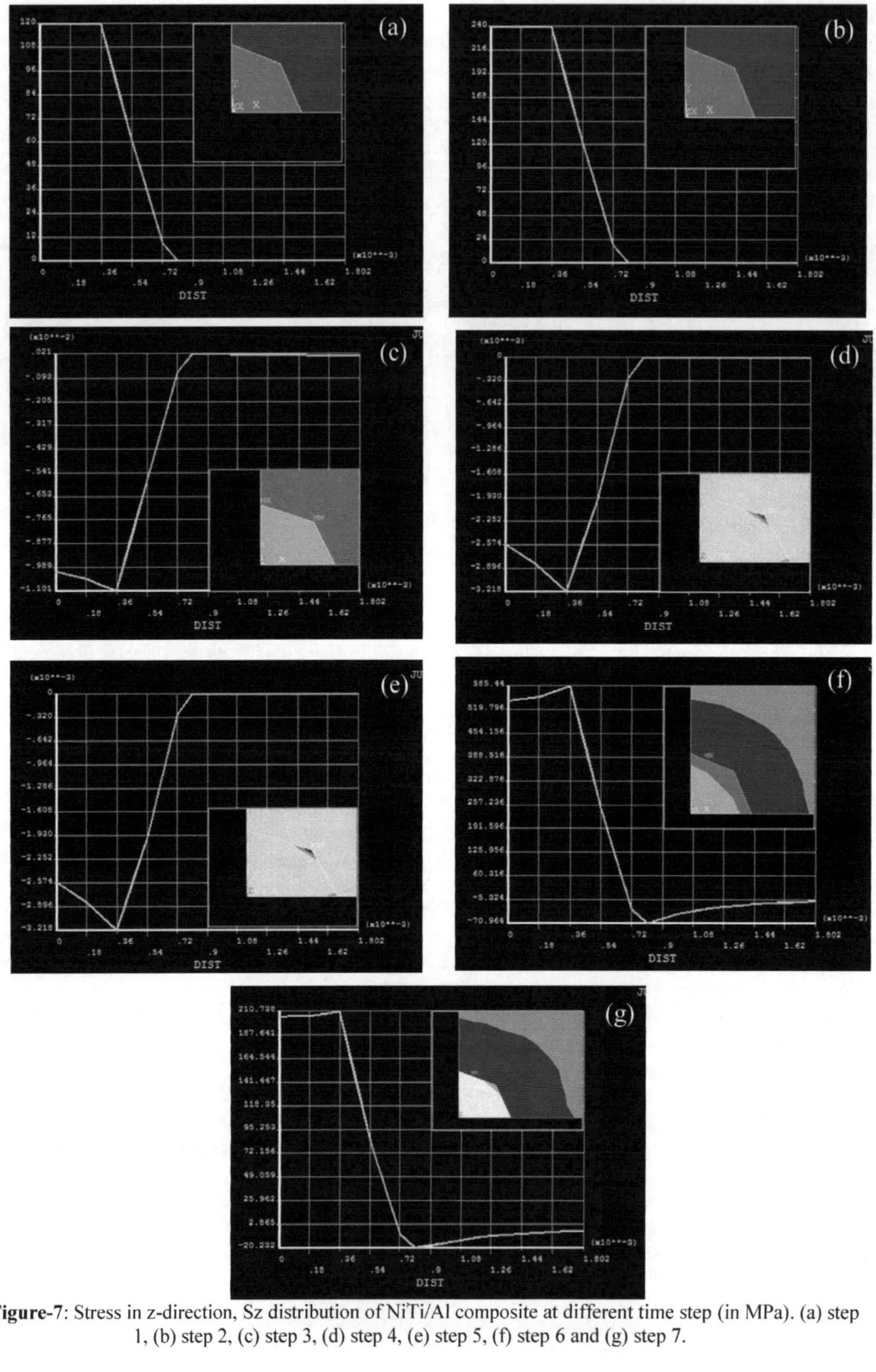

Figure-7: Stress in z-direction, Sz distribution of NiTi/Al composite at different time step (in MPa). (a) step 1, (b) step 2, (c) step 3, (d) step 4, (e) step 5, (f) step 6 and (g) step 7.

Then, the model undergo load step 5 to activate the Al matrix elements surrounding of NiTi wire. The Al matrix is modified to low Young modulus material to reduce simulation failure in load step 4. Hence, in this step Al matrix have converted back to original material property after activated the element. The stress distribution as shown in Figure-7(e) is same as load step 4 since no external load is applied on composite. Then, the simulation is proceeded to load step 6 where the temperature of the model increased from 289K to 363K. The residual stress is generated in this load step where there are no external force acting on it.

Figure-7(f) shows the residual stress distribution of NiTi/Al composite at time=6s. The residual stress generated in this load step is due to shape memory effect and mismatch of thermal expansion between Al matrix and NiTi fiber reinforcement. Based on Figure-7(f), the residual stress is around 520 MPa at origin and increased to 585 MPa at 0.36mm which is in the region of NiTi wire. Then, the residual stress starts to decrease drastically from 585MPa to a maximum compressive residual stress value -70 MPa around 0.78mm which is at region on Al matrix. The shrinkage of NiTi is due to shape memory effect and expansion of Al matrix due to increase of thermal load. This phenomenon has caused the residual stress is decreased along the interfacial between matrix and fiber due to the pulling action of NiTi on Al matrix. This pulling action generated a compressive residual stress on Al matrix that surrounded the NiTi wire. However, the compressive residual stress is slowly decrease for the region far away from the NiTi wire and eventually reach a saturated value -5.324 MPa.

Then, for load step 7 the model temperature is decreased from 363K to 300K (room temperature). Based on Figure-7(g), the residual stress distribution pattern is still the same but the residual stress value is decreased if compare with Figure-7(g). The maximum residual stress is about 210.8 MPa which is decreased about 64% and the minimum residual stress is about -2MPa which is decreased about 62%.

Beside simulation stress distribution of NiTi/Al composite is simulated, stress distribution of Ni/Al composite and Ti/Al composite also has been simulated. Then, the maximum compressive stress generated in Al matrix in Ni/Al and Ti/Al composite then would be compared with NiTi/Al composite at 363K and 300K.

Table-4: The maximum compressive stress generated

Temperature	NiTi/Al	Ni/Al	Ti/Al
363 K	-70.964 MPa	-13.483 MPa	-13.537 MPa
300 K	-30.732 MPa	-2.004 MPa	-2.012 MPa

Table-4 indicates the NiTi/Al has highest compressive residual stress compared to Ni/Al and Ti/Al. The compressive residual stress of NiTi/Al generated during heating process about 425 % higher than Ni/Al and Ti/Al. This is because the residual stress of NiTI/Al is generated with combination shape memory effect of NiTi and mismatch of thermal expansion of between matrix and fiber. In contrast, Ni/Al and Ti/Al do not have shape memory effect during heating process so they formed weaker compressive residual stress compared with NiTi/Al. Furthermore, the compressive residual stress of these composites is higher at higher temperature. This is because of the degree of thermal expansion of Al matrix is higher when the temperature of composite is higher.

4. Conclusion

The process induced residual stress of NiTi/Al has been simulated by ANSYS APDL version 16.2. Furthermore, two others composite, Ni/Al and Ti/Al, also has been simulated.

After simulation, the three types of composites able to generate compressive residual stress on Al matrix. Compressive residual stress on Al matrix able to enhance the material property of composite by closed the micro crack of the Al matrix of the composite. However, after compared the maximum residual stress generated by difference composites. The NiTi/Al, SMA composite, is the superior because it has about 425% of compressive residual stress value higher than Ni/AL and Ti/Al. The higher value of compressive residual stress is caused by the residual stress of NiTi/Al is induced with combination of shape memory effect and mismatch of thermal expansion of matrix and fiber. Meanwhile, the residual stress of Ni/Al and Ti/Al is only induced by the mismatch of thermal expansion.

Acknowledgement

The authors wish to thank Ministry of Higher Education (MOHE) and Universiti Tun Hussein Onn Malaysia for their financial support.

References

[1] Jani M., Leary J., Subic M., A. & Gibson M. A. (2014). A review of shape memory alloy research, applications and opportunities. Materials and Design 56, 1078-1113.

[2] Shimamoto A., Ohkawara H., & Nogata F. (2004). Enhancement of mechanical strength by shape memory effect in TiNi fiber-reinforced composites. Engineering Fracture Mechanics 71 (4-6), 737-746

[3] Watanabe Y., Miyazaki E., & Okada H. (2002). Enhanced mechanical properties of Fe-Mn-Si-Cr shape memory fiber/plaster smart composite. Materials Transactions 43(5), 974–983.

[4] Ni, D. R., & Ma, Z. Y. (2014). Shape Memory Alloy-Reinforced Metal-Matrix Composites: A Review. Acta Metallurgica Sinica (English Letters) 27(5), 739-761.

[5] Tan W.C., Salleh A.S., Jamian S.,& Ghazali M.I. (2007). Phase trasnformation temperatures for shape memory alloy wire. World Academy of Science, Engineering and Technology 25, 304-307.

[6] 1100 (Al99.0Cu, A91100) Aluminum. (2016, May 8). Retrieved July 02, 2016, from http://www.makeitfrom.com/material-properties/1100-Al99.0Cu-A91100-Aluminum

[7] Special Metals | Home. (2016). Specialmetals.com. Retrieved 25 June 2016, from http://www.specialmetals.com/

[8] Nickel 200 - Alloy Wire International. (2016). Alloywire.com. Retrieved 25 June 2016, from http://www.alloywire.com/nickel_alloy_200.html

[9] Petrini, L., Migliavacca, F., Massarotti, P., Schievano, S., Dubini, G., & Auricchio, F. (2005). Computational Studies of Shape Memory Alloy Behavior in Biomedical Applications. Journal of Biomechanical Engineering 127(4), 716-725.

[10] Jamian S., Sato H., Miura-Fujiwara E., & Watanabe Y. (2011). Finite element analysis of severe plastic deformation of difficult-to-work material by equal-channel angular pressing at ambient temperature. Japanese Journal of Applied Physics 50 (1S1), 01AJ06.

[11] Jamian S., Sato H., Miura-Fujiwara E., & Watanabe Y. (2012). Experimental study on severe plastic deformation of Ti by novel equal-channel angular pressing. Japanese Journal of Applied Physics 51, 01AK04.

Computational Analysis on Stent Geometries in Carotid Artery: A Review

Muhammad Sufyan Amir Paisal[1], Ishkrizat Taib[2], Al Emran Ismail[3]

Faculty of Mechanical and Manufacturing Engineering, Universiti Tun Hussein Onn Malaysia, Parit Raja, 86400 Batu Pahat, Johor, Malaysia.

[1]muhammadsufyanamir@gmail.com, [2]iszat@uthm.edu.my, [3]emran@uthm.edu.my

Abstract. This paper reviews the work done by previous researchers in order to gather the information for the current study which about the computational analysis on stent geometry in carotid artery. The implantation of stent in carotid artery has become popular treatment for arterial diseases of hypertension such as stenosis, thrombosis, atherosclerosis and embolization, in reducing the rate of mortality and morbidity. For the stenting of an artery, the previous researchers did many type of mathematical models in which, the physiological variables of artery is analogized to electrical variables. Thus, the computational fluid dynamics (CFD) of artery could be done, which this method is also did by previous researchers. It lead to the current study in finding the hemodynamic characteristics due to artery stenting such as wall shear stress (WSS) and wall shear stress gradient (WSSG). Another objective of this study is to evaluate the nowadays stent configuration for full optimization in reducing the arterial side effect such as restenosis rate after a few weeks of stenting. The evaluation of stent is based on the decrease of strut-strut intersection, decrease of strut width and increase of the strut-strut spacing. The existing configuration of stents are actually good enough in widening the narrowed arterial wall but the disease such as thrombosis still occurs in early and late stage after the stent implantation. Thus, the outcome of this study is the prediction for the reduction of restenosis rate and the WSS distribution is predicted to be able in classifying which stent configuration is the best.

1. Introduction

Carotid artery is a blood vessel from aorta that circulates blood from neck to the brain. It divides into two which are internal carotid artery and external carotid artery. The abnormality of hemodynamics in carotid artery caused by high blood pressure leads to the implantation of stent. A stent is a device with circular section used to reinforce the internal wall of the artery [1]. However, there are still a few complications that occur after the implantation which are restenosis of blood vessel, thrombosis and embolization. These complications give high mortality and morbidity [2, 3 and 4]. Figure 1 shows the basic anatomy of carotid artery in human body [5].

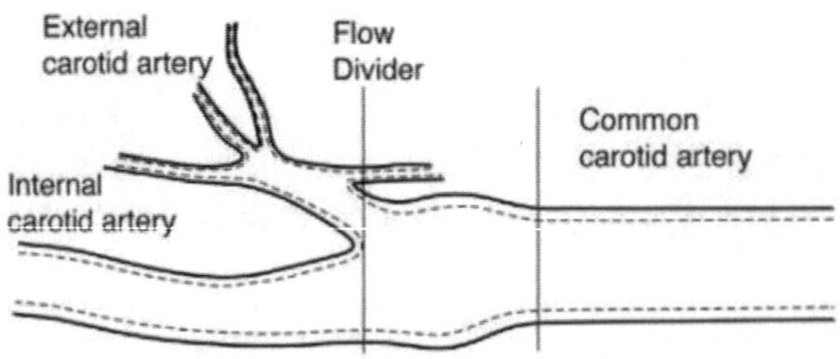

Figure 1. Basic anatomy of carotid artery [5].

The criteria in choosing the right configuration of stent also need to be focused on, because each type of stent has its advantages and disadvantages. There are two type of stent which are opened and closed type. Opened type stent is chosen because the closed type is not suitable to be used in bendable muscular with torsional effect. Opened type stent is then categorized into two which are self-expendable and balloon expendable. Self-expendable is not chosen due to sedative medications that lead into over-sedation, drug reaction and aspiration. Balloon expendable type is selected although it may cause cell proliferation [6].

Stent geometry and its strut configuration also play the main role in maintaining the efficiency of the blood flow in the artery. This is because the flexibility and stiffness are controlled by the parameters which are the material and geometry of the stent. The geometry of the stent has the structure which having a repetitive number of cell or geometrical pattern. Usually, the diameter of stent is two or four times larger of the original one after the implantation [1]. Besides that, the stent strut position also influences the stress distribution and the maximum compressive stress which increase by 10 percent due to changes in the final struts configuration alone when the stent strut is very distant from its neighbor [7].

The development of Computational Fluid Dynamics (CFD) nowadays made it easier to predict the blood flow through the stent. The inlet of this flow is the common carotid artery and the outlets of this flow are internal carotid artery and external carotid artery. However, the CFD analysis considers the outlets as free flow that lead to inaccurate result. The outlets have its own blood pressure accordingly to the cardiac cycle of the heartbeat. The blood pressure in the carotid artery is calculated by using lumped parameter mathematical model. This mathematical model is common in use nowadays for studying the factors that affect pressure and flow waveforms [8].

2. Patient Condition

The condition of patient is important in determining the pressure which to be used in the CFD computation. For this research, the considered four conditions of patient are normal, prehypertension, stage 1 hypertension and stage 2 hypertension. Hypertension is a disease of medical term for high blood pressure [9].

For normotensive condition, the mean pressure in left carotid artery is 100 mmHg and its pulse pressure is 48 mmHg [10, 11]. To get the hypertensive blood pressure in the carotid artery, following formula is used by using the aortic pulse pressure [11].

$$PP_{car} = (0.98 \pm 0.11) \times PP_{ao} + (0.13 \pm 8.14) \quad (1)$$

Where PP_{car} is a carotid pulse pressure and PP_{ao} is a pulse pressure for aortic artery. Since pulse pressure is difference between systolic and diastolic blood pressure, these pressures could be get from the Table 1 [12]. Systolic blood pressure means the highest pressure amount of the heart beats while diastolic blood pressure is the lowest pressure amount of the beating heart. Table 1 shows the classification of normotensive and hypertensive blood pressure in adult patient [12].

Table 1. Classification of normotensive and hypertensive blood pressure [12].

Condition	Systolic Blood Pressure (mmHg)	Diastolic Blood Pressure (mmHg)
Normal	90-119	60-79
Prehypertension	120-139	80-89
Hypertension (Stage 1)	140-159	90-99
Hypertension (Stage 2)	160-179	100-109

The high blood pressure in the artery could cause blocking effect of embolism which the blood flow becomes narrow gradually [2]. Embolism is the effect of embolus formation where the blood clots due to irregular linings of artery. The hardening of arterial walls which called as atherosclerosis is also occurred due to inflammation of inner lining of arteries and deposition of fatty material as plaques as shown in Figure 2 [3]. The rupture of atherosclerotic plaque triggers the formation of blood clot which could also be called as thrombosis that blocks the normal hemodynamics of blood flow in artery [4].

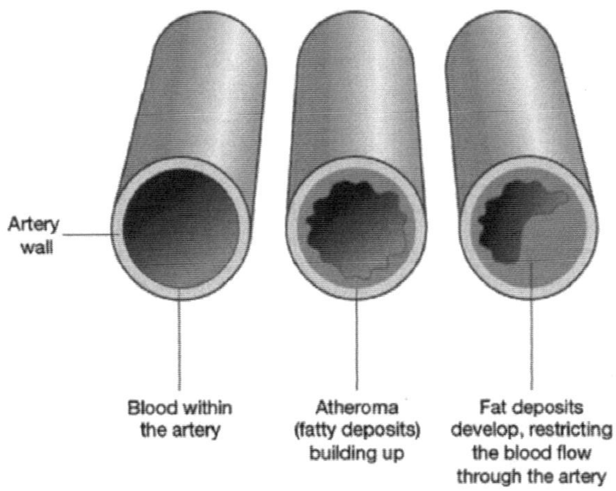

Figure 2. The narrowing blood vessel due to atherosclerosis [3].

3. Artery Stenting

The following topics explain the effects of stent design to the hemodynamic and its arterial diseases, by relating to the previous research that has been done.

3.1. Effect of Stent Configuration to the Artery

Uncontrollable of hypertensive blood pressure could lead to arterial disease which is stroke due to damaged and weakened blood vessel [13]. Stenting of carotid artery in this research could treat the re-stenosis, embolization and thrombosis that occurred.

For this research, balloon expandable of opened type stent is suitable to be applied in the carotid artery. This is due to its good properties in the fully annealed, well adapted mechanical properties, biocompatibility and resistance to the corrosion as well as fatigue performances [14].

Since this research has four conditions, five different designs of stent for each condition are studied. Thus, the blood flow performance of each design is predicted by using CFD method. Figure 3 shows some example of different commercial design of stents.

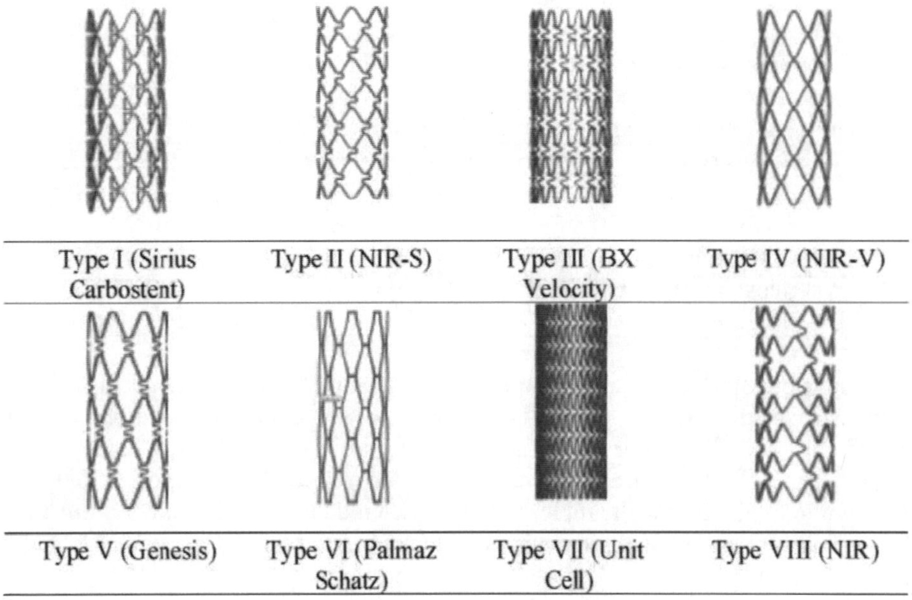

Figure 3. Examples of commercial stent design [15].

In the implantation of stent, each artery has different condition that need to be considered. Thus, there are many designs of stent implemented according to its desired configurations. However, there are still a few complications occurred although a lot of stentings are improved as shown in the Table 2.

Table 2. Significance along with the complications of the previous study.

Researcher	Stent Configuration	Significance	Complication
[16,17]	Bare Metal Stent	Allow shorter dual-antiplatelet therapy duration for high bleeding risk or need of immediate surgery.	High rate of restenosis.
[18]	Drug-Eluting Stent (First Generation)	Risk of Stenosis after percutaneous coronary intervention is reduced.	High risk of late stent thrombosis.
[19]	Drug-Eluting Stent (Second Generation)	More safety and efficiency than the first generation of drug-eluting stent.	Stent thrombosis still occurs.
[20]	Biodegradable Polymer Stent	High efficiency in reducing very late stent thrombosis.	Not reducing the risk of mortality.
[21]	Balloon Expandable Stent	Immediate substantial improvement in lumen diameter after implantation and low rate of re-stenosis.	High potential of early thrombosis and relatively high rate of peripheral vascular complications.
[22]	Self-Expandable Stent	Lower the occlusion rate, less therapeutic failure, less need of re-intervention and lower cholangitis incidence.	Mortality within 30 days.

3.2. Thrombosis Versus Stent Strut Geometry

Different design of stent strut geometry alters different early thrombosis rate. Early thrombosis is a stage which is within 30 days after the stent implantation. Different geometry shape gives different rate of early thrombosis.

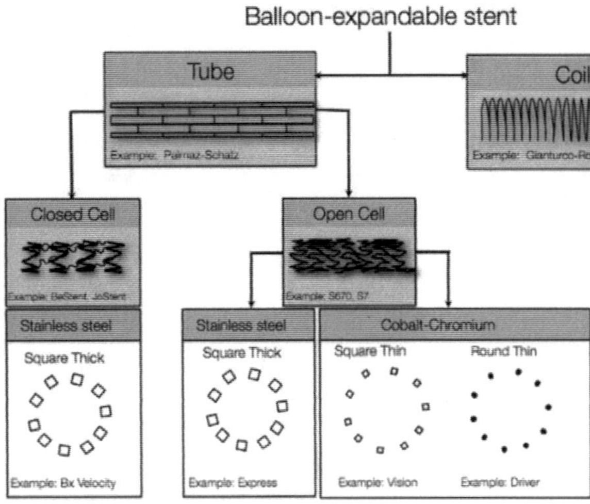

Figure 4. Basic Strut Geometry of Balloon-Expandable Stent [23].

For self-expandable with mesh type stent, the early stent thrombosis is very high which is almost 5 percent. Since the early thrombosis rate is very high, the usage of stent is move onto the balloon-expandable type stent. As shown in Figure 4, the balloon-expandable type stent have two which are tube and coil. The coil type stent has also high early thrombosis rate which is almost 4 percent. For the tube stent type, the early thrombosis is low which is 2 percent and below [23].

If Drug-Eluting Stent (DES) and Bare Metal Stent (BMS) are to be compared, the long-term rate of stent thrombosis for DES is lower than BMS. This is based on the patients treated with DES have a difference about 1.1 percent of weighted risk from the patients treated with BMS through 33 months [24].

3.3. The Hemodynamic Effect of Stent Strut Geometry

The changes of parameters for stent design such as decreasing the strut width, increasing the strut-strut spacing and decreasing the strut-strut intersection, are proved to be in the reduction of restenosis rate. This is due to the mechanical properties of fluid dynamics near to the wall such as intrastrut flow patterns and Wall Shear Stress (WSS) [25].

A higher normalized Wall Shear Stress enhances the hemodynamic performance of the stent. Furthermore, the most optimum stent design is defined which have the most minimum area of low intrastrut time averaged WSS [26].

Wall Shear Stress Gradient (WSSG) is a parameter to represent the non-uniformity in the flow patterns. WSSG that having the value of 20 dynes/cm^3 and above is likely to have a cellular proliferation and Neointimal Hyperplasia (NIH) in the artery. The value of WSSG is high near the strut and remains distinctly non-zero for most of the region between the struts of the stent.

The previous study that has been done by Ladisa et al. [27] shows that WSS is lower at the stented region for all simulation and the stagnation zones occurred around stent strut. The maximum value of WSSG is occurred at four-strut polygonal which it is higher than 20 dynes/cm^3. However, the maximum WSSG did not occur in the eight-strut polygonal stent. This can be seen in Figure 5 that investigating the value of WSS and WSSG which affected by different design of stent configuration.

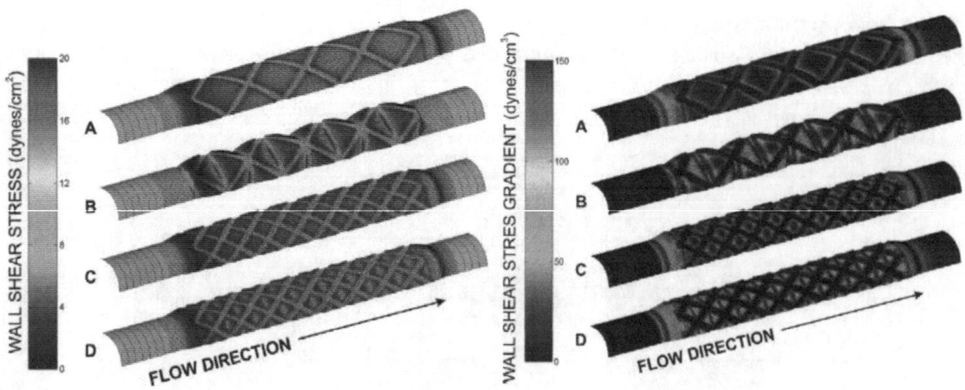

Figure 5. CFD analysis of WSS (left) and WSSG (right) for four different design of stent configuration [27].

3.4. Flow Characteristic on Single Strut and Complete Pattern Strut

The following topics explain the hemodynamic characteristic based on single strut stent and complete pattern strut.

3.4.1. Single Strut

A study on a single strut is a concern because it affects the spatial distribution of WSS. The change in spatial WSS distribution is also a change to the progression of endothelialization, neointimal hyperplasia, and restenosis.

The stent design with intrastrut of approximate 40 degree angle minimize the area of low time-averaged WSS regardless of intrastrut area or vessel size since the design of stent that have minimum area of low time-averaged WSS is considered as optimum stent design. Figure 6 shows the parameters that affecting the WSS [26].

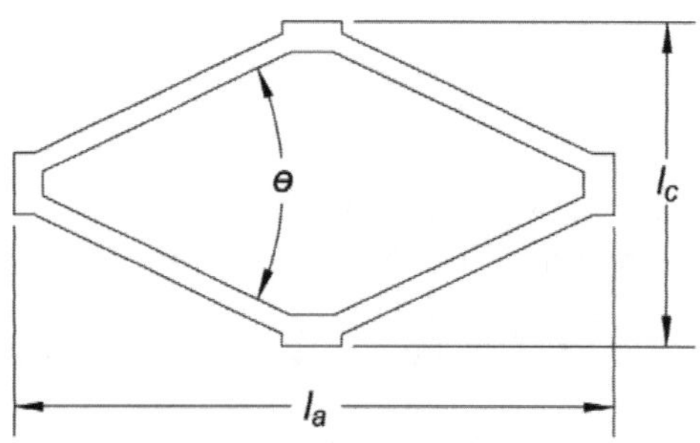

Figure 6. Parameter on single strut which Θ is angle, l_a is length and l_c is width [26].

3.4.2. Complete Pattern Strut

Compared to the study of single strut, the analysis of complete pattern strut plays the main role of restenosis by using CFD method. The hemodynamic behavior and WSS distribution could be obtained in overall.

Based on the study that has been done by Hsiao et al. [28], the simulated result shows the minimum WSS occurs at the circulation zones located at the downstream or backside of each stent struts. At the regions near the backside of each stent strut, the flow velocity and shear rate are fairly low, but a large disparity in viscosity is exist in Newtonian and non-Newtonian flow models. As shown in Figure 7, the WSS value is investigated in two models which are Newtonian and Non-Newtonian flow.

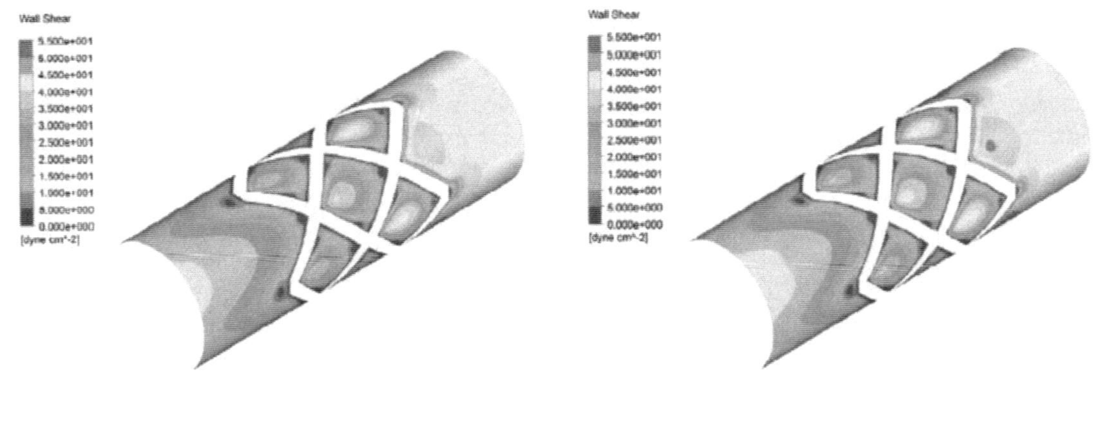

(a) (b)

Figure 7. WSS contours for (a) Newtonian flow and (b) non-Newtonian flow model [28].

3.5. Different Shape of Stent Geometry

Shape of stent geometry is commonly dictated by manufacturing limitations but it can play a fundamental role in the improvement of hemodynamic performance [29].

Circular shape stent is the most commonly used in the artery. However, Kim et al. [30] states that the efficiency of stent is related to several parameters including porosity and stent strut shapes. Thus, the concept of flow reduction is used in characterizing the flow efficiency in rectangular shape stent. The rectangular shape stent is observed to be optimal and to decrease the magnitude of the velocity by 53.92% in the 3D model and 89.25% in the 2D model in the aneurysm sac.

In other hand, Mejia et al. [29] also did a study on the evaluation of stent strut profile on shear stress distribution using statistical moments. The result shows that, the streamlined profile exhibits better hemodynamic performance than the standard circular and rectangular profile. For tear drop profile, up to 96% of the area between strut is exposed to the wall shear stress levels above the critical value for the onset of restenosis while for the square profile, the analogous value is 19.4%. Figure 8 shows the investigated stent strut profile that affecting shear stress distribution.

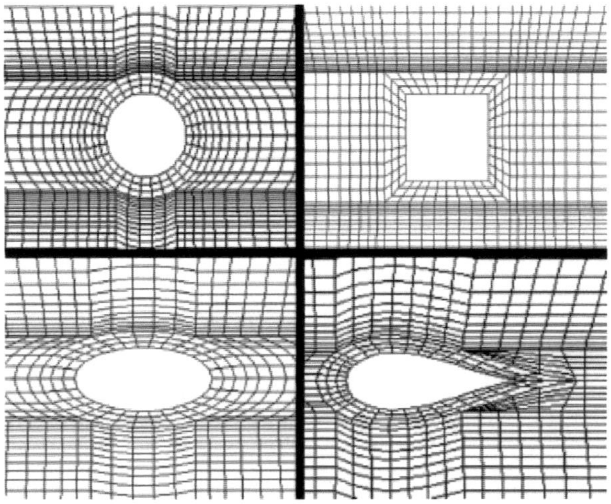

Figure 8. Stent strut profile of circular (top-left), square (top-right), elliptical (bottom-left) and tear-drop (bottom-right) [29].

Dumoulin et al. [31] who studied the mechanical behavior modeling of balloon-expandable stents states that the lowest value of critical pressure by the radial compression or geometric nonlinearity influence is important to maintain the long-term usage. The geometry mode of stents being studied are elliptical, triangle and rectangular. As shown in the figure below, the elliptical mode has the lowest critical pressure which is 0.116 MPa among the three modes to withstand the radial compression. Figure 9 shows the obtained critical pressure according to the shape of stent geometry.

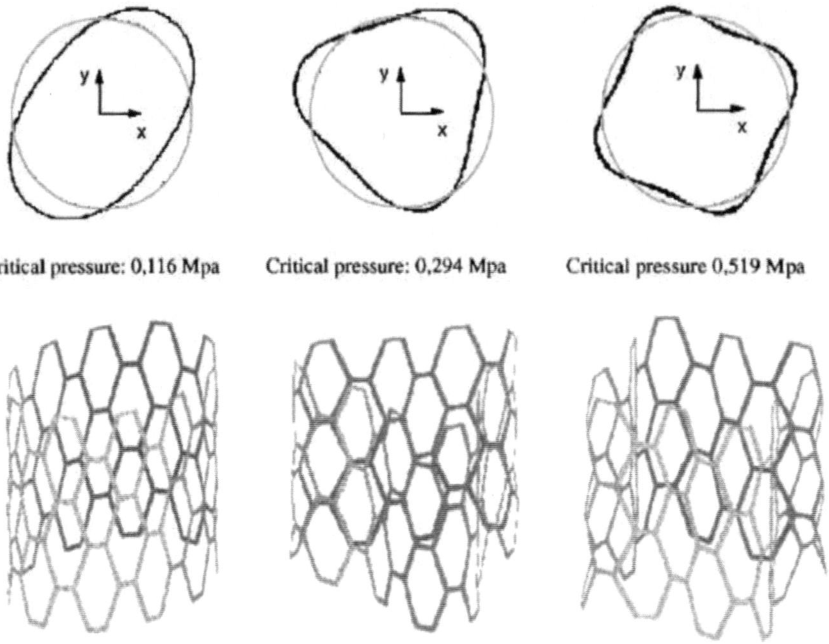

Figure 9. Critical pressure according to the shape of stent geometry [31].

4. Influence of Lumped Parameter Mathematical Modelling to the Artery

The first step in doing the CFD is to have all the required parameters. To do so, mathematical model of a carotid artery need to be done. The use of this mathematical model is to simplified the uneven and complexity of biological condition into computable data [32]. Most of the mathematical model of an artery use electrical schematic model as an analogy for the flow in a tube.

To be in detail, Table 3 shows the physiological parameters of artery which respectively describe the parameters of fluid dynamic as electrical analogue [33].

Table 3. The quantities of fluid dynamics and their analogues [33].

Fluid dynamics	Physiological variables	Electrical analogue
Pressure P [Pa=J/m^3]	Blood pressure [mmHg]	Voltage U [V=J/C]
Flow rate Q [m^3/s]	Blood flow rate [L/s]	Current I [A=C/s]
Volume V [m^3]	Blood volume [V]	Charge q[C]
Viscosity η	Blood resistance $\square = \frac{8\eta Nl}{\pi r^4}$	Electrical resistance R
Elastic coefficient	Vessel's wall compliance	Capacitor's capacitance C
Inertance	Blood Inertia	Inductor's inertance L
Poiseuille's law:	$Q = \frac{\Delta P}{\square} = \frac{\Delta P \pi r^4}{8\eta\ell}$	Ohm's law: $I = \frac{\Delta U}{\square}$

In this research, the mathematical model used is lumped parameter. This model describes the physics of the problem in an easier way and changing the model parameters give insight to the behavior of the carotid artery [34].

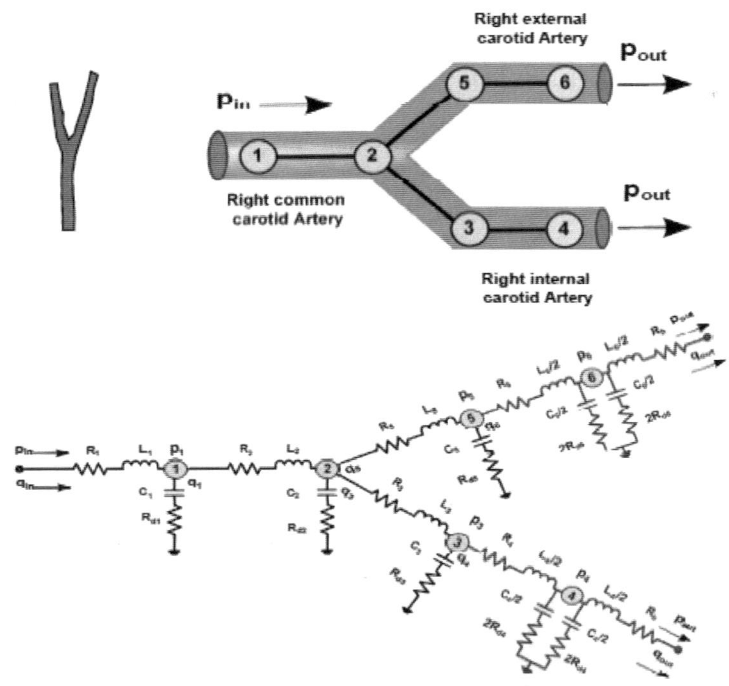

Figure 10. Simplified anatomy (top) and electrical schematic diagram (bottom) of a right carotid artery [34].

Figure 10 illustrates on how the anatomy of a right carotid artery is changed into electrical schematic diagram. To find the pressure at each node, following equations are used [8]:

$$P_{(i-1)} - P = q_i R_i + L_{V_i} \frac{dq_i}{dt} \quad (2)$$

$$q_{C_i} = C_{V_i} \frac{dP_i}{dt} \quad (3)$$

Table 4 shows the findings by the previous researchers [34, 35, 36, 37, 38] on how they modeled the artery into computable form.

Table 4. Type of mathematical model used for artery by previous researchers.

Researcher	Type of mathematical model used	Findings
[34]	Lumped parameter mathematical model on human cardiovascular system.	Able to measure pressure and flow of the hemodynamics.
[35]	Finite difference method with the governing equations of motion of incompressible steady couple stress fluids in the absence of body force and body moment.	It leads the shear stress, flow rate and impedance to be changed suddenly with all the parameters of both sides of apex in the artery.
[36]	One dimensional wave propagation theory and assumes the pressure waveform is a superposition of forward propagating wave and backward waves from many reflection sites.	This model shows the age-related changes in blood pressure waveform. If the pressure wave form is impossible to be obtained due to moving subject, this model is suitable to be used.
[37]	A stenosed artery could be obtained by using Herschel-Bulkley fluid which the blood flow is considered axially symmetric, laminar, pulsatile, fully developed flow and assumed incompressible.	The variation of pressure gradient corresponding to the axial distance is obtained.
[38]	A Windkessel model which the small blood vessels or capillaries are considered as resistance R and the large blood vessels are considered as compliance C.	This method able to estimate the blood pressure in the artery because of its simplest form using 2 element equations but conflicts with other modern methods in terms of wave propagation and reflection.

5. CFD of Blood Flow

Next step is doing the CFD method for the blood flow through stented carotid artery. Computational Fluid Dynamics (CFD) nowadays could be applied to simulate and analyze the flow pattern of hemodynamic in artery. Thus, there a few assumptions made to simplify the computation.

Uemiya et al. [39] did a study on analysis of restenosis after carotid artery stenting. In the work that has been made, the blood is assumed to be Newtonian, incompressible and laminar flow with a constant dynamic viscosity of 0.0035 Pa s and density of 1050 kg/m^3. Besides that, the wall of the patient is considered as rigid. The inflow of blood is ranged from 250 ml/minute to 380 ml/minute.

Conti et al. [40] also did a study on carotid artery hemodynamics of before and after stenting. For the CFD analysis, the outflow sections consisting external and internal carotid artery, are prescribed by using Windkessel model. The peripheral impedance for both outflows is represented by two resistances R_1, R_2 and one compliance C.

6. Validation and Verification

The computation of lumped parameter mathematical model needs a validation by referring to the previous researcher in order to know whether the method is correct or not with the percentage of accuracy. If the method is accurate enough, the work can be proceed to the CFD of carotid artery stenting.

The involved case for the validation is the pressure waveform of healthy patient condition in carotid artery. Figure 10 below shows the pressure waveform of carotid artery which to study about the validation of carotid artery as a means of estimating augmentation index of ascending aortic pressure by Chen et al. [41]. Solid curve is carotid AI (AI_t) while dotted curves is aortic AI (AI_m). Figure 10(a) shows carotid waveform has a clear inflection point on the upstroke, but no inflection point could be found on the aortic pressure curve. Figure 10(b) shows no inflection point could be found on either waveform. For Figure 10(c), the inflection points were on the upstrokes of both waveforms while for Figure 10(d), the inflection points were on the downstrokes of both waveforms. In Figure 10(e), the inflection points were on the downstroke of carotid and on the upstroke of aortic waveforms. Lastly for Figure 10(f), the inflection points were on the upstrokes of both waveforms, but AI_t was greater than AI_m [41].

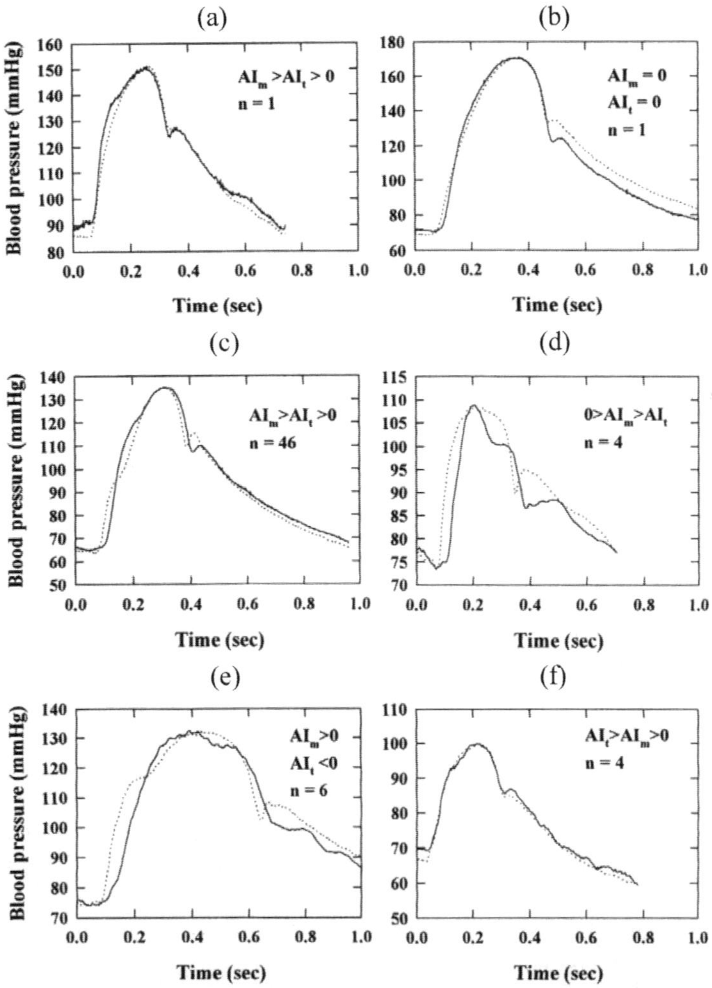

Figure 11. Summary comparisons of waveforms and augmentation indexes (AIs) at baseline in adult patient [41].

7. Conclusion and Future Direction

Over the years, the CFD of hemodynamic in artery stenting become more relevance in predicting the mechanical properties and performance of blood. However, this work requires a high level of understanding of bio-fluid mechanism in relating to the schematic diagram of lumped parameter circuit.

At the end of this study, the resulting of WSS distribution is predicted to be able in classifying which stent design is the best in overcoming the arterial disease of hypertension such as restenosis, thrombosis and embolization. Thus, the rate of failure in the carotid artery could be reduced. The calculated WSS distribution could also reduce the try-and-error method which has been done by previous clinicians.

In the new paradigm of surgical field, the community is hoped to have a better healthcare with the use of today's computational and evaluation method along with the stent technologies.

References

[1] Guimaraes T A, Duarte M A V and Oliveira S A G 2005 Topology Optimization of the Stents Cells Plane Model with Maximum Hardening and Flexibility *Proceedings of the Inverse Problems, Design and Optimization (IPDO - 2004) Symposium, Rio de Janeiro, Brazil, March 17-19, 2004* ed Colaço M J et al (Rio de Janeiro: E-papers Pub. House) p 192.

[2] Ramdass and Shivakumar A 2005 *A Guide to High Blood Pressure* (New Delhi: Lotus Press).

[3] Aziz S and Aziz Z 2015 *Understanding High Blood Pressure* (London: SPCK).

[4] Sanders T and Emery P 2003 *Molecular Basis Of Human Nutrition*. (London: Taylor & Francis).

[5] Polak J F, Pencina M J, Meisner A, Pencina K M, Brown L S, Wolf P A and D'Agostino R B 2010 *Journal of Ultrasound in Medicine: Official Journal of the American Institute of Ultrasound in Medicine* **29**(12) 1759-1768.

[6] Casserly I P, Sachar R and Yadav J S 2011 *Practical Peripheral Vascular Intervention* (Market Street: Lippincott Williams & Wilkins).

[7] Amatruda C M, Hose D R, Lawford P and Narracott A 2012 *Journal of Biomechanics* **45** S25.

[8] Waite L and Fine J M 2007 *Applied Biofluid Mechanics* (New York: McGraw-Hill).

[9] Brewer S 2014 *Overcoming High Blood Pressure: The Complete Complementary Health Program* (Oxford: Duncan Baird Publishers).

[10] Khamdaeng T, Luo J, Vappou J, Terdtoon P and Konofagou E 2012 *Ultrasonics* **52**(3) 402-411.

[11] Boutouyrie P, Bussy C, Lacolley P, Girerd X, Laloux B and Laurent S 1999 *Circulation* **100**(13) 1387-1393.

[12] Chobanian A V, Bakris G L, Black H R, Cushman W C, Green L A, Izzo J L and Jones D W 2003 *Hypertension* **42** 1206-1252.

[13] Siddiqui A, Morr S and Lin N 2014 *Medical Device: Evidence and Research* **7** 343-355.

[14] Auricchio F, Constantinescu A, Conti M and Scalet G 2015 *International Journal of Fatigue* **75** 69-79.

[15] Taib I 2016 *Improvement of Haemodynamic Stent Strut Configuration for Patent Ductus Arteriosus Through Computational Modelling* (Johor: Universiti Teknologi Malaysia).

[16] Wiebe J, Colleran R and Kastrati A 2016 *JACC: Cardiovascular Interventions* **9**(12) 1256-1258.

[17] Mauri L, Silbaugh T S, Garg P, Wolf R E, Zelevinsky K, Lovett A and Varma M R 2009 *New England Journal of Medicine* **360**(3) 300-302.

[18] Kaiser C, Galatius S, Erne P, Eberli F, Alber H, Rickli H and Pedrazzini G 2011 *New England Journal of Medicine* **364**(12) 1178-1180.

[19] Shi H, Chu H, Gu W, Cai X, Guo J, Ding Z and Wen S 2016 *International Journal of*

Cardiology **214** 393-397.

[20] Yang Y, Lei J, Huang W, and Lei H 2016 *International Journal of Cardiology* **222** 486-493.

[21] Savage M P, Fischman D L, Schatz R A, Teirstein P S, Leon M B, Baim D and Goldberg S 1994 *Journal of the American College of Cardiology* **24**(5) 1207-1212.

[22] Sawas T, Al-Halabi S, Parsi M A and Vargo J J 2015 *Gastrointestinal Endoscopy* **82**(2) 256-267.

[23] Mukherjee D 2010 *Cardiovascular Catheterization and Intervention: A Textbook of Coronary, Peripheral, and Structural Heart Disease* (New York: Informa Healthcare).

[24] Kereiakes D J, Yeh R W, Massaro J M, Driscoll-Shempp P, Cutlip D E, Steg P G and Mauri L 2015 *JACC: Cardiovascular Interventions* **8**(12) 1552-1562.

[25] Duraiswamy N, Schoephoerster R T and Moore J E 2009 *Journal of Biomechanical Engineering* **131**(6) 061006.

[26] Gundert T J, Marsden A L, Yang W and LaDisa J F 2012 *Journal of Biomechanical Engineering* **134**(1) 011002.

[27] LaDisa J F 2004 *Journal of Applied Physiology* **98**(3) 947-957.

[28] Hsiao H, Lee K, Liao Y and Cheng Y 2012 *Procedia Engineering* **36** 128-136.

[29] Mejia J, Ruzzeh B, Mongrain R, Leask R and Bertrand O F 2009 *BioMedical Engineering OnLine* **8**(1) 8.

[30] Kim Y H, Xu X and Lee J S 2010 *Annals of Biomedical Engineering* **38**(7) 2274-2292.

[31] Dumoulin C and Cochelin B 2000 *Journal of Biomechanics* **33**(11) 1461-1470.

[32] Abdi M, Navidbakhsh M and Razmkon A 2016 *J Biomed Phys Eng* **6**(1) 33-40.

[33] Kokalari I, Karaja T and Guerrisi M 2013 *Journal of Biomedical Science and Engineering* **6**(1) 92-99.

[34] Raheem G 2016 *Mathematical Modelling and Sensitivity Analysis of Lumped-Parameter Model of the Human Cardiovascular System* (Abbottabad: Freien Universitat Berlin).

[35] Srinivasacharya D and Rao G M 2016 *Mathematical Biosciences* **278** 37-47.

[36] Lazović B, Mazić S, Zikich D and Žikić D 2015 *Wave Motion* **56** 14-21.

[37] Sankar D and Hemalatha K 2006 *International Journal of Non-Linear Mechanics* **41**(8) 979-990.

[38] Tsanas A, Goulermas J Y, Vartela V, Tsiapras D, Theodorakis G, Fisher A C and Sfirakis P 2009 *Medical Engineering & Physics* **31**(5) 581-588.

[39] Uemiya N, Lee C, Ishihara S, Yamane F, Zhang Y and Qian Y 2013 *Journal of Clinical Neuroscience* **20**(11) 1582-1587.

[40] Conti M, Long C, Marconi M, Berchiolli R, Bazilevs Y and Reali A 2016 *Computers & Fluids* **000** 1-13.

[41] Chen C, Ting C, Nussbacher A, Nevo E, Kass D A, Pak P and Yin F C 1996 *Hypertension* **27**(2) 168-175.

Axial quasi-static crushing behaviour of cylindrical woven kenaf fiber reinforced composites

Noorhaslinda Abdullah, Al Emran Ismail*

Faculty of Mechanical and Manufacturing Engineering, Universiti Tun Hussein Onn Malaysia, Batu Pahat, 86400 Johor, Malaysia.

*emran@uthm.edu.my / al_emran@hotmail.com

Abstract. This paper presents the crushing responses of cylindrical woven kenaf fibre reinforced composites under quasi-static compression. Kenaf fiber in the form of yarn is weaved into woven mat. It is then submerged into polyester bath before it is wrapped into a cylindrical shape. There are two important parameters investigated such as number of layers and fiber orientations. According to the experimental results, as expected increasing the number of layers increased the energy absorption performances. However, increasing the fibre orientations from 0° to 45° capable to decrease the energy absorption capability. It is also observed that during progressive collapses, localized buckling is the dominant failure mechanism where there is no large composite fragmentation occurred.

1. Introduction

The utilizations of natural fibers in fabricating the composite structures increased especially for non-load bearing applications. Due to environmental awareness and the potential of natural fibers, some researchers [1-10] investigated the capability of these fibers to withstand under quasi-static and dynamic forces. One of the important research areas for composite applications is crashworthiness where the composite is used to manage the crushing behavior in order to minimize the injuries or fatalities.

Crushing performances of engineering materials for an example metal [1] and synthetic fiber reinforced composites [2] can be found tremendously in open literature. However, due to environmental awareness among researchers, synthetic fibers are increasingly replaced with natural fibers. One of the aspect of research works conducted using natural fiber is crashworthiness [3]. According to literature survey, lot of works conducted to study the crushing behavior of natural fiber and most of them used randomly oriented fibers in fabricating the composites. It is hard to find the works deal with the woven natural fiber reinforced composites. For an example Alkbir et al. [4] conducted an investigation on the effect of geometries of hexagonal tubes fabricated using natural kenaf fibers. In their studies, angles played an important role in determining the crushing performances.

On the other hand, Yang and Chouw [5] performed an investigation on the crashworthiness characteristics of flax composite tubes. In fabricating their composite tubes, three crucial important factors are used for example inner diameter, number of thickness and length-to-diameter ratios. They stated that flax fiber has a great potential to be used as an energy absorbing fiber.

Additionally, most of the natural fibers used are randomly oriented [6]. Recently, several research works conducted in [3-9] concentrating on the applications of oriented fibers in producing the composite tubes. Ismail and Sahrom [10] investigated the lateral crushing energy absorption of cylindrical kenaf fiber reinforced composites. Three fiber orientations and number of layers are used in making the composite tubes and they are quasi-statically crushed. It is found that crushing response is strongly related with the number of layers and unfortunately fiber orientation is not played an important role in affecting the energy absorption. Some works on axial energy absorption since the tubes are crushed laterally of kenaf fiber composites can also be found in [11-20].

In this work, kenaf fiber in the form of yarn is weaved into a plain woven mat. The woven mat is firstly immersed into a polyester resin bath before it is wrapped around the cylindrical mold. There are three type of thicknesses are considered such as two, three and four layers. Each layer is aligned to the specific directions for example 0°, 15°, 30° and 45°. Once the composite tubes hardened, they are quasi-statically crushed to obtain the responses of force versus displacement. Then, the crashworthiness characteristics are extracted and discussed with the relation with of number of layers and orientations.

2. Experimental Program

In this work, kenaf fiber in the form of yarn is used and then weaved into a plain woven type structure where there are two sets of fiber such as warp and weft. There fibre are perpendicular to each other and thus creating an unbalanced plain woven mat. The number of warp and weft are determined with 6x4 where there are 6 yarns in warp and 4 yarns in weft directions per $10mm^2$. An in-house weaving machine is used as shown in Figure 1(a) and the completed woven mat is revealed in Figure 1(b).

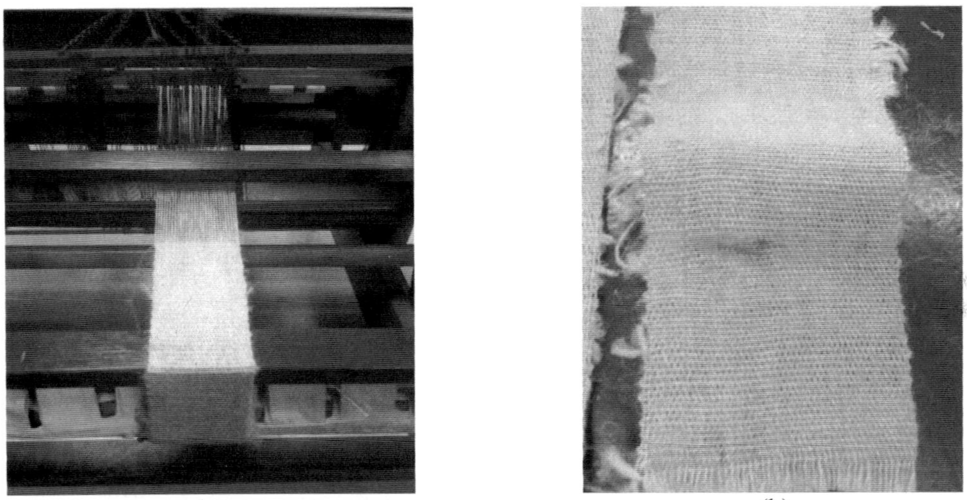

(a) (b)

Figure 1 (a) Weaving process using in-house machine and (b) Completed woven kenaf mat.

Kenaf woven mats are then submerged into polyester resin and a special attention is paid to ensure resin uniformly distributed into the fibres. Then, wet kenaf mats are wrapped around the cylindrical mould before they are circumferentially compressed to squeeze out any excessive resin. After 24 hours of curing time, the mould is removed and any unwanted composites especially at both edges are trimmed smoothly to ensure the total length is 70mm. In this work, two, three and four layers are used and each layer has its own orientation as tabulated in Table 1.

Cylindrical-shaped woven kenaf reinforced composites are quasi-statically compressed using a constant cross-head displacement of 5mm/min. All composite tubes are crushed up to 60% of the original length in order to obtain the curves of force versus displacement. The area under the curves represented the energy absorption performances. On the other hand other crashworthiness parameters are also determined such as peak force, mean force and force ratio. Peak force is defined as the maximum elastic force, mean force is the average fluctuating forces occurred during the collapsing processes and force ratio is the ratio between the peak with mean forces. The effect of fibre orientations and number of layers on the energy absorption performances and the force ratios are discussed.

Table 1 Types of sample used in this work.

Samples	Orientation Θ°	Thickness (cm)	Inner Diameter (cm)	Length (cm)
C1	[0°/0°]	0.5	5.0	7.0
C2	[0°/15°]	0.5	5.0	7.0
C3	[0°/30°]	0.5	5.0	7.0
C4	[0°/45°]	0.5	5.0	7.0
C5	[0°/0°/0°]	0.7	5.0	7.0
C6	[0°/15°/0°]	0.7	5.0	7.0
C7	[0°/30°/0°]	0.7	5.0	7.0
C8	[0°/45°/0°]	0.7	5.0	7.0
C9	[+15°/0°/-15°]	0.7	5.0	7.0
C10	[+30°/0°/-30°]	0.7	5.0	7.0
C11	[+45°/0°/-45°]	0.7	5.0	7.0
C12	[0°/0°/0°/0°]	1.0	5.0	7.0
C13	[0°/+15°/-15°/0°]	1.0	5.0	7.0
C14	[0°/+30°/-30°/0°]	1.0	5.0	7.0
C15	[0°/+45°/-45°/0°]	1.0	5.0	7.0

3. Results and Discussion

Figure 2 shows the force-displacement curves of different layers cylindrical tubes under axial compression. It is found that as expected, the crushing responses are significantly related with the thicknesses of the cylindrical wall where the peak forces increased as the thickness increased. In general, the crushing responses can be divided into three stages, the first stage is the elastic deformation. In this stage, the displacement is proportional to the force. Once the tubes experienced the peak force, the first collapse occurred. Localized buckling is the responsible mechanism in collapsing the tubes normally initiated at the either at the upper or lower edges of the tubes. It is also observed that for four-layered tubes, large force drop occurred indicating that probably the tubes collapsed catastrophically as shown in Figure 2(c) compared with two-layered composite shown in Figure 2(a). Then, the crushing mechanism entered the second stage where the tube wall experienced progressive collapses. It is continued until all the composites are crushed and the force is gradually increased since there is no composites to be collapsed. It is then entered the last stage which is the densification stage. According to the final observation, composite wall still intact together and no large or small composite fragmentations can be observed. This is contradictory mechanisms as reported in [4, 5] where during crushing processes large composite fragmentations contributing to lower the crushing performances.

Figure 3 reveals the effect of number of layers on the crushing responses. Each composite tubes are fabricated using identical orientations but different in number of layers. It is also indicated that similar behaviour of crushing patterns can be observed where thicker the composites higher the responses of force-displacement curves. Even though higher peak forces are produced, the composites experienced catastrophic failure since the peak force dropped gradually. This behaviour contributed to lower the specific energy absorption.

Figure 4 indicates the effect of number of layers with different fibre orientations on the force-displacement curves under axial compressions. It is revealed that increasing the number of layers increased the responses of force-displacement. No strong effect is also observed on the force-displacement curves if fibre orientations increased from 15° to 45°. However, the composites containing 15° fibre orientations sustained higher responses especially the peak forces. This is due to the fact that during crushing process, circumferential stress is the most dominant force. Then, 15° fibre orientation capable to resist the circumferential deformation and therefore strengthening the composite walls.

According to Table 2, it is obvious that increasing the angles from 0° to 45° have decreased the specific energy absorption performances regardless of their thicknesses. This is due to the fact that 0° fibre alignments capable to strengthen the composite wall circumferentially. During compression, fibres not only experienced axial but radial deformations then [0°/0°], [0°/0°/0°] and [0°/0°/0°/0°] capable to provide a maximum strengthening mechanism in axial and radial directions.

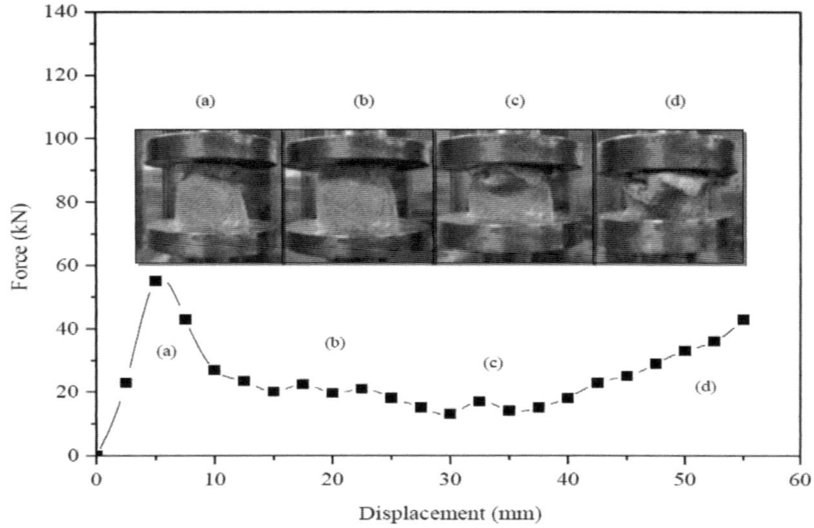

(a)

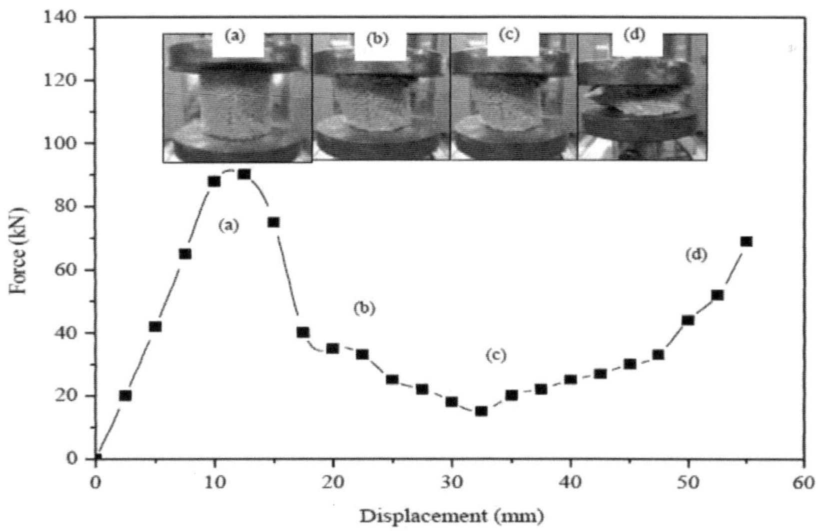

(b)

Figure 2 Continued….

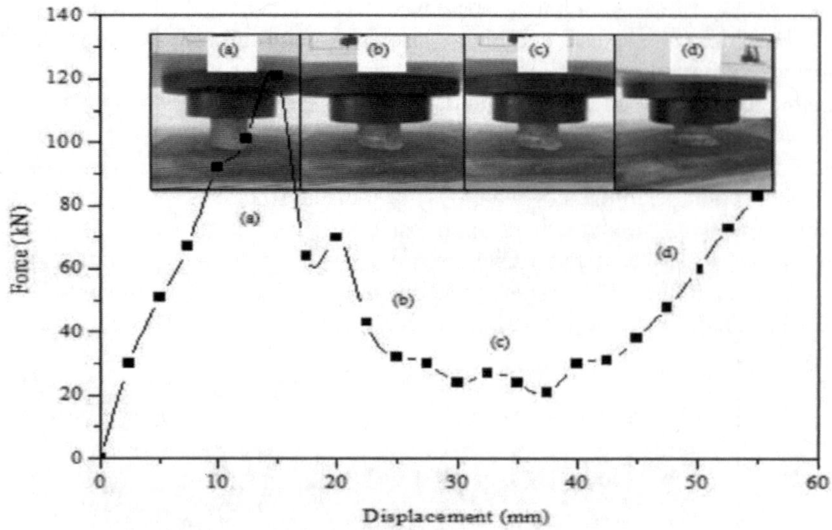

(c)

Figure 2 Crushing responses of (a) two-, (b) three- and (c) four-layered cylindrical composite tubes under axial quasi-static compression.

Table 2 Effect of fibre orientations and number of layers on the peak force, mean force and energy absorption.

	Specimens	Peak force (kN)	Mean force (kN)	Energy absorption (kJ)	Specific energy absorption (kJ/kg)
2-layers	[0°/0°]	62.4	20.7	1.025	15.530
	[0°/15°]	55.0	18.0	0.754	11.781
	[0°/30°]	42.1	11.2	0.687	10.905
	[0°/45°]	46.7	14.3	0.561	9.350
3-layers	[0°/0°/0°]	101.0	32.1	2.189	22.567
	[0°/15°/0°]	91.8	28.4	1.371	14.581
	[0°/30°/0°]	69.2	21.6	1.173	12.748
	[0°/45°/0°]	73.4	25.2	0.957	10.638
	[15°/0°/15°]	71.2	25.9	1.070	12.442
	[30°/0°/30°]	65.4	18.5	0.903	11.153
	[45°/0°/45°]	69.7	21.7	0.804	10.180
4-layers	[0°/0°/0°/0°]	123.7	24.2	6.414	53.450
	[0°/15°/15°/0°]	121.0	22.5	5.248	47.280
	[0°/30°/30°/0°]	93.1	19.5	3.583	33.489
	[0°/45°/45°/0°]	105.6	20.4	2.163	21.2111

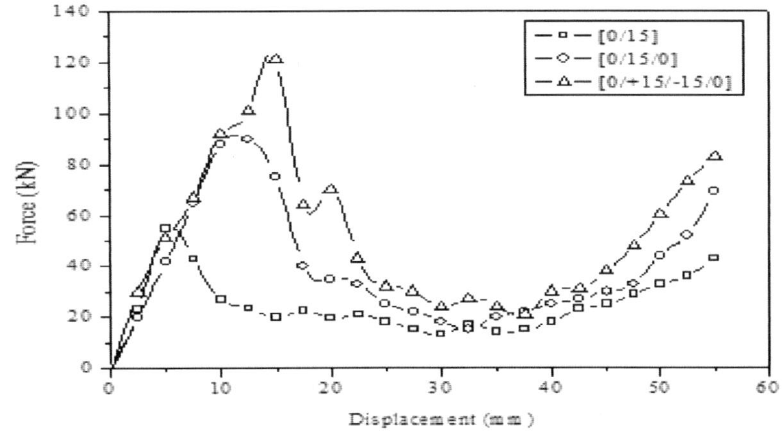

(a)

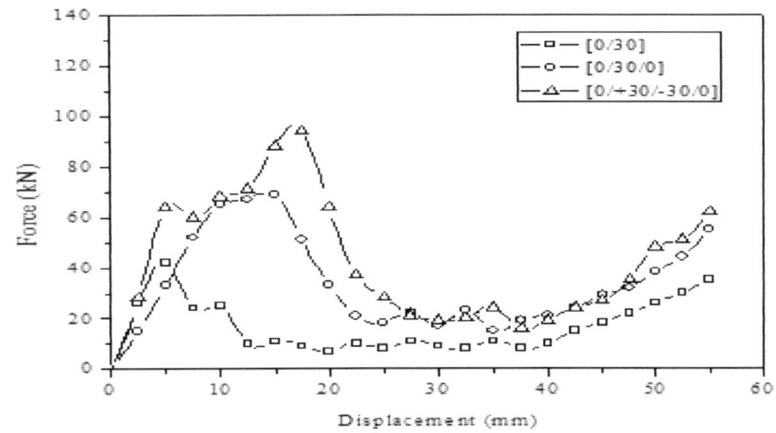

(b)

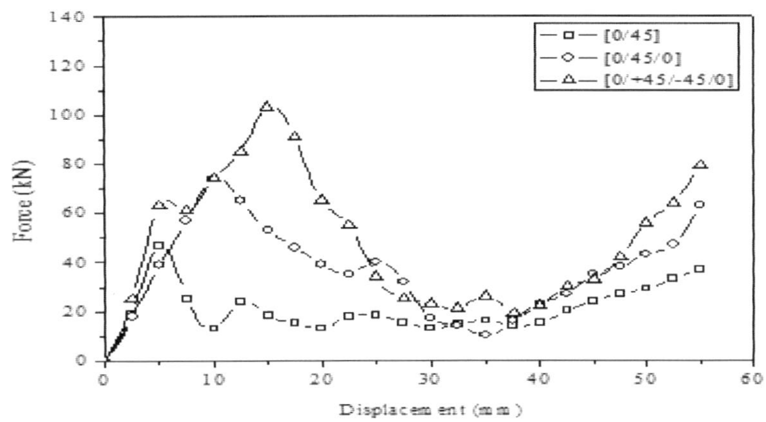

(c)

Figure 3 Effect of number of layers on the force-displacement curves for composites contained (a) 15°, (b) 30° and (c) 45°.

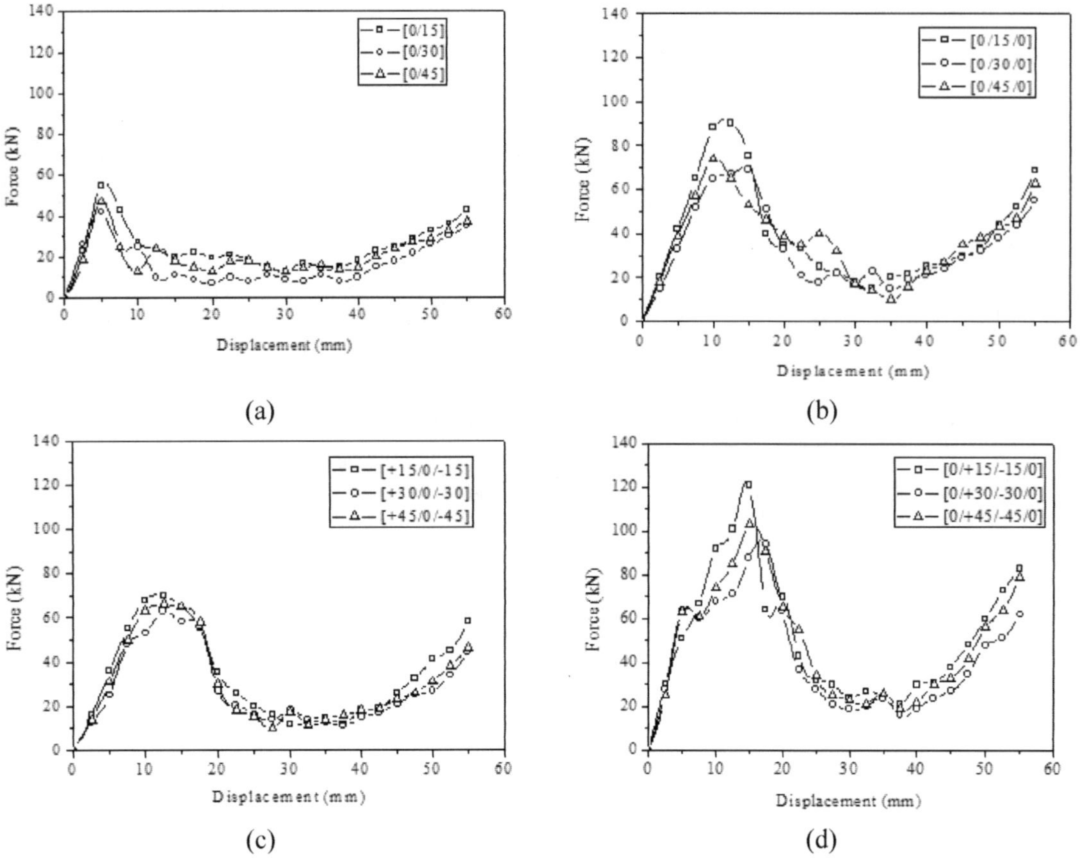

Figure 4 Effect of fibre orientations on the force-displacement curves, (a) two-layered, (b) three-layered, (c) three-layered (different arrangements) and (d) four-layered cylindrical composite tubes.

4. Conclusion

An experimental program is conducted on cylindrical composite tubes fabricated using woven kenaf fibers. There are three different thicknesses are used depending on the number of layers. Each layer has its own orientation. The composite tubes are quasi-statically compressed and crushing response is automatically recorded. Then, several conclusions can be drawn such as:

1. Crushing responses of these composites are significantly depend on the number of layers and fiber orientations.
2. As expected increasing number of layers increasing the specific energy absorptions.
3. Increasing the fiber orientations from 0° to 45° capable to decrease the specific energy absorptions.
4. Most of the composites collapsed through localized buckling especially at the upper edge/moving plate boundary. Then, the buckled wall progressive collapsed without showing large composite fragmentations.

Acknowledgement

Authors acknowledged Universiti Tun Hussein Onn Malaysia and Center for Graduate Studies for sponsoring this work.

References

[1] El-Hage H, Mallick PK and Zamani N 2006 *Composite Structures* **73** 505-514.
[2] Mahdi E, Hamouda AMS and Sebaey TA 2014 *Materials and Design* **56** 923-928.

[3] Misnon MI, Islam MM, Epaarachchi JA and Lau KT 2014 *Materials and Design* **59** 359-368.
[4] Alkbir MFM, Sapuan SM, Nuraini SM and Ishak MR 2014 *Materials and Design* **60** 85-93.
[5] Yan L and Chouw N 2013 *Materials and Design* **51** 629-640.
[6] Akil HM, Omar MF, Mazuki AAM, Safiee S, Ishak ZAM and Abu Bakar A 2011 *Materials and Design* **32**(8-9) 4107-4121.
[7] Abdullah NH, Irwan MN, Ismail AE 2015 *Applied Mechanics and Materials* **773-774** 123-128.
[8] Khalid SNA, Ismail AE and Zainulabidin MH 2105 *Applied Mechanics and Materials* **773-774** 134-138.
[9] Hassan MA and Ismail AE 2015 *Applied Mechanics and Materials* **773-774** 149-153.
[10] Ismail AE and Sahrom MF 2015 *International Journal of Applied Engineering Research* **10**(8) 19277-19288.
[11] Ismail AE 2015 *International Journal of Engineering and Technology* **6**(6) 2662-2668.
[12] Ismail AE 2015 *International Journal of Engineering and Technology*, **7**(3) 1098-1104.
[13] Hedayati R, Ziaei-Rad S, Eyvazian A and Hamouda AM 2014 *Journal of Mechanical Science and Technology* **28**(4) 1381-1392.
[14] Eyvazian A, Habibi MK, Hamouda AM and Hedayati R 2014 *Materials and Design* 54 1028-1038.
[15] Eyvazian A, Akbarzadeh I and Shakeri M 2011 *Proc. IMechE Part L: J. Materials Design and Applications* 226 109-118.
[16] Ismail AE 2016 *Jurnal Teknologi* 78(6) 45-50.
[17] Hamdan A, Mustapha F, Ahmad KA, Mohd Rafie AS, Ishak MR and Ismail AE 2016 *International Journal of Polymer Science* 6514041 2016.
[18] Ismail AE 2016 *International Journal of Mechanical and Mechatronics Engineering* 16(1) 49-53.
[19] Ismail AE 2016 *International Journal of Mechanical and Mechatronics Engineering* 16(4) 65-69.
[20] Ismail AE and Mohd Tobi AL 2016 *ARPN Journal of Engineering and Applied Sciences* 11(4) 8668-8672.

Effect of High Velocity Ballistic Impact on Pretensioned Carbon Fibre Reinforced Plastic (CFRP) Plates

Kamarul Azhar KAMARUDIN[1*], Iskandar ABDUL HAMID[2]

[1]Department of Engineering Mechanics, Faculty of Mechanical and Manufacturing Engineering, Universiti Tun Hussein Onn Malaysia.
[2]Crash Reconstruction Unit, Vehicle Safety & Biomechanics Research Centre, Malaysian Institute of Road Safety Research, Kajang, Selangor.
*corresponding author: kamarula@uthm.edu.my

Abstract. This work describes an experimental investigation of the pretensioned thin plates made of Carbon Fibre Reinforced Plastic (CFRP) struck by hemispherical and blunt projectiles at various impact velocities. The experiments were done using a gas gun with combination of pretension equipment positioned at the end of gun barrel near the nozzle. Measurements of the initial and residual velocities were taken, and the ballistic limit velocity were calculated for each procedures. The pretension target results in reduction of ballistic limit compared to non-pretension target for both flat and hemispherical projectiles. Target impacted by hemispherical projectile experience split at earlier impact velocity compared to target by flat projectile. C-Scan images analysis technique was used to show target impact damaged by hemispherical and flat projectiles. The damage area was shown biggest at ballistic limit velocity and target splitting occurred most for pretention plate.

1. Introduction

For decades, large number of studies had been conducted to investigate the impact behavior of composites at low and high velocities using different projectile nose shapes[1][2]. At low velocity impact, the damage is not visible by naked eye but may affect the internal mechanical properties strength. However, at high velocity impact, the damage is relatively much severe and contributes to instant catastrophic.

Ulven et al.[3] performed a series of ballistic impact tests on eight-harness satin weave carbon/epoxy laminates with four different projectile geometries. The perforation mechanism, ballistic limit and damage evolution of each laminate has been analyzed. At high velocity impact conical projectile resulted in the greatest amount of energy absorbed followed by flat, hemispherical, and fragment simulating projectile impact. During impact, flat projectile creates shear zone, which results in plugging while conical and hemispherical result in fibres stretching.

A review on the literatures conclude that the data of impact on pretension targets is limited and therefore it is difficult to make comparisons between experimental results[4][5]. Hence the effect of pretension on plates during impact is not conclusive and require further works. Several authors have studied the effects of pretension on the dynamic response of composite structures/panels at low velocity impact[6][7][8]. Whittingham et al.[9] presented experimental data of pre-strained composite materials under drop weight low velocity impact. Three different strain conditions; 500, 1000 and 1500 $\mu\varepsilon$, were applied to the samples. The results showed that the pre-strain did not affect the peak force, impact energy, depth, penetration, or perforation of the specimens. The relationship indicated that the strains value is small when a pretension of less than 15% of the ultimate strength is applied. Low strain value is seen contribute to reduce effects on the results.

Different findings were produced by Chiu et al.[10], who reported a much more convincing experimental results for pretension test. A higher pretension of 20% of the ultimate tensile strength was used with a drop weight impact tester on a composite laminate. The pretension contribution shows a decrease in impact resistance and ballistic limit value while increasing its damage area.

This argument was supported by Bhavani and Sun [11], who investigated the low velocity impact on composite laminates with an existing preload. An approximation of 22 KN load (about one-third of the ultimate strength of the sample) preload was applied to study its impact behaviour. The results showed that the pretension did reduce the tensile strength of the material. However, results from the low velocity impact cannot be extrapolated to a high velocity impact.

To date, little works have been published on the impact of structures subjected to pretensions under high velocity impact[12][13]. Few studies are available with regard to metal structures under pretensions. Garcia-Castillo et al.[12] performed an experiment using glass/polyester woven plates under pretensions of 31% of ultimate tensile strength. Its ballistic limit shows difference of only 10% with non-pretension samples.

Later in 2011, Garcia-Castillo et al.[14] conducted an experiment on the aluminium alloys under pretension and non-pretension conditions. A tensile pretension of 243 MPa and at about 38% of the ultimate strength of the aluminium alloy was applied axially to the plate. Impacted by hemispherical projectile, it revealed that there is only a small difference of less than 5% on the measured ballistic limit.

An almost related study were conducted by Garcia-Castello et al.[15] which investigate pretension damage at high velocity impact with existence of pretension using hemispherical type projectile. Studies shows that higher damage occurred at higher impact velocity near ballistic limit however damage reduction occurred when velocity above than ballistic limit.

It is clear that ballistic impact focussing on the effect of pretension has been studied widely. However, less attempts have been made on damages caused by hemispherical and flat projectiles. The purpose of this research is to observe the ballistic limit and damage influence of uniaxial pretension on the CFRP plate under impact using two different projectile shapes.

2. Experimental Methods

The Carbon Fibre Reinforced Plastic(CFRP) material with the commercial name HEXCEL AS4 (12K) was supplied by a company dealing with fabrication of aircraft components. The individual layers of the unidirectional CFRP prepreg, known as ply, have a thickness of 0.125 mm, and the overall density of the material is approximately 1600 kg/m^3. The elastic and strength properties are shown in Table 1. The CFRP rectangular panel was manufactured using a hand layup technique, consisting of 24 plies of layup forming a bidirectional composite. The test panels were cut into required sizes using a diamond disc cutter with coolant to avoid any heating and stress concentration, which might have caused delamination and damage. The target dimension was 145x45x3 mm for length, width, and thickness respectively.

Table 1: Material properties of unidirectional CFRP.

Description and (unit)	Symbol	Value
Young's modulus in fibre direction 1 (GPa)	E_{11}	145
Young's modulus in fibre direction 2 (GPa)	E_{22}	11
Poisson's ratio	v_{12}	0.3
Shear modulus, 1-2 plane (GPa)	G_{12}	4.5
Tensile failure stress in fibre direction 1 (MPa)	X_{1T}	1620
Compression failure stress in fibre direction 1 (MPa)	X_{1C}	1200
Shear strength, 1-2 plane (MPa)	S_{12}	120

Figure 1 shows the flat and hemispherical nose geometry of projectile with a mass of approximately 3 grams. The silver steel projectiles were machined using a lathe, then subjected to a hardening process to increase their strength. Both projectiles were 5 mm in diameter, with the hemispherical nose a little longer in order to have the same weight with the flat projectile.

Figure 1: Flat-end (left) and hemispherical (right) nose shape projectile

Light sensors and high speed camera were used to measure the impact and residual velocity respectively. A tensile rig has been designed specifically to run the pretension job. It is attached near the gas gun testing machine as shown in Figure 2. A couple of hydraulic cylinders were used to pull the upper and lower frame in the opposite direction to create tensile force to the specimen. To measure the strain, a simple 'load cell' is attached on the tensile rig position in-line with the tensile direction. The load cell is attached with reading indicator and when the strain reached the required value, the hydraulic cylinders are locked to maintain a constant pretension before the impact test take place on the required velocity.

Tests for CFRP specimens were performed under two axial loading conditions; without pretension (0%), and with a pretension of 30% of the maximum tensile strength. The ultimate tensile strength of the bidirectional CFRP was approximately 705 MPa, therefore 30% of pretension is 211.5 MPa. After the impact test, the specimens were inspected by a non-destructive inspection with C-Scan technique using Olympus Phase Array C-Scan MX2. Here an ultrasonic sound will inspect any defect in the specimens.

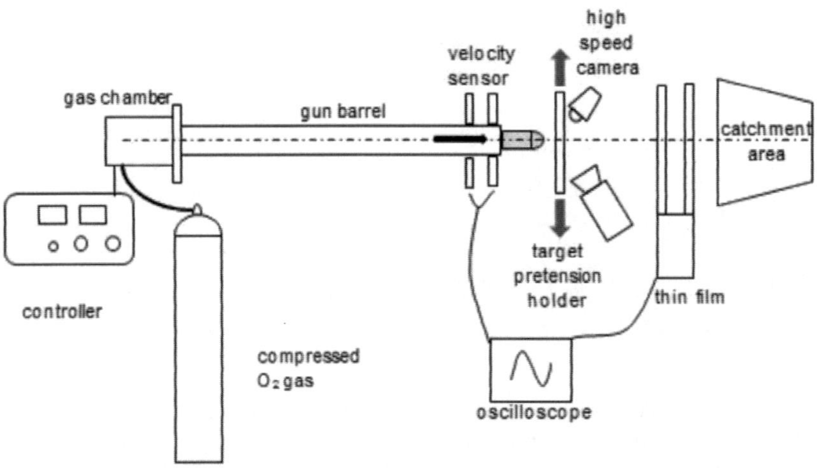

Figure 2: Schematic illustration of the high velocity impact test arrangement

3. Results and Discussion

3.1 Ballistic Limit Prediction

Figure 3 presents the ballistic test result under impact of flat and hemispherical projectiles with target pretension of 0% and 30% from material ultimate strength. Using the modified-Navy ballistic limit, four tests consisting of two complete and two partial penetrations, were used to predict the ballistic limit(V_{50})[16]. For a target with 0% pretension, an average ballistic limit from flat projectile was approximately 102 m/s. While for a hemispherical-nose projectile, the average ballistic limit was approximately 118 m/s. The ballistic limit for a flat-nose projectile is 15% lower than that for a hemispherical-nose projectile. Figure 3 shows detail of impact tests with plate under 30% pretension. Similarly, using the modified-Navy definition, the ballistic limit for a flat-nose projectile was approximately 90 m/s, whereas the ballistic limit for a hemispherical-nose projectile was approximately 99 m/s. At 30% pretension, the ballistic limit for a flat-nose projectile was found to be 10% lower than a hemispherical-nose projectile. For a flat-nose projectile, plate splitting only occurred after a complete penetration has occurred. However, for a hemispherical-nose projectile, plate splitting occurred for both partial and complete penetrations.

Existence of pretension led to a reduction in the ballistic limit for both projectile types. Under 0% and 30% target pretension, flat projectile results in 12% reduction. On the other hand, at impact velocity of 85 m/s with 30% target pretension, the results show that the flat projectile perforated through the sample while the hemispherical only dented the sample. This indicates that hemispherical projectile require a higher velocity to reach ballistic limit. This also proved that shear and plugging behaviour by flat projectile are easily to perforate compared to hemispherical projectile.

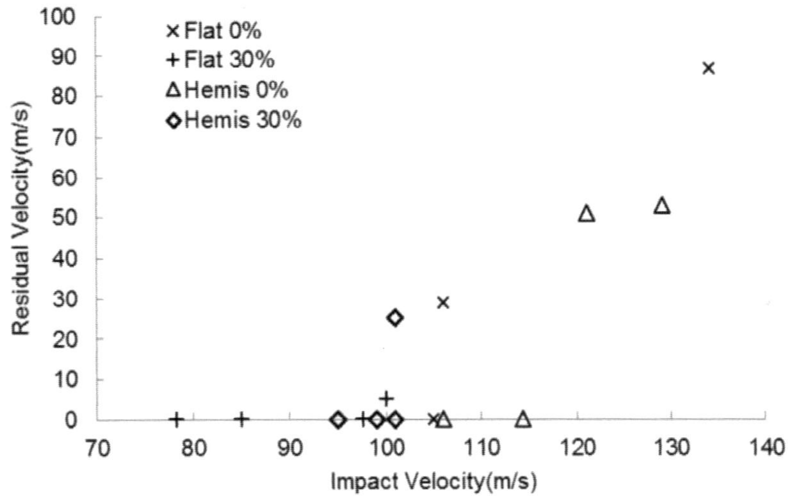

Figure 3: Variations of the residual velocity with impact velocity for CFRP panels with and without pretension.

In addition, the hemispherical-nose-shape projectile is subjected to higher resistance upon impact, compared to flat-nose projectile, regardless of the existence of pretension. This increased resistance was due to the increased of contact surface area, as the hemispherical nose shape allows the target to stretch sideways to create more plastic deformation, while the sample failed by shear plugging for the flat-nose projectile.

3.2 Damage Assessment of impacted CFRP plate

Figure 4 shows damage images of all plates inspected by the C-Scan at three different speed categories: (i) below than ballistic limit, (ii) ballistic limit(V_{50}) and (iii) above the ballistic limit. Most images show that the damage areas are localized in the centre. The C-Scan inspection reveals only the projection image of the area. However, the lamination between plies is not been observed. It can be seen that the damage area was smaller at velocity lower than ballistic limit, as shown in Figure 4a, b, c and d. The damage area becomes larger when approaching ballistic limit velocity (Figure 4e, f, g and h). As the impact velocity increases higher than ballistic limit, the perforation threshold shows the reduction. For pretension target which was impacted at higher velocity than ballistic limit, it shows split on target (Figure 4j and l). In overall, although the size and shape are differed, the results are having some similarities to the works of Garcia-Castillo et al.[15]. In addition, with existence of pretension, hemispherical projectile tends to split the target at early impact velocity of 95 m/s compared to flat projectile at 105 m/s.

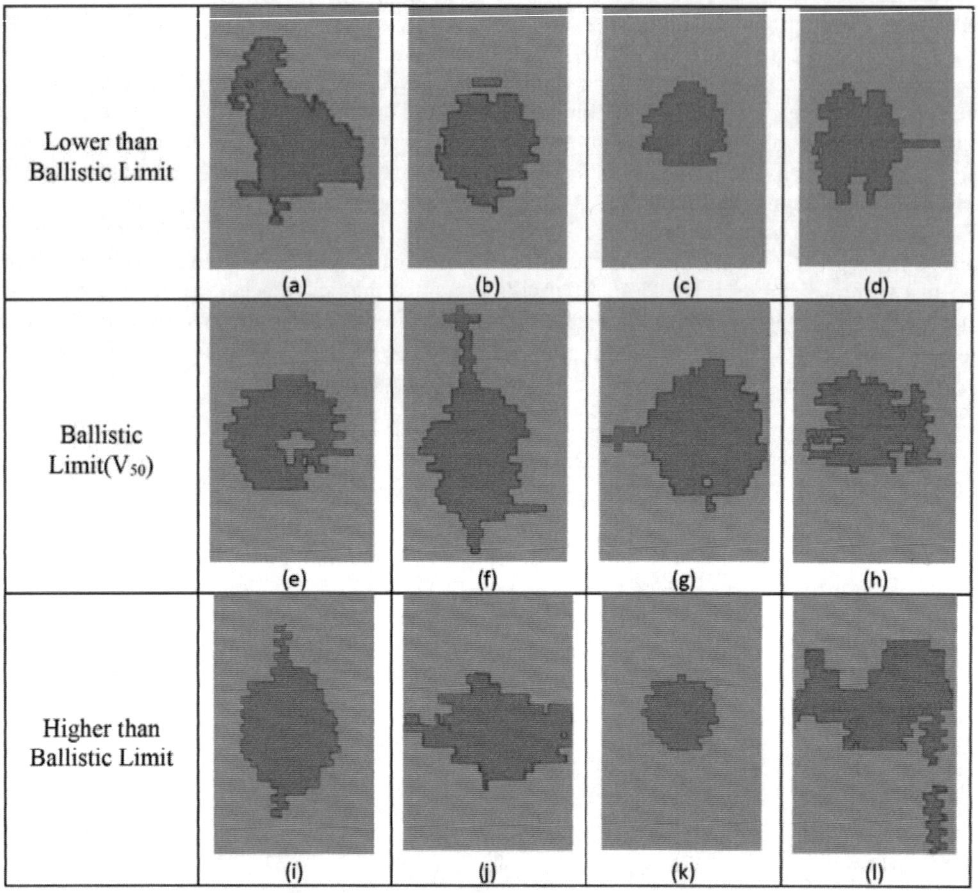

Figure 4: C-Scan images of CFRP target at impact velocity below, average and higher than ballistic limit.

4. Conclusions

Two projectiles types, flat and hemispherical shapes have been used to evaluate the ballistic performance of CFRP plate subjected to high velocity impact with the effect of axial pretension. The impact velocity, residual velocity, the ballistic limit and the damages were obtained. The damage for pretensioned target behaves differently due to sample splitting. In high velocity impact, the damaged area increases until the ballistic limit is reached and experience reduction at velocity higher than ballistic limit. It is also observed that hemispherical projectile split target plate at lower impact velocity compared to flat projectile. It was due to tensile damage by hemispherical projectile compared to shear failure by flat projectile. Using the C-Scan, it shows clear images of internal damage due to the pretension target being impacted by flat and hemispherical projectiles.

5. Acknowledgement

The authors would like to thank Ministry of Higher Education (Malaysia) and Universiti Tun Hussein Onn Malaysia for financial support of this work.

References

[1] C. S. Lopes, P. P. Camanho, Z. Gürdal, P. Maimí, and E. V. González, "Low-velocity impact damage on dispersed stacking sequence laminates. Part II: Numerical simulations," *Compos. Sci. Technol.*, vol. 69, no. 7–8, pp. 937–947, 2009.

[2] F. Jiying, G. Zhongwei, and J. C. N. O. Westley, "Modeling Perforation in Glass Fiber Reinforced Composites Subjected to Low Velovity Impact Loading," *Polym. Polym. Compos.*, vol. 16, no. 2, pp. 101–113, 2008.

[3] C. Ulven, U. K. Vaidya, and M. V. Hosur, "Effect of projectile shape during ballistic perforation of VARTM carbon/epoxy composite panels," *Compos. Struct.*, vol. 61, pp. 143–150, 2003.

[4] H. M. Wen, "Predicting the penetration and perforation of FRP laminates struck normally by projectiles with different nose shapes," *Compos. Struct.*, vol. 49, no. 3, pp. 321–329, 2000.

[5] M. Garcia-Avila, M. Portanova, and A. Rabiei, "Ballistic performance of composite metal foams," *Compos. Struct.*, vol. 125, pp. 202–211, 2015.

[6] M. D. Robb, W. S. Arnold, and I. H. Marshall, "The damage tolerance of GRP laminates under biaxial prestress," *Composite Structures*, vol. 32. pp. 141–149, 1995.

[7] R. Mines, Q. Li, and R. Birch, "Static Behaviour of Transversely Loaded CFRP Laminate Panels Subject to In-Plane Tension," *Strain*, vol. 36, no. 2, 2000.

[8] T. Mitrevski, I. H. Marshall, R. S. Thomson, and R. Jones, "Low-velocity impacts on preloaded GFRP specimens with various impactor shapes," *Compos. Struct.*, vol. 76, pp. 209–217, 2006.

[9] B. Whittingham, I. H. Marshall, T. Mitrevski, and R. Jones, "The response of composite structures with pre-stress subject to low velocity impact damage," *Compos. Struct.*, vol. 66, no. 1–4, pp. 685–698, 2004.

[10] S. T. Chiu, Y. Y. Liou, Y. C. Chang, and C. long Ong, "Low velocity impact behavior of prestressed composite laminates," *Materials Chemistry and Physics*, vol. 47. pp. 268–272, 1997.

[11] B. V. Sankar and C. T. Sun, "Low-velocity impact damage in graphite-epoxy laminates subjected to tensile initial stresses," *AIAA J.*, vol. 24, no. 3, pp. 470–471, 1986.

[12] S. Garcia-Castillo, S. Sanchez-Saez, J. Lopez-Puente, E. Barbero, and C. Navarro, "Impact behaviour of preloaded glass/polyester woven plates," *Compos. Sci. Technol.*, vol. 69, no. 6, pp. 711–717, 2009.

[13] K. Kamarudin and A. E. Ismail, "Modelling High Velocity Impact on Aluminium Alloy 7075-T6 under Axial Pretension," *Appl. Mech. Mater.*, vol. 629, pp. 498–502, Oct. 2014.

[14] S. García-Castillo, S. Sánchez-Sáez, and E. Barbero, "Behaviour of uniaxially preloaded aluminium plates subjected to high-velocity impact," *Mech. Res. Commun.*, vol. 38, no. 5, pp. 404–407, Jul. 2011.

[15] S. Garcia-Castillo, C. Navarro, and E. Barbero, "Damage in preloaded glass/vinylester composite panels subjected to high-velocity impacts," *Mech. Res. Commun.*, vol. 55, pp. 66–71, 2014.

[16] M. E. Backman and W. Goldsmith, "The mechanics of penetration of projectiles into targets," *Int. J. Eng. Sci.*, vol. 16, no. 1, pp. 1–99, Jan. 1978.

Colloquium of Advanced Mechanics (CAMS2016) IOP Publishing
IOP Conf. Series: Materials Science and Engineering **165** (2017) 012006 doi:10.1088/1757-899X/165/1/012006

A review on fracture prevention of stent in femoropopliteal artery

Bainun Akmal Mohd Atan[1,*], Al Emran Ismail[2], Ishkrizat Taib[3], Zulfaqih Lazim[4]

Faculty of Mechanical and Manufacturing Engineering, Universiti Tun Hussein Onn Malaysia, 86400 Parit Raja, Batu Pahat, Johor, Malaysia

[1]bainun89@gmail.com, [2]emran@uthm.edu.my, [3]iszat@uthm.edu.my, [4]zulfaqihlazim@gmail.com

Abstract. Heavily calcific lesions, total occlusions, tortuous blood vessels, variable lengths of arteries, various dynamic loads and deformations in the femoropopliteal (FP) arterial segment make stenosis treatments are complicated. The dynamic forces in FP artery including bending, torsion and radial compression may lead to stent fracture (SF) and eventually to in-stent restenosis (ISR). Stent design specifically geometrical configurations are a major factor need to be improved to optimize stent expansion and flexibility both bending and torsion during stent deployment into the diseased FP artery. Previous studies discovered the influence of various stent geometrical designs resulted different structural behaviour. Optimizing stent design can improve stent performances: flexibility and radial strength to prevent SF in FP arterial segment.

1. Introduction

Peripheral artery disease (PAD) has affecting over 200 million individuals worldwide making PAD the third leading cause of morbidity and mortality in the world [1–3]. In fact, 60-80% of patients with PAD have coronary artery disease (CAD) in at least one vessel, and 25% have carotid stenosis [4]. Through PAD encompass a several vascular beds, it is most often occured in the lower body limb, involving aortoiliac (AI), femoropopliteal (FP), and infrapopliteal (IP) vascular beds. The FP arterial segment composed of the superficial femoral artery (SFA) in the proximal region and popliteal artery (PA) in the distal region [5] (Figure 1), often has high atherosclerotic plaque represent a challenge for endovascular revascularization [6]. Atherosclerosis occurred when excessive cholesterol, calcium and fibrin attached to the wall of blood vessel leading to the thickening and hardening of the vessel wall. As a result, the forming fatty substances called plaques capable to narrow the arteries.

Percutaneous transluminal ballon angioplasty (PTA) followed by stenting have proven superior to prevent abrupt artery closure [7], restore blood flow and vessel patency [8] in FP vascular beds. In this process, the stent is mounted on a balloon. The stent is expanded by inflating the balloon at the blockage region to compress the plaque against the artery wall. When the balloon is deflated the stent remains in place and support the artery open [9]. However, FP arterial segment is often found to have calcific and occlusive [10] atherosclerosis due to its length, the longest in the body. Heavily calcific plaque can affect the expansion of stent leading to less luminal gain that is resulted with higher rates of restenosis, known as in-stent restenosis (ISR) [6] leading to late thrombosis and stents fractures (SF) probably due to low radial strength of the stent. From previous study, PTA through effective for shorter (<5 cm) non-calcified lesions [11], fails when lesions are longer, calcified, and more complex. The restenosis rates vary, but are between 35% and 65% at 6-12 months for lesions more than 100 cm [12–14]. The main reason of restenosis is smooth muscle cell and intima hyperproliferation resulting adverse remodelling, caused by endothelium injury, inflammation of the persistent atherosclerosis plaque, hemodynamic factors, and mechanical stress from a permanent stent [15,16]. Indeed, the 12-month rates of ISR after implantation of modern bare metal nitinol stents in the FP segment range from 18% to 37% [11].

FP arterial segment is subjected to various dynamic forces including radial compression, bending and torsion [8] due to daily physical movements have been associated with high rates of SF [17–19]. This failure is reliable attributed by low flexibility, contributed to increased rates of restenosis [20]. The dynamic mechanical loading generated within tortuous blood vessels during deliverability of the stents to the target lesion that are going to be cured, might also lead to SF. Flexibility and radial strength of the stent in FP artery may be improved by optimizing the stent design parameter. Previous studies have reported the influence of various stent design configuration resulted different structural behaviour of the stent under numerical studies. This paper will be focused on FP arterial stenting commonly known as most anatomic locations of lower extremity atherosclerosis and discussed the effect of stent geometrical design on stent flexibility and expansion to prevent SF. Previous numerical studies of stent structural behaviour: bending, torsion and expansion will be reviewed aiming to know the influence of stent geometry design on the stent performances. These data may assist in the future modification process of stent design.

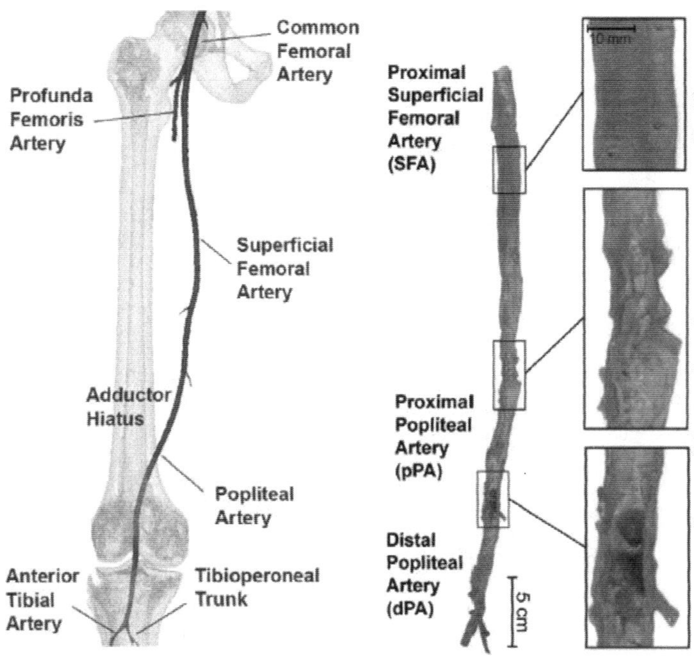

(a) (b)

Figure 1. Anatomy of (a) FP arterial segment and (b) severe PAD in the distal SFA and PA [5].

2. The influence of stent design on fracture

Heavily calcific lesions, total occlusions, tortuous blood vessels and numerous dynamic loads on FP artery may lead to SF and eventually to severe ISR and thrombosis. The mechanical forces involved are radial compression, bending and torsion due to limbs movements. SF can result unwanted event and life threatening [21–23]. Failure may potentially occur when the stress is locally high at certain locations [24]. High radial strength and flexibility is characterized as significant properties during the

IOP Conf. Series: Materials Science and Engineering **165** (2017) 012006 doi:10.1088/1757-899X/165/1/012006

implementation of FP stent. Radial strength and flexibility can be improved by optimizing the stent strut and stent geometry configurations. An ideal stent should be considered to have the following design requirements: high radial strength, good flexibility, good fatigue properties, low elastic radial and longitudinal recoil and optimum scaffolding. Stent geometry dictates many of the delivery system and contributes to the performance of the sten [25,26]. The optimization of stent geometry and stent strut design will enhance stent structural behaviour: bending, torsion and radial strength, thereby reduce the possibility of SF and eventually ISR in FP arterial segment.

2.1. Flexibility

Flexibility of FP stent is often associated with the capacity of an expanded stent to assume the natural curve of a vessel specifically during the movements of the lower extremity without unnatural straightening. Flexibility also signifies deliverability or the capacity of the unexpanded stent to navigate tortuous passage to the stenosed region [24]. Torsion and bending ability of the stent are important characteristics to withstand the dynamic mechanical forces in the FP artery as the knee will flexes and twists and during delivery of the stent into tortuous blood vessels. Improving flexibility demands the ability for expandable stent to pull apart from each other on the outer radius of a curved vessel and push together on the inner radius. Optimizing of stent geometrical design has led to improvements of stent performance subsequently minimize the frequency of SF. The structural behavior of stent has been examined by previous studies such as Azaouzi et al. [24], Petrini et al. [27], Mori and Saito [28] and Junlei et al. [29]. These findings may be discussed in the following paragraphs.

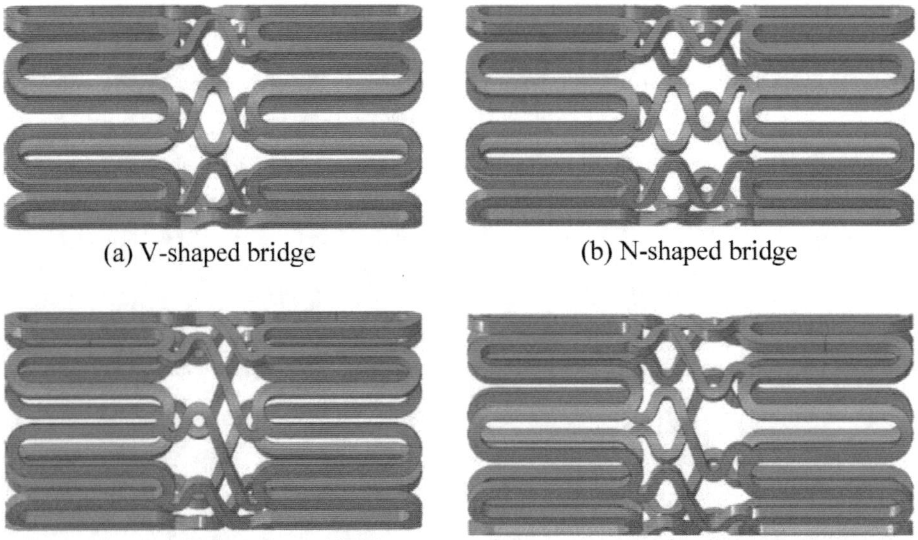

(a) V-shaped bridge (b) N-shaped bridge

(c) Unsymmetrical V-shaped bridge (c) Unsymmetrical N-shaped bridge

Figure 2. Geometries of the bridges [24].

2.1.1. Bending

The structural behavior of a stent in terms of bending can be demonstrated by constraining one of the extremities of the stent unit cell, and moment of bending is applied on the other extremity in stent simulation tests. Azaouzi et al. [24] has investigated the effect of the different connecting element geometry known as 'bridge' depicted in Figure 2 on the structural behaviour of balloon expandable (BE) stent including bending, torsion and expansion demonstrated using FEA. Basically, most of BE stents have two fundamental elements: expandable ring elements and connecting elements (bridge) [24]. Expandable rings consist of some number of struts in a zigzag pattern arrangement and offer excellent combination of flexibility and strength during in service. Petrini et al. [27] stated that the expandable ring element is mainly function to scaffold the wall of the artery.

Various bridge geometry configurations aim to upgrade stent flexibility however influence foreshortening characteristics of the stent. During stent expansion, a peak-to-peak bridge connection will tend to pull adjacent expanding rings apart from each other as strut angles increase resulting in reduction in length (foreshortening). Based on Azaouzi et al. [24] simulation result, the location and magnitude of the maximum stress within the bridge is different for each bridge geometry design. As a result, at the beginning of bending, unsymmetrical bridges are remarkably more flexible than symmetrical bridges. The unsymmetrical V-shaped bridge is more flexible than the unsymmetrical N-shaped bridge. This study indicates the geometry modification of bridge elements using numerical analysis is crucial to enhance the flexibility of the stent.

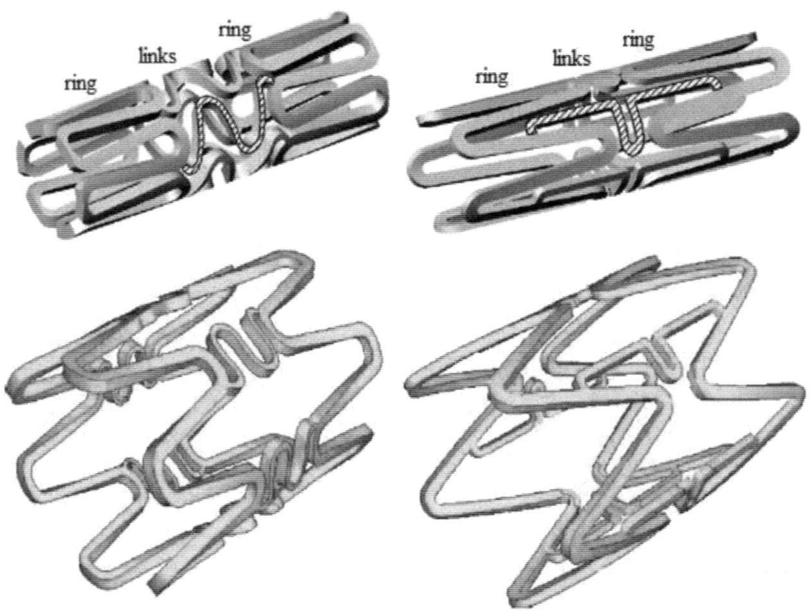

(a) (b)

Figure 3. Geometries of (a) CV and (b) SC models in unexpanded (top) and expanded (bottom) configurations [27].

Petrini et al. [27] has studied the performance of stent by analysing three-dimensional models of two generation of stent: Cordis BX-Velocity (CV) and Sirius Carbostent (SC) (Figure 3) using FEA and demonstrated interference between link struts at the compressive side and the stent thickness. Bending tests in unexpanded and expanded configurations were carried out. During unexpanded bending analysis, CV links show a high ability to independently deform from the rings compare to SC model. For SC model, rings and links do not deform independently. This can be described based on the links geometries shown in Figure 3. The links of CV stent have two crests and are connected with the top of the ring (Figure 4(a)), while the links of the SC model have only one crest and are connected with the center of the ring (Figure 4(b)). The expanded bending responses reveal a tendency similar to the unexpanded stent response, but with a lower flexibility. Different thicknesses of the stent models would also influence the stent response during bending. Indeed, this study showed the influence of link geometry and stent thickness on stent performance in terms of flexibility.

Mori and Saito [28] have evaluated stent flexibility using FEM with four different link structures shown in Figure 5. From the results, N-stent was the most flexible however W-stent was the most rigid stent. This can be explained by the link configuration that deforms flexibly associates from symmetric or asymmetric link configuration and link strut length. The FEM analysis demonstrated that the W-stent rigidity with a non-point symmetric link configuration resulted from interference between the struts. The narrow distance between the link struts subsequent interference between struts at the compressive side leading to reduction of flexibility and promotion of buckling deformation.

Closed-cell and open-cell stent design is vital to be considered during optimization of stent performance. The primary advantage of closed-cell design (Figure 6(a)) is supply uniform vessel scaffolding of the stenosed region probably reduce overall stent flexibility leading to fracture [29–31]. In contrast, an open-cell design (Figure 6(b)) allows fewer internal inflection points and connection to bridging elements, thus allowing for less vessel scaffolding but offer greater flexibility reducing the probability of SF.

The selection of a particular type of stent design relies on the possible performance of stent in the target lesion that is going to be cured. For example, laser-cut tube stent made of stainless steel are mostly used on coronary heart disease treatment, while stents made of Nickel-Titanium alloy are more suitable for the treatment of peripheral vascular disease positioned in femoral, abdominal, carotid or renal arteries because of their shape memory and super-elasticity properties [33].

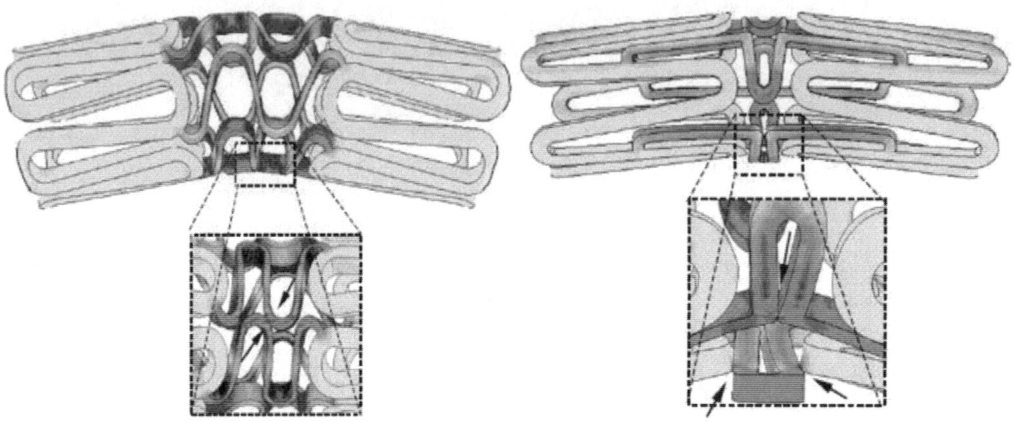

(a) (b)

Figure 4. Von Mises stresses for the (a) CV and (b) SC models in the unexpanded confguration [27].

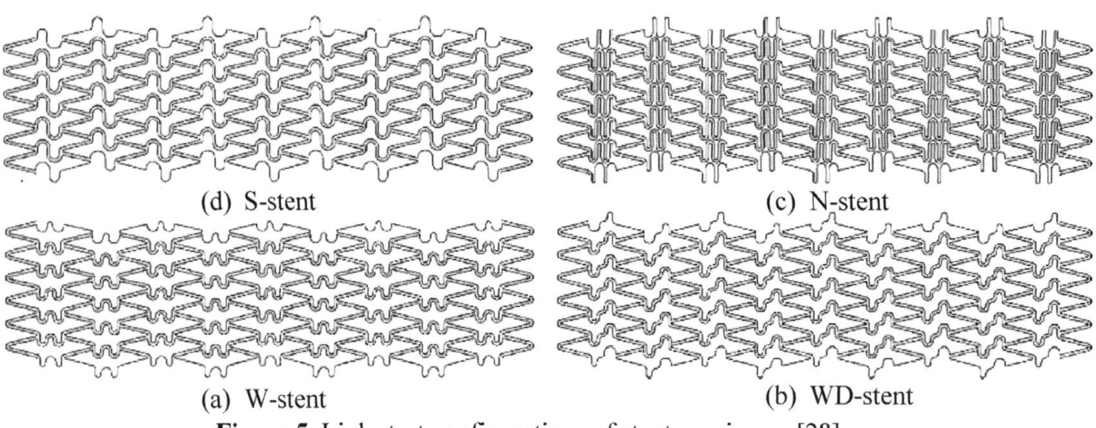

Figure 5. Link strut configurations of stent specimens [28].

2.1.2. Torsion

Since FP artery located in the largest segment of the artery branches, composed of SFA in the proximal region and PA in the distal region, stent need to twists around the femur as it through behind the knee during deliverability described by its tortuous geometry [8], thus torsion ability of a stent is an important factor need to be considered in this review. Similarly to bending, torsion is measured by constraining one extremity of the stent and moment of torsion is applied at the constrained extremity of the stent unit cell during simulation test analysis. Azaouzi et al. [24] studied the influence of bridge geometry on the torsion ability of the stent. It is resulted the unsymmetrical N-shaped bridge (Figure 2) is obtained as the most impressive in terms of torsion ability. These results are very useful to understand the influence of bridge geometrical configuration on torsion ability of stent during insertion along tortuous FP artery.

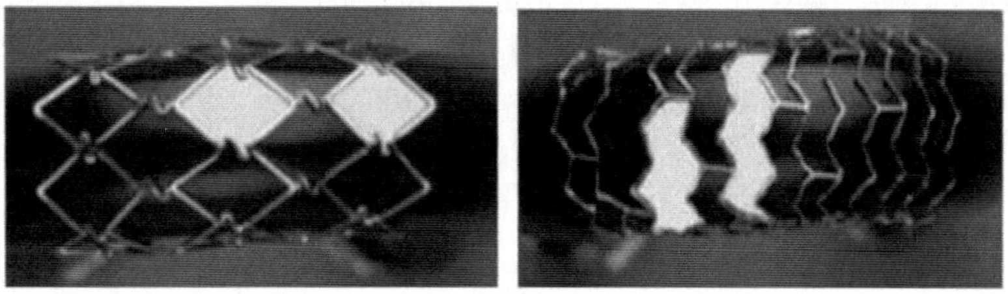

(a) (b)

Figure 6. Stent design: (a) closed-cell design and (b) open-cell design [32].

2.2. Radial strength

Radial strength is characterized very important to prevent SF due to high calcific plaque in stenosed FP arterial segment. The strength of a stent is interpreted as the amount of recoil the vessel experiences after the stent is expanded inside the artery. The radial strength describes the capability of stent to withstand collapse under short-term or long-term external forces [24] and is the compressive resistance force of a stent to maintain the diameter and the vessel lumen after implantation [31]. The radial strength is imperative to the treatment of ostial lesions and highly calcified lesions where diameter maintenance over time is critical [34]. The choices of strut and bridge geometry, dimensions and materials are the main factors that influence stent strength leading to a better long-term efficacy of the device. Azaouzi et al. [24] studied the influence of bridge geometry on the radial strength of stent. It is observed that symmetrical N-shaped bridge (Figure 2) has high expansion due to low radial force and stresses, therefore capable to extend the life of stent.

Li et al. [29] demonstrated the influence of stent geometry on the performance of the magnesium alloy stent (MAS). The strut width of the optimized stent was designed gradually decreased from the curved end to the middle segment of the strut (Figure 7). The size distribution of the struts width is proven a superior to reduce the maximum principle strain after expansion and the maximum principle stress of the stent after recoil. The decreased size of the strut width at the middle segment caused the plastic deformation resistance weakened. Therefore, the stent is expended not only by the plastic deformation at the curved ends of the circumferential rings but also by the plastic deformation of the middle segment of the struts. Meaning that the strain is not concentrated on curved ends of the circumferential rings only, but evenly distributed on the whole struts of the stent. Moreover, optimized stent showed similar radial strength with common structure although the optimized structure has reduced gradually the strut width.

Additionally, the fundamental width of the struts has different effects. Thinner struts may incorporate more deeply into the arterial wall and cause fewer angulations and stretching of the arterial wall [35]. Although thicker struts result in greater radio-opacity and radial force for improved arterial support, the larger struts result in abnormal apposition, greater arterial injury, and neointimal proliferation. However, previous studies have observed that a reducing strut thickness (from 100 μm to about 65 μm) capable to increase radial strength of the stent, therefore improving the deliverability and life of the stent [36]. Thus, bridge geometry and stent struts thickness should be considered during modification of FP stent.

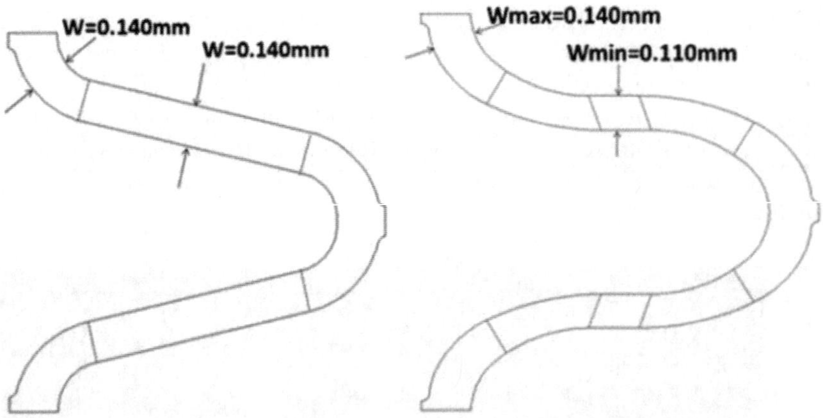

Figure 7. Stent struts width size of the common (a) and optimized (b) stent [29].

3. Fenite Element Analysis

FEA is an useful method and has proven to be powerful and superior providing more detailed understanding of design and fatigue [34,35]. FEA is required because prototypation process is costly, instead of traditional trial and error technique to verify and design new models of stent. FEA provides an access to extensive information under highly controlled conditions, thus, making it possible to screen different and competing stent design alternatives before to expansive stent prototype fabrication [27]. The structural behavior of the stent is normally investigated under three loading conditions: bending, torsion and radial strength. The stent flexibility: bending and torsion is normally evaluated in the expanded and unexpanded state. The bending is measured by constraining one extremity of the stent unit cell and imposing a bending moment on the other extremity. Similarly to bending, the torsion is measured by constraining one extremity of the stent and moment of torsion is applied at the constrained extremity of the stent unit cell during simulation test analysis. Petrini at al. [27] demonstrated the response of one unit of a stent is similar to the full length stent for flexibility analysis. Hence, simulation analysis to optimize stent flexibility can be performed for a unit structure only.

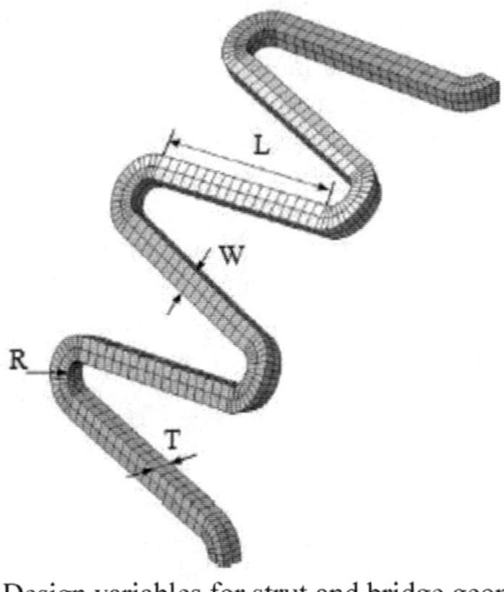

Figure 8. Design variables for strut and bridge geometry [24].

The radial strength of the stent is examined by expanding the stent unit cell in the radial direction using a cylinder. The expansion of stent is often examined by a virtual test that is developed where the stent unit cell is subjected to this loading scenario: radial displacement driven expansion of a cylindrical balloon revealing the expanded stent shape and the foreshortening reduction of the diameter of the cylindrical balloon revealing the elastic radial recoil of the stent [24]. In order to perform numerical simulations of the bending and torsion, some nodes of the stent unit cell are fixed on its constrained extremity (the translational and rotational degree of freedom of each node are constrained).

4. Discussion

Flexibility is an important property of stents during deliverability along tortuous blood vessel. FP artery associated with various mechanical loadings (bending, torsion and radial compression). Based on stress-concentration feature of stent, as the local stress is high at certain locations, fracture may potentially occur. Fracture may occur during initial deployment of the stent, which involves large amounts of plastic deformation as the stent expanded inside the stenosed artery. Optimization on strut and bridge geometry may improve the stent performances during insertion and subsequently reduce the residual stresses generated during stent placement. From the previous studies reviewed in this paper, bridge geometry configuration (symmetry and asymmetry) and stent thickness indicated a very important factor for stent flexibility (bending and torsion) and expansion. The interference between struts results in the rigidity of the stent with symmetry link configuration. In contrast, the stent with asymmetry link configuration has high flexibility because the interference between struts was geometrically averted. Besides that, stent strut width possesses a gradient size distribution influenced the amount of strain and residual stress of the stent. These findings are very useful and contribute valuable information about the structural integrity of stent purposely to minimize percentage of SF.

According to Lim et al. [39], the strut and bridge geometry may be parameterized using four design variables: length (L), curvature radius (R), width (W) and thickness (T) shown in Figure 8. These design variables differently affect the performance of the stent such as on foreshortening, elastic recoil, fatigue resistance and more importantly flexibility [39]. Reducing the strut thickness will reduce catheter system and prevent excessive flow disturbances. Reducing the strut length will upgrade the capacity of an expanded stent to assume the natural curve of a vessel without unnatural straightening. Increasing the strut width will improve the vessel scaffolding and decrease the vessel tissue prolapse between the struts. Other than that, closed-cell and open-cell design, stent material also should be considered prior stent fabrication. Thus, the specified parameters may be considered in future modification studies.

5. Conclusion

Strut and bridge geometrical configurations of stent may influence on stent performance. FEM analysis of stent is crucial to provide data of stresses generated by mechanical loadings: bending, torsion and expansion for stent modification prior to prototype fabrications. Further investigation may be conducted by combining FEA with various stent design optimization strategies aiming to improve the structural behaviour of the commercial stent in FP arterial segment to prevent SF.

6. Acknowledgement

This paper was partly sponsored by the Centre for Graduate Studies UTHM. Authors also acknowledged Office for Research, Innovation, Commercialization and Consultancy Management (ORICC) UTHM through FRGS (Vot 1592).

References

[1] Fowkes FGR, Rudan D, Rudan I, Aboyans V, Denenberg JO, McDermott MM, Norman PE, Sampson UKA, Williams LJ, Mensah GA and Criqui MH 2013 *Lancet* **382**(9901) 1329–1340.

[2] Litsky J, Chanda A, Stilp E, Lansky A and Mena C 2014 *Med. Devices Evid. Res.* **7**(1) 149–

156.
[3] Roger VL, Go AS Lloyd-Jones DM, Benjamin EJ, Berry JD, Borden WB, Bravata DM, Dai S, Ford ES, Fox CS, Fullerton HJ, Gillespie C, Hailpern SM, Heit JA, Howard VJ, Kissela BM, Kittner SJ, Lackland DT, Lichtman JH, Lisabeth LD, Makuc DM, Marcus GM, Marelli A, Matchar DB, Moy CS, Mozaffarian D, Mussolino ME, Nichol G, Paynter NP, Soliman EZ, Sorlie PD, Sotoodehnia N, Turan TN, Virani SS, Wong ND, Woo D and Turner MB 2012 *Circulation* **125**(1) e2–e220.
[4] Cheng SW, Wu LL, Ting AC, Lau H and Wong J 1999 *Cardiovasc. Surg.* **7**(3) 303–309.
[5] Kamenskiy AV, Pipinos II, Dzenis YA, Phillips NY, Desyatova AS, Kitson J, Bowen R and MacTaggart JN 2015 *Acta Biomater.* **11**(1) 304–313.
[6] Klein AJ and Ross CB 2016 *Trends Cardiovasc. Med.* **26**(6) 495–512.
[7] E Y, He NS, Wang Y and Fan HL 2008 *J. Vasc. Interv. Radiol.* **19**(4) 499–503.
[8] Ní Ghriallais R and Bruzzi M 2013 *Med. Eng. Phys.* **35**(11) 1620–1628.
[9] Moore J and Zouridakis G 2004 Boca Raton, London, New York, Washington, D.C: CRC Press.
[10] Klein AJ, Chen SJ, Messenger JC, Hansgen AR, Plomondon ME, Carroll JD and Casserly IP 2009 *Catheter. Cardiovasc. Interv.* **74**(5) 787–798.
[11] Krankenberg H, Schlüter M, Steinkamp HJ, Bürgelin K, Scheinert D, Schulte KL, Minar E, Peeters P, Bosiers M, Tepe G, Reimers B, Mahler F, Tübler T and Zeller T 2007 *Circulation* **116**(3) 285–292.
[12] Klein AJ, Pinto DS, Gray BH, Jaff MR, White CJ and Drachman DE 2014 *Catheter. Cardiovasc. Interv.* **84**(4) 529–538.
[13] Rocha-Singh KJ, Jaff MR, Crabtree TR, Bloch DA and Ansel G 2007 *Catheter. Cardiovasc. Interv.* **69**(6) 910–919.
[14] Schillinger M, Sabeti S, Loewe C, Dick P, Amighi J, Mlekusch W, Schlager O, Cejna M, Lammer J and Minar E 2006 *N. Engl. J. Med.* **354**(18) 1879–1888.
[15] Kearney M, Pieczek A, Haley L, Losordo DW, Andres V, Schainfeld R, Kenneth R and Isner JM 1997 *Circulation* **95**(8) 1998–2002.
[16] Chaabane C, Otsuka F, Virmani R and Bochaton-Piallat ML 2013 *Cardiovasc. Res.* **99**(2) 353–363.
[17] Schlager O, Dick P, Sabeti S, Amighi J, Mlekusch W, Minar E and Schillinger M 2005 *J. Endovasc. Ther.* **12**(6) 676–684.
[18] Müller-Hülsbeck S, Schäfer PJ, Charalambous N, Yagi H, Heller M and Jahnke T 2010 *J. Endovasc. Ther.* **17**(6) 767–776.
[19] Jonker FHW, Schlösser FJV, Moll FL and Muhs BE 2008 *Endovasc. Today* 53–58.
[20] Scheinert D, Scheinert S, Sax J, Piorkowski C, Bräunlich S, Ulrich M, Biamino G and Schmidt A 2005 *J. Am. Coll. Cardiol.* **45**(2) 312–315.
[21] Carrozza M, Santoro G, Giovanna Russo M, Caianiello G and Calabrò R 2006 *Int. J. Cardiol.* **113**(1) 127–128.
[22] Marrey RV, Burgermeister R, Grishaber RB and Ritchie RO 2006 *Biomaterials* **27**(9) 1988–2000.
[23] Lee SH, Park JS, Shin DG, Kim YJ, Hong GR, Kim W, Shim BS 2007 *Am. J. Cardiol.* **100**(4) 627–630.
[24] Azaouzi M, Makradi A and Belouettar S 2013 *Comput. Mater. Sci.* **72** 54–61.
[25] Garasic JM, Edelman ER, Squire JC, Seifert P, Williams MS and Rogers C 2000 *Circulation* **101**(7) 812–818.
[26] Sullivan TM, Ainsworth SD, Langan EM, Taylor S, Snyder B, Cull D, Youkey J, Laberge M 2002 *J. Vasc. Surg.* **36**(1) 143–149.
[27] Petrini L, Migliavacca F, Auricchio F and Dubini G 2004 *J. Biomech.* **37**(4) 495–501.
[28] Mori K and Saito T 2005 *Ann. Biomed. Eng.* **33**(6) 733–742.
[29] Li J, Zheng F, Qiu X, Wan P, Tan L and Yang K 2014 *Mater. Sci. Eng. C* **42** 705-714.

[30] Wholey MH and Finol EA 2007 *Endovasc. Today* 25–34.
[31] Patel SM, Li J and Parikh SA 2016 *Interv. Cardiol. Clin.* **5**(3) 365–380.
[32] AL-Mangour B, Mongrain R, and Yue S 2013 *Mater. Sci. Appl.* **4**(10) 606–621.
[33] Azaouzi M, Makradi A and Belouettar S 2012 *Mater. Des.* **41** 410–420.
[34] Price MJ 2014 2nd ed. Philadelphia: Elsevier 238.
[35] Moscucci M 2013 8th ed. Philadelphia: Lippincott Williams & Wilkins.
[36] O'Brien B and Carroll W 2009 *Acta Biomater.* 5(4) 945–958.
[37] Argente dos Santos HAF, Auricchio F and Conti M 2012 *J. Mech. Behav. Biomed. Mater.* **15** 78–92.
[38] Li N, Zhang H and Ouyang H 2009 *Finite Elem. Anal. Des.* **45**(6) 468–475.
[39] Lim D, Cho SK, Park WP, Kristensson A, Ko JY, Al-Hassani STS, Kim HS 2008 *Ann. Biomed. Eng.* **36**(7) 1118–1129.

IOP Conf. Series: Materials Science and Engineering **165** (2017) 012007 doi:10.1088/1757-899X/165/1/012007

A review of stent's failure on patent ductus arteriosus

Zulfaqih Lazim[a*], Al Emran Ismail[b], Ishkrizat Taib[c], Bainun Akmal Mohd Atan[d]

Faculty of Mechanical and Manufacturing, Universiti Tun Hussein Onn Malaysia, 86400, Parit Raja Batu Pahat, Johor, Malaysia.

[a*]zulfaqihlazim@gmail.com, [b]emran@uthm.edu.my, [c]iszat@uthm.edu.my, [d]bainun89@gmail.com

Abstract. This paper presents a review of stent's failure on patent ductus arteriosus (PDA). Ductus arteriosus (DA) is an opening for newborn babies and some patient that experienced cynotic congenital heart disease (CCHD) should maintain the duct opening for survival. To date, there are no specific research on mechanical stent failure study at DA. The challenging of the stent implantation on PDA is the PDA morphology. The failure of stent in term of stent fracture have been reported and reviewed in this paper. Furthermore, the failure prediction of stent is important for further stent design development. The morphology of PDA, stent type and material used in PDA and method for accessing the failure of stent is reviewed.

1. Introduction

The adaption of coronary stent has been establish on 1986. Since then, the scientist have been developing more efficient stent design to treat coronary artery disease. On 2006, there's been reported that more than 1 million stents are implanted in human arteries [1]. The purpose of stent is keeping the coronary artery opened in the location where it's deployed. However the failure of stent preventing its main objective to allow blood flow to the targeted artery location. The stent structure should be capable in order to hold or keep coronary arteries open and reduce the chance of a heart attack. However, human body is dynamic organism and the effect on implanted stent by mechanical fatigue is inevitable. Mechanical fatigue on stent structure lead to fracture and failure.

The Ductus arteriosus (DA) is a small vessel connecting between two major blood vessels, the aortic arch (AA) and pulmonary artery (PA) that permit blood to bypass the pulmonary circulation. Originally, DA is a normal part of every foetal to allow the exchange of oxygenation blood with maternal blood through placenta. After a few days of birth, the DA usually close, the closure does not occur on some people and lead to the condition known as Patent Ductus arteriosus (PDA) [2].

The purpose of stenting on PDA mainly because of typical lesions such as pulmonary and tricuspid atresia, critical pulmonary stenosis and tetralogy of Fallot. Implantation of stent in PDA is needed for survival in these patients [2, 3]. Tricuspid atresia is a type of congenital heart disease where the valve between the right atrium and right ventricle fails to develop. The pulmonary atresia on the other hand is defect of the pulmonary valve, which is the valve that controls blood flow from the right ventricle to the main pulmonary artery. Pulmonary stenosis is a condition of obstruction to blood flow from the right ventricle to the pulmonary artery. Tetralogy of Fallot is a congenital heart defect. This is a problem with the heart's structure that's present at birth and change the normal flow of blood through the heart. It occurs in about 5 out of every 10,000 babies [3].

Other method to connect the AA and PA by modified Blalock-Taussig (mBT) which an open surgery that cause high morbidity and mortality compare to PDA stenting. Traditionally, mBT surgery is the first treatment option, but due to some factor such as very severe disease high risk patient, very low birth weight, or lack of service's experience, mBT can be limited [4, 5].

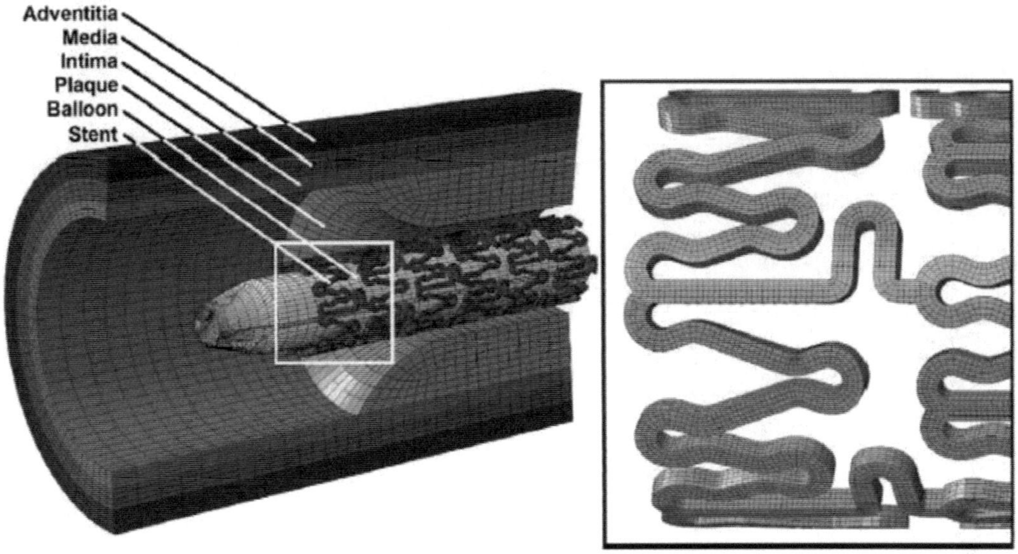

Figure 1. Finite element mesh for the stent–artery–balloon system [6]

Figure 1 above show the finite element mesh for the stent–artery–balloon system. Normally there are two type of expanding stent inside human artery which are balloon-expanding and self-expanding stent. The wall of an artery consists of three layers. The tunica intima is innermost layer, which is simple squamous epithelium surrounded by a connective tissue basement membrane with elastic fibers. The tunica media is middle layer, which is primarily smooth muscle and is usually the thickest layer. The tunica adventitia is outermost layer, which attaches the vessel to the surrounding tissue. This layer is connective tissue with varying amounts of elastic and collagenous fibers.

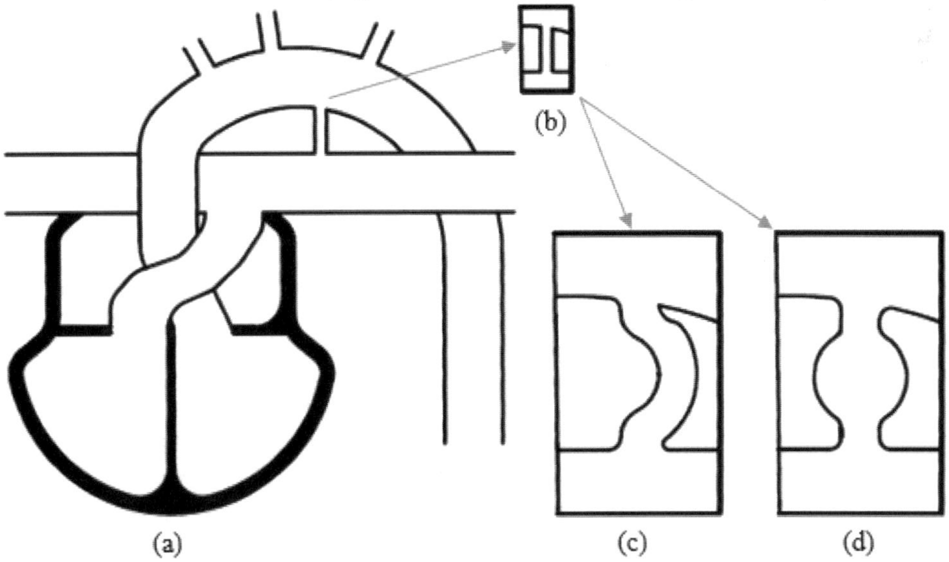

Figure 2. (a) Location of PDA, (b) Straight tubular vessel DA, (C) Curved vessel DA, (d) Bulging shape DA [7]

Figure 1 show the location of PDA near to human heart. There are straight tubular vessel DA, curved vessel DA and bulging shape duct DA. The opening of DA supply blood flow to the lungs from critical right heart obstructive lesion and also supplies to systemic and parallel circulation from critical left heart lesion. The bidirectional flows exist in DA from mixing of oxygenated and deoxygenated blood.

2. Stent placement on PDA

Patent ductus arteriosus (PDA), one of the most common congenital heart defects (CHD). Some PDA morphology is straight duct while some are too big in diameter, curved or tortuous, tubular, funnel shaped, long and meandering, short or have no length or also known as window duct [8]. This paper only review on straight duct type of PDA. Morphology in this paper refer to the shape, form and structure of patent ductus arteriosus. The selection of stent placement on PDA based on the morphology of PDA. The common size of straight duct PDA varies from 7 to 11mm long and 4 to 5mm in diameter [8]. Table 1 show Patient and Procedural Characteristics on PDA, two types of stents are used and two from three patients are survive.

Table 1. Patient and Procedural Characteristics on PDA [9]

	Patient 1	Patient 2	Patient 3
Clinical data			
Age, y/sex	9.7/male	3.3/male	2.3/male
Weight, kg/height, cm	24.6/134	14/94	12.5/88
Stent data			
Number of stents	1	1	2
Type of stent	Valéo	Valéo	Genesis
Diameter/length, mm	9×26	7×26	6×18+7×18
Follow-up, mo/outcomes	8/alive	10/alive	24 mo/death

The materials selection for coronary stents preferably consist of criteria such as flexible, supportive, capable of expansion, and biocompatible. The using of material such as metallic based have been used for coronary stents [10]. Ishkrizat et al. [11] used two types of materials for finite element analysis (FEA) of stent in PDA which is stainless steel 316L and cobalt–chromium L605 and their parameters showed in Table 2. The most commonly used material for Bare-metal stent (BMS) platforms is 316L stainless-steel, the advantages of using BMS stainless steel are adequate mechanical properties and high corrosion resistance [12]. However, the cobalt–chromium alloy has an excellent radial strength, which allows the development of stents with ultrathin struts while preserved radial force. Other favorable the cobalt–chromium alloy characteristics was adequate radio-opacity and better deliverability [13].

Table 2. The Material Properties for Stents [11]

Material of the stents	Parameters				
	Density (kg/m^3)	Young's Modulus (Mpa)	Poisson's Ratio	Yield Strength (σ_y)	Tangent Modulus
Stainless Steel 316L	7800	193000	0.27	207	692
Cobalt-Chromium L605	9700	243000	0.3	476	680

Thin-strut stents tend to have lower profiles than thick-strut stents. This lower profile strut lead to less contact between stent and arterial wall surface and decrease the in stent restenosis (ISR). Clinical reports reported that strut thickness may affect restenosis rates [14]. Radial strength describes the external pressure that a stent is able to withstand without incurring 'clinically significant damage', the radial strength of the stent should be enough to support the hemodynamic and arterial wall force. Stents must have sufficient radial strength to treat diseased artery [15].While the radio-opacity is the advantage of stent visibility and deliverability refer to the flexibility of the stent on deliver phase using catheter. On the deliver phase of stent, the stent is compressed into catheter and deliver through curvy blood vessel to the stent implantation site.

In contrast to BMS material, biodegradable stent offer some advantage. The polymers and metal alloys are example of biodegradable stent. Some advantage of polymers that it could predict the degradation time however having unsatisfied mechanical characteristic while magnesium alloys is the other way around [16]. At some point, for temporary stenting in PDA, biodegradable stent is suitable device compare to BMS stent. Advantage of stent decaying proses poses by degradable stent solved the main concern in pediatric stenting as suitable for baby's growth. The ideal characteristics of a biodegradable stent have been defined as (1) sufficient radial strength to prevent vascular recoil, (2) minimal thrombotic and inflammatory response (3) avoidance of intimal proliferation (4) reabsorption of stent components within months (5) no release of toxic products or embolic material during breakdown and (6) easy processing and sterilization [17].

3. Failure of Stent

Failure of stent on PDA is a key point to address before stent optimization process. The previous work by previous researchers regarding the stent's failure are reviewed in this section. Mechanical stent failure (MSF) is seldom occasion however the failure is very dangerous [18]. The most common categories are longitudinal stent deformation, stent fracture, stent under expansion and stent recoil [18]. The MSF rate reported that from 136 patients undergoing stent implantation, 17 patients (12.5%) endure mechanical stent failure [19]. From 17 patients, 16 having focal and 1 having in stent restenosis (ISR) [19].

The focal ISR known as lesion that less than 10mm and occur at unscaffolded segment, while diffuse ISR is lesion that more than 10mm [20]. There are three types of diffuse ISR, which are confined to stent margin, beyond stent margin, and total occlusion or have a flow rate 0 [21]. Complete stent fracture occur in one patient (5.9%), partial stent fracture endure by 3 patients (17.6%) while 13 (76.5%) more experienced longitudinal deformation and stent's strut fracture [19].

The stent's fracture caused by vessel tortuosity, longer stent length, overlapping stenting, calcified and ostial lesions and high deployment pressure while stent's longitudinal deformation caused by guideliner use, use of post dilatation balloons and multiple stent implantation [18]. The pre and routine post dilation balloon is useful for facilitate optimal stent deployment (SD) and stent optimization (STOP). Pre-dilatation with the correct size and type of balloon will cause less trauma to the vessel than direct stenting [22, 23]. Pre-dilatation also: (1) Increases the rate of successful stent delivery by creating a pathway for the delivery system; (2) May prevent damage to the drug coating during delivery; (3) Can help to estimate lesion length and diameter using the balloon markers; (4) Aids full stent apposition by modifying the lesion before the practitioner places the stent.

The fatigue of stent structure is one of the stent failure. The stent fatigue is the weakening of the stent material and structure caused by repeatedly applied loads. It is localised structural damaged that occurs when stent is subjected to cyclic loading. These loads must be tested before stent are used in blood vessel. Every stent have fatigue limit that define the highest stress that material can withstand for infinite number of cycles without breaking. Marrey et al. [24] mention that fatigue life of 10^8 cycles is required for a 10 year life of stent to design against premature mechanical failure. This approach to fatigue design also used in most other fields of engineering, e.g., in automobile or aerospace applications [24]. According to Marrey et al. [24] the fatigue design using two methodologies which are the traditional stress/strain-life (S-N) approach, and the damage-tolerant or fracture-mechanics approach. Damage tolerant approaches relying on fracture mechanics based crack growth data have been proposed and successfully applied to larger implant devices such as metallic and pyrolytic carbon mechanical heart valve prostheses [25, 26]

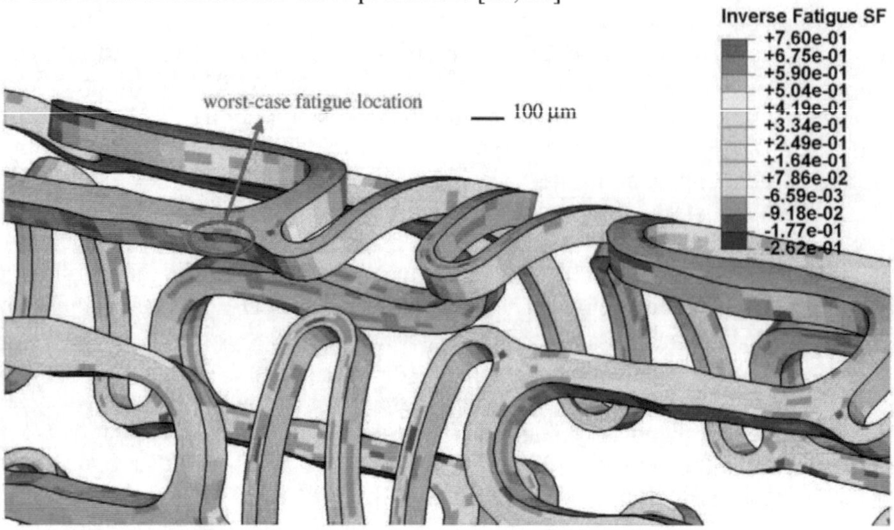

Figure 3. The simulation of estimated location cracks initiation [24]

Figure 3 showed the Finite element methods (FEM) simulation contour of estimated location of cracks initiation. FEM are considered to be an efficient way of testing and improving stent designs to minimize In-stent restenosis (ISR) from biomechanical aspects [27]. The inverse fatigue safety factors (SF) indicate the worst-case fatigue location where fatigue cracks probably initiate. The worst-case fatigue location mean that the location of stent element with the lowest SF or highest inverse fatigue SF. The safety factor means load carrying capacity of structure beyond or under the expected or actual load. The red colour location from the simulation indicate the possible location of structure failure and important for stent structure design optimization.

The investigation of stent's failure stent also have been done experimentally by Pelton et al, [28]. According to Pelton et al. [28] stent experimental testing is reliable as a stent's failure test and its results are very important as reference for stent design optimization. The testing specimen, variables, and load generator are controlled following some standardization testing. Figure 4 showed Diamond subcomponent of stent and fixture of cyclic testing. It was designed to determine stent's pulsatile fatigue properties. The tests were conducted at 30 Hz per ASTM recommendations [28].

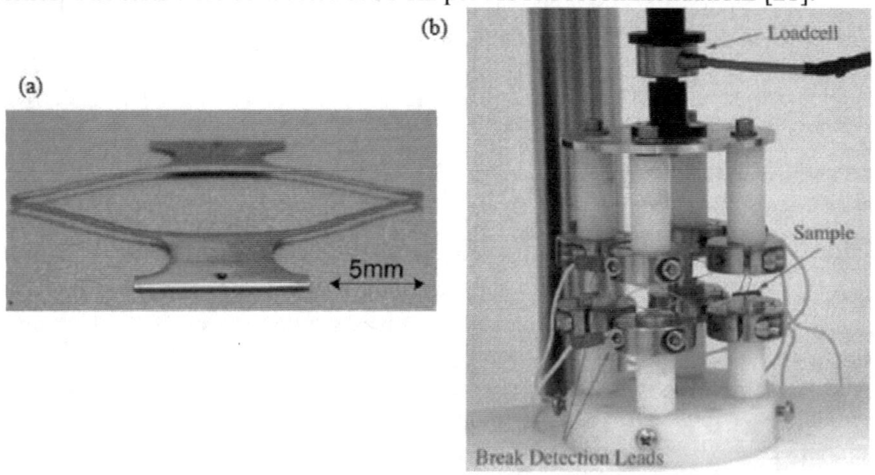

Figure 4. (a) Diamond subcomponent of stent (b) Fixture of cyclic testing [28]

A total of 432 subcomponents was tested [28]. This diamond stent's subcomponent consists of strut length, width and thickness by 8mm, 330μm, and 350μm [28]. The subcomponent were designed by top and bottom holder to provide proper alignment, structural stability, reduction of axial buckling, and secure gripping during laboratory testing [28]. The producing of nitinol test specimens by laser machined, processed at 500°C about 10 min and water quenched [28]. Although experimental subcomponent was scale up geometrically, but the material properties (nitinol) and material machining (laser-machined) was keeping nearly as actual product [28]. The effect of controlling environment during testing, high frequency of stent's subcomponent testing in air resulted the same fatigue reaction as in vivo conditions [29]. The tests were conducted until the stents fractured or achieve 10 million cycles [28].

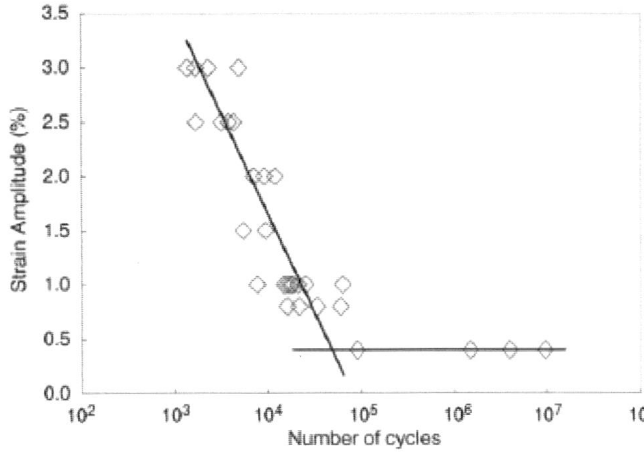

Figure 5. Strain Amplitude (%) versus Number of cycles [28]

Figure 5 above showed graph of percentage strain amplitude versus number of cycles of experimental stent fatigue failure test. The graph indicate that higher percentage of strain amplitude resulted lower number of cycles and vice versa. The number of cycles represent the stent's life cycle before fatigue failure. The stent's design optimization lifetime could be predicted by referring to the tabulated graph on figure above.

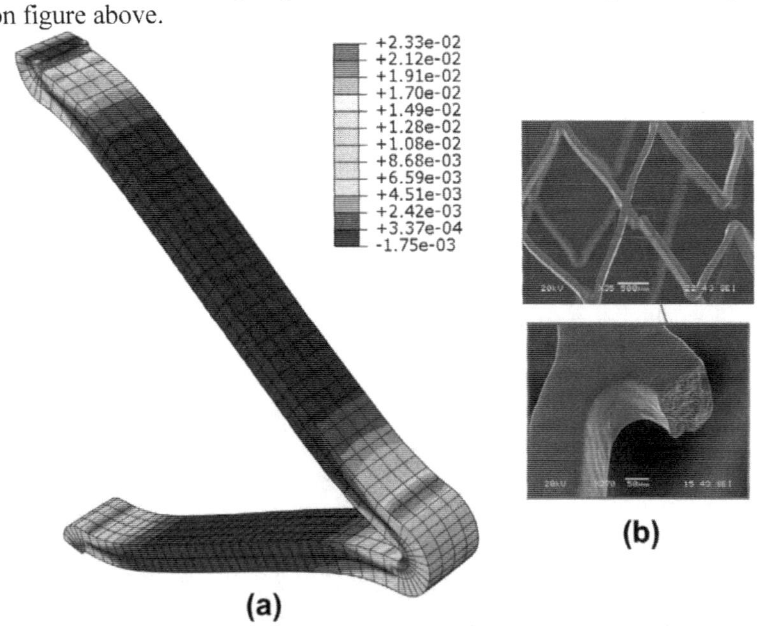

Figure 6. (a) Distribution of equivalent strain (b) Fatigue fracture on bridge tip region [30]

Figure 6 show the distribution of equivalent strain and fatigue fracture on bridge tip region. Simulation of stent's represent the motion of stent when in service and indicate the critical region of stent's failure. Critical region for equivalent strain show identical with fatigue fracture on bridge tip region. The external work done on an elastic member in causing it to distort from its unstressed state is transformed into strain energy.

The plastic strain energy from to low or high cycle stress can be predicted by simulation contour and failure location of stent can be estimated for further stent optimization. Energy principles in structural mechanics express the relationships between stresses, strains or deformations, displacements, material properties, and external effects in the form of energy or work done by internal and external forces. Since energy is a scalar quantity, these relationships provide convenient and alternative means for formulating the governing equations of deformable bodies in solid mechanics.

4. Conclusion

The stent structure failure by fatigue load have been a major concern for the researcher all over the world. This human implant device is very important for example stenting on patent ductus arteriosus to save patients having cynotic congenital heart disease (CCHD). PDA stenting safer than open surgery known as modified Blalock-Taussig to treat CCHD. PDA morphology varies such as bulging shape, curve or tortuous, and straight duct, for simplicity this paper reviewed the straight duct stent implantation on PDA. The reviewed of investigation about the fatigue stent failure in this paper are executed by simulation and experiment, however the simulation is preferable due to experimental fabrication cost and time. The experimental investigation of fatigue failure on stent strut following ASTM standardization. Comparing the material selection, cobalt–chromium alloy is more favorable than BMS stainless steel considering ultrathin stent strut structure for decreasing in stent restenosis effect. To design better stent structure, part of important feature are ultrathin strut, sufficient radial strength, radio-opacity and deliverability. The simulation of stent's subcomponent illustrated the lowest safety factor by red colour contour location which was similar location on experimental testing. The fatigue failure of stent structure can be concluded mostly occur at bridge tip region or strut's edge of stent.

Acknowledgement

This paper was partly sponsored by the Centre for Graduate Studies UTHM. Authors acknowledged Universiti Tun Hussein Onn Malaysia (UTHM), Office for Research, Innovation, Commercialization and Consultancy Management (ORICC) through FRGS 2016/1 Vot. 1592, Center for Graduate Studies UTHM.

References

[1] Marrey R V, Burgermeister R, Grishaber R B and Ritchie R O 2006 Fatigue and life prediction for cobalt-chromium stents: A fracture mechanics analysis *Biomaterials* **27** pp 1988-2000

[2] Setchi A, Mestel A J, Siggers J H, Parker K H, Tan M W and Wong K 2012 Mathematical model of flow through the patent ductus arteriosus *Journal of mathematical biology*

[3] Kaskurthy S, Veeduluri A R and Nidadavolu V K 2015 Tetralogy of fallot : A rare disease of heart defect **4** pp 305–312

[4] Petrucci O, O'Brien S M, Jacobs M L, Jacobs J P, Manning P B and Eghtesady P 2011 Risk Factors for Mortality and Morbidity After the Neonatal Blalock-Taussig Shunt Procedure **2** pp 642–652

[5] C. Mariño et al. 2016 Immediate and medium-term outcomes of ductal stenting in neonates and infants **23** pp 211–215

[6] Schiavone A and Zhao L G 2016 A computational study of stent performance by considering vessel anisotropy and residual stresses *Materials Science and Engineering C* **62** pp 307–316

[7] Krichenko A, Benson L N, Burrows P, Möes C A F, McLaughlin P and Freedom R M 1989 Angiographic classification of the isolated, persistently patent ductus arteriosus and

implications for percutaneous catheter occlusion *The American Journal of Cardiology* **63** pp 877-880

[8] Matsui H, McCarthy K and Ho S 2008 Morphology of the patent arterial duct: features relevant to treatment *Images in paediatric cardiology* **10** pp 27-38

[9] Boudjemline Y, Patel M, Malekzadeh-Milani S, Szezepanski I, Lévy M and Bonnet D 2013 Patent ductus arteriosus stenting (transcatheter potts shunt) for palliation of suprasystemic pulmonary arterial hypertension a case series *Circulation: Cardiovascular Interventions*

[10] Gewillig M, Boshoff D E, Dens J, Mertens L and Benson L N 2004 Stenting the neonatal arterial duct in duct-dependent pulmonary circulation: new techniques, better results *Journal of the American College of Cardiology* **43** pp 107-112

[11] Taib I, Shea K, Shafii N S and Zahran A 2015 Finite Element Analysis for Stent in Patent Ductus Arteriosus (PDA)

[12] Karjalainen P P, Nammas W and Airaksinen J K 2014 Optimal stent design: past, present and future *Interventional Cardiology* **6** pp 29-44

[13] Kereiakes D J, Cox D A et al. 2003 Usefulness of a cobalt chromium coronary stent alloy *American Journal of Cardiology* **92** pp 463-466

[14] Briguori C, Sarais C et al. 2002 In-stent restenosis in small coronary arteries *Journal of the American College of Cardiology* **40** pp 403-409

[15] Bedoya J, Meyer C A, Timmins L H, Moreno M R and Moore J E 2006 Effects of stent design parameters on normal artery wall mechanics *Journal of biomechanical engineering* **128** pp 757-765

[16] Debusschere N, Segers P, Dubruel P, Verhegghe B and De Beule M 2015 A finite element strategy to investigate the free expansion behaviour of a biodegradable polymeric stent *Journal of biomechanics* **48** pp 2012-8

[17] Peters B, Ewert P and Berger F 2009 The role of stents in the treatment of congenital heart disease: Current status and future perspectives *Annals of pediatric cardiology* **2** pp 3-23

[18] Kübler P, Tomasiewicz B, Johansson M, Szczepański A and Reczuch, K 2016 Mechanical stent failure as a cause of life-threatening left main restenosis *Advances in Interventional Cardiology* **3** pp 271-273

[19] Inaba S, Mintz G S et al. 2014 Mechanical complications of everolimus-eluting stents associated with adverse events: an intravascular ultrasound study *EuroIntervention* **11** pp 1301-1308

[20] Ragosta M, Samady H, Gimple L W, Sarembock I J, Fenster M and Powers E R 2004 Percutaneous treatment of focal vs. diffuse in-stent restenosis: a prospective randomized comparison of conventional therapies *Catheterization and Cardiovascular Interventions : Official Journal of the Society for Cardiac Angiography & Interventions* **61** pp 344-349

[21] Mehran R, Dangas G et al. 1999 Angiographic Patterns of In-Stent Restenosis *Circulation* **100** pp 1872 LP - 1878

[22] Wilson S H, Berger P B et al. 2000 Immediate and late outcomes after direct stent implantation without balloon predilation *Journal of the American College of Cardiology* **35** pp 937-943

[23] Pentousis D, Guerin Y et al. 1998 Direct stent implantation without predilatation using the MultiLink stent *The American journal of cardiology* **82** pp 1437-1440

[24] Marrey R V, Burgermeister R, Grishaber R B and Ritchie R O 2006 Fatigue and life prediction for cobalt-chromium stents: A fracture mechanics analysis *Biomaterials* **27** pp 1988-2000

[25] Ritchie R and Lubock P 1986 Fatigue Life Estimation Procedures for the Endurance of a Cardiac Valve Prosthesis : Stress / Life *Journal of Biomechanical Engineering* **108** pp 1-8

[26] Ritchie R 1996 Fatigue and fracture of pyrolytic carbon: a damage-tolerant approach to structural integrity and life prediction in "ceramic" heart valve prostheses *Journal of Heart Valve Disease*

[27] Wu W, Wang W Q, Yang D Z and Qi M 2007 Stent expansion in curved vessel and their interactions: A finite element analysis *Journal of Biomechanics* **40** pp 2580-2585

[28] Pelton A R, Schroeder V, Mitchell M R, Gong X Y, Barney M and Robertson S W 2008 Fatigue and durability of Nitinol stents *Journal of the Mechanical Behavior of Biomedical Materials* **1** pp 153-164
[29] Robertson S W and Ritchie R O 2007 In vitro fatigue-crack growth and fracture toughness behavior of thin-walled superelastic Nitinol tube for endovascular stents: A basis for defining the effect of crack-like defects *Biomaterials* **28** pp 700-709
[30] Azaouzi M, Lebaal N, Makradi A and Belouettar S 2013 Optimization based simulation of self-expanding nitinol stent *Lecture Notes in Mechanical Engineering* **1** pp 423-450

IOP Conf. Series: Materials Science and Engineering **165** (2017) 012008 doi:10.1088/1757-899X/165/1/012008

Mode I stress intensity factors of slanted cracks in plates

Al Emran Ismail*,1, Mohd Zubir Mohd Ghazali[2], Nik Hisyamudin Muhd Nor[1]

[1]Faculty of Mechanical and Manufacturing Engineering, Universiti Tun Hussein Onn Malaysia, Batu Pahat, 86400 Johor, Malaysia.

[2]Malaysia Marine and Heavy Engineering Sdn. Bhd., 81700 Pasir Gudang, Johor, Malaysia

*emran@uthm.edu.my / al_emran@hotmail.com

Abstract. This paper presents the roles of slanted cracks on the stress intensity factors (SIF) under mode I tension and bending loading. Based on the literature survey, lack of solution of SIFs of slanted cracks in plain strain plates are available. In this work, the cracks are modelled numerically using ANSYS finite element program. There are two important parameters such as slanted angles and relative crack length. SIFs at the crack tips are calculated according to domain integral method. Before the model is further used, it is validated with the existing model. It is found that the present model is well agreed with the previous model. According to finite element analysis, there are not only mode I SIFs produced but also mode II. As expected the SIFs increased as the relative crack length increased. However, when slanted angles are introduced (slightly higher than normal crack), the SIFs increased. Once the angles are further increased, the SIFs decreased gradually however they are still higher than the SIFs of normal cracks. For mode II SIFs, higher the slanted angels higher the SIFs. This is due to the fact that when the cracks are slanted, the cracked plates are not only failed due to mode I but a combination between both modes I and II.

1. Introduction

The concept of fracture mechanics is successfully used to characterize the elastic behavior of cracks in mechanical components. There are two important tools generally used for such purposes such as stress intensity factor (SIF) and J-integral approaches. The first approach is the traditional method where it is relatively difficult to use as compare with the second tool. Therefore, this work focuses more on the implementation of J-integral and then convert it to SIFs. The advantages of J-integral over traditional SIF can be found in [1]. On the other hand, some basic applications of SIF method can also be found in [2].

Another important fracture criterion can be found in [3]. They have reviewed the developments and application of crack tip opening angle or displacement (CTOA/CTOD). This paper also discussed on the current issues regarding to CTOA/CTOD. Based on the literature, the works on the edge cracks are tremendously found especially on the normal cracks. However, the solution of SIFs for the slanted cracks are considerably difficult to find. For example Albinmousa et al. [4] studied slanted cracks under mode I tension. They demonstrated the applicability, simplicity and flexibility of the new approach where it is then validated through comparison with the existing analytical and numerical solution. It is found that their finding are well agreed with experimental results. However, they are assumed that the cracks are in the plane stress but not in plain strain.

Dempsey and Mu [5] used an approach of weight function for calculating the stress intensity factors of in-situ edge-cracked rectangular plated under arbitrary crack face loading. Various crack geometries are investigated and stress intensity factor (SIF), crack opening displacement (COD) and crack opening angle (COA) are determined and then compared with the existing results. It is found that a good accuracy is obtained for all crack geometries considered.

Matsumto et al. [6] computed the SIFs of interfacial cracks based on the interaction energy release rates and boundary element method (BEM) sensitivity analysis. Several crack configurations are considered. In their analyses, the present method can give more accurate results with a coarse mesh than the method based on the displacement extrapolation. Sensitivity analysis is conducted and found that the computational cost are comparatively small. Hammond and Fawaz [7] reviewed the solutions of SIFs of various size single edge-cracked tension specimens. Comparison are made between the literature reported data and the present effort's results with generally satisfactory correlation. Significant differences are also found for the short crack with the plate aspect ratio of less than 3.

Kuang and Chen [8] used a displacement extrapolation method for analyzing the problem of mixed mode loading of single edge cracks. According to their reports, the proposed modified extrapolation method associated with collapsed triangular point element is the most reliable method for determining the SIFs for both normal and inclined cracks.

Ismail et al. [9-12] investigated the surface cracks in round bars under bending, tension and their combination of stresses. They found that the stress intensity factors significantly affected by the crack geometries and for combined stresses conditions, the combined stress intensity factors can be predicted using a superposition technique. While the stress intensity factors from different loading modes can also be predicted using an equivalent von-Mises stress however there some discrepancies when compared with the values predicted using finite element method. He is also studied the stress intensity factors of three-dimensional slanted cracks in round bars. The detail of the solution of stress intensity factors can be found in [13-15].

According to the literature above, it is found that lack of the solutions of stress intensity factors are available especially for slanted cracks under mode I loading. Therefore, this paper presents the solutions of SIFs for various single edge crack geometries and angles embedded in the plain strain plates.

2. Numerical Modelling

Single edge crack is placed at one edge of the plain strain plate subjected to mode I loading (tension and bending forces). Two types of crack configurations are considered such as normal and slanted cracks. The schematic diagram of the cracked plate is shown in Figure 1. The width of the plate is $W = 50$mm while the total height is $2H = 100$mm. There are seven relative crack length, a/L are used such as 0.1, 0.2, 0.3, 0.4, 0.5, 0.6 and 0.7 where a is a crack length and L is a plate width. The normal crack condition, $W = L$. Another important parameter is slanted angle, θ then they are normalized against 90^0. Therefore, the normalized angles used are 0.00, 0056, 0.111, 0.167, 0.222, 0.278, 0.333, 0.389, 0.444 and 0.500. The crack geometries are then modelled using ANSYS finite element program through the use of ANSYS Parametric Design Language (APDL). Since the problem is elastic then modulus of elasticity, $E = 200$GPa and Poisson's ratio, $\nu = 0.3$ is used. The cracked plated is also assumed to fulfil the plain strain condition and iso-parametric 8-node quadrilateral element called PLANE183 is used to model the plate. A special attention is paid to the crack tip where the nodes around it is shifted to the location of ¼ close to the tip. This is to ensure the singularity around the crack tip.

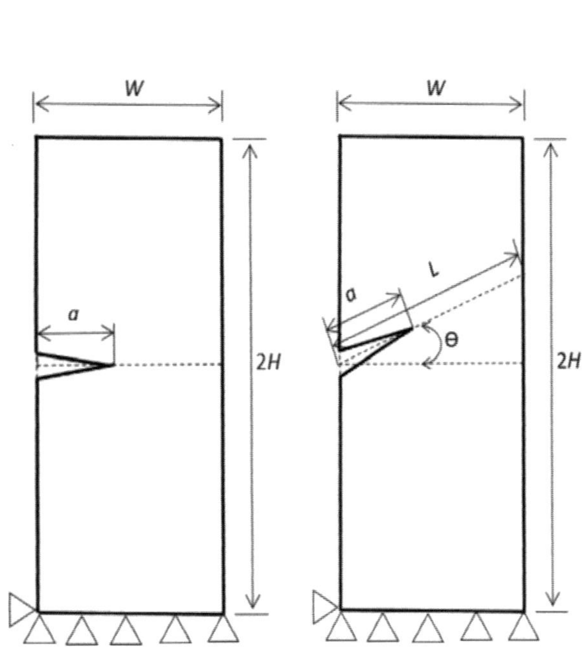

Figure 1 Schematic diagram of (a) normal and (b) slanted cracks.

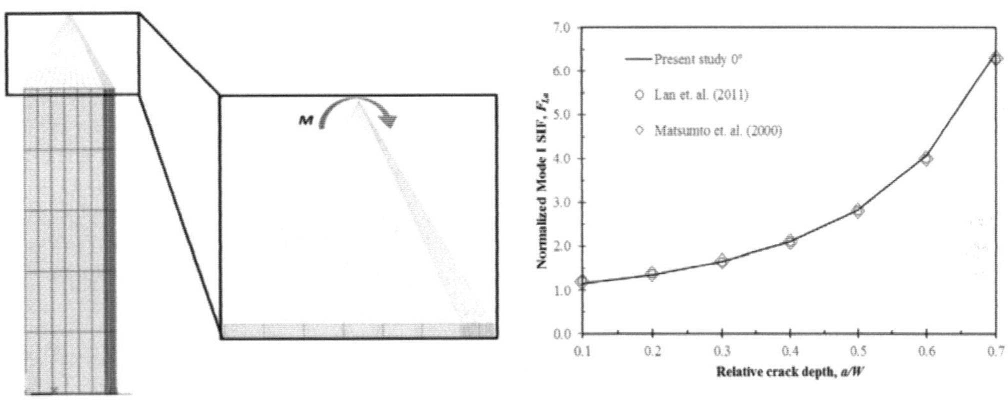

Figure 2 (a) MPC184 element is used for remotely applied bending moment and (b) Validation of the present with the existing models.

Figure 2(a) shows the finite element model of the crack plate. In order to properly manage the mechanical deformation of the plate, it must be constrained properly. At the bottom edge of the plate is fixed in y-direction. This is to ensure that the plate is capable to elongate in both x- and y-directions. In order to prevent model rotation, the plate is also constrained at bottom left point. Tension stress is applied directly at the upper edge while the bending moment is applied remotely through the use multi-point constraint element called MPC184 as shown in Figure 2(a). Both bending moment and tension force are applied to the independent node. Stress intensity factor (SIF) based on the J-integral (domain integral method). In this work, the SIFs are normalized as function of crack length, a and applied stress, σ using the following expression:

$$F = \frac{K}{\sigma\sqrt{\pi a}} \tag{1}$$

where, F is a normalized stress intensity factor, K is a stress intensity factor, σ is an applied stress and a is a crack length. It is a paramount important aspect to validate the present with the existing models (Lan et al. [14] and Matsumto et al. [6]) when the normal crack is used. Figure 2(b) reveals the validation results and it is found that the present model is well agreed with the existing results. In this work, it is also assumed that the normal crack is sufficiently validated the slanted cracks since the solutions of SIFs of slanted cracks are difficult to obtain.

3. Results and Discussion

Normalized SIFs for slanted cracks under tension forces is shown in Figure 3 for both modes I and II. They are plotted against relative crack depth, a/L. In general, the mode I SIFs gradually increased with the increased of a/L. The lowest SIFs are obtained from normal or 90° cracks. It is observed that slightly increased the slanted angle to 5° has significantly increased the SIFs compared with the normal cracks. No significant different of SIFs obtained from the cracks slanted using 5°, 10°, 15° and 20° where the SIFs are almost identical. However, mode I SIFs (Figure 3(a)) decreased as the slanted angles are increased. Similar SIFs trend is observed from the works conducted by Lan et. al. [13] and Matsumto et. al. [6]. In this work, when the slanted angles are increased to up to 45°, the SIFs seemed to approach the SIFs obtained from the normal cracks.

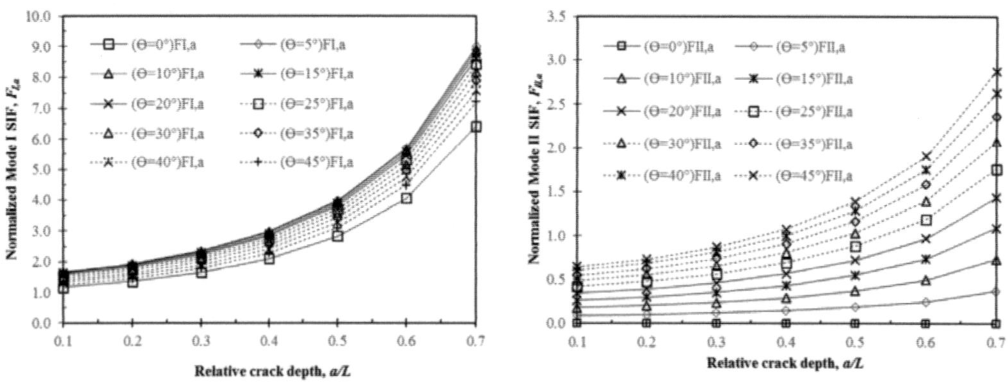

(a) (b)

Figure 3 Effect of relative crack depth on the stress intensity factor of (a) mode I and (b) mode II under tension stress.

Figure 3(b) shows the mode II SIFs plotted against the relative crack depth, a/L for different slanted angles. It is observed similarly with mode I SIFs, mode II SIFs are also increased when a/L increased since the ligament of the plate decreased when crack length increased and therefore increasing the stress intensity around the crack tip. It can also be seen that mode II SIFs increased when slanted angles are increased. Contrarily with mode I SIFs where the SIFs decreased when higher slanted angels are used. Figure 4 reveals the effect of slanted angles on the mode I SIFs for different relative crack length. As expected longer the crack length higher the SIFs can be obtained. However, it is interested to note that slightly slanted the crack for an example 5° has the great effect on the mode I SIFs. It is also observed that the maximum SIF occurred when 5° angle is used. However when the angles increased, their effect are diminished where the SIFs reduced to approach the SIFs for normal cracks.

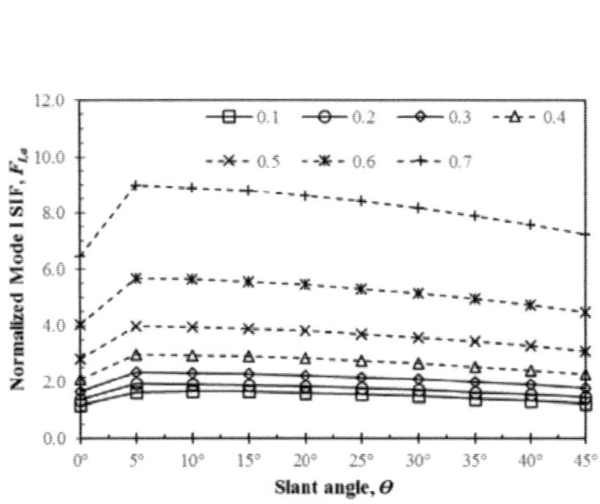

Figure 4 Effect of slanted angles on the SIFs when relative crack depth are varies under tension stress.

Figure 5 shows the stress intensity around crack tip of different slanted cracks. Figure 5(a) reveals the crack behaviour for normal crack. Since the problem is symmetrical only half of crack is modelled. It is observed the stress distribution is also symmetrical where the crack face is almost parallel with the symmetrical plane. On the other hand, Figures 5(b) and 5(c) indicate the crack deformations when the cracks are slanted at 25° and 45°, respectively. Once the angles are introduced, the stress intensity around the crack tip is disturbed where the bottom side of the crack the stress distribution is almost similar with the normal cracks. However at the upper side, the stress intensity seemed to shear the crack to the right of the plate and thus creating mode II SIFs. The shear severity increased as the slanted angles increased and therefore increasing the mode II SIFs.

Figure 6 shows the effect of relative crack depth, a/L on the modes I and II SIFs under remote bending moment, respectively. In general both modes of SIFs increased as crack length are increased due to the fact that the ligament cross-section decreased in length and therefore increasing the stress intensities. Figure 6(a) indicates the effect of a/L on the mode I SIFs when slanted angles are varied. In comparison, the SIFs obtained from the normal cracks are also presented. It is found that the introduction of slanted angles have decreased the SIFs significantly. Under bending moment, there is insignificant effect of slanted angles on the mode I SIFs. However, the introduction of slanted angles have slightly decreased the mode I SIFs. Figure 6(b) reveals the influence of relative crack depth on the mode II SIFs. It is found that the introduction of slanted angles are strongly affected the mode II SIFs where increasing the angles also increase the SIFs. It is also observed that for relatively shallow cracks ($a/L \leq 0.4$), the a/L has insignificant role on the mode II SIFs. This is due to the ligament of the plate is strong enough to resist the sliding deformation of the cracks.

Figure 7 reveals the effect of slanted angles on the SIFs when relative crack length, a/L are varied for modes I and II SIFs, respectively. In this work, 0° crack is defined as a normal crack where the crack faces are perpendicular to the axial forces. For the relatively shallow cracks ($a/L \leq 0.4$), the effect of slanted angles on the both modes I and II SIFs are insignificant. Increasing the crack length, slightly increased the SIFs however the increments can be ignored if compared with other SIFs obtained using $a/L \leq 0.4$.

IOP Conf. Series: Materials Science and Engineering **165** (2017) 012008 doi:10.1088/1757-899X/165/1/012008

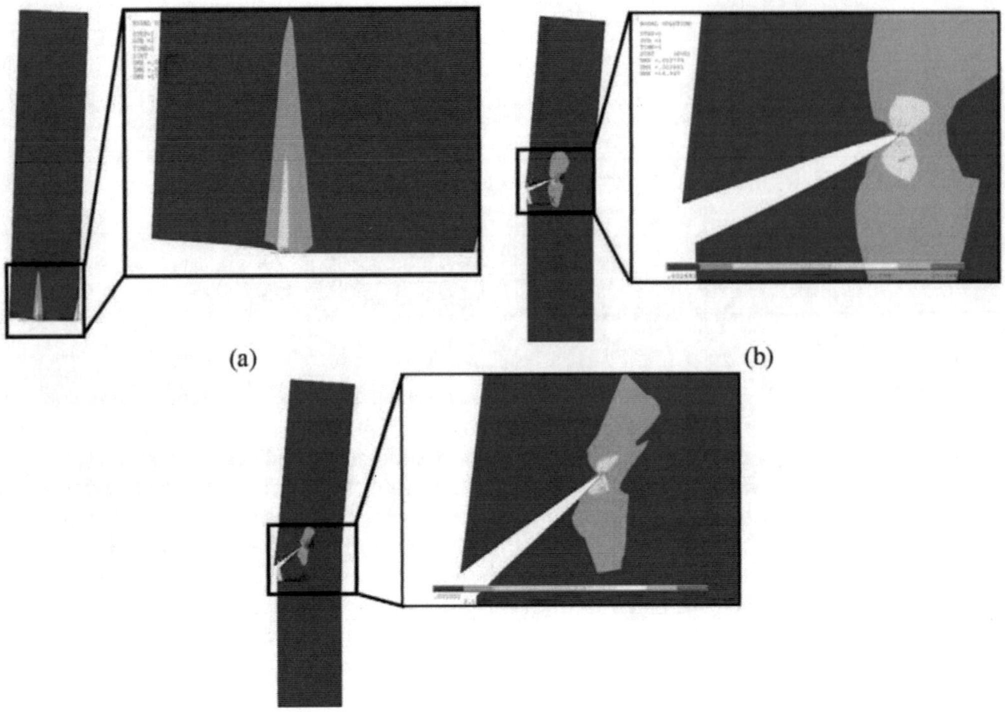

(a) (b)

(c)

Figure 5 Crack deformation under mode I tension stress of different slanted cracks, (a) 0°, (b) 25° and (c) 45° under tension stress.

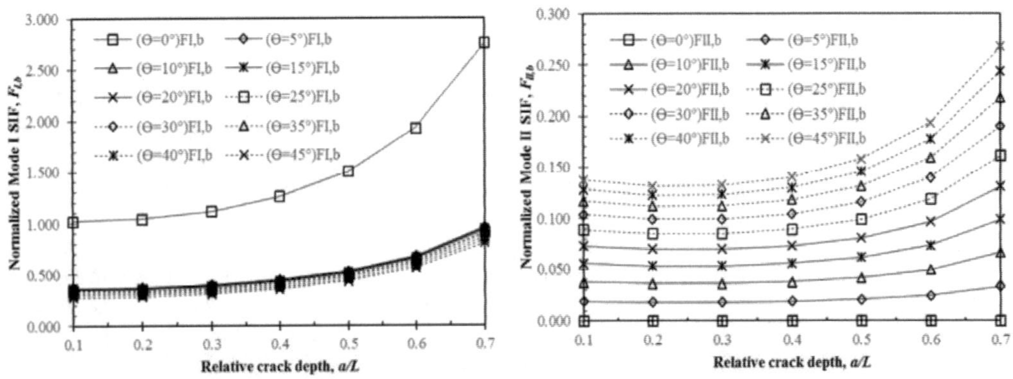

(a) (b)

Figure 6 Effect of relative crack depth on the SIFs of (a) mode I and (b) mode II under bending moment.

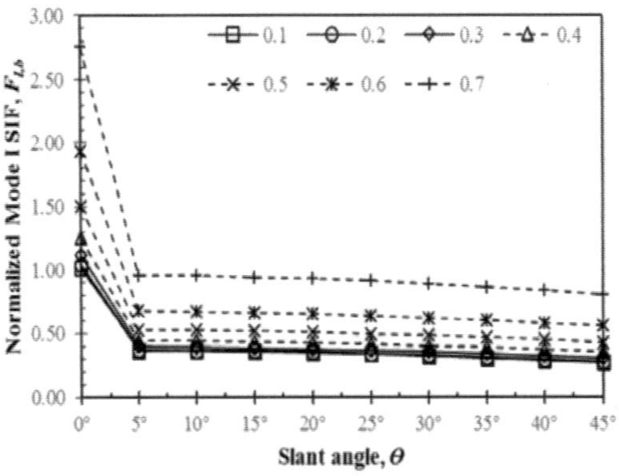

(a)

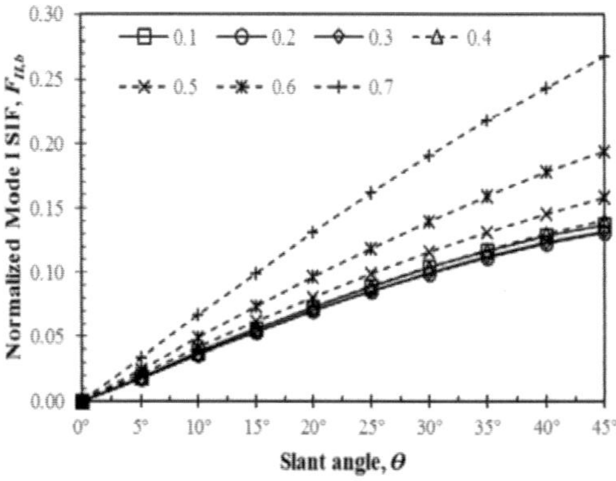

(b)

Figure 7 Effect of slanted angles on the SIFs when relative crack depth are varies under bending moment.

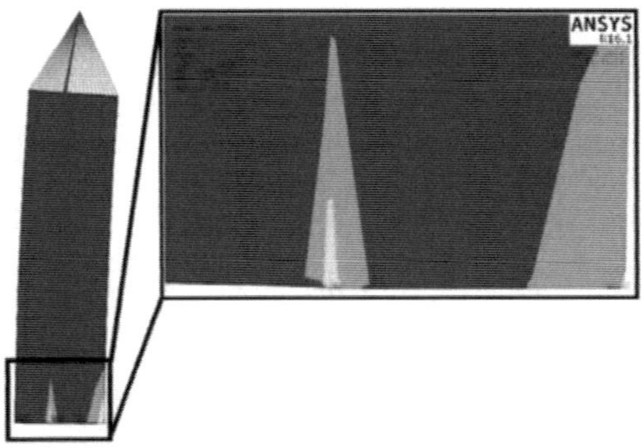

(a)

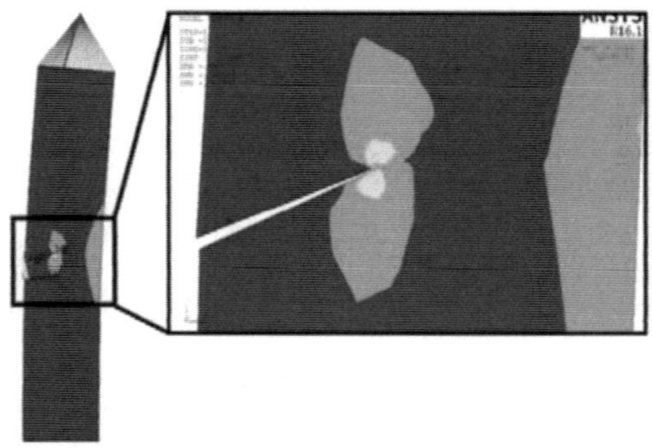

(b)

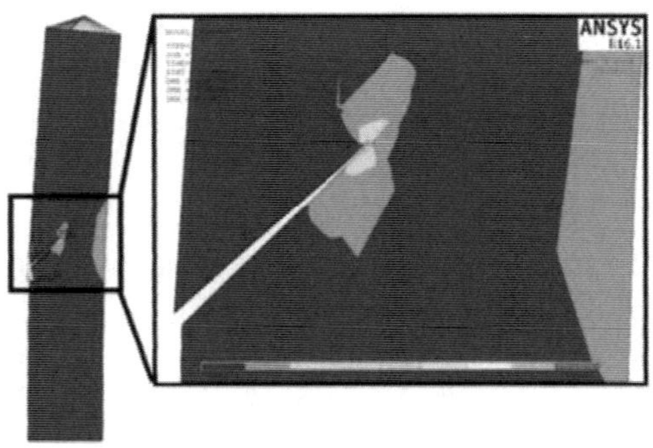

(c)

Figure 8 Crack deformation under mode I tension stress of different slanted cracks, (a) 0°, (b) 25° and (c) 45° under bending moment.

Figure 8 shows the stress distribution around crack tip located at plate edge. The crack was subjected to bending moment. The SIF for slant crack subjected to single load was shown in Figure 7. Slant crack at angle 25° and 45° as shown in Figures 8(b) and 8(c) produces smaller opening while slant crack at angle 0° produces larger crack opening as shown in Figure 8(a). As the slanted angles are increased, the symmetrical pattern of stress intensity distributions are disturbed where the upper distribution is shifted to the right while the lower distribution is then shifted to the left. This is due to the fact that the degree of mode II stress intensity factors increased when the slanted angles are increased.

4. Conclusion

This work numerically presents the slanted crack behavior in plain strain plate under bending and tension forces. In general tremendous amount of works on the normal cracks are available. However, lack number of stress intensity factor solutions are obtained especially when the cracks are slanted. Based on the numerical works, several conclusion can be stated as below:

1. Both under bending and tension forces, stress intensity factor increased as a function of crack length.
2. Under tension force, increasing the slanted angles lowered the stress intensity factors. However, the mode I stress intensity factors obtained from slanted angles are higher than the stress intensity factors for normal cracks. On the other hand, mode II stress intensity factors increased when slanted angles are increased due to the capability of plate to slide along the crack lines.
3. Under bending moment, the effect of slanted angles on the mode I stress intensity factors are insignificant where increasing the slanted angles slightly reduced the stress intensity factors. The effect of slanted angles on mode II stress intensity factors are significant where increasing the slanted angles gradually increased the mode II stress intensity factors.

Acknowledgement

Authors acknowledged Universiti Tun Hussein Onn Malaysia and Kementerian Pegajian Tinggi for sponsoring this work through Fundamental Research Grant Scheme (FRGS) 1/2016 Vot. 1592.

References

[1] Courtin S, Gardin C, Bezine G and Ben Hadj Hamouda H 2005 *Engineering Fracture Mechanics* **72** 2175-2185.
[2] Guinea GV, Planas J and Elices M 2000 *Engineering Fracture Mechanics* **66** 243-255.
[3] Newman JC Jr, James MA and Zerbst U 2003 *Engineering Fracture Mechanics* **70** 371-385.
[4] Albinmousa J, Merah N and Khan SMA 2011 *Engineering Fracture Mechanics* **78** 3300-3307.
[5] Dempsey JP and Mu Z 2014 *Engineering Fracture Mechancs* **132** 93-103.
[6] Matsumto T, Tanaka M and Obara R 2000 *Engineering Fracture Mechanics* **65** 683-702.
[7] Hammond MJ and Fawaz SA 2016 *Engineering Fracture Mechanics* **153** 25-34.
[8] Kuang JH and Chen LS 1993 *Engineering Fracture Mechanics* **46**(5) 735-741.
[9] Ismail AE, Ariffin AK, Abdullah S and Ghazali MJ 2014 *Research Journal of Applied Sciences, Engineering and Technology* **7**(10) 1985-1993.
[10] Ismail AE, Ariffin AK, Abdullah S and Ghazali MJ 2014 *Jurnal Teknologi (Sciences & Engineering)* **68**(1) 7–18.
[11] Ismail AE, Ariffin AK, Abdullah S and Ghazali MJ 2012 *Meccanica* **47** 1141-1156.
[12] Ismail AE, Ariffin AK, Abdullah S and Ghazali MJ 2012 *Journal of Zhejiang University-SCIENCE A (Applied Physics & Engineering)* **13**(1) 1-8.
[13] Ismail AE, ALM Tobi and NHM Nor 2015 *AIP Conference Proceedings* 1660 070027.
[14] Lan X, Noda NA, Zhang Y and Michinaka K 2012 *Acta Mechanica Solida Sinica* 25 404-416.
[15] Ismail AR, Azlan MA, Mohd Tobi AL and Ahmad MH 2016 *IOP Conf. Series: Materials Science and Engineering* 152 012049.

Stress intensity factors and interaction of two parallel surface cracks on cylinder under tension

MK Awang[1], AE Ismail[1], AL Mohd Tobi[1] and MH Zainulabidin[1]

[1]Faculty of Mechanical and Manufacturing Engineering, Universiti Tun Hussein Onn Malaysia, Batu Pahat, 86400 Johor, Malaysia.

Email: mdkhair@uthm.edu.my

Abstract. This paper presents three-dimensional (3D) finite element solution on multiple surface cracks. The cracks on solid cylinder are similar in sizes, parallel each other, assumed to grow in semi-elliptical shape and subjected to remotely tension loading (mode I). A wide range of parametric study involving crack depth ratios ($0.1 \leq a/D \leq 0.4$), crack aspect ratios ($0.2 \leq a/b \leq 1.2$), normalized coordinates on crack front ($0.0 \leq x/h \leq 0.93$) and inter-crack distance ratios ($0.005 \leq c/l \leq 0.32$) are considered for numerical estimation of stress intensity factors (SIFs) along crack front. For multiple surface cracks under axial loading, the stress intensity factors along crack front decreased when c/l decreased. When multiple cracks approach one another, the stress intensity changes due to interaction of the stress field. The results show that it produces a stress shielding effects.

Keywords: Stress Intensity Factor, Crack Interaction, Shielding

1. Introduction

Cylinders under tension loading are widely used in mechanical and structural engineering. They are normally used as a shaft, rods and so on and frequently subjected to cyclic loadings and tends to fail due to nature of said loadings. In most cases, multiple cracks are likely initiated due to corrosion [1] and fatigue [2]. Conventional assessments for multiple crack problems are derived from fracture mechanics principle based on single crack configuration. Even by using current design codes, such as BS7910, multiple cracks are usually recharacterized as one larger crack, following certain rules and condition. This may lead to over-conservative and incorrect predictions of the service life of cracked component. On top of that, crack interaction criterion in this failure assessment guideline, are extensively validated by past research work on crack interaction in plate model. However, similar validation based on cylinder model is still lacking.

This paper focused on crack interaction for surface cracks in a cylinder under different crack spacing. The interaction of surface cracks is examined using a calibrated three dimensional finite element procedure, for linear elastic analysis. Crack interactions are evaluated based on the crack driving force, measured by the stress intensity factor values, determined from an interaction integral approach implemented in a commercial finite element code. The finite element results are validated against available SIF results [3] for cracked cylinder with a single semi-elliptical crack. Despite of availability of efficient techniques and computer to solve three-dimensional crack problems, solution for interactive multiple cracks on cylinder are not much in the literature.

2. Review of existing solutions

The problem involved interacting cracks receive increased attention since the last two decade. Chudnovsky et al. [4] have developed different solution techniques that use the method of dislocations. By using the body force method, Murakami and Nemat-Nasser [5] have presented Mode I SIF solutions for a number of interacting coplanar surface cracks in a half-space. These solutions were given for a variety of crack shapes and sizes under tension and bending loads. Using the same technique, Isida et al. [6] studied an array of parallel semi-elliptical surface cracks in a 3D semi-infinite solid under tension. Employing the boundary element method, Tu and Cai[7] have also studied the behavior of 2D non-coplanar cracks contained in an infinite body. Miyazaki et al. [8] used a combined line-spring and boundary element method to obtain mode I solutions for embedded and surface coplanar cracks interacting in a plate. Applying the alternating finite element methods, Stonesifer et al. [9] calculated the SIFs for two symmetric non-coplanar surface cracks interacting in a plate subjected to remote uniaxial tension. Further, classical finite element methods were also used by Jiang et al. [10] to analyze two parallel surface cracks in 3D. In many of the cases studied, the cracks were treated as being symmetric.

3. Finite element modelling

The geometry of the crack shown in Figure 1 can be described by the dimensionless parameters a/D and a/b which are crack depth ratio and crack aspect ratio, respectively. Different crack shapes were considered i.e a/D of 0.1, 0.2, 0.3 and 0.4. Meanwhile a/b ranging from 0.2 to 1.2 with each increment of 0.2, where the definitions of geometrical parameters agree with those of Shin and Cai [3] and Carpinteri [11]. In this work, there are 14 points along crack front being considered through normalized coordinates x/h with special attention given to most outer point, deepest point and one point in between. Only a quarter model are required due to symmetry of the case. The diameter of the cylinder is 50mm and 200mm in length. The Poisson ratio is assumed to be 0.33 and Young's modulus is 70GPa.

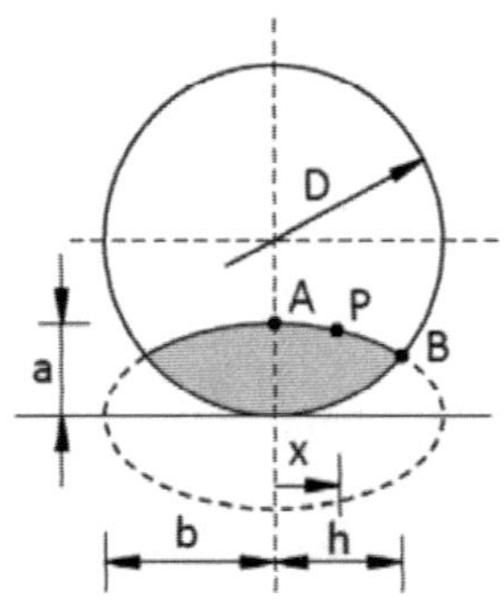

Figure 1: Crack characterization [3]

The layout of the problem is shown in Figure 2 where both cracks take place at the middle of the cylinder and the inter-crack distance ratio, c/l ranges of 0.005, 0.01, 0.02, 0.04, 0.08, 0.16 and 0.32. Interaction integral method is used to determine the stress intensity factor at the crack front. Since the crack faces are normal to direction of forces, only mode 1 of SIFs is produced. The SIFs are also normalized as below:

$$F_{I,a} = \frac{K_I}{\sigma\sqrt{\pi a}} \tag{1}$$

where K_I is the modes I stress intensity factors, while F_I is their corresponding geometrical correction factors or normalized stress intensity factors, σ is an axial stress and a is a crack depth.

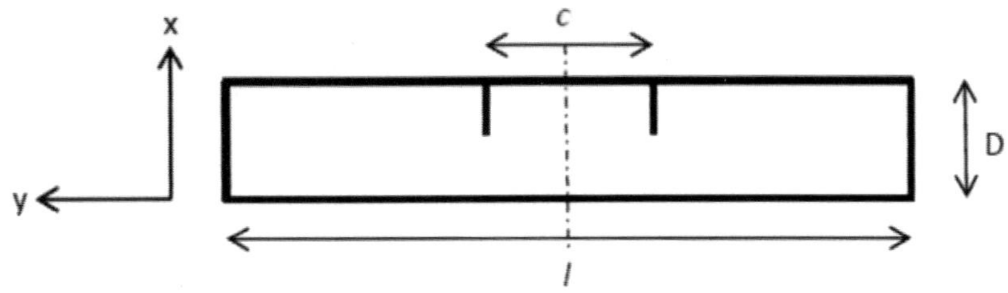

Figure 2: The layout of crack

4. Results and discussion

4.1 Stress intensity factors for single crack

In order to employ the appropriate finite element model, it is necessary to compare the proposed model with published data. Figure 3 shows a comparison of the SIFs values of a single crack under tension loading as well as those predicted by Shin and Cai [3]. It is found that the results are in a good agreement. Therefore, the proposed model can be used with confident to determine the SIF values of multiple cracks.

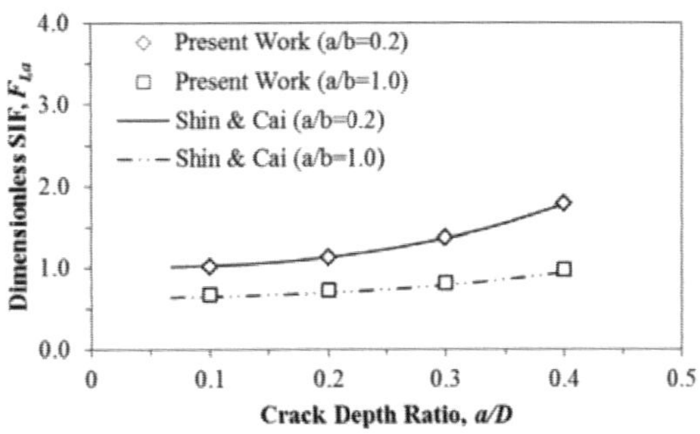

Figure 3: Validation of present model

4.2 Double crack interaction

Schematic diagram of the interactive cracks is shown in Figure 2. The cylinder is subjected to a remote tension load in y-direction. A finite element model is developed using ANSYS Parametric Design Language (APDL) code and parallel interacting surface crack are modelled using 20-noded isoparametric quadratic brick elements. The interaction process is quantified by calculating the interaction factor, γ which is defined by ratio of normalized SIF between a cylinder with two cracks and a cylinder with a single crack. Since the problem is symmetrical, only a quarter of cylinder has been modelled. Wide ranges of crack configuration, i.e ($0.1 \leq a/D \leq 0.4$), ($0.2 \leq a/b \leq 1.2$), ($0.0 \leq x/h \leq 0.93$) and ($0.005 \leq c/l \leq 0.32$) have been analyzed and the results are displayed both in graphical and tabular forms.

Figures 4 – 7 show the variation of SIF along crack front for various inter-crack distance when the crack aspect ratio, a/b is kept constant at 0.6. It is apparent that various inter-crack distance produce shielding effect rather than amplification effect. The closer the two cracks are, the greater the interaction is, the stress field is relaxed thus the more the dimensionless SIF decreases. The same pattern goes for the rest of crack aspect ratio. The dimensionless SIF values almost identical with single crack when *c/l* reaches 0.32. This means there is no more interaction beyond this ratio and two interacting crack can be treated as a single separate crack. It can be seen that the plotted graph in Figure 7 have more gap each other than Figure 4 which indicates the dimensionless SIF not only depends on *c/l*. In this regards, *a/D* plays a significant roles. The dimensionless SIF always achieve its maximum at outer point (x/h = 0.93). This is due to state of stress changes slowly from plain strain at deepest point to plain stress at outer point. Therefore, to estimate the critical load for any crack propagation, determination of the stress intensity factor at the free surface of the cylinder is the most crucial.

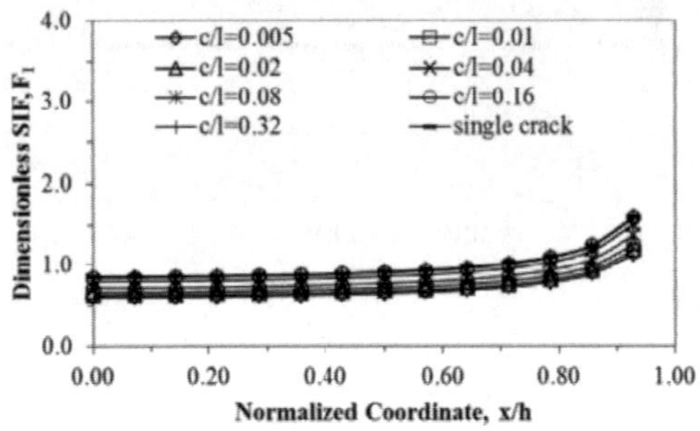

Figure 4: The variatons of SIF along crack front for various intercrack distance, *c/l* when *a/D*= 0.1, *a/b*=0.6

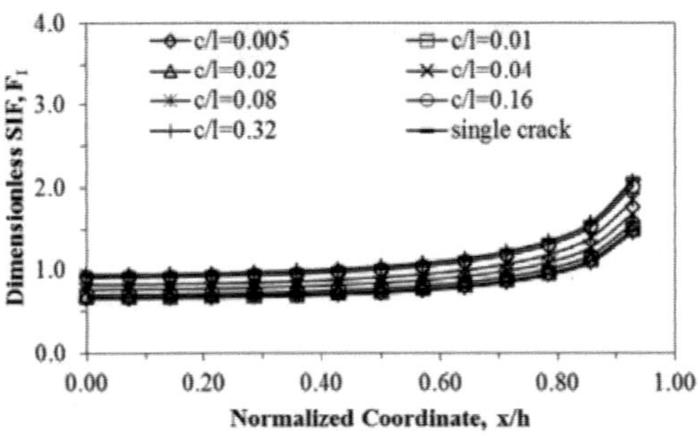

Figure 5: The variatons of SIF along crack front for various intercrack distance, *c/l* when *a/D*= 0.2, *a/b*=0.6

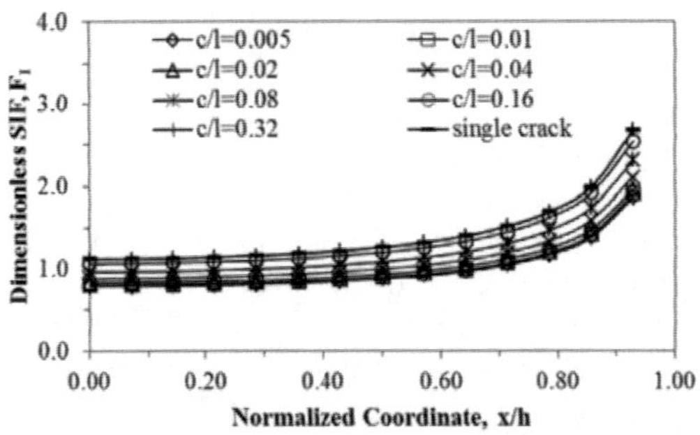

Figure 6: The variatons of SIF along crack front for various intercrack distance, *c/l* when *a/D*= 0.3, *a/b*=0.6

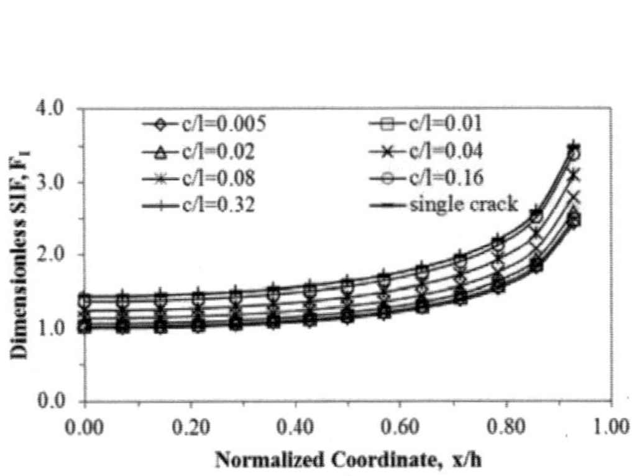

Figure 7: The variatons of SIF along crack front for various intercrack distance, *c/l* when *a/D*= 0.4, *a/b*=0.6

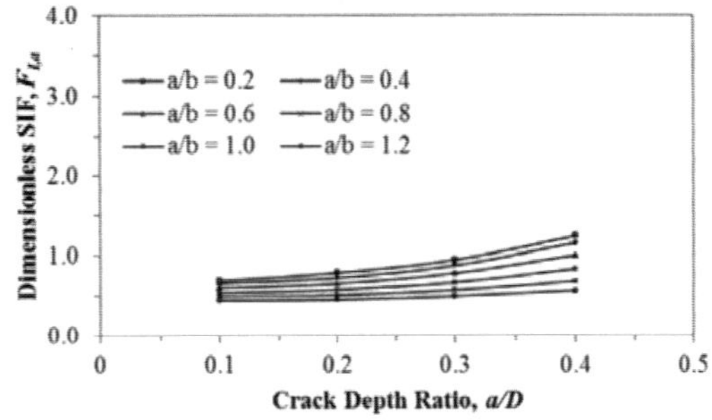

Figure 8 : Dimensionless SIF vs *a/D* at deepest point when *c/l*=0.005

To further investigate the normalised SIF effect due to the size of cracks, parametric study with varying *a/D* and *a/b* is carried out. Figure 8 shows the variation of the normalised SIFs with respect to crack depth ratio, *a/D*. The latter plays a significant role at lower *a/b* but less effect at *a/b*=1.2. Meanwhile, *a/b* between 0.4-0.8 has a biggest effect on value of normalised SIF as shown in Fig 9.

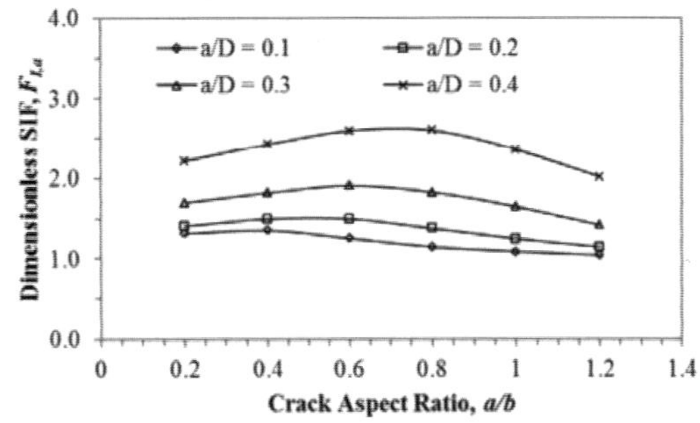

Figure 9 : Dimensionless SIF vs *a/b* at outer point when *c/l*=0.02

Tables 1 – 6 show the variation of interaction factor, measured at three different locations along crack front. These point represent deepest point x/h=0, outer point x/h=0.93 and x/h=0.5. It is noted that the values of γ always less than unity, which serves as a reference as it represents the magnification or shielding effects. Then only when *c/l* approaches 0.32, there is no more shielding effect at all locations along crack front. The maximum reduction of 33% in interaction factor occurs when *a/b*=0.2, *a/D*=0.1 at deepest point. The variations in the values of γ can be compared with the variations of normalized SIF.

Table 1: Interaction factor for two parallel surface cracks under tension when $a/b = 0.2$

					c/L			
a/D	x/h	0.005	0.01	0.02	0.04	0.08	0.16	0.32
	0.00	0.67	0.71	0.75	0.80	0.89	0.97	0.99
0.1	0.50	0.66	0.70	0.74	0.81	0.89	0.95	0.99
	0.93	0.75	0.80	0.86	0.92	0.96	0.99	1.00
	0.00	0.69	0.71	0.73	0.78	0.86	0.96	1.00
0.2	0.50	0.70	0.72	0.74	0.79	0.87	0.97	1.00
	0.93	0.71	0.74	0.78	0.85	0.93	0.98	1.00
	0.00	0.69	0.71	0.74	0.78	0.86	0.95	1.00
0.3	0.50	0.69	0.71	0.74	0.79	0.86	0.96	1.00
	0.93	0.70	0.73	0.77	0.83	0.92	0.99	1.00
	0.00	0.69	0.71	0.74	0.79	0.87	0.95	1.00
0.4	0.50	0.69	0.71	0.74	0.79	0.87	0.96	1.00
	0.93	0.70	0.72	0.76	0.82	0.91	0.99	1.00

Table 2: Interaction factor for two parallel surface cracks under tension when $a/b = 0.4$

					c/L			
a/D	x/h	0.005	0.01	0.02	0.04	0.08	0.16	0.32
	0.00	0.69	0.72	0.76	0.82	0.91	0.98	0.99
0.1	0.50	0.69	0.73	0.77	0.83	0.92	0.98	1.00
	0.93	0.73	0.77	0.83	0.89	0.96	1.00	1.00
	0.00	0.69	0.71	0.74	0.79	0.86	0.96	1.00
0.2	0.50	0.69	0.71	0.74	0.79	0.87	0.96	1.00
	0.93	0.70	0.73	0.78	0.84	0.91	0.98	1.00
	0.00	0.69	0.71	0.74	0.79	0.86	0.95	1.00
0.3	0.50	0.69	0.71	0.74	0.79	0.86	0.95	1.00
	0.93	0.69	0.72	0.76	0.82	0.90	0.95	1.00
	0.00	0.70	0.71	0.74	0.79	0.87	0.96	1.00
0.4	0.50	0.69	0.72	0.74	0.80	0.87	0.96	1.00
	0.93	0.70	0.72	0.76	0.82	0.91	0.99	1.00

Table 3: Interaction factor for two parallel surface cracks under tension when $a/b = 0.6$

					c/L			
a/D	x/h	0.005	0.01	0.02	0.04	0.08	0.16	0.32
	0.00	0.70	0.73	0.77	0.84	0.93	0.98	0.99
0.1	0.50	0.70	0.73	0.78	0.85	0.93	0.99	0.99
	0.93	0.71	0.74	0.79	0.86	0.94	0.99	1.00
	0.00	0.69	0.71	0.75	0.80	0.87	0.96	1.00
0.2	0.50	0.69	0.71	0.74	0.79	0.87	0.96	0.99
	0.93	0.69	0.71	0.74	0.80	0.88	0.96	0.99
	0.00	0.69	0.71	0.74	0.79	0.86	0.95	1.00
0.3	0.50	0.69	0.71	0.74	0.79	0.86	0.95	1.00
	0.93	0.69	0.71	0.73	0.78	0.86	0.95	1.00
	0.00	0.69	0.71	0.74	0.79	0.86	0.95	1.00
0.4	0.50	0.70	0.71	0.74	0.79	0.87	0.96	1.00
	0.93	0.70	0.71	0.75	0.80	0.89	0.97	1.00

Table 4: Interaction factor for two parallel surface cracks under tension when $a/b = 0.8$

					c/L			
a/D	x/h	0.005	0.01	0.02	0.04	0.08	0.16	0.32
	0.00	0.70	0.74	0.79	0.86	0.94	0.98	0.99
0.1	0.50	0.71	0.74	0.79	0.86	0.94	0.99	0.99
	0.93	0.69	0.72	0.77	0.85	0.94	0.99	1.00
	0.00	0.69	0.72	0.75	0.81	0.88	0.97	0.99
0.2	0.50	0.70	0.72	0.75	0.81	0.88	0.97	1.00
	0.93	0.69	0.71	0.74	0.80	0.88	0.97	1.00
	0.00	0.69	0.71	0.74	0.79	0.87	0.95	1.00
0.3	0.50	0.69	0.71	0.74	0.79	0.87	0.95	1.00
	0.93	0.69	0.71	0.74	0.79	0.86	0.96	1.00
	0.00	0.69	0.71	0.74	0.79	0.87	0.95	1.00
0.4	0.50	0.70	0.71	0.74	0.79	0.87	0.95	1.00
	0.93	0.69	0.70	0.74	0.79	0.87	0.96	1.00

Table 5: Interaction factor for two parallel surface cracks under tension when $a/b = 1.0$

					c/L			
a/D	x/h	0.005	0.01	0.02	0.04	0.08	0.16	0.32
	0.00	0.73	0.75	0.81	0.88	0.96	0.99	0.99
0.1	0.50	0.73	0.75	0.80	0.88	0.96	0.99	1.00
	0.93	0.72	0.72	0.77	0.86	0.95	0.99	1.00
	0.00	0.70	0.72	0.76	0.82	0.90	0.97	1.00
0.2	0.50	0.70	0.72	0.76	0.82	0.90	0.97	1.00
	0.93	0.69	0.71	0.74	0.79	0.88	0.98	1.00
	0.00	0.70	0.72	0.75	0.80	0.88	0.96	1.00
0.3	0.50	0.70	0.72	0.75	0.80	0.87	0.96	1.00
	0.93	0.70	0.71	0.73	0.78	0.85	0.95	1.00
	0.00	0.70	0.72	0.74	0.79	0.87	0.96	1.00
0.4	0.50	0.70	0.72	0.75	0.80	0.87	0.96	1.00
	0.93	0.70	0.71	0.74	0.78	0.86	0.95	1.00

Table 6: Interaction factor for two parallel surface cracks under tension when $a/b = 1.2$

					c/L			
a/D	x/h	0.005	0.01	0.02	0.04	0.08	0.16	0.32
	0.00	0.73	0.76	0.83	0.90	0.97	0.99	0.99
0.1	0.50	0.72	0.75	0.82	0.90	0.97	1.00	1.00
	0.93	0.68	0.72	0.78	0.87	0.96	0.99	1.00
	0.00	0.70	0.73	0.78	0.84	0.92	0.98	1.00
0.2	0.50	0.70	0.73	0.77	0.83	0.91	0.98	1.00
	0.93	0.70	0.71	0.73	0.79	0.89	0.98	1.00
	0.00	0.70	0.72	0.76	0.82	0.90	0.97	1.00
0.3	0.50	0.70	0.72	0.75	0.81	0.89	0.97	1.00
	0.93	0.69	0.69	0.72	0.78	0.85	0.96	1.00
	0.00	0.70	0.72	0.75	0.80	0.88	0.96	1.00
0.4	0.50	0.70	0.72	0.75	0.80	0.88	0.96	1.00
	0.93	0.70	0.71	0.73	0.78	0.85	0.95	1.00

Figure 10 shows the interaction factor along crack front. It can be seen that there is a slight increase of interaction factor as graph approaches outer point for the most of the considered cases. On the other hand, there are the cases where the interaction is higher at outer point. In other words, the location on crack front does not show the significant effect on interaction factor as far as parallel crack is concern. Unless the crack configuration is different which is interesting to be investigates.

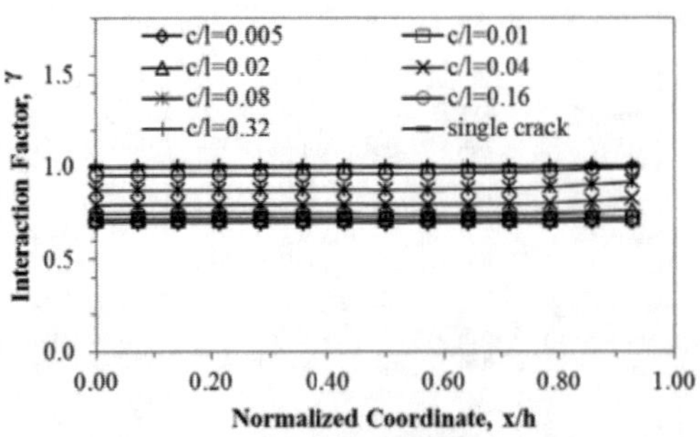

Figure 10: Interaction factors along crack front *at* $a/D = 0.4$, $a/b = 0.2$

5. Conclusions

Based on the investigations conducted numerically using ANSYS finite element program on the parallel cracks subjected to tension stress, the following remarks are drawn from this study:-

1. As expected, higher crack depth ratio produced higher SIFs for mode 1 stress intensity factors.
2. The maximum SIF always occurs at outer point on crack front. This is due to state of stress changes gradually from plain strain at deepest point to plain stress.
3. When there is two parallel interacting cracks, the flexibility of cylinder increases, the SIF decrease.
4. The interaction diminishes along with the increase of inter-crack distance.

References

[1] Leis BN 1997 *Proceedings of the Seventh ISOPE, Vol. 4, Quebec, Canada* pp. 607–613.
[2] Soboyejo WO, Knott JF 1990 *Eng. Fracture Mech.* 37:323–340.
[3] Shin CS, Cai CQ 2004. *Int Journal of Fracture* 129:239–64.
[4] Chudnovsky A, Dolgopolsky A, Kachanov M 1987. *Int. J. Solids Structures* 23:1–10.
[5] Murakami Y, Nemat-Nasser S. 1982 *Eng. Fracture Mech* 16:373–386.
[6] Isida M, Yoshida T, Noguchi H.1991. *Eng. Fracture Mech.* 39:845–850.
[7] Tu ST, Cai RY. 1988 *Computational Mechanics Publications*, pp. 239–247.
[8] Miyazaki H, Kaneko H, Murakata T.1989. *Int. J. Press. Vess. Piping* 38:1–14.
[9] Stonesifer RB, Brust FW, Leis BN. 1993 *Eng. Fracture Mech.* 45:357–380.
[10] Jiang ZD, Petit J, Akid R, Bezine G. 1991 *Eng. Fracture Mech.* 40:345–354.
[11] Carpinteri A. 1992 *Fatigue Fract Eng Mater Struct* 15(11):1141–53.

Optimization on Impact Strength of Woven Kenaf Reinforced Polyester Composites using Taguchi Method

S.N.A Khalid[a] , A.E Ismail[b], M.H Zainulabidin[c]

Faculty of Mechanical and Manufacturing Engineering, Universiti Tun Hussein Onn Malaysia, 86400 Batu Pahat, Johor.

[a]sitinorazilabintikhalid@yahoo.com, [b]emran@uthm.edu.my, [c]hafeez@uthm.edu.my

Abstract. This paper focuses on the effect of weaving patterns and orientations on the energy absorption of woven kenaf reinforced polyester composites. Kenaf fiber in the form of yarn is weaved to produce different weaving patterns such as plain, twill and basket. Three woven mats are stacked together and mixed with polyester resin before it is compressed to squeeze out any excessive resin. There is nine different orientations are used during stacking processes by following Taguchi orthogonal arrays method. The hardened composites are cured for 24 hours before it is shaped according to specific dimensions for impact tests. The composites are perforated with 1m/s blunted projectile. According to the experimental findings, weaving pattern and orientation have distinct potential effects on value of energy absorption. The optimization using Taguchi method reveals preferable orientation of each weaving pattern composites. Based on the fracture observation, the fragmentations after optimization indicating lower distance surface fracture perforated obtained.

1. Introduction

During the recent years, the attentions are given to crashworthiness and energy absorption management on composite structures under oblique impact loading [1,2]. Nowadays, the kenaf natural fibers have the potential to be used as a replacement for glass or other traditional reinforcement materials in composites. There have been many researchers involved in this field of kenaf reinforced plastics under perforated impact [3,4,5]. Most of these findings focus on unidirectional, short fiber and little research on woven especially in kenaf natural fiber. The ability of woven fabric is better mechanical properties compare to non-woven [3,4]. However, there are several issues such as their poor fire resistance, variable quality, depending on unpredictable influences such as weather and lower strength properties. In the light and due to the versatility and enormous potential, it is currently being explored by researchers by some aspect such as woven and orientation performance [6,7,8].

In the recent years, researchers have investigated a variety of approach in applied Taguchi orthogonal array on composite [9,10]. In this research, glass fibers and epoxy at several compositions are applied to determine which led to minimization of erosion rate of the composite. From previous research, normally Taguchi method are apply in determining parameters such as percentages material composition, temperature, pressure load and curing time. By evaluating the performance of the product in several environmental conditions, there would be a realistic data to calculate the real word variance. The classical experimental design method are too complex, time consuming and not easy to use. A large number of experiments have to be carried out when the number of process parameters are

IOP Conf. Series: Materials Science and Engineering **165** (2017) 012010 doi:10.1088/1757-899X/165/1/012010

more. To solve this problem, Taguchi method uses to study the entire parameter space with the minimum number of experiments [11].

Several of previous studies using natural fiber as the main ingredient in assessing the strength of composite, but not really focus on angle orientations. In this paper, Taguchi orthogonal array is applied to facilitate the arrangement of orientation and indirectly optimize the orientation that proposed higher energy absorption. On the other hand, optimization using Taguchi methods are more suitable tools. This systematic experiment has led to determination of significant process parameters and orientations variables that predominantly influence the energy absorption respectively.

2. Methodology

As-received kenaf yarn with a size of 1mm as shown in Figure 1(a) is used and weaved to form a woven mat as reveal in Figure 1(b). During weaving processes, three different fiber architectures are used such as plain, twill and basket. Woven mats are then stacked (four layers) together using different fiber orientations followed by Taguchi method $L_9 3^4$as listed in Table 1. The $L_9\ 3^4$ orthogonal array was used based on the number of parameters and levels that has been selected. These parameters are four layered were investigated and three angles each composite. Taguchi method are uses a special design of orthogonal arrays to study the entire parameter space of orientation with the minimum number of experiments.

Then, the stacks are mixed with polyester resin before it is compressed to spread the resin uniformly across the fiber mats. Once the composites hardened and cured for 24 hours (Figure 1(c)), they are shaped into shape and geometry as specified in ASTM D 3763 standard of 100x100x3 mm.

Impact test is performed using Hydroshot HITS-T10 Shimadzu machine. The samples are positioned and clamped in all-degree of freedom. Before the tests, ASTM D 3763 is properly followed to ensure the results are reliable. The blunted-shaped projectile is used to perforate the samples where the speed of projectile is fixed to 1m/s and the mass of projectile is 5kg. During tests, the responses of force-displacement diagrams are recorded. The energy absorptions are determined using the area under the curves of force versus displacement.

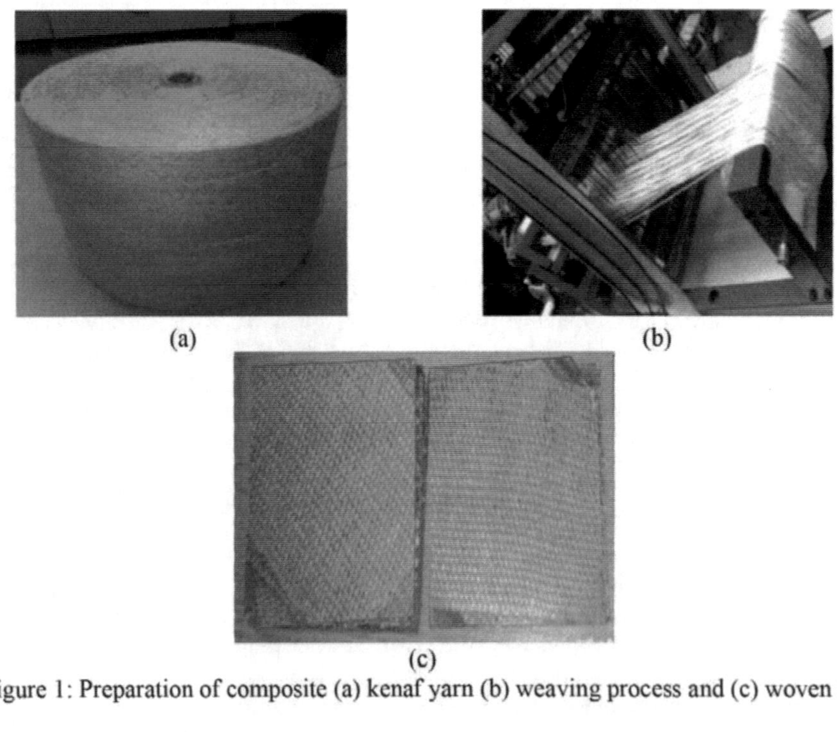
(a) (b) (c)

Figure 1: Preparation of composite (a) kenaf yarn (b) weaving process and (c) woven mats

3. Result and Discussion

In Taguchi Method, optimizations refer to determination of the best level for control factors. In other words, the best levels of control factors are maximize the S/N (Signal to Noise) ratios value. However the experiments that conducted to determine the best levels are based on Orthogonal Array (OA) that concentrated on all factors in minimum number also minimize the cost of material besides the time in complete the experimental. In this experimental, the variables of angle were chosen as input factor. The result of response measures are converted into S/N ratio using Minitab software. The bigger value of response denoted better, such as residual energy absorption, is called 'bigger is better'. It seems that a larger response is better and the S/N ratio will be defined as follows:

$$\frac{S}{N} = -10\ log\left(\frac{1}{n}\sum_{i=1}^{n}\frac{1}{{y_i}^2}\right)$$

Where n is the trial repetition and y_i is the result of the ith experiment for each trial

Figure 2 shows the main effects plot (data means) for S/N ratios of composites (a) Plain (b) Twill (c) Basket. Regarding to this graph pattern on Figure 2, the optimized factor of energy absorption can be express by selecting the highest level of S/N ratios value for each factor of orientation. The result for angle arrangement from Figure 2 is represented the best orientation for plain woven is [+40/-15/+40/-15], twill woven is [+40/+75/+40/+75] and basket woven is [-15/+40/+75/+75].

Figures 3 shows value of energy absorption after optimization with different type of woven and orientation. From the result of plain woven in Figure 3(a), the value after optimization is 27.910 Joule. The percentage error compare with actual result is 9.307%. For Twill woven, value of energy absorption after optimization is 31.946 Joule and the error is 9.038%. Meanwhile for basket woven the value of energy absorption after optimization is 29.893 and the error is 8.048%. Furthermore, the value produces still under acceptable range even though it is below the value of actual result due to material yarn size under control and sensitivity to dynamic loading. Hence the result is still can be used for the research purposed although the energy absorption is lower than after optimization.

From the results that have been obtained, shows the best arrangement of angle is suggested by using Minitab software for each type of woven which is plain is [+40/-15/+40/-15], twill is [+40/+75/+40/+75] and basket is [-15/+40/+75/+75]. The analysis of data shows the influence of various process variable angles on value of energy absorption. It has been noticed that residual angles give the energy absorption enhanced. Due to the result, it demonstrated that a fiber orientation for each layers is significantly affected to value of energy absorption.

Table 1: The orientation arrangement of samples.

Trial (Sample)	A	B	C	D
1	-15	-15	-15	-15
2	-15	40	40	40
3	-15	75	75	75
4	40	-15	40	75
5	40	40	75	-15
6	40	75	-15	40
7	75	-15	75	40
8	75	40	-15	75
9	75	75	40	-15

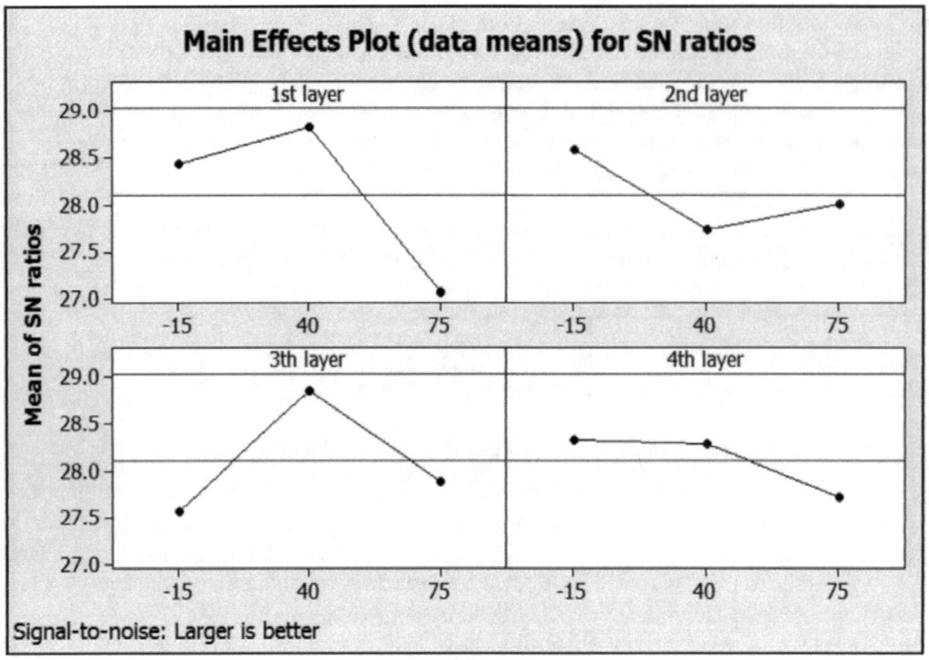
Main Effects Plot (data means) for SN ratios
1st layer
2nd layer
3th layer
4th layer
Mean of SN ratios
29.0
28.5
28.0
27.5
27.0
-15
40
75
Signal-to-noise: Larger is better

(a)

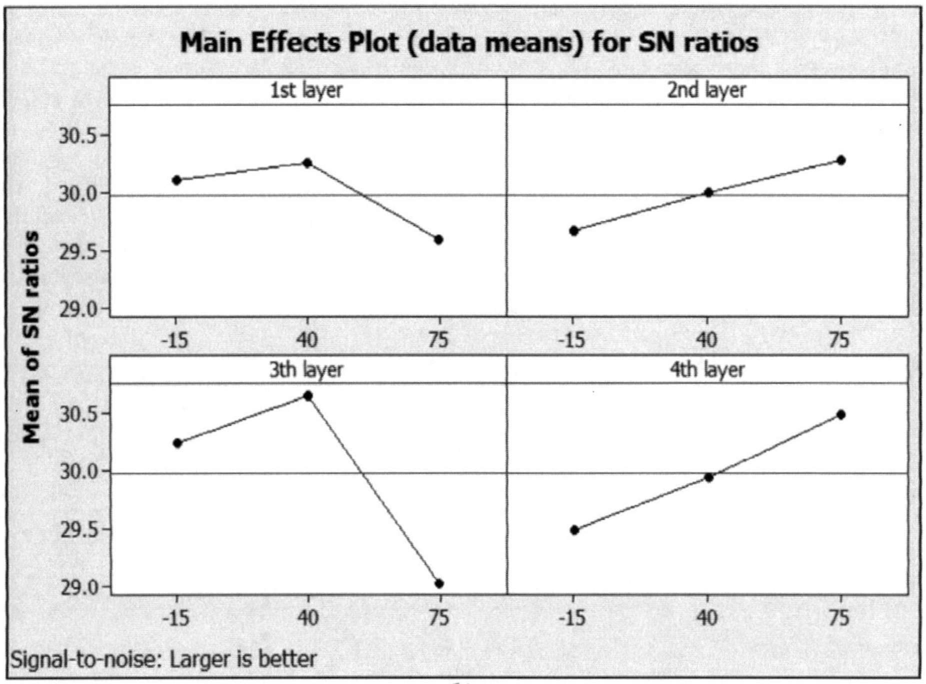
Main Effects Plot (data means) for SN ratios
1st layer
2nd layer
3th layer
4th layer
Mean of SN ratios
30.5
30.0
29.5
29.0
-15
40
75
Signal-to-noise: Larger is better

(b)

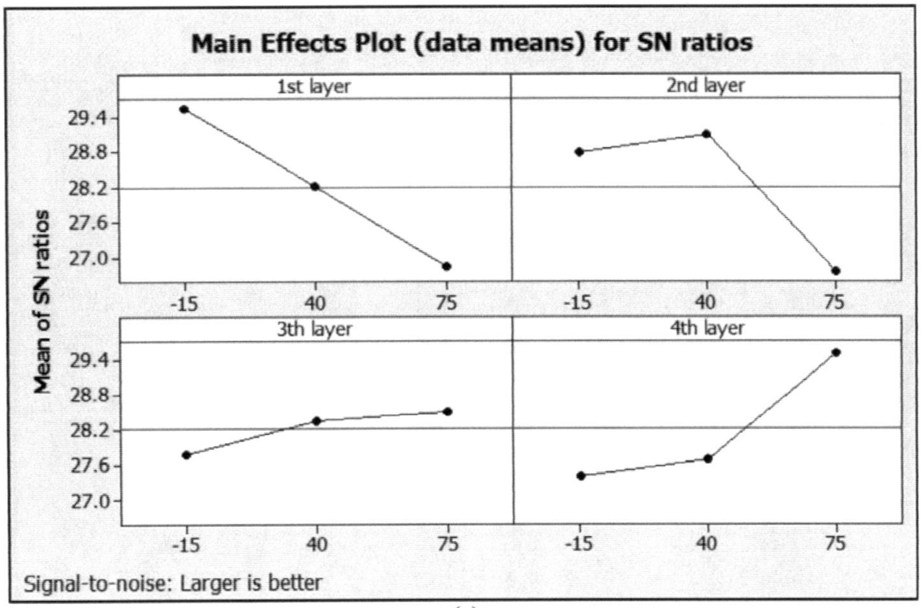

(c)

Figure 2: Main effect plot (data mean) for S/N rations (a) Plain (b) Twill (c) Basket

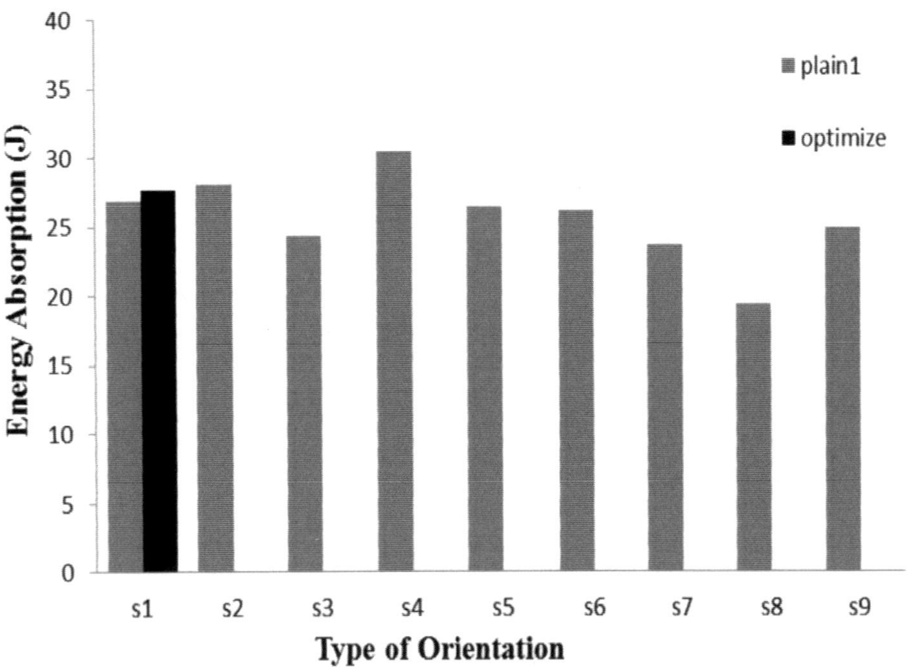

(a)

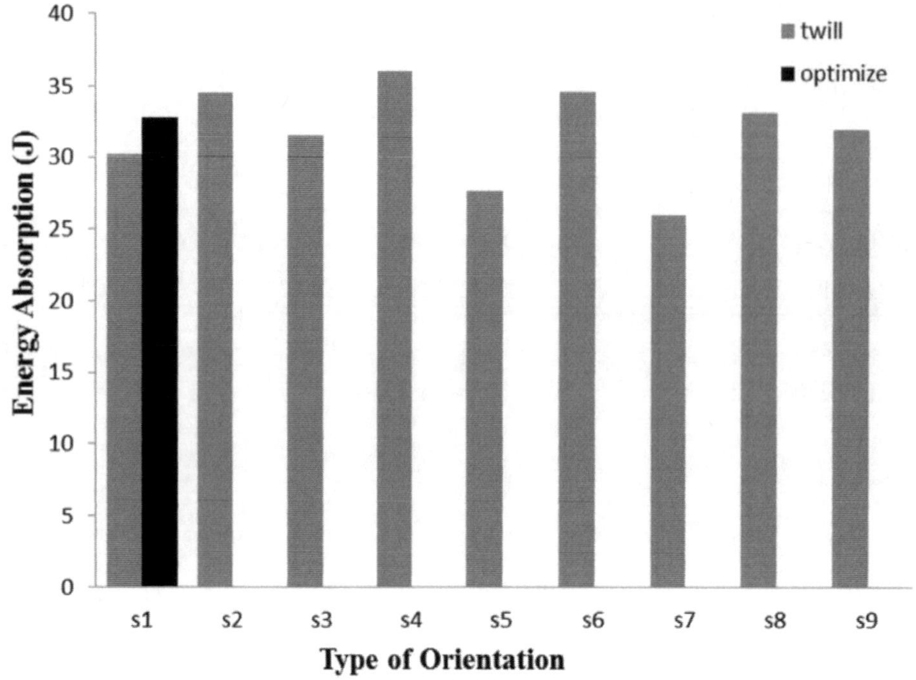

(b)

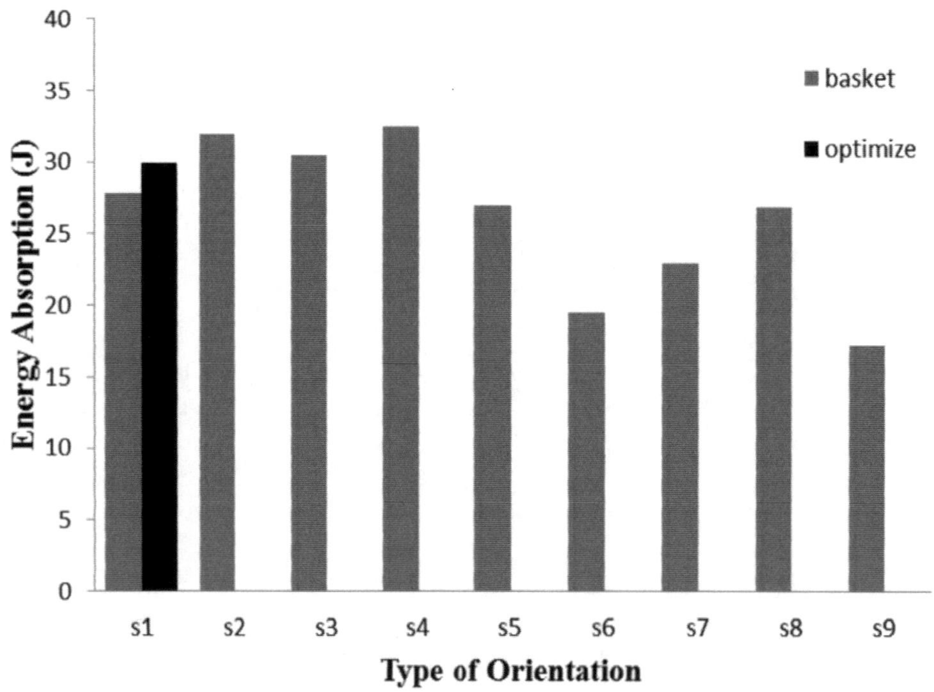

(c)

Figure 3: Energy absorption after optimization (a) Plain (b) Twill (c) Basket

In general, the fracture mechanisms are almost similar where the fragmentation surface after impact effect. Figure 4 shows fragmentations of three type of woven such as plain, twill and basket. Overall observation, it can see that fragmentation after optimization is smaller than before. For the plain woven it can be seen that fragmentation on the back is around distance 3.0 cm. Besides that, for twill woven the fragmentation is 2.5 cm and for basket is 6 cm. After optimization the fragmentation for plain is 2.0 cm, twill is 2.1 cm and for basket 2.0 cm. Overview from the fragmentation after optimization, it shows that, not only increase energy absorption, but also decrease the length of fragmentation. This study has found that the highest energy absorption and less perforated distance on twill woven. It can be explained that when f fabric is stretched in one direction which is load direction, it amplifies in the reverse direction. Owing to the crimp, the fabric strength is less than the strength of twisted yarns, because of the twisted yarns; the yarn strength is less than the strength of fibres.

Type woven	Before optimization	After optimization	
Plain	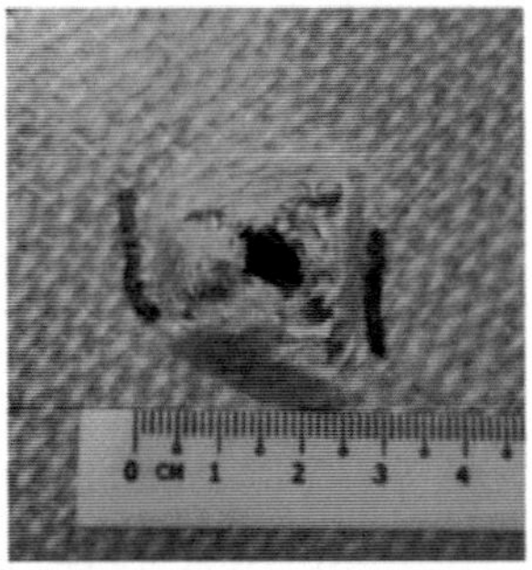	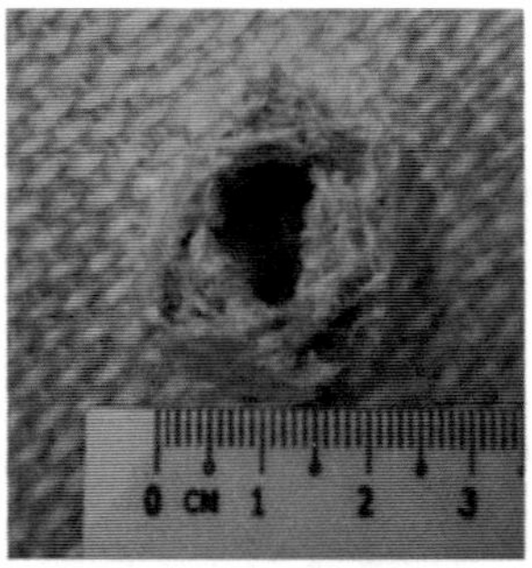	
Twill	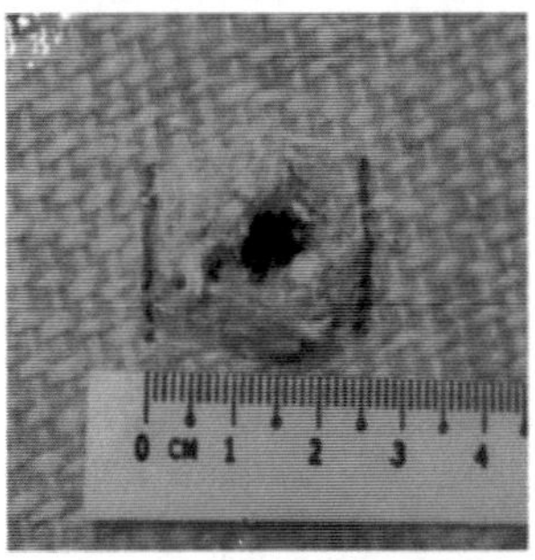	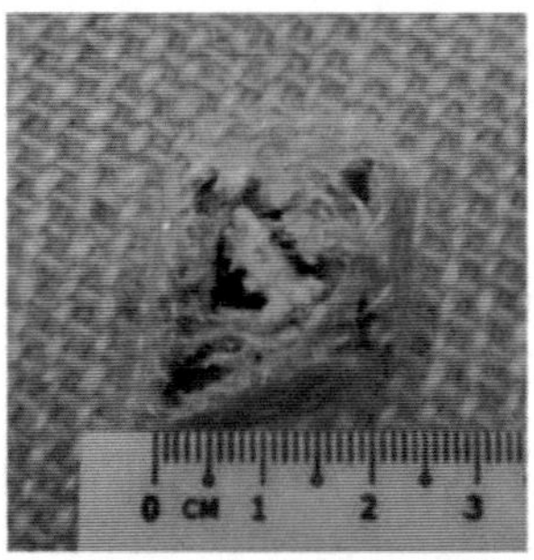	
Basket	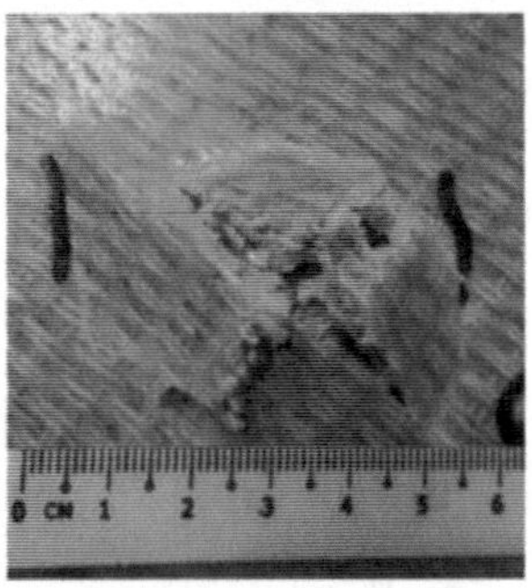	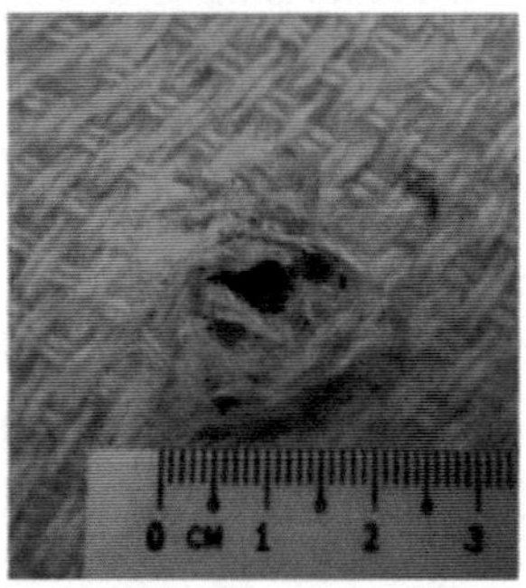	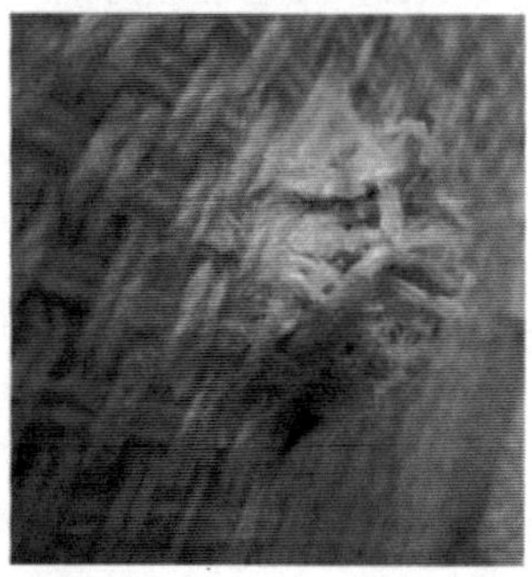

Figure 4: Fragmentation of composite before and after optimization

4. Conclusion

Experimental study was carried out on behaviour of composite with different type of woven and orientations are investigated. This result presents the best performance on mechanical strength and energy absorption using Taguchi method. Based on the results, the following conclusions are drawn:

i. The result is represented that the best orientation for plain woven is [+40/-15/+40/-15], twill woven is [+40/+75/+40/+75] and basket woven is [-15/+40/+75/+75].
ii. The value of energy absorption after optimization still accepted and can be used because percentage of error is below than 10%, due to material yarn size under control and sensitivity to dynamic loading.
iii. The fragmentation after optimization of three type of woven plain, twill and basket shows the smaller perforated compare to before optimization.

5. Acknowledgements

Authors acknowledged Universiti Tun Hussein Onn Malaysia (UTHM), MyBrain15 Scholarship (MyPhd), Office for Research, Innovation, Commercialization and Consultancy Management (ORICC) through RAGS Vot. R019, Center for Graduate Studies UTHM.

6. References

[1] Othman A, Abdullah S, Ariffin AK, Mohamed NAN 2016 *Composites Part B* doi: 10.1016/j.compositesb.2016.04.027.
[2] Roslan MN, Ismail AE, Hashim M.Y, Zainulabidin MH, Khalid SNA 2014 *Applied Mechanics and Materials* 465 1324-1328
[3] Khalid SNA, Jamaluddin MR, Roslan MN, Ismail AE, Zainulabidin MH 2015 *Journal of Applied Science and Agriculture* 10(7)
[4] Ismail AE, Hassan MA, Kamaruddin KA 2014 Vol. 1, no. c, pp. 1-5
[5] Khalid SNA, Ismail AE and Zainulabidin MH 2015 *Applied Mechanics and Materials* 773-774 149-153
[6] Mahdi E, Hamouda AMS, Sedaey TA 2014 *Material and Design* 56 923-928.
[7] GM Khan Arifuzzaman, M Terano, MA Gafur, Alam Shamsul M 2013 *Engineering Science*
[8] Ferdousa Nasrin, Sadiqur Md, Rahman b, Kabir Reashad, Arif Eftekhar Ahmed. 2014 *International Journal of Scientific Research Engineering & Technology (IJSRET)*, ISSN 2278 – 0882, Volume 3, Issue 9
[9] Rahul Kumar, Kaushik Kumar, Sumit Bhowmik 2014 *International Conference on Advance in Manufacturing and Material Engineering, AMME* 2014. Procedia Materials Science 5 688 – 696
[10] B. sutharson, Dr.M. Rajendran, A. Karapagaraj *International Journal of Materials and Biomaterials Applications*, ISSN 2249–9679
[11] Chandramohan, D. and J. Bharanichandar, *American Journal of Environmental Science* 9 (6): 494-504, 2013, ISSN: 1553-345X

Impact Strength of Different Weaving Patterns of Woven Kenaf Reinforced Polyester Composites

S.N.A Khalid[a] , A.E Ismail[b], M.H Zainulabidin[c]

Faculty of Mechanical and Manufacturing Engineering, Universiti Tun Hussein Onn Malaysia, 86400 Batu Pahat, Johor.

[a]sitinorazilabintikhalid@yahoo.com, [b]emran@uthm.edu.my, [c]hafeez@uthm.edu.my

Abstract. This paper focuses on the effect of weaving patterns and orientations on the energy absorption of woven kenaf reinforced polyester composites. Kenaf fiber in the form of yarn is weaved to produce different weaving patterns such as plain, twill and basket. Three woven mats are stacked together and mixed with polyester resin before it is compressed to squeeze out any excessive resin. There is 9 different orientations are used during stacking processes. The hardened composites are cured for 24 hours before it is shaped according to specific dimensions for imp act tests. The composites are perforated with 1m/s blunted projectile. According to the experimental findings, both weaving patterns and orientations have distinct potential effects on the force-displacement diagrams. However, fiber orientations have insignificant effect for plain woven especially in the first stage of deformations. Energy absorption performances for each composite condition are calculated and then plotted against fiber orientations for different weaving patterns. It is found there is no strong relationship between energy absorption and fiber orientations. However for each case of composites, higher energy absorption is found for the composites orientated using [+40°/-15°/+40°/+75°]. Based on the fracture observation, both plain and basket-type woven composites reveal large fragmentations occurred indicating lower energy absorption performances. While for twill condition, no obvious fragmentation is observed where the impact damage around the perforated hole is uniformly distributed leading to higher capability of energy absorptions.

1. Introduction

During the recent years, the attentions are given to crashworthiness and energy absorption management on composite structures. The main advantages of fibre reinforced composite material over conventional isotropic materials, however the specific strength and stiffness which can be archived [1]. Nowadays, natural fiber is significantly used to fabricate various parts in automotive applications. Considering the high performance of composite materials in term of cost effectiveness, durability and maintenance, the application of kenaf fiber reinforced composites as construction material has great potential and is important to achieve sustainability. Moreover with composites the designer of fiber, matrix, and fiber orientation are used in produce composite to improved material properties.

Several works had been done on investigating the strength of composite under perforated impact [2,3]. Most of these findings focus on unidirectional, short fiber and little research on woven especially in kenaf natural fiber. The ability of woven fabric is better mechanical properties compare to non-woven [4,5]. Difference of woven and non-woven gives different response on the impact due to the ability to displace and withstand before failure [6]. Therefore, the present work investigates the effect of weaving types and fiber orientation on energy absorption under impact loading. Several researchers mentioned in their research that orientation of 75°/-15° is recommended for better load capacity and energy absorption [3,7]. Three type of weaving which is plain, twill and basket are reoriented using best angle that are recommended by previous research.

2. Methodology

As-received kenaf yarn with a size of 1mm as shown in Figure 1(a) is used and weaved to form a woven mat as reveal in Figure 1(b). During weaving processes, three different fiber architectures are used such as plain, twill and basket. Woven mats are then stacked (four layers) together using different fiber orientations as listed in Table 1. The stacks are mixed with polyester resin before it is compressed to spread the resin uniformly across the fiber mats. Once the composites hardened and cured for 24 hours (Figure 1(c)), they are shaped into shape and geometry as specified in ASTM D 3763 standard of 100x100x3 mm.

Impact test is performed using Hydroshot HITS-T10 Shimadzu machine. The samples are positioned and clamped in all-degree of freedom. Before the tests, ASTM D 3763 is properly followed to ensure the results are reliable. The blunted-shaped projectile is used to perforate the samples where the speed of projectile is fixed to 1m/s. During tests, the responses of force-displacement diagrams are recorded. The energy absorptions are determined using the area under the curves of force versus displacement.

(a)

(b)

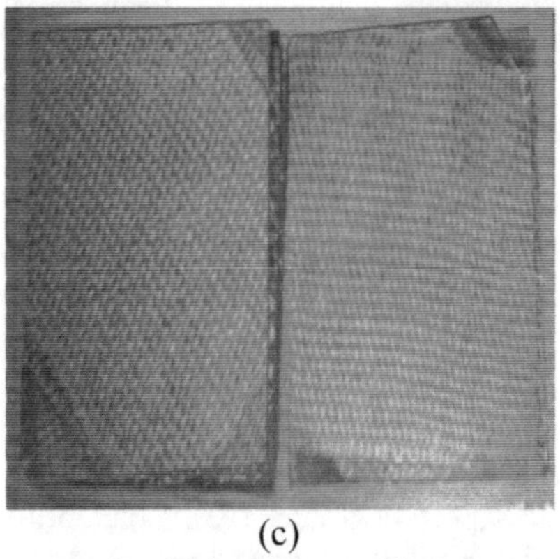

(c)

Figure 1: Preparation of composite (a) kenaf yarn (b) weaving process and (c) woven mats

3. Result and Discussion

Result from impact test showed the maximum force of penetration, total energy at the maximum penetration and energy absorption. Areas under the graph are analysis to obtain the total energy absorption in composite samples with different orientation and different type of weaving. Total force versus displacement with varying type of weaving as shown in Figure 1. From Figures 1(a),(b) and (c) it is shows the similar pattern which is upward and after touching the sample, points of impact are change the curved, but the effect to the force of the sample have a fraction. This situation is almost similar to each other for all samples. From curve that are plotted, there have 4 stages before perforated. At stage 1, there is no damage accusing. As the load increased, matrix cracking occurs at stage 2. The size and extend of the matrix cracking may progress such that interfacial debonding can occurs at stage 3. This in turn, leads to delamination, then fibre breaking, and finally perforation of the impacted specimen at stage 4. The different of each pattern are depending on value of peak force.

Table 1: The orientation arrangement of samples.

Sample	Orientation (θ°)
S1	-15/-15/-15/-15
S2	-15/+40/+40/+40
S3	-15/+75/+75/+75
S4	+40/-15/+40/+75
S5	+40/+40/+75/-15
S6	+40/+75/-15/+40
S7	+75/-15/+75/+40
S8	+75/+40/-15/+75
S9	+75/+75/+40/-15

However, the composite fibres with twill woven shows higher energy absorption compare to plain and basket. Figure 3 shows the energy absorption of plain, twill and woven using nine type of orientation referring to Table 1. From the result that are obtain, the higher energy for three type of woven indicate on sample four which is using orientation [40/-15/40/75]. Due to the result, it demonstrated that a fiber orientation for each layers is significantly affected to value of energy absorption.

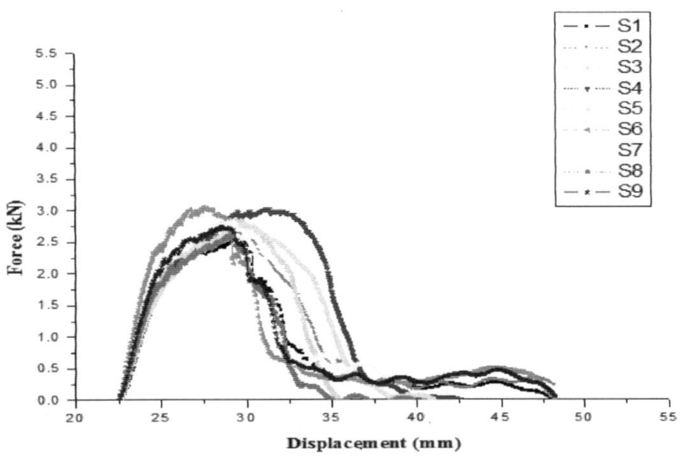

(a)

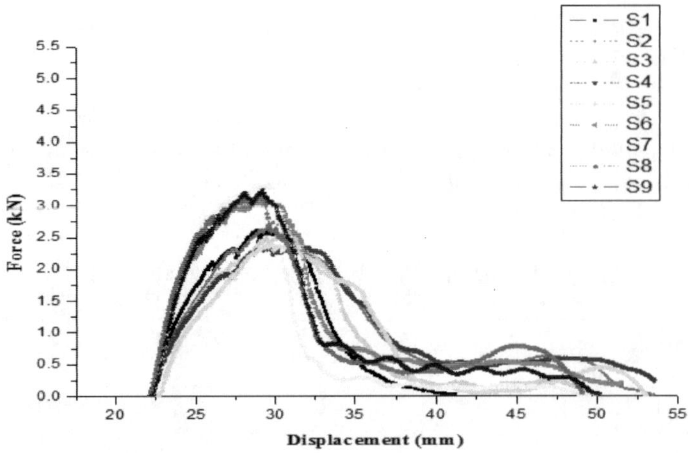

(b)

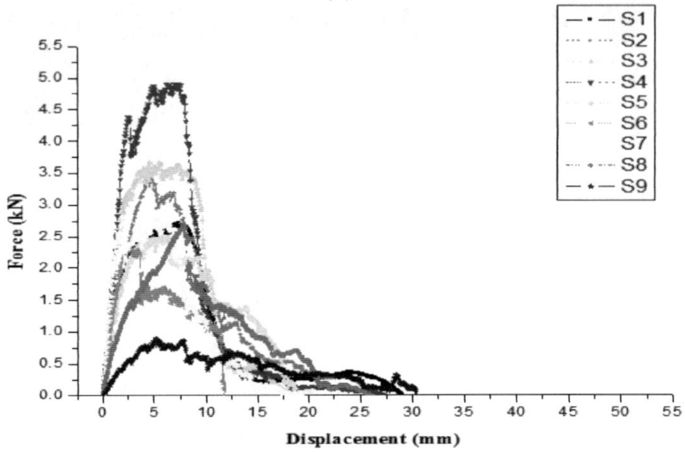

(c)

Figure 2: Force-displacement on (a) Plain (b) Twill and (c) Basket

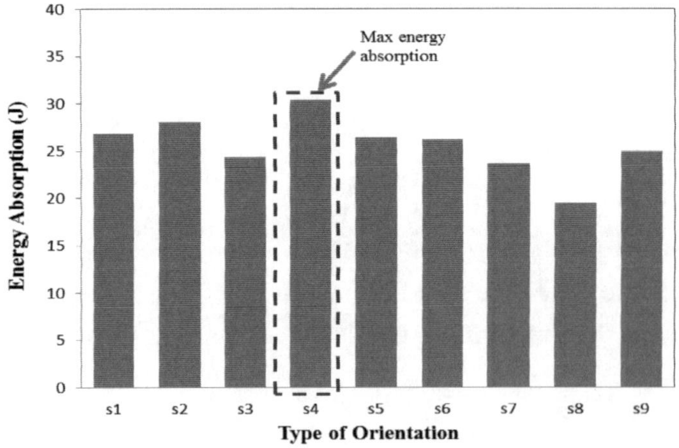

(a)

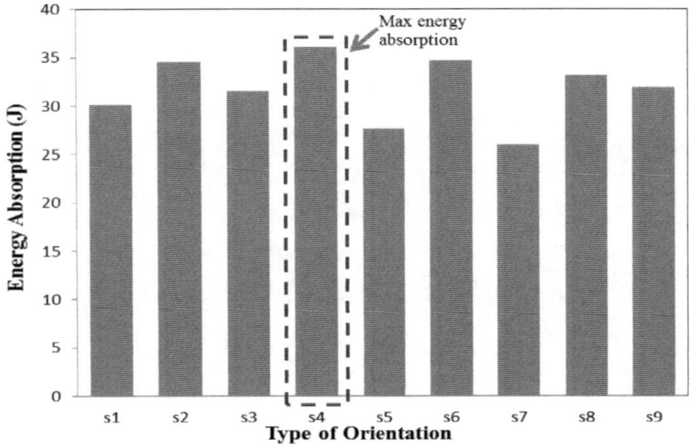

(b)

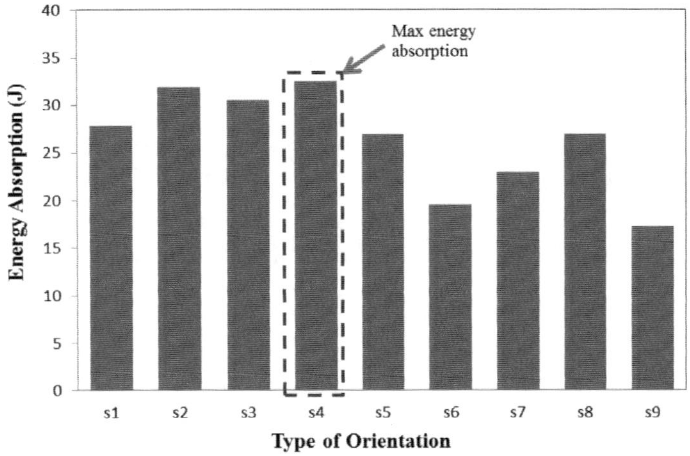

(c)

Figure 3: Energy absorption (a) Plain woven (b) Twill woven (c) Basket woven

In general, the fracture mechanisms are almost similar where the surfaces are perforated after impacted. Figure 4 shows fragmentation of three type of woven such as plain, twill and basket. For the plain woven it can be seen that fragmentation that happen at the back is around distance 2.5 cm to 3.0 cm. Besides, for twill woven the fragmentation of fracture surface with distance 2.0 cm to 2.4 cm and for basket woven, fragmentation is 3 cm to 6 cm. Whether the size of perforated on basket woven is greater than plain woven, but still have higher capability to absorb energy absorption. From the result it can see that the highest energy absorption is on twill woven. It can be explained that when fabric is stretched in one direction which is load direction, it amplifies in the reverse direction. Owing to the crimp, the fabric strength is less than the strength of twisted yarns, because of the twisted yarns; the yarn strength is less than the strength of fibres [8].

IOP Conf. Series: Materials Science and Engineering **165** (2017) 012011 doi:10.1088/1757-899X/165/1/012011

Type woven	Front	Back
Plain	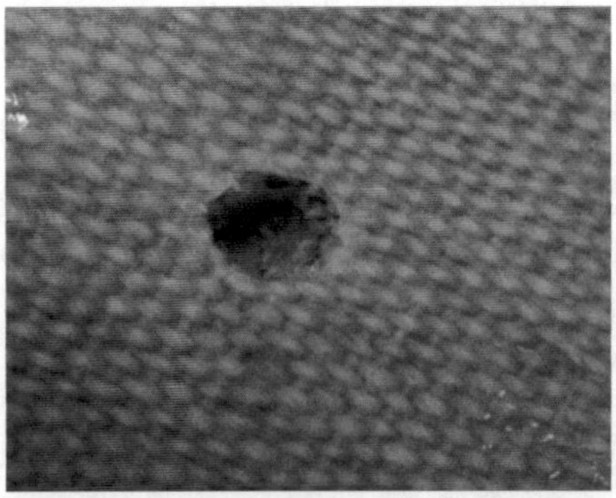	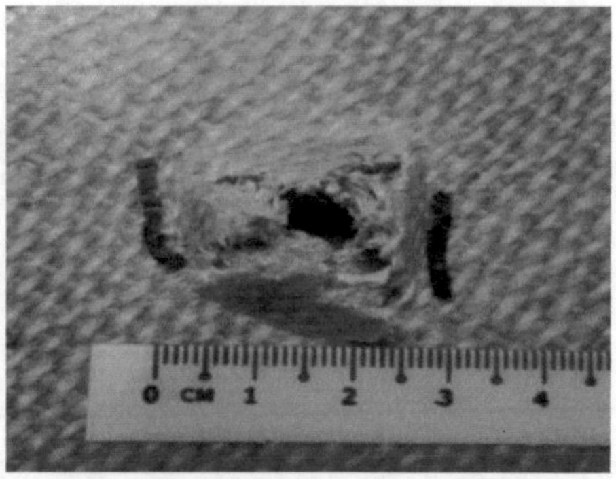
Twill	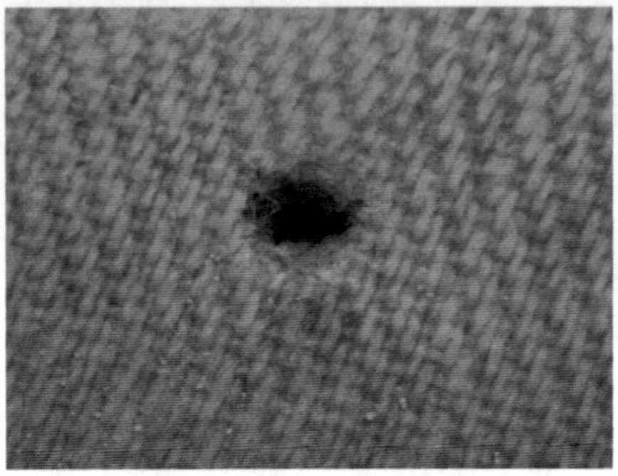	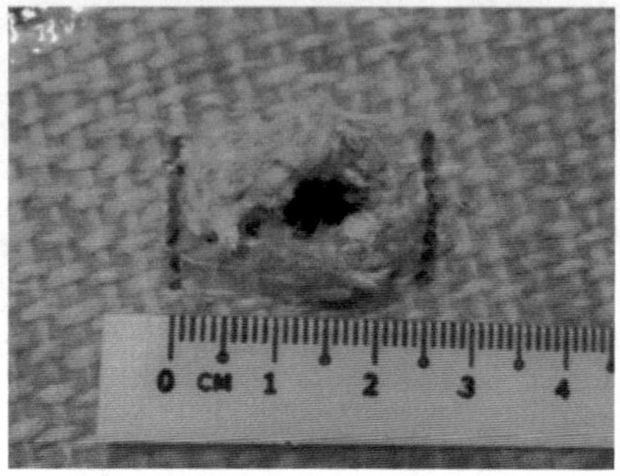
Basket	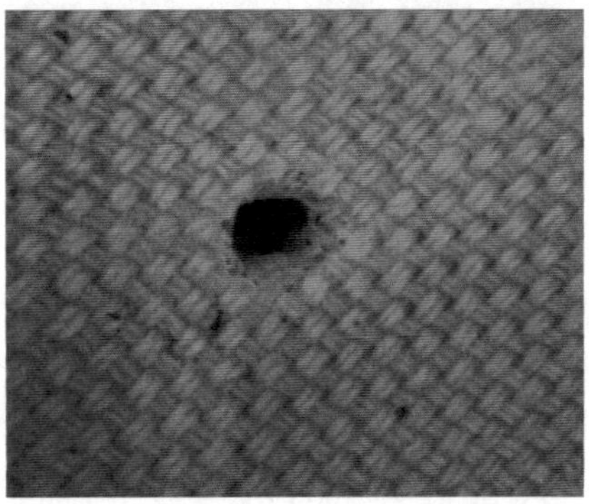	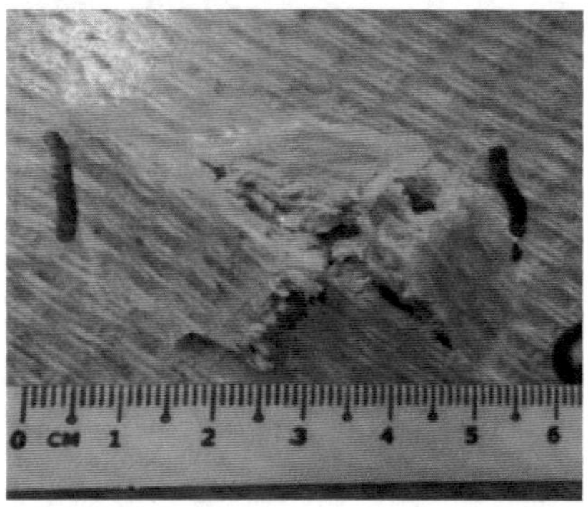

Figure 4: Fragmentation of composite after impacted

4. Conclusion

The present study is carried out on the behaviour of composite with different type of woven and orientation. The result presents the perforated impact performance on deformation, peak force, mechanical strength and energy absorption. Following conclusions are drawn;

i. Energy absorbed of twill woven was higher than those of plain and basket. The best orientation for three type of woven is [+40/-15/+40/+75].
ii. Value of peak force of basket woven is higher compare to other weaving which is reach 5.0 kN.
iii. The fragmentation of basket produces greater distance which is 3 cm to 6 cm of absolute penetration. Referring to energy absorption result, whether have greater fragmentation but still have higher capability to absorbed energy compare to plain woven.

5. Acknowledgements
Authors acknowledged Universiti Tun Hussein Onn Malaysia (UTHM), MyBrain15 Scholarship (MyPhd), Office for Research, Innovation, Commercialization and Consultancy Management (ORICC) through RAGS Vot. R019, Center for Graduate Studies UTHM.

6. References
[1] Mamalis AG, Robinson M, Manolakos DE, Demosthenous GA, Ioannidis MB, Carruthers J. 1997 *Composite Structures* 37 109-134
[2] Roslan MN, Ismail AE, Hashim M.Y, Zainulabidin MH, Khalid SNA 2014 *Applied Mechanics and Materials* 465 1324-1328
[3] Mahdi E, Hamouda AMS, Sedaey TA 2014 *Material and Design* 56 923-928
[4] Khalid SNA, Jamaluddin MR, Roslan MN, Ismail AE, Zainulabidin MH 2015 *Journal of Applied Science and Agriculture* 10(7)
[5] Khalid SNA, Ismail AE and Zainulabidin MH 2015 *Applied Mechanics and Materials* 773-774 149-153
[6] GM Khan Arifuzzaman, M Terano, MA Gafur, Alam Shamsul M 2013 *Engineering Science*
[7] Ismail AE, Hassan MA, Kamaruddin K.A 2014 Vol. 1, no. c, pp. 1-5
[8] Ferdousa Nasrin, Sadiqur Md, Rahman b, Kabir Reashad, Arif Eftekhar Ahmed. 2014 *International Journal of Scientific Research Engineering & Technology (IJSRET),* ISSN 2278 – 0882, Volume 3, Issue 9

Evaluation of hardness strength and microstructures of recycled Al chip and powder AA6061 fabricated by cold compaction method.

M. I. A. Kadir[1], M. S. Mustapa[1]*, A. S. Mahdi[1], S. Kuddus[1], and M. A. Samsi[1]

[1]Structural Integrity and Monitoring Research Group (SIMReG), Faculty of Mechanical and Manufacturing Engineering, Universiti Tun Hussein Onn Malaysia (UTHM) Malaysia

*Email: sukri@uthm.edu.my

Abstract. The present study is aimed at investigating Mechanical properties of the milled recycling Aluminium type AA6061 according to the change of the volume fraction for the continents and the reinforcement material. In the study, high speed millings were used for producing the chip sizes of AA6061 and followed by a cold press forging process. Milling process was used to produce smaller particles of aluminium chips by the planetary ball mill. Three groups of specimens were taken according to the percentage of chip and powder, while, two groups were taken according to the mixture between different particle sizes of powders. Where, four types of particle size were chosen (25, 63, 100, mix of particles) μm. The results were showing that the Microhardness increased with increasing the percentage of chip, then the density was decreased due to a large amount of pores were found. By the experiments, it can be concluded, the mixture of powders is the best choice for all groups. On the other hand, the type D3 (78.5% (25μm) + 21.5% (100μm)) gave the best result for microhardness compare with others according to particle.

Keyword: Aluminium Alloy AA6061, recycling chip, cold compaction method, microhardness.

1. Introduction

The melting process of reuse for aluminium metal gives many pollution effects to the environment. Aluminium is the most heavily used nonferrous metals in the world [1, 2]. Aluminium alloys are very favourable for structural applications in military, aerospace and transportation industries due to their high specific strength, low density and resistance to corrosion, and especially as regards high energy cost [3]. Aluminium hybrid composites are a new generation of metal matrix composites that have the potentials of pleasing the recent requests of advanced engineering applications which have a high specific modulus, a tailorable coefficient of thermal expansion and good wear resistance [4, 5]. The optimization of the mechanical properties of aluminium matrix composites is

IOP Conf. Series: Materials Science and Engineering **165** (2017) 012012 doi:10.1088/1757-899X/165/1/012012

largely dependent on the microstructure of the metal. Some parameters characterize this structure, i.e. size, shape and particle distribution, and to the matrix, such as precipitates, grain size, texture, and dispersoids [7]. Thorough characterization of the composite microstructure is required to understand the mechanical properties of aluminium matrix.

In metal industries, scrap and waste metals that residue after manufacturing processes are chip and discards. These waste metals are recycled by retreating them to smelters. However, during melting processes of metals for reusing, many metals are lost due to the costs of labor and occurring oxidation, energy and environmental protection expenditures.

The current research work focuses on understanding the effect of the volume fraction of aluminium alloy chip and powder for getting optimum mechanical properties of aluminum. There are some researches for powder metallurgy but there are not for recycling aluminum AA6061 until now.

2. Experimental setup

The main raw materials in this study were aluminium AA6061, zinc stearate ($C_{36}H_{70}O_4Zn$). In this study, AA6061 was mixed with zinc stearate as a binder. The mechanical properties for Aluminium AA6061 are shown in Table.1 were used this material in some research.

Table 1. Mechanical properties of Aluminum AA6061 (ASTM B308/B308M)

Yield Strength (MPa)	Tensile Strength (MPa)	Density g/cm^3	Hardness (Vickers)
240	260	2.7	107

The first part explains the experimental design to prepare the chip from the bulk. High-speed milling machine (Sodick-MC 430L) at UTHM was used to produce the chip from AA6061 ingot using the following constant set up parameter to obtain a chip with uniform size as shown in Table 2.

Table 2. Chip production specification

Machine name	HSM (SODICK – MC430l)
Tool	Ø 10 mm carbide 2 flute
Feed rate	1100 mm/min
Depth of cut	1.0 mm
Number of cycles	11000

AA6061 chips need to be clean in order to remove oil and other contaminants that stick to the chips, so that it would not affect the obtained results. The cleaning process of the chips was performed by using an ultrasonic bath cleaner with duration of (1hr). During the cleaning process, acetone (ch_3coch_3) was used as cleaning agent.

After the cleaning process the chip will be drying undergone by furnace type (KUITTHO Linn High Therm). The temperatures of drying and duration time were 75°C and 1h respectively. The next steps are chip milling process or production of particle size. Smaller Aluminum particle size was formed by this process. The aluminum chip was milled by using the planetary ball mill type (Retsch PM 100).

In this process, the aluminum chip was milled in a stainless steel jar for the total duration of 20 hours. Aluminum chip was added to the machine for the rotation speed of 350 RPM. During this process, the machine was set to pause for every 15 minutes to ensure that the aluminum chips in the ball mill were not cold welded due to high energy collision subjected in the ball mill. The ball to powder ratio of this process was 20:1. A preventive measure such as ensuring the steel cup was clamped properly was taken into account in order to prevent the machine from damaging.

The sieving process was used to separate the various particle sizes. Three sizes were separated in this study (25, 63, 100 μm). The duration time was 30 min and the amplitude of vibration was 4.

The second part explains the experimental design to produce a recycled aluminum sample. The aluminum particle size from the previous part was then compacted into standard ASTM (E9-89a) shaped and compacted at pressure 9 ton and held at holding time 20 min and sintered at 552 ° C due to optimum sintering temperature [8].

3. Result and Discussion

The Aluminium alloy AA6061 was produced as a chip under constant parameter. It was then milled under speed 350 RPM and classified according to the particle size (25, 63, 100, mix) μm. After that, some specimens were fabricated for several testing to obtain enough data for microhardness behaviour for compaction process.

3.1. Chip and powder production

This paragraph discusses on producing of the aluminium chip AA6061. The chips were produced from aluminium block by using high speed CNC milling machine (SODICK – MC430l). Chip length, chip width and the shape of the chip were observed using scanning electron microscopy at lowest magnification possible. Figure 1 illustrates the several of chips sizes and shapes according to the parameters used for milling process

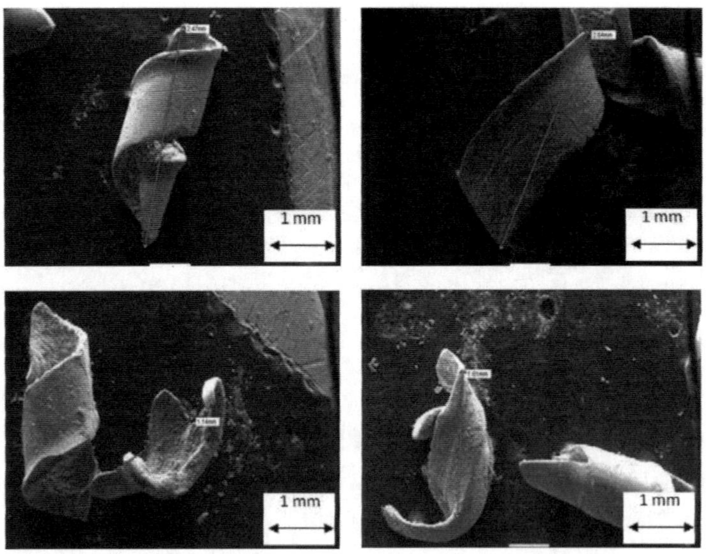

Figure 1. Image of different chip shapes.

Meanwhile, ball milling process was used to produce smaller particle size of aluminium chips into powder form. One speeds of milling was used which was 350 RPM. The duration time for milling process was 20 hours. The ball to powder ratio (b.p.r) of this process was 20:1. Smaller particle size was obtained by this process and three particle sizes were sieved at 25 μm, 63 μm and 100 μm.

3.2 Classification of the specimens

There are many variables considered in the selection, in this paper parameter that considered were the different volume fraction of aluminium chip and particle sizes of powder. The denotations of each composition were as follows in Table.3.

Table 3. Classification of specimens

Symbol	Sub-sym	Chip %	Powder %
A	A1	90	10 (one size 25 μm)
B	B1	70	30 (one size 25 μm)
C	C1	50	50 (one size 25 μm)
D	D1	0	21.5% (25μm) + 78.5% (100μm)
	D2	0	50% (25μm) + 50% (100μm)
	D3	0	78.5% (25μm) + 21.5% (100μm)
E	E1	0	21.5% (25μm) + 78.5% (63μm)
	E2	0	50% (25μm) + 50% (63μm)
	E3	0	78.5% (25μm) + 21.5% (63μm)

3.3 Effect of volume fraction on microhardness

Hardness gives a good idea of the durability and coherence of the material mass by using small loads. In this test which is non-destructive, the instrument consists of a hard and accurate head for easing the penetrating in the material. The material will usually suffering from elastic deformation and follows by the plastic deformation. In this test, two groups were taken, the first group was mixture between aluminium chip and aluminium powder (25 μm). The second group was mixture between aluminium powders 25 μm, 63 μm and 100 μm.

Eight values for each sample were inspected to calculate the average. Figure 2 shows the relationship between volume fraction aluminium chip and microhardness for the sample A, B, and C. It can be seen, that the biggest value of microhardness was sample A1 due to it has a little amount of pores formation (5.67%), therefore it has higher value of hardness which was 61.003 Hv. While sample C1 has a lower value of microhardness 49.031Hv due to it has large percentage of porosity (7.69%). Whereas, sample B1 between A1 and C1 gave hardness value of 56.87Hv, whilst, resulting porosity at 6.33%. Generally, the relationship between volume fraction of aluminium chip and powder and hardness is directly proportional.

On the other hand, the value decreased from A1 to C1 due to it has insufficient amount of powder and thus gives the possibility of it particles to rearrange in the voids formation from the large size of chips. By the previous data, it can be concluded that the using of powder will improve the performance of the mixture and next the mixture of powder was studied in this paper. Figure 1 shows the relationship between the type of specimen and microhardness.

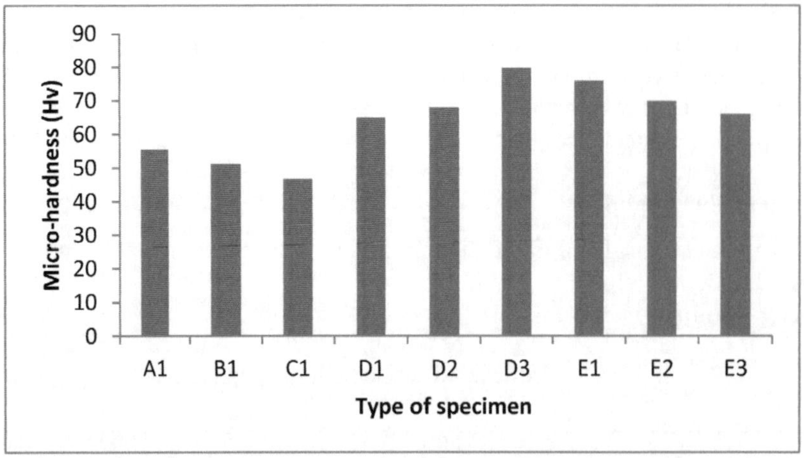

Figure 2. Relationship between type of specimen and microhardness

It can be seen that the specimens with mixture between powders gave higher values of micro-hardness due to these specimens have little amount of pores. The higher value obtained was for the specimen D3 (78.5% 25μm + 21.5% 100μm), while the lower value was from the specimen C1 (50% chip +50% powder with size of 25 μm) due to it has a large amount of porosity.

3.4 Microstructural analysis

The microstructure was obtained by using optical microscopy to for better analysis between chip-powder and powder-powder mixtures. The images were magnified at 50 and 200 times for good comparison between chip-powder and 200 magnifications were used at powder-powder mixture samples (D and E). Thus, Figure 3 shows microstructure for the specimen C1 and Figure 4 shows micro-structure for the specimen D3. The specimen sample of C1 illustrated the different structure of chips and powder. The image shows clear view and separation of chip and powder, whereby, the powder resulted in filling up voids formation due to the large size of chips.

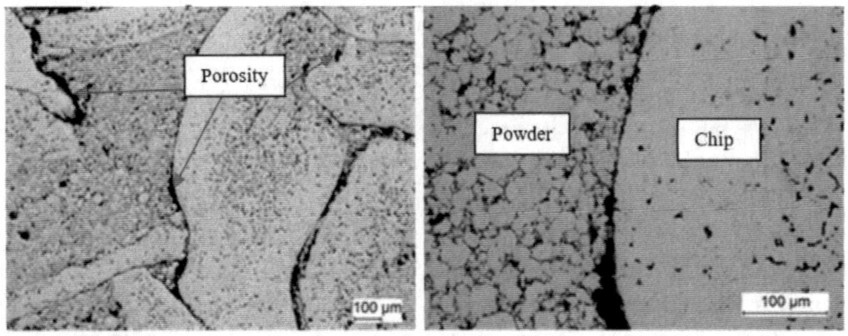

Figure 3. Microstructure for the specimen C1 (50X and 200X magnifications)

Meanwhile, Figure 4 shows the microstructure between powder-powder specimens (D and E). The microstructures illustrate that specimen D1, D2 and D3 show larger grain sizes due to the particles mixture between 25μm (smallest particles used) and 100μm (largest particles used). Besides, the specimen of E1, E2 and E3 provide better bonding between each particles due to smaller size of pore formation. But both groups illustrates fair amount of porosity due to the implementation of powder metallurgy route in the study. With smaller particle size powder use may help in the formation of grain boundary thus relatively filling up the pores surrounded by the larger particle size.

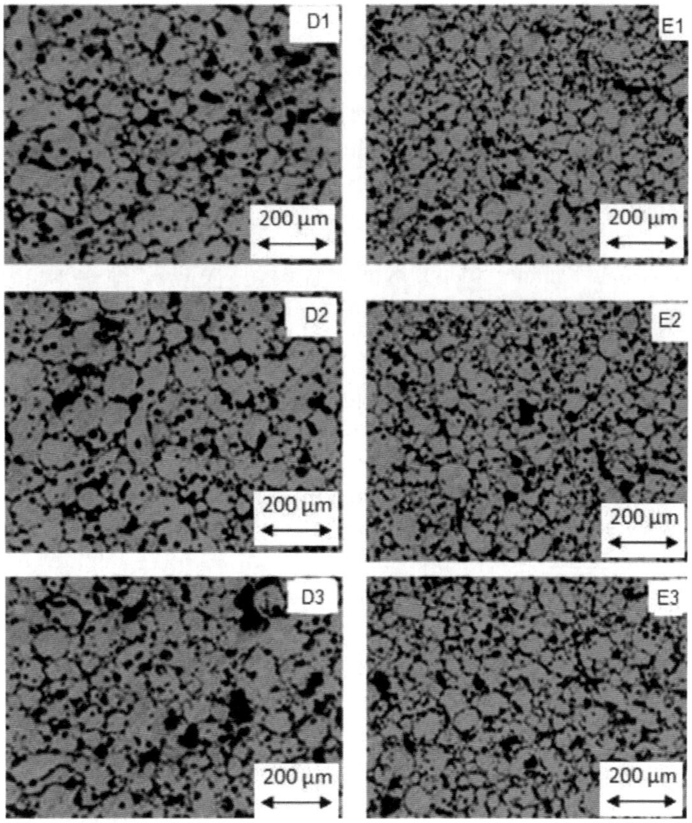

Figure 4. Microstructure for the specimen D and E

4. Conclusion

In this work, the study of recycling aluminium chip was successfully done. The chip sizes of AA6061 that produced from high speed milling were used and the mechanical property of the recycled chip of AA6061 was studied. The smaller size of powder particles was produced by using ball milling was achieved at 350RPM speed. Micro-hardness test was used to the specimen for different particle size and different mixture. Five groups were used for this study, three groups were mixed between chip and powder and two groups were mixed between powder and powder. It can be shown that the mix group D (78.5% 25µm + 21.5% 100µm) has a bigger value for this study due to relatively smaller porosity and well distributed mixture compared to others. Direct proportional relationship was found between the hardness and the percentage of chip.

ACKNOWLEDGMENT

This work is supported by the fundamental research grant scheme (FRGS) under vot 1596. The authors would like to thank the Office for Research, Innovation, Commercialization and Consultancy Management (ORICC) and University Tun Hussein Onn Malaysia (UTHM) for assistance provided in this work.

References

[1] S.S. Khamisa, M.A. Lajisb, R.A.O. Albert, 2015, A Sustainable Direct Recycling of Aluminum Chip (AA6061) in Hot Press Forging Employing Response Surface Methodology, science direct, Procedia CIRP 26, 477 – 481.

[2] Ahmed Sahib Mahdi , Mohd Sukri Mustapa ,Mohd Amri lajis , Mohd warikh abd Rashid, 2015, Effect Of Compaction Pressure On Mechanical Properties Of Aluminium Particle Sizes Aa6061 Through Powder Metallurgical Process, ARPN Journal of Engineering and Applied Sciences.

[3] Devaraju Aruri, Kumar Adepu, Kumaraswamy Adepu, Kotiveerachari Bazavada, 2013, Wear and mechanical properties of 6061-T6 aluminum alloy surface hybrid composites [(SiC + Gr) and (SiC + Al2O3)] fabricated by friction stir processing, jmaterres technol ;2(4):362–369.

[4] Michael Oluwatosin Bodunrin, Kenneth Keneth Alaneme, Lesley Heath Chown, (2015)" Aluminium matrix hybrid composites: a review of reinforcement philosophies; mechanics, corrosion and tribological characteristics" j mater res Technol.

[5] K. Umanath, S. T. Selvamani, K. Palanikumar and R. Sabarikreeshwaran (2014)" Dry Sliding Wear Behavior of AA6061-T6 Reinforced SiC and Al2O3 Particulate Hybrid Composites" ScienceDirect**,** Procedia Engineering 97,694 – 702.

[6] Mohd. Suhail, Mahmood Alam, Reyaz Ur Rahim (2015), The Effect of Process Parameter on Metal Matrix Composite (Al+4%Cu+5%Sic) By Stir Casting, *International Journal of Engineering Trends and Applications (IJETA) – Volume 2 Issue 1, Jan-Feb.*

[7] R. Song, D. Ponge, D. Raabe, R. Kaspar (2005), Microstructure and crystallographic texture of an ultrafine grained C–Mn steel and their evolution during warm deformation and annealing, Acta Materialia 53, 845–858.

[8] M. S. Mustapa, A. S. Mahdi, and M. Amri, "Physical Properties of Recycling Milled Aluminium Chip (AA6061) for Various Sintering Temperature," vol. 3, no. 2, pp. 33–40, 2016.

Preliminary Investigation to Determine the Suitable Mixture Composition for Corn Starch Matrix

Nazri Huzaimi Zakaria[1*], Zamani Ngali,[2] Mohd Zulkefli Selamat[1]

[1]Faculty of Mechanical Engineering, Universiti Teknikal Malaysia Melaka, Hang Tuah Jaya Durian Tunggal Melaka

[2]Faculty of Mechanical & Manufacturing Engineering, Universiti Tun Hussien Onn, Batu Pahat,

*E-mail: nazrihuzaimi@utem.edu.my

Abstract. The use of natural fiber as reinforcement in polymeric composites has been seen a dramatically increase over the last decades. The surge in the interest of natural fiber composite or biodegradable composite is mainly due to the attractive cost of production, improved of hardness, better fatigue endurance and good thermal and mechanical resistivity. In this work, corn starch in the form of powder is utilized as the matrix of the composite. However, starch is brittle and has low strength make it inappropriate candidate for matrix binder. The main objective of this study is to modify the mechanical properties of pure corn starch by mixing it with water, glycerol and vinegar. The composition ratio of water is 60~80%, corn starch 10~35%, glycerol is 5~15% and vinegar is 0~5%, ten samples (A-J) have been manufactured and the best mixture composition is selected based on few selection criteria. The selection criteria are visual impaction, hardness and density. From the results, the samples without vinegar are not suitable to be used because of the fungus availability on the surface. Meanwhile the results from the samples with 5 ml vinegar have no fungus on their surface even has been exposed to the ambient air. While the sample C has shown the best sample based on the visual, hardness and density test.

1. Introduction

Composite materials are materials made from two or more constituent materials with significant different physical and chemical properties [1]. The combination of two different materials produces a new type of material that has different structural properties from the individual component [2]. The new material generated is generally preferred due to many reasons. One of the reasons is that the new material is generally has higher strength, lighter, and less expensive than the old one [1]. The use of natural fibers as reinforcement in polymeric composites has seen a dramatic increase over the last decades. The surge in the interest of natural fibers is mainly due to the attractive cost of production, improved of hardness, better fatigue endurance and good thermal and mechanical resistivity [3]. Apart from that, natural fibers are also considerably environmentally friendly [4]. In order to give the unique ability for the natural fibers, binder/matrix such as starch, epoxy and polypropylene are added to enhance the existing mechanical properties. During the ancient civilization, clay was reinforced by straw to build strong walls. The analogy is that the clay becomes the binder that holds the straw together resulting in stronger building structure.

Matrix or binder is an important component that should be thoroughly consider in order producing composite material. This consideration is needed due to the varying properties of the matrix such as the durability in various temperature, electricity conductivity, moisture sensitivity, weight, and ease of handling depending on the usage purpose of the matrix [5]. Besides that, the selection of matrix also will influenced by the fabrication method and the effect of matrix on the properties of reinforcement itself. However, the prime consideration is that the constituent should be chemically inert non-reactive. The matrix can be future classified into few categories, namely polymer matrix, metal matrix, ceramics matrix and graphic matrix materials. Even though the strength of the materials on the composites is basically due to the fiber reinforcement but, the function of the matrix cannot be underestimated as it also provides support for the fibers by helping to withstand the applied loads [6]. Moreover, the matrix also provides stability to the composite materials that is by holding the fibers together, and protects the fibers from the harsh environment. Distribution of even load between the fibers is achieved with the aid of the matrix. Due to all these reasons, the selection of matrix is critical as the mechanical properties of the composites such as the inter-laminar shear (bending loads) and the in plane shear (torsion loads) is heavily influenced by the type of matrix utilized.

In green category of composites materials, the matrix can be future classified into three types. The first type is named as partial biodegradable with non–biodegradable polymers matrix such as thermoplastic polymers (polypropylene, polyethylene) and thermoset polymers (epoxy, polyster). Meanwhile, the second type is name as fully biodegradable with biodegradable polymers matrix such as the renewable biopolymer matrix. The examples of this type are soy plastics, cellulosic plastics and starch plastic. The third type of matrix is defined as petro-based bio-degradable polymer matrix such as aliphatic co-polyster and polyesteramides [7, 8]

Starch matrix is classified under carbohydrate that is abundantly available in plants where it is stored in granule forms and acts as an energy reserve. The starch is composed of two types of polymers, which are named as amylose and amylopectin. The properties of hydrophilicity in the starch make the components incompatible with most of hydrophobic polymers. This incompatibility results to a condition where the molecules component of the starch is easily consumed by microorganism when it is exposed to soil. In order to improve the ductility of the starch, the addition of the glycerin can be performed as it lower down the transition and melting point temperature [9]. Besides that, starch has been the popular research subject for the past decade studied owing to to the availability, biodegradability and low in cost of production. Starch can be easily found in corn, rice or potatoes. One unique characteristics of starch is that it becomes thermoplastic if water, glycerol or sorbitol is added as the plasticizer. However, several researchers reported that the main disadvantages of starch are poor of process ability, highly brittle, high water resistance and poor stability in the long term duration [10].

Previous study reported that starch has huge potential to be a good matrix or binder to produce a fully biodegradable composite [10]. However, the mechanical properties of the starch need to be improved through modification process. The improvement process starts with the determination of the composition of water, starch, glycerol and also vinegar. The best composition of the sample is then utilized in mechanical testing.

2. Methodology

In this work, corn starch in the form of powder is utilized as the matrix of the composite. However, starch is known to be brittle and has low strength that makes it inappropriate candidate for matrix binder [10]. Therefore, the strength properties of starch need to be enhanced. It is proposed in this paper that the enhancement of the starch structural properties can be enhanced by modifying its chemical composition. The modification of the chemical composition is conducted by adding water, glycerol and vinegar. Glycerol enhances the ductility of the corn starch while vinegar breaks the branched of amylopectin into straight chained amylose molecules. The composition volume of the mixture is depicted in Table 1. The powder form of starch is first mixed together with water, glycerol and vinegar. The combination of glycerol and water and vinegar results to liquidification of the starch powder. Then, the mixture is heated up so that the liquid form of starch, water and glycerol can be converted into a new form of liquid that has high viscosity. Next, this liquid is poured into a mould of which the solidification will take place.

The samples of the new starch are then prepared for mechanical testing. Two types of mechanical testing were performed, namely hardness and density test. The hardness test has been conducted using Hardness Shore Scale. The whole process of modification of the corn starch is presented in the flow chart depicted in Figure 1.

Table 1. Composition mixture of corn starch, glycerol and vinegar in weight %

Sample	Water (g)	Starch (SH) (g)	Glycerol (g)	Vinegar (g)
A	60	25	10	5
B	60	20	15	5
C	70	20	5	5
D	70	15	10	5
E	60	35	5	0
F	70	25	5	0
G	75	15	10	0
H	75	20	5	0
I	80	10	10	0
J	80	15	5	0

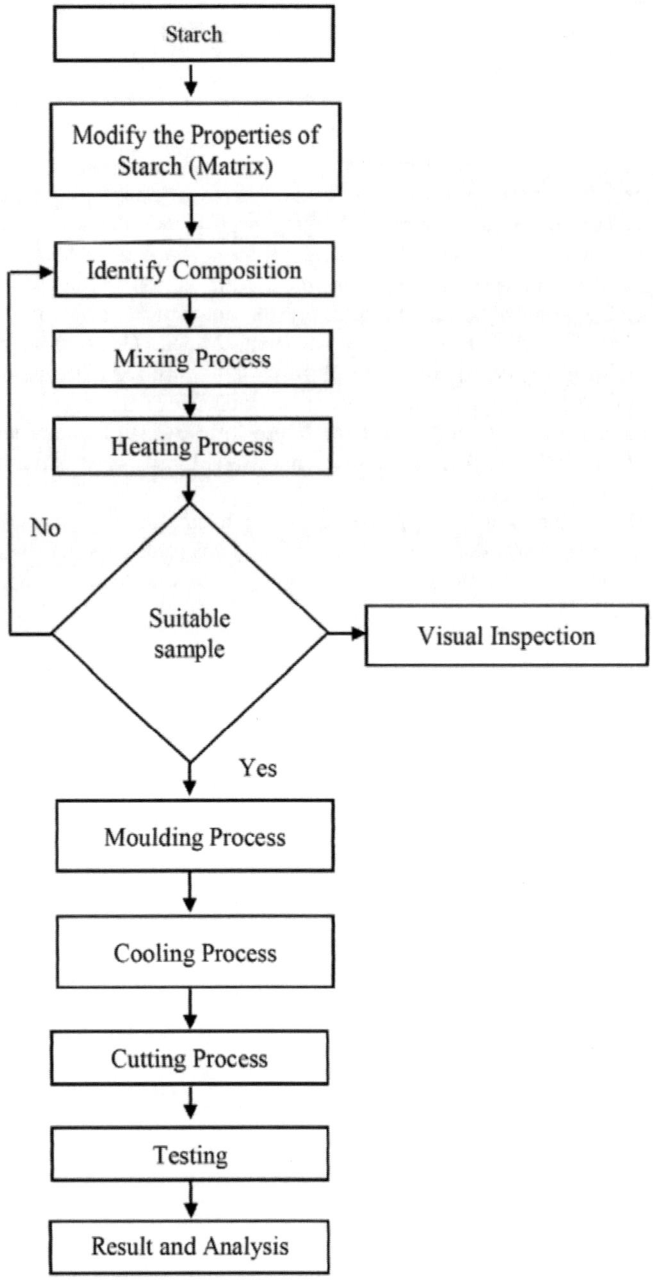

Figure 1. Flow chart of corn starch modification process

3. Results and Discussion

Figure 2 (Sample A up to D) and Figure 3 (Sample E up to J) below present ten samples after the completion of the cooling process. After doing the visual inspection of each sample, as seen in Figure 2, the surface of the samples from A-D are in good condition and have the same colour throughout the sample surface. On the other hand, samples E-J as presented in Figure 3 indicate that there is fungus on the surface. It clearly can be seen the dark area on the surface of each sample. Since the main difference in term of composition between samples A-D and samples E-J, is the availability or present of vinegar in the mixture composition is about 5ml. The present of vinegar as one of mixture components, it gives the acidic effect such as resisted of the fungus to be from and spreader throughout on the surface.

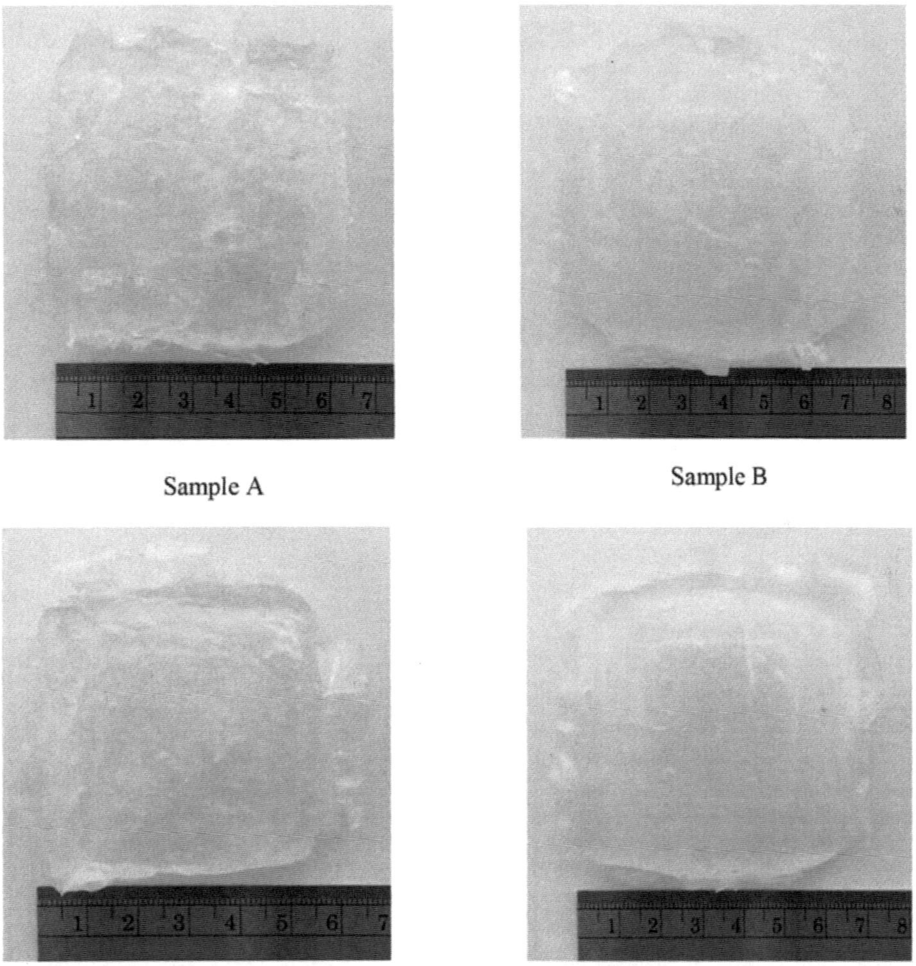

Sample A

Sample B

Sample C

Sample D

Figure 2. Sample of A to D (no fungus on surface)

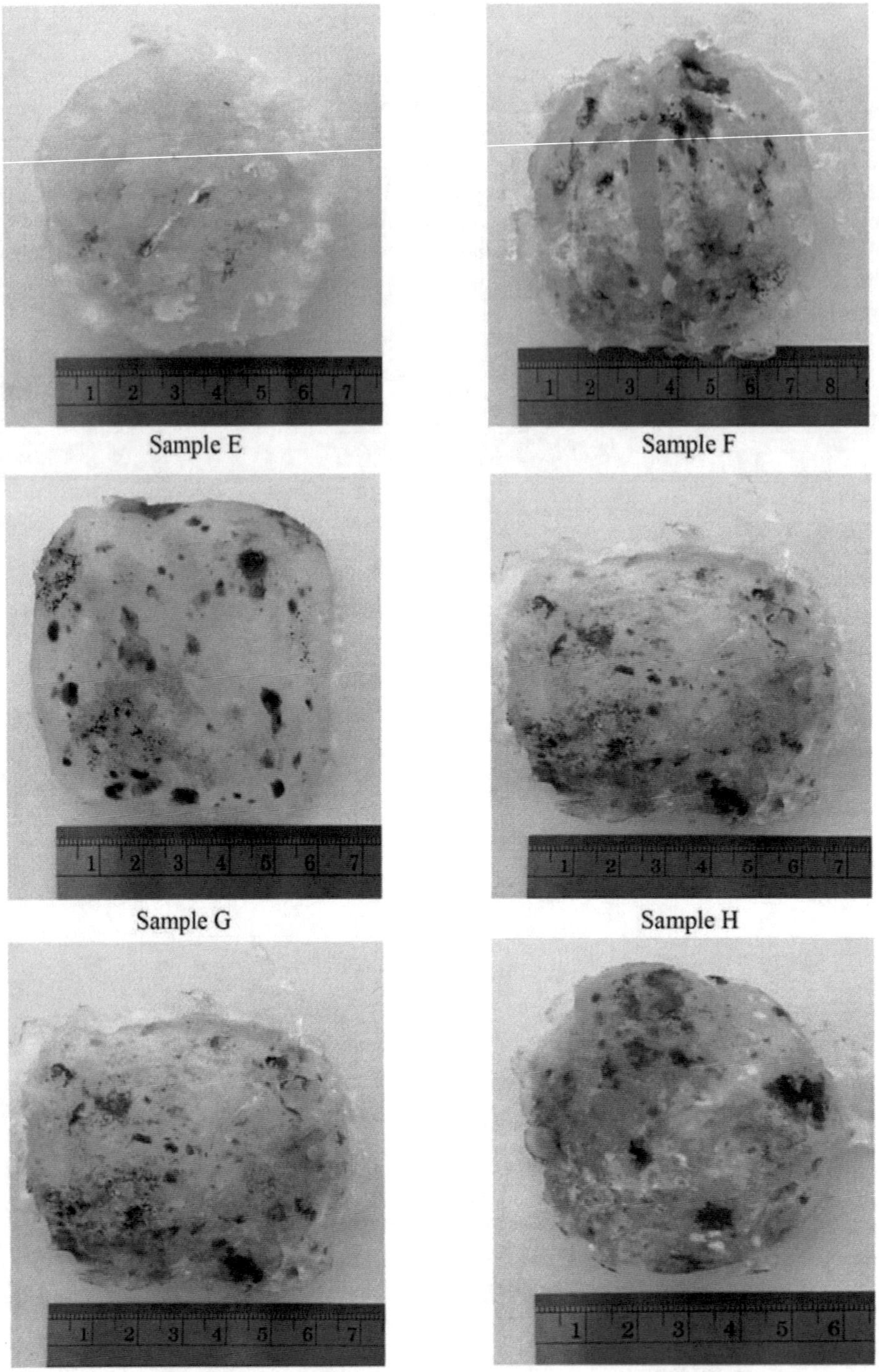

Sample E Sample F

Sample G Sample H

Sample I Sample J

Figure 3. Sample of E to J (with fungus on the surface)

Meanwhile Table 2 and Table 3 present the results of density and hardness tests respectively. For both tests, the tests have performed only for samples A, B, C and D since these samples have no fungus on their surface. The results in Table 2 show the density value of an average of five readings. Based on data collected, generally, the density values remain almost constant for A,B,C and D samples. This result indicates that the mixture composition has no significant effect on the density. Meanwhile from Table 3, sample C has the highest hardness number. As for the comparison purpose, the content of glycerol in sample C is the lowest with 5 gram. Thus, it is suggested that the content of glycerol significantly affects the sample hardness.

Table 2. Density Testing Results for Sample A-D (unit is in g/cm^3)

No	Sample A	Sample B	Sample C	Sample D
1	1.305	1.268	1.367	1.287
2	1.299	1.259	1.363	1.287
3	1.297	1.261	1.370	1.290
4	1.298	1.274	1.336	1.282
5	1.288	1.279	1.379	1.297
Average	1.30	1.27	1.36	1.29

Table 3. Hardness Testing Results for Sample A-D (dimensionless)

No	Sample A	Sample B	Sample C	Sample D
1	9.5	3	8	1
2	8	4	6	2
3	6	3.5	10	1
4	8	5	11.5	0.5
5	3	1.5	17	1
6	4	4	14	1.5
7	6	1	13	0.5
8	8.5	3	5	0.5
9	9	1.5	11	0.5
10	6	1	7	1
11	4.5	4	18	0.5
12	9	1	6	1
Average	6.79	2.71	10.54	0.92

4. Conclusion

The process of modifying the mechanical properties of pure corn starch by mixing with water, glycerol and vinegar has been successfully demonstrated. It shows from the result that the sample without vinegar is prone to have fungus over some time. So that this samples with fungus on the surface is not suitable for future use. Apart from that, the hardness and density testing shows that Sample C has the best characteristics in terms of mechanical properties. Sample C has the composition of 70% water, 20% starch, 5% glycerol and 5% vinegar. Based on this result, it's have generate potential to be used as matrix or binder with natural fibers to produce biodegradable composite. Therefore, further investigation is needed such as to obtain the best composition and orientation between the modified starch with the natural fiber to produce the best result of composite.

Acknowledgement
The first author would like to thank Universiti Teknikal Malaysia Melaka and the Malaysian Ministry of Higher Education for sponsoring his study scholarship.

References

[1] Amini MHM, Hashim R, Hiziroglu S, Sulaiman NS, Sulaiman O. 2013. *Properties of particleboard made from rubberwood using modified starch as binder.* Composites Part B: Engineering.50:259-64.
[2] Mohd Zulkefli Selamat, Mashitah Razi, Ayu Natasya Kasim, Sivakumar Dharmalingam, Azma Putra, Mohd Yuhazri Yaakob and Mohd Ahadlin Mohd Daud. 2016. *Mechanical Properties of Starch Composite Reinforced by Pineapple Leaf Fiber (PLF) From Josapine Cultivar.* ARPN Journal of Engineering and Applied Sciences. Vol. 11, No. 16, ISSN 1819-6608.
[3] Poulikidou S, Jerpdal L, Björklund A, Åkermo M. 2016. *Environmental performance of self-reinforced composites in automotive applications — Case study on a heavy truck component.* Materials & Design.103:321-9.
[4] Rubio-López A, Hoang T, Santiuste C. 2016. *Constitutive model to predict the viscoplastic behaviour of natural fibres based composites.* Composite Structures.155:8-18.
[5] Zain NM, Roslin EN, Ahmad S. 2016. *Preliminary study on bio-based polyurethane adhesive/aluminum laminated composites for automotive applications.* International Journal of Adhesion and Adhesives. 71:1-9.
[6] Le AT, Gacoin A, Li A, Mai TH, Rebay M, Delmas Y. 2014. *Experimental investigation on the mechanical performance of starch–hemp composite materials.* Construction and Building Materials. 61:106-13.
[7] Breche Q, Chagnon G, Machado G, Nottelet B, Garric X, Girard E, et al. 2016. *A non-linear viscoelastic model to describe the mechanical behavior's evolution of biodegradable polymers during hydrolytic degradation.* Polymer Degradation and Stability. 131:145-56.
[8] Pittayavinai P, Thanawan S, Amornsakchai T. 2016. *Manipulation of mechanical properties of short pineapple leaf fiber reinforced natural rubber composites through variations in cross-link density and carbon black loading.* Polymer Testing. 54:84-9.
[9] Jamaludin AR, Kasim SR, Abdullah MZ, Ahmad ZA. 2014. *Sago starch as binder and pore-forming agent for the fabrication of porcelain foam.* Ceramics International. 40:4777-84.
[10] Soykeabkaew N, Thanomsilp C, Suwantong O. 2015. *A review: Starch-based composite foams.* Composites Part A: Applied Science and Manufacturing. 78:246-63.

IOP Conf. Series: Materials Science and Engineering **165** (2017) 012014 doi:10.1088/1757-899X/165/1/012014

THERMAL CHARACTERISTIC OF WASTE-DERIVED HYDROXYAPATITE (HA) REINFORCED ULTRA HIGH MOLECULAR WEIGHT POLYETHYLENE (UHMWPE) COMPOSITES FOR FUSED DEPOSITION MODELING (FDM) PROCESS

Mohamad Helmi Bin Md Ansari[1, 2, a], Mohd Halim Irwan Bin Ibrahim[1, 2, b]

[1] Advanced Manufacturing and Materials Centre, ORICC UTHM, [2]Faculty of Mechanical and Manufacturing, University Tun Hussein Onn Malaysia, Johor, Malaysia

[a]helmi_ansari@yahoo.com, [b]mdhalim@uthm.edu.my

Abstract— The present study provides a hydrothermal synthesis to obtain Hydroxyapatite (HA) powder from waste eggshells. This waste-derived HA has been characterized by X-ray diffraction, scanning electron microscopy and energy dispersive spectroscopy analysis. Waste-derived HA will be reinforced the Ultra-High Molecular Weight Polyethylene (UHMWPE) to develop a material composite for biomedical applications because of impressive mechanical properties owned by UHMWPE. Main challenger is UHMWPE has an ultra-high viscosity that renders continuous melt-state processes including one of the additive manufacturing processes which is Fused Deposition Modeling (FDM). To develop this material as feedstock in FDM process, it has been overcome by blending UHMWPE with waste-derived HA as filler. It exhibit the inclusion of 50wt% HA has reduced the degradation temperature in TGA and DSC thus enhances the processability in FDM process.

1.0 Introduction

In a recent years, additive manufacturing (AM) has been used extensively in biomedical applications such as bone implantation and dental application. AM processes have shown significant potential for developing patient-specific scaffolds with different structural properties using several kinds of materials. With this AM processes, it can make a porous scaffolds to make an augment in tissue regeneration and bone repair. AM normally refers to techniques that have the ability to fabricate physical objects automatically through continual build-up or creation of solid material through additive manufacturing. Applications of additive manufacturing (AM) in tissue engineering offer the pledge of growing these regenerative tissues and functional organs. AM has been effective in integrating structural architecture and assimilation of hormones in scaffolds. In these recent years, there are many additive manufacturing (AM) technologies and techniques have been developing due to extensively used in many industries.

One of additive manufacturing (AM) technologies is fused deposition modeling (FDM) process by layered deposition of polymer plastic on the platform in three dimensional objects [1]. Application of fused deposition modeling (FDM) in biomedical applications has shown its potential compared to the conventional manufacturing process where conventional manufacturing techniques process no control over the pore sizes, their distribution or their interconnectivity. FDM process as has shown in Figure 1 is one of AM technologies where 3D objects are built layer-by-layer from CAD file on a computer-controlled fixtureless platform. By using wire filament has been passes through a heated liquefier and continuous bead or road of materials through a nozzle and deposit onto a platform. Most challenge in the FDM process when to develop a new material of feedstock for this process is thermal characteristic of filament and surrounding environment [2]. Besides FDM process, there are others techniques that are most popular which are Stereo Lithography (SL), Laminated Object Manufacturing (LOM), Selective Laser Sintering (SLS), Solid Ground Curing (SGC), and Ink Jet Printing (IJP) [3] based on layered deposition of polymer plastic on the platform in three dimensional objects. Despite on each type of techniques are based on same concept to build a product, there are have different aspects and own characteristic which can indicates the user to choose of every technique on certain applications more than others [3]. The selection of the right technique for any field leads to better results. There has been a study that presenting about a systematic approach for selection of AM technology to help the user to select the applicable process technique technology. Selection of technology is a complex task because it depends on many criteria such as cost, flexibility, complexity, user friendliness, environmentally green as well as technical capabilities [4].

This paper selects the FDM process because it is potential to produce prototypes with locally controlled properties (porosity and mechanical properties). FDM is functional for use in design verification, prototyping, development and manufacturing. This kind of method also involves into making of three dimensional objects based upon design data which was provided from a computer aided design (CAD) system. Stratasys Inc. was known as a leading company to fabricate the FDM machine and also a manufacturer of 3D production systems which are based on rapid prototyping and solid free-form fabrication. In 1999, several inventors which were been assigned Stratasys Inc. has patterned first their FDM machine [5]. Nowadays, many industrial and manufacturer use Stratasys system to develop and produce parts which are having complex dimensions and shape in variety thermoplastic materials such as ABS, PPSF and PC. FDM process has been widely used in many industries including automotive, aerospace, biomedical, construction and also in manufacturing industries [6]. This FDM process is economical and quite. Thermoplastic extrusion is perfectly adequate fast for minor parts and those parts that have simple geometries.

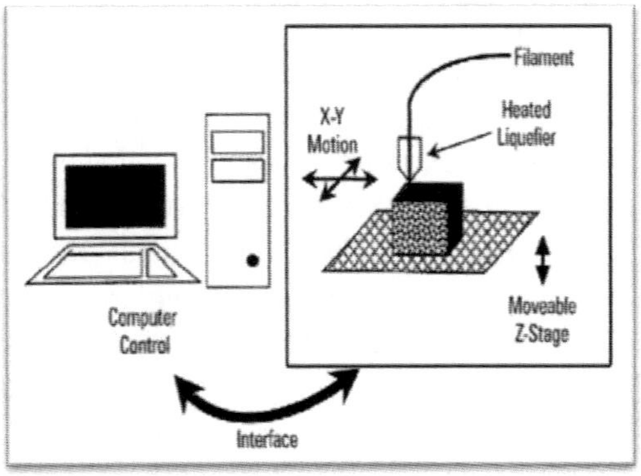

Figure 1. Schematic of fused deposition modeling (FDM) process. [5]

1.1 FDM Material

As we known, feedstock material for FDM process is in filament wire formed which need to deposit in a semi-melted phase and solidifies under jurisdiction of the platform temperature. In generally, the researchers in this field believe that the filament material for FDM process needs to consider two factors which are nozzle temperature and platform temperature. This parameter plays a significant role that affected the properties of the final product. A process parameter optimization for FDM process was conducted by Peng et al. (2014) inappropriate temperature applied in the nozzle temperature can cause the nozzle clogging [7]. Therefore, a melting point temperature of material needs to know and concern. It applies in all material including metal, polymer, ceramic and composite for the engineering application or biomedical application because it is influenced the interlayer bonding thus determine the strength of the material. Nowadays, there are only few material for FDM technology that has been commercialized regardless for engineering applications or biomedical applications due to heating temperature are available ranges within around 120 until 210 °C [8]. This caused FDM technologies suffer limitations of new materials. However with this kind of circumstances, polymers and some low melting point of metals are more suitable for the FDM technologies thus attract many researchers to explore the possibility to process new materials by using FDM process. Most problem faces by the researchers are to study the durability of the material during heating process in FDM. To produce a polymer and some low melting point suitable for FDM process, they have to be thermoplastics that become molten at reasonably low heating temperature usually not above 250 °C [8] and solidify fast enough (decently high glass transition temperature) so that they becomes hardened [9] and hold their shape rather than remaining soft and droopy. This is a very different requirement as compared to injection molding which could literally accept any molten plastics, as long as they can be removing from the mold relatively easily after cooling.

Type of FDM materials application can be divided to two type which are engineering and biomedical applications. In the engineering applications, FDM technology produces strong parts and prototypes and is known to accurately produce feature details [10]. The final parts usually renowned because it is well to being drilled threaded and painted. This technology has entered various types of industries such as aerospace, automotive and manufacturing. Thus, the expanding material will give benefit for the FDM process to evolve and improve. Stratasys's founder, Scott Crump invented FDM Technology more than 20 years ago, and Stratasys has continued to study ever since, developing a range of systems that appeal to large manufacturers, designers, engineers, educators and other professionals [5]. Many researchers usually have focused on processing polymers using FDM process but to expand FDM benefits; there are also has been reported work that has been carried out using low melting alloys [11]. It has been mentioned that it has posses more strength than polymers. Alloy is a mixture of metals or with another element and has been characterized by metallic bonding character. According to Mirelles et al. (2013), the main factors controlling the depositing metal alloys using FDM were the melting temperature and phase transition characteristics based on alloy composition [11]. Besides of polymer and some of the low melting point metal alloys, there are also researchers who are attempts to mix between of metal and polymer [12] and nowadays it has been used in several applications.

In the scope of FDM material for biomedical application, the researchers need to emphasize the characteristic selection of material by selecting the suitable implant materials that have good biocompatibility and supporting mechanical properties with human body. Although metals became regular choice as biomaterial in biomedical applications by manufacturer and designer but polymer has been explored by many researchers and scientist as another option [13]. By using FDM process as manufacturing process, polymer or thermoplastic polymer can become a significant biomaterial for biomedical applications with high biocompatibility and suitable strength of mechanical properties. In this paper, researcher are intend to make HA/UHMWPE as feedstock material for FDM process. UHMWPE as we know is ultra high molecular weight polyethylene that has extremely long chains of polyethylene with molecular mass which has proven has impressive mechanical strength. The main challenge to use UHMWPE as biomedical application material is it difficult to process and needed the incorporation of the other material as binder system. Therefore, the presence of HA in UHMWPE will be expected shows the better thermal stability of the composites. Further discussion about HA, UHMWPE and previously study on HA/UHMWPE will be discussed in the next topic.

1.2 HA/UHMWPE Composites as Feedstock for FDM Process

1.2.1 Hydroxyapatite (HA). For a numbers of years, a various synthetic implant materials have been routinely utilized as replacement for biomedical applications. Bioactive apatite such as Hydroxyapatite (HA) has been extensively investigated for biomedical applications due to its excellent biocompatibility and tissue bioactivity properties. HA is a natural occurring mineral from calcium apatite with the usually written as $Ca_{10}\,(PO_4)_6(OH)_2$ and has similarity in chemical composition to the inorganic matrix of bone [20]. HA is a natural occurring mineral from calcium apatite with the usually written as $Ca_{10}\,(PO_4)_6(OH)_2$. HA has been widely used as an implant material because its close similarity in composition to natural bone. Based on the previous study, biologically derived natural material such as corals [14], fish bone [15] and eggshells [16] have been converted into useful biomaterials like HA. One of these natural materials is eggshell has most composition similar with the human bone. Everyday millions of tons of eggshells are produced as biowaste around the world. Most of this waste is disposed of in landfills without any pretreatment. Eggshells in landfills produce odors and promote microbial growth as they biodegrade. The eggshell contributes by 11% of the total weight of the egg and has been composed of calcium carbonate

(94%), calcium phosphate (1%), organic matter (4%) and magnesium carbonate (1%). The natural biological origin of eggshells results in HA with a crystalline structure and composition similar to that of human bone.

1.2.2 HA reinforced UHMWPE. By using only Hydroxyapatite (HA) as biomaterial to make scaffolds or load-bearing applications is not much suitable due to its poor mechanical properties. Therefore, a study conducted by Bonfield et al., in 1981 shown about HA reinforced polyethylene for bone replacement, it shown that this composite of HA and polymer has potential to improve the mechanical properties and has both the porous structure and the chemical structure of bone [17]. Composite blending with a polymeric material such as ultra-high-molecular-weight polyethylene (UHMWPE) indicate its potential to improve the mechanical and biological properties of bio-composites which is can be adapted to use as implant material for biomedical applications. UHMWPE has extremely long chains of polyethylene with molecular mass which all align in the same direction. UHMWPE has been long used in biomedical applications as biomaterial due to its impressive mechanical strength. Recently, many applications were directed towards the development of polymer composites for load-bearing orthopedic applications. The advantage of using composite material for such application is due to the fact that by varying the amount and type of reinforcing material such as Hydroxyapatite (HA), the mechanical and biological properties can be tailored for specific applications [18]. Furthermore, the low processing temperature of the composites prevents phase and compositional changes of the HA.

Among the wide choice of polymer matrix that have been widely promoted, UHMWPE and its composites have been highlighted as having the potential for such application [18]. Thermal and chemical stability of ceramics, high strength, wear resistance and durability. These characteristic contribute to making ceramics good candidate materials for bones implant and UHMWPE as a composite matrix has these entire characteristic. UHMWPE is a high performance thermoplastic which have to date been associated with high performance devices, structural and load-bearing biomaterials in various fields including electronic and automotive industries [19]. It is best suited for medium mechanical duties in water, oil hydraulics, pneumatics and unlubricated applications. It has a good abrasion resistance but is better suited to soft mating surfaces. It is anticipated that by incorporating particulate HA into UHMWPE matrix, the mechanical performance of the material, particularly the Young's modulus could be improved considerably [20]. Reinforcing UHMWPE with HA could potentially reduce the risk for infection due to wear particles. Because of it, this could help to increase the osseointegration (bone growth onto the acetabular cup) [21]. This HA/UHMWPE composite shown many benefit and can help the growth of cells of bones. Therefore, the presence of bioactive HA would subsequently enhance implant fixation with proper provision for their interactions with the host tissue [22]. A study by Panin et al. in 2012 shows that the role of HA in HA/UHMWPE composite to increase the wear resistance in 3.5 times compared to the pure UHMWPE polymer [23]. Thus, in the FDM process it will be advantages to solve the problem with the conventional manufacturing is that it takes an amount of processing time for the whole process to complete and producing a part. Besides that, FDM process has control over the pore sizes, their distribution or their interconnectivity with the help of HA within UHMWPE composites towards cell culture of bones. Thus, introduction of HA/UHMWPE composites as a new biomaterial shows a promising potential to be study for biomedical applications like bone grafting and orthopedic implants. As mentioned previously, this paper has been proposed a suggestion to use HA/UHMWPE composites as feedstock for FDM process. FDM process is an easy as long the material made in form of filament wire. Due of certain circumstances, FDM process suffer a limitation of material and it is worth extending the FDM material list with materials/composites. Although FDM process has extensively used in engineering application such as manufacturing industry and construction but it has proven that FDM process can make a significant useful in biomedical application. In this present paper, waste-derived-HA from eggshell and ultra high molecular weight polyethylene (UHMWPE) has been used and formed as feedstock for FDM process. Therefore, a quick brief review about HA and previous study HA reinforced UHMWPE as biomaterial in biomedical applications has been taken.

2.0 Methodology

2.1 Synthesis of HA

Hydrothermal synthesis can be defined as a method of synthesis of single crystals that depends on the solubility of minerals in hot water under high pressure. A study by Hui et al proves that HA powder has been successfully synthesized from waste chicken eggshell [24]. First steps to the method of synthesis of HA from waste chicken eggshell by using hydrothermal synthesis are waste of chicken eggshell was collected and crushed into small piece. After that, surface of eggshell was mechanically cleaned by distilled water and boiled in water about 20 minutes. Next, they were placed in a three-stage thermal treatment of furnace machine. In first-stage thermal treatment, the temperature was set from 0°C until 450°C with the heating rate per minute was 7°C and the soaking time was 2 hours. For the second-stage thermal treatment, the temperature was set from 450°C till 600°C with the heating rate per minute was 5°C and the soaking time was 2 hours. Finally in the third-stage, temperature was set from 600°C till 900°C with the heating rate per minute was 4°C and the soaking time was 1 hour. At the temperature 900°C, calcium oxide will be formed by releasing carbon dioxide (CO_2) from the waste chicken eggshells. The chemical equation is shown in equation (1) and by obtained CaO from the eggshells, it will transform into HA powder by mixing with Tri-Calcium Phosphate ($Ca_3(PO_4)_2$) to following procedure by Rivera et.al [25]. The equation will be expected reacted like equation (2) below:

$$CaCO_3 \rightarrow CaO + CO_2 \qquad (1)$$

$$3Ca_3(PO_4)_2 + CaO + H_2O \rightarrow Ca_{10}(PO_4)_6(OH)_2 \qquad (2)$$

2.2 Blend Formulations

In this research, ultra high molecular weight polyethylene (UHMWPE) will been incorporating with derived pure Hydroxyapatite (HA) powder to become HA/UHMWPE composites and investigate on their characteristic as a candidate for wire filament feedstock of fused deposition modeling (FDM) machine. Step of procedure for this mixing process will be follow as Mirsalehi et. al., 2015 which are resulting the increase in HA content leads to an increase of the interfacial bonding between HA and UHMWPE thus as a result can reduce viscosity of UHMWPE and help the improvement of UHMWPE polymer as a bone substitute material rather than pure UHMWPE alone. At first a sufficient amounts of UHMWPE will be determine by their weight and hand-mixer it with paraffin oil at 100°C. Paraffin oil will act as surfactant and lubricant to the UHMWPE for improving processing the powder [26] After a certain time, the compounding of HA powder and swelled UHMWPE will be mixing by using Brabender Plastograph mixer type W50. UHMWPE will be feed into mixer at range of 230°C with rotation speed of 10rpm for 5 minutes. After that, HA will be adding into mixer for 10 minutes. Then, the compounding samples are cooling down to the room temperature. Besides that, it is important to remove the paraffin oil for HA/UHMWPE further analysis. The removing paraffin oil will by carry out by using hot press at 100°C at pressure 3.8 MPa for an hour. Then, the samples will be extracting by soaking in hexane for 24 hours to remove the remaining paraffin oil after hot press process. Subsequently, the samples then dry in an oven at 80°C overnight. The characterization of each sample was shown in the Table 1.

Table 1.Blend formulations of HA/UHMWPE composites samples

Designation	**UHMWPE (%wt)**	**HA(%wt)**
U100	100	0
U90H10	90	10
U80H20	80	20
U70H30	70	30
U60H40	60	40
U50H50	50	50

2.3 Morphology Properties on waste-derived HA from eggshell

2.3.1 SEM/EDS Analysis. To study the material composition of the samples, Scanning Electron Microscope (SEM) tool has been chosen because the ability to perform on-site analysis of the selected point in the specimen to determine the chemical composition of the specimen through the use software Energy Dispersive Spectroscopy (EDS) in the SEM tool. In the SEM tool has the electron beam which is functional to scan across a sample's surface. When the electrons beam focus on the samples, a variety of signals can be generated and detection of specific area of samples can resulting an image or composition of the samples. The three signals which provide the greatest amount of information in SEM are the secondary electrons, backscattered electrons and X-rays. In a previous study, SEM/EDS analysis has been used almost for every research and experiment to study the relation of viscosity and shear rate inside the element [27]. Scanning electron microscopy SEM studies have demonstrated that the morphology of structure inside the materials or element. Therefore, the application of SEM/EDS analysis was used in this experiment to analysis the morphological properties of synthesized of HA. Thus, the application of SEM/EDS analysis could be important because the analysis can determined the successfully research to achieve the objective that has been assigned.

2.3.2 X-ray powder diffraction (XRD) Analysis. X-ray diffraction (XRD) is a tool used for recognized the atomic and molecular structure of a crystal, which by the crystalline atoms can result a beam of incident X-ray to cause an undergo diffraction into many specific area of directions. In this research, XRD analysis needed to study the synthesized HA powder produced from waste eggshell.

2.4 *Thermal analysis on different formulation of HA/UHMWPE composites*

2.4.1 Differential Scanning Calorimeter (DSC). Differential scanning calorimeter is a thermo-analytical tool in which the difference in the amount of heat needed to increase the temperature of a sample. The instrument used in this study was TGA4000-Perkin Elmer. The references and the samples are sustained at closely the same temperature during the experiment. The basic principle in this technique is that when the sample undergoes physical transformation, heat is needed to flow to the sample than the reference to maintain both at same temperature. By observing the difference in heat flow, differential scanning calorimeter is able to detect the amount of heat absorbed or released throughout the transitions. DSC can be applied in measure a number of characteristic properties of a sample. This technique is possible to observe fusion and crystallization as well as glass transition temperature. This ability of examine the transition temperatures and enthalpies makes DSC a valuable tool in producing phase diagrams for various chemical systems. The heat of fusion pure UHMWPE, ΔH°_{f} is 291 J/g, therefore by using below equation (3), we can obtained the degree of crystallinity. ΔH_f is the enthalpy of each sample.

$$\chi = \frac{\Delta H_f}{\Delta H^{\circ}_{f}} \times 100\% \qquad (3)$$

2.4.2 Thermo-gravimetric Analysis (TGA). Thermo-gravimetric analysis (TGA) is commonly used to determine selected characteristics of materials that display either mass loss or gain due to decomposition. In this study, TGA4000-Perkin Elmer was used in examination of materials characterization through analysis of characteristic decomposition patterns and determination of organic content of a sample. The methods of using this instrument is by continuously weighs a sample as it is heated to temperatures for coupling with FTIR and mass spectrometry gas analysis. As the temperature increases, various components from the sample decomposed and weight percentage for resulting change in mass can be measured. The thermo-gravimetric analysis data results in the effect of temperature change affect the relatively weight loss rate. Thermo-gravimetric analysis (TGA) was performed from the ambient temperature of 50 °C – 700°C of under a controlled environment with a heating rate of 10 °C /min by using TGA4000-Perkin Elmer to determine the thermal degradation behavior of the composites.

2.5 *Pre-processing on FDM*

The samples of HA/UHMWPE composites that has best thermal properties will be chosen to undergo a preliminary test in FDM process. First, the sample will go through the extruder process by using single screw extruder machine. By having the sample in the form of filament, a preliminary test was carried out using Flashforge 3D Printer machine to evaluate the extrudability through the print head nozzle of 0.4 mm diameter. The nozzle temperature and platform temperature will be set at 230°C and 110°C respectively.

3.0 Result and Discussions

3.1 Morphology properties of waste-derived HA from eggshell

3.1.1 XRD Analysis. The structural characterization of HA was done by using X-ray Diffraction with Cu-Kα of 1.5406A radiation that generated at generator setting of 40kV and 40mA. The samples of HA with calcined temperature of 900°C. The data was collected in the range of 10-70°, with step size of 0.020°. Phase identification for the sample was observed by comparing the diffraction pattern of HA obtained in laboratory with JCPDS standards. Figure 2 shows the XRD pattern of HA after calcined at 900°C. The phase of derived HA was confirmed with XRD analysis. The result was obtained from mixing of CaO and tri-calcium phosphate under several process followed by calcination at 900°C. The peaks were very sharp and clearly reveal the identity and strong broadening of lines. The formation of hydroxyapatite $Ca_5 (PO_4)_3(OH)$ was confirmed by study of X-ray diffraction (XRD). From the results obtained, the sample was recognized as hydroxyapatite in monoclinic phase with JCPDS standard pattern 01-089-4405 and the spectrum matched with the standard JCPDS pattern with reference code of 00-055-0592. From the figure shown, the peak start rising at 10.27° and attained highest peak at 31.71°.

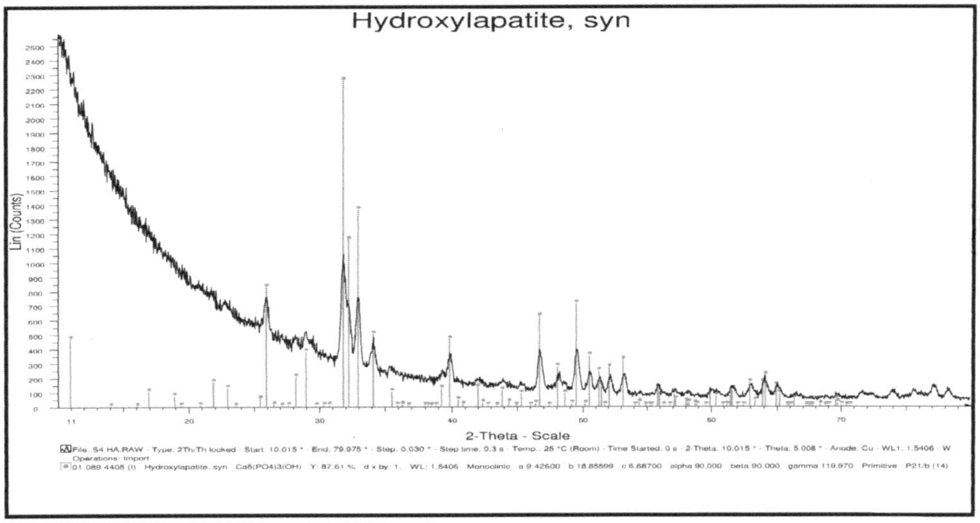

Figure 2. XRD pattern of HA

3.1.2 SEM/EDS Analysis. The surface morphology and crystal size of the derived HA which have been milled, sieved and calcined were investigated under SEM. Figure 3 shows the SEM images of waste-derived HA at 900°C. The HA samples were scaled for resolution of 30µm at 10kV. The results of the images were magnified to1000X for 30µm for the whole field to view. The z-axis of the between the sample holder and detector was set to 10mm before scanning. The SEM images of the HA powder in

Figure 3 revealed that the produced HA microstructures are less spherical. The HA microstructure ascertain high porous in calcined HA powder. Before calcined, HA particle size was extremely small and the agglomerates are hard to disperse due to the small particle size. Formation of the calcined HA microstructures contributed to the ability of the particles to crystallize and form agglomeration at high temperature [28]. Electron dispersive X-ray spectroscopy was used for the elemental analysis of chemical composition for the HA samples. Based on the EDS analysis, the Ca/P ratio for the calcined HA was calculated and found to be 1.69 at 900 °C as shown at the Figure 4 below. The previous study reports that Ca/P ratio for HA is 1.67 thus the results of HA synthesized from eggshell could be considered successful because the result is approximately close to the standard Ca/P ratio.

Figure 3. SEM images of waste-derived HA powder. The blue form box indicates the area of the element analysis.

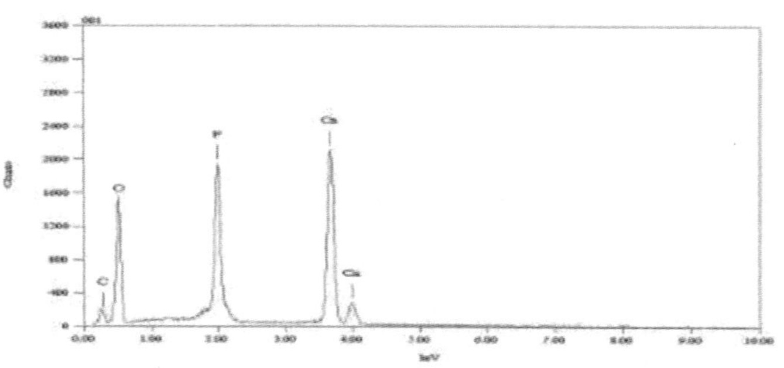

Element	(keV)	mass%	Error%	At%	Compound	mass%	Cation	K
C K	0.277	6.94	0.09	7.34			1.3389	
O K	0.525	44.47	0.26	60.71			26.2238	
P K	2.013	18.09	0.12	11.61			20.7048	
Ca K	3.690	30.50	0.24	20.35			51.7325	
Total		100.00		100.00				

Figure 4. EDS results of HA powder eggshell.

3.2 Thermal properties of HA/UHMWPE composites

3.2.1 Thermo-gravimetric Analysis (TGA). TGA of pure UHMWPE and HA/UHMWPE blends was carried out to assess the effects of blend composition on the thermal stability of the polymers. The TG curves are shown in Figure 5 where as the degradation temperature of each sample blends obtained are shown in Table 2. The initial thermal stability is categorized by temperatures at 10% weight loss, which referred as $T_{10\%}$. The incorporation of HA into UHMWPE as polymer matrix has the ability to reduce the decomposition temperature of pure UHMWPE and HA/UHMWPE composites. In the Table 2 shown the initial decomposition temperature ($T_{10\%}$) for U50H50 has start at 232.255 °C and the peak of decomposition temperature (T_p) is 320.472°C. It shown the significant changes from result obtained by the pure UHMWPE by adding HA as filler. This results was also consistent with a previous study that adding HA reinforced UHMWPE up until 50 wt% has reduces UHMWPE chain mobility allowing HA particle to penetrate inside the composites thus can enhanced the processability of the HA/UHMWPE composites by the degradation of melting temperature.

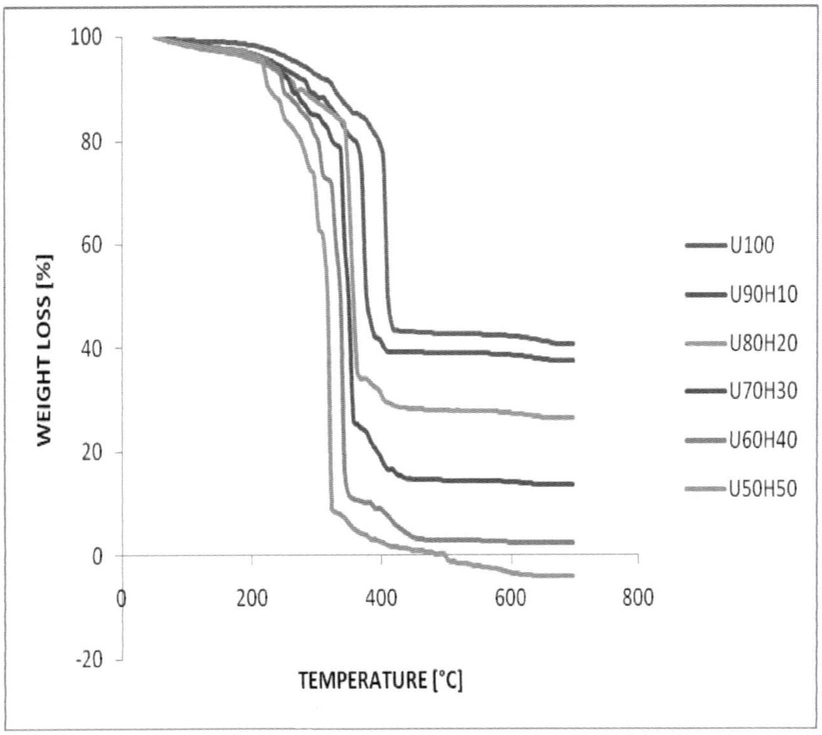

Figure 5. Thermograms of HA/UHMWPE Composites

Table 2. Degradation temperature of components and UHMWPE/PEG blends obtained from TG and DTG curves.

Formulation	Percentage composition (wt %)		Initial decomposition temperature, $T_{10\%}$ (°C)	Peak of decomposition temperature, T_p (°C)
	UHMWPE	**HA**		
U100	100	0	333.151	421.352
U90H10	90	10	292.446	394.898
U80H20	80	20	272.462	361.296
U70H30	70	30	265.929	354.378
U60H40	60	40	252.735	340.746
U50H50	50	50	232.255	320.472

3.2.2 Differential Scanning Calorimeter (DSC) Analysis. Table 3 below shown the result obtained by DSC analysis which has melting temperature, heat fusion and degree of crystallinity. By addition of HA, it induce of the decreasing of the melting point, heat fusion and crystallinity. Therefore the incorporation of HA can help the processability of HA/UHMWPE composites for FDM process. It also exhibit by introducing of HA as filler can also decrease the degree of crystallinity (χ) of the composites. It shown that increasing HA reinforced UHMWPE can cause the composites hinder the polymer macromolecules to rearrange regularly to from crystals. These results are consistently with the previous study by increasing HA particles in the polymer matrix decreasing the decomposition of crystal formation [29]. However, it shown that degree of crystallinity of HA/UHMWPE composites don't show significant of changes thus the incorporation of HA in UHMWPE can't affect the mechanical properties of the composites.

Table 3. The melting temperature, heat fusion and degree of crystallinity for pure UHMWPE and HA/UHMWPE composites

Designation	**Melting point, T_m (°C)**	**Heat fusion, ΔH_f (J/g)**	**Crystallinity, χ (%)**
U100	326.3	174.89	60.1
U90H10	312.7	170.82	58.7

U80H20	305.1	166.74	57.3
U70H30	285.3	155.98	53.6
U60H40	271.4	152.48	52.4
U50H50	244.7	147.54	50.7

4.0 Conclusion

The present study successfully synthesis of Hydroxyapatite (HA) from waste eggshell by using hydrothermal synthesis method and exhibit the result of Ca/P ratio is approximately with the standard Ca/P ratio of HA. By using this waste-derived HA to incorporate with UHMWPE, the degradation temperature of HA/UHMWPE composites is successfully reduced significantly rather than pure UHMWPE with the incorporation of waste-derived HA as filler. Paraffin oil also plays a significant role to reduce the viscosity of UHMWPE before mix with the waste-derived HA. By result from TGA and DSC, this research successfully reduces the HA/UHMWPE composites of melting point by increasing the waste-derived HA as filler thus enhances the processability in FDM process. By increasing HA content up till 50wt%, it will be expected that the HA /UHMWPE composites allows the HA content to the level of cortical bone without the brittleness problems. In the pre-processing on FDM process, HA/UHMWPE composites by using U50H50 formulation is showing a promising results where the composites able to extrude with a constant diameter of 1.80mm. Further study will be about optimization parameter in extruder process to fabricate wire filament feedstock for FDM process with application of U50H50 formulation HA/UHMWPE composite by using Taguchi method because Taguchi method has been proved by its ability to improve the quality of product with the lesser experiment [30]. Overall, the viscosity of UHMWPE is successfully reduced to become a feedstock material in FDM process with the degradation temperature by using waste-derived HA as filler and also with the help of paraffin oil.

.5.0 References

[1] S. J. Kalita, "13 - Rapid prototyping in biomedical engineering: structural intricacies of biological materials," in *Woodhead Publishing Series in Biomaterials*, 2010, pp. 349–397.

[2] K. Torres, G. Staśkiewicz, M. Śnieżyński, a Drop, and R. Maciejewski, "Application of rapid prototyping techniques for modelling of anatomical structures in medical training and education.," *Folia Morphol. (Warsz).*, vol. 70, no. 1, pp. 1–4, 2011.

[3] J.-P. Kruth, M. C. Leu, and T. Nakagawa, "Progress in Additive Manufacturing and Rapid Prototyping," *CIRP Ann. - Manuf. Technol.*, vol. 47, no. 2, pp. 525–540, 1998.

[4] K. Lokesh and P. K. Jain, "Selection of Rapid Prototyping Technology," *Adv. Prod. Eng. Manag.*, vol. 5, no. 2, pp. 75–84, 2010.

[5] E. Stratasys, Inc., "RAPID PROTOTYPING SYSTEM WITH FILAMENT SUPPLY SPOOL

MONITORING," 1999.

[6] C. K. Chua, K. F. Leong, C. M. Cheah, and S. W. Chua, "Development of a tissue engineering scaffold structure library for rapid prototyping. Part 1: Investigation and classification," *Int. J. Adv. Manuf. Technol.*, vol. 21, no. 4, pp. 291–301, 2003.

[7] A. Peng, X. Xiao, and R. Yue, "Process parameter optimization for fused deposition modeling using response surface methodology combined with fuzzy inference system," *Int. J. Adv. Manuf. Technol.*, vol. 73, pp. 87–100, 2014.

[8] L. Marcus, J. J. St, W. Barlow, L. B. David, and H. Crawford, "Solid Freeform Fabrication Proceedings," 1994.

[9] R. H. Abdul Haq, "Characterization and Development of Polycaprolactone (PCL)/Montmorillonite(MMT)/Hydroxapatite(HA) Nanocomposites for fused deposition modelling (FDM) process," Universiti Tun Hussein Onn Malaysia, 2015.

[10] P. Gu and L. Li, "Fabrication of Biomedical Prototypes with Locally Controlled Properties Using FDM," *CIRP Ann. - Manuf. Technol.*, vol. 51, no. 1, pp. 181–184, 2002.

[11] J. Mireles, D. Espalin, D. Roberson, B. Zinniel, F. Medina, and R. Wicker, "Development of a Fused Deposition Modeling System for Low Melting Temperature Metal Alloys," *Univ. Texas El Paso, USA*, pp. 836–845, 2013.

[12] M. Trojanowska-Tomczak, R. Steller, J. Ziaja, G. Szafran, and P. Szymczyk, "Preparation and Properties of Polymer Composites Filled with Low Melting Metal Alloys," *Polym. Plast. Technol. Eng.*, vol. 53, no. 5, pp. 481–487, 2014.

[13] L. T. Kuhn, "Biomaterials," in *Introduction to Biomedical Engineering*, 2005, pp. 255–312.

[14] M. Sivakumar, T. S. Sampath Kumar, K. L. Shantha, and K. Panduranga Rao, "Development of hydroxyapatite derived from Indian coral," *Biomaterials*, vol. 17, no. 17, pp. 1709–1714, 1996.

[15] M. H. I. Ibrahim, N. Mustafa, A. Amin, and R. Asmawi, "Processability Study of Natural Hydroxyapatite and SS316L Via Metal Injection Molding," pp. 358–362, 2015.

[16] A. R. Ibrahim, W. Wei, D. Zhang, H. Wang, and J. Li, "Conversion of waste eggshells to mesoporous hydroxyapatite nanoparticles with high surface area," *Mater. Lett.*, vol. 110, pp. 195–197, 2013.

[17] M. Wang, R. Joseph, and W. Bonfield, "Hydroxyapatite-polyethylene composites for bone substitution: effects of ceramic particle size and morphology.," *Biomaterials*, vol. 19, no. 24, pp. 2357–2366, 1998.

[18] S. L. Evans and P. J. Gregson, "Composite technology in load-bearing orthopaedic implants," vol. 19, no. February 1997, 1998.

[19] L. Fang, Y. Leng, and P. Gao, "Processing of hydroxyapatite reinforced ultrahigh molecular weight polyethylene for biomedical applications.," *Biomaterials*, vol. 26, no. 17, pp. 3471–8, 2005.

[20] L. M. Fang, Y. Leng, and P. Gao, "Effect of HA Content on Mechanical Properties of Hot Drawn HA/UHMWPE Nanocomposites for Bone Substitutes," *Key Eng. Mater.*, vol. 334–335, pp. 701–704, 2007.

[21] D. G. Reynolds, M. Takahata, A. L. Lerner, R. J. O'Keefe, E. M. Schwarz, and H. A. Awad, "Teriparatide therapy enhances devitalized femoral allograft osseointegration and biomechanics in a murine model.," *Bone*, vol. 48, no. 3, pp. 562–70, 2011.

[22] R. A. Surmenev, M. A. Surmeneva, and A. A. Ivanova, "Significance of calcium phosphate coatings for the enhancement of new bone osteogenesis – A review," *Acta Biomater.*, vol. 10, no. 2, pp. 557–579, 2014.

[23] S. V. Panin, L. a. Kornienko, N. Sonjaitham, M. V. Tchaikina, V. P. Sergeev, L. R. Ivanova, and S. V. Shilko, "Wear-resistant ultrahigh-molecular-weight polyethylene-based nano- and microcomposites for implants," *J. Nanotechnol.*, vol. 2012, 2012.

[24] P. Hui, S. L. Meena, G. Singh, R. D. Agarawal, and S. Prakash, "Synthesis of Hydroxyapatite Bio-Ceramic Powder by Hydrothermal Method," *J. Miner. Mater. Charact. Eng.*, vol. 9, no. 8, pp. 683–692, 2010.

[25] E. M. Rivera, M. Araiza, W. Brostow, V. M. Castaño, J. R. Díaz-Estrada, R. Hernández, and J. R. Rodríguez, "Synthesis of hydroxyapatite from eggshells," *Mater. Lett.*, vol. 41, no. 3, pp. 128–134, 1999.

[26] M. H. . Ibrahim, "Optimization of Micro Metal Injection Molding with Multiple Performance Characteristics using Grey Relational Grade Optimization of Micro Metal Injection Molding with Multiple Performance Characteristics using Grey Relational Grade," no. November 2016, 2010.

[27] M. R. Harun, N. Muhamad, A. B. Sulong, N. H. M. Nor, and M. H. . Ibrahim, "Rheological Investigation of ZK60 Magnesium Alloy Feedstock for Metal Injection Moulding using Palm Stearin Based Binder System," vol. 47, pp. 4126–4130, 2011.

[28] Y. Han, S. Li, X. Wang, L. Jia, and J. He, "Preparation of hydroxyapatite rod-like crystals by protein precursor method," vol. 42, pp. 1169–1177, 2007.

[29] M. E. Babiker, W. Guangcheng, Z. Sen, T. Y. Fei, P. Yuqing, R. Huaiping, and Y. Muhuo, "The Effect of Pressure-Induced Flow (PIF) Processing on the Thermal Stability and Mechanical Properties of Ultra-High Molecular Weight Polyethylene (UHMWPE) Gel," *Science (80-.).*, no. 2006, pp. 22–34, 2008.

[30] N. H. M. Nora, N. Muhamad, M. H. I. Ibrahim, M. Ruzia, and K. R. Jamaludin, "OPTIMIZATION OF INJECTION MOLDING PARAMETER OF Ti-6Al-4V POWDER," vol. 6, no. 1, pp. 126–132, 2011.

IOP Conf. Series: Materials Science and Engineering **165** (2017) 012015 doi:10.1088/1757-899X/165/1/012015

PARAMETER OPTIMIZATION OF NATURAL HYDROXYAPATITE/SS316L VIA METAL INJECTION MOLDING (MIM)

N.Mustafa[1] , M.H.I Ibrahim1[2] A.M.Amin[3], R.Asmawi[4]

Advanced Manufacturing & Material Center (AMMC) Universiti Tun Hussein Onn Malaysia, 86400, Parit Raja, Batu Pahat, Johor, Malaysia,

E-mail: nj.wawa@gmail.com[1] , mdhalim@uthm.edu.my[2] , azriszul@uthm.edu.my[3], roslias@uthm.edu.my[4]

Abstract. Metal injection molding (MIM) are well known as a worldwide application of powder injection molding (PIM) where as applied the shaping concept and the beneficial of plastic injection molding but develops the applications to various high performance metals and alloys, plus metal matrix composites and ceramics. This study investigates the strength of green part by using stainless steel 316L/ Natural hydroxyapatite composite as a feedstock. Stainless steel 316L (SS316L) was mixed with Natural hydroxyapatite (NHAP) by adding 40 wt. % Low Density Polyethylene and 60 %wt. Palm Stearin as a binder system at 63 wt. % powder loading consist of 90 % wt. of SS316 L and 10 wt. % NHAP prepared thru critical powder volume percentage (CPVC). Taguchi method was functional as a tool in determining the optimum green strength for Metal Injection Molding (MIM) parameters. The green strength was optimized with 4 significant injection parameter such as Injection temperature (A), Mold temperature (B), Pressure (C) and Speed (D) were selected throughout screening process. An orthogonal array of L_9 $(3)^4$ was conducted. The optimum injection parameters for highest green strength were established at A_1, B_2, C_0 and D_1 and where as calculated based on Signal to Noise Ratio.

Introduction

Metal injection molding is a new technologies that offers to produce metal and ceramic in small and intricate part geometry with more economically and large volume of production which is consist of four major process in MIM includes of mixing, molding, debinding and sintering [1,2]. In metal injection molding, stainless steel has become as a most popular materials and at the mean time broadly used as metallic implant material in medical implant application. In term of medical application, SS316L have an excessive mechanical property such as fracture toughness, fatigue strength and cost effectiveness [1, 3, 4]. SS316 L water atomized was selected as a metal powder since owing to their superior characterization in MIM such as, high corrosion resistance, high packing densities and non spherical particle shape plus resulted to the higher interparticle friction [5,6, 7,].

Lately, hydroxyapatite (HAP) was revealed to be an attractive material for biomedical applications since it offerings a chemical composition close to the bone mineral [8]. Hydroxyapatites have resembled properties to bone structure and teeth based on the chemical and structural observation. HA with the stoichiometric formula $Ca_{10}(PO_4)_6(OH_2)$ and a Ca/P molar ratio=1.67 respectively. HA is a promising biomaterials for bone substitutes and have an excellent properties such as biocompatible, osteoconductive, non-toxic, non inflammatory, non immunogenic agent, and bioactive. Bioactive denotes to the ability to form a direct chemical bond with living tissues [9, 10]

The green part of SS316l/ NHAP feedstock were fabricate by using the high class features of the MIM process that can be practical to manufacture the metal matrix composite (MMC) or ceramic matrix composite (CMC) parts [11] and established by earlier researcher such as [4,12,13,14] but they laboring the synthetic HA.

Sample that manufactured from MIM process are recognized as a green part. Usually, the properties of green part have to be evaluating in terms of density, strength and defect. Even though the MIM process compromises many advantages, it requires the proper molding condition to produce better quality of green part. Regarding to this concern, an optimization of injection molding parameter could be increased or decreased the quality of green part hooked on a tolerable assessment [15,6].

Taguchi's parameter design is an imperative tool for robust design. It promises a simple and systematic approach to optimize design for performance, quality and cost. Robust design applied two major tools which is Signal to noise ratio and orthogonal array. Signal to noise ratio (S/N), concentrate on measures quality with stress on variation, and orthogonal arrays, which accommodates various design factors simultaneously [16].

Taguchi method applied the Design Experiment (DOE) techniques owing a systematic and efficient methodology and outcomes the optimum combination input parameter of process [17, 18, 19].The beneficial of Taguchi Method includes simplification of experimental plan and feasibility study of interaction between different parameters. Previous researchers that work with Taguchi technique are [20, 6]. Taguchi Methods also popular as the best method in order to optimize the design, performance and costing issue [17].

Taguchi method is supreme technique in order to optimize the green part in metal injection molding [21]. There are various injection parameters were optimized in determination the highest green strength which is injection pressure (A), injection parameter (B), mold temperature (C), injection time (D) and holding time (E) and the outcome proves that Taguchi Method could be the best methods in order to solve the problem with a minimum number of trials [6,22]. Other researcher that used Taguchi method as a tool in injection molding optimization parameter is [23, 24].

The main objective of this paper is to optimize the injection parameter for highest green strength by using Taguchi ($L_9 3^4$) Orthagonal Array. As addition information, NHAP was derived from Tilapia fish bone used to apply as ceramic powder in metal injection molding with the combination of metal powder SS316 L and Low density Polyethylene (LDPE) and Palm Stearin as binder system.

Materials and Methods

SS316L water atomized (Epson Atmix Corp) with irregular shapes and high corrosion resistance was selected as a metal powder and NHAP powder was categorized as a ceramic powder. The Tilapia Fish bone was subjected to calcination process in order to produce NHAP powder. Table 1 presented Combination characteristic of SS316L water atomized powder/NHAP characteristic. Low Density Polyethylene (LDPE) and Palm stearin (PS) were used as binder component in this research. LDPE and PS are well known as an established binder system in MIM and the detailed properties described in Table 2. The importance of binder system are plays as transportation agent for feedstock homogeneity and packing powder into desired shaped before sintering phases [25]

Table 1. Combination characteristic of SS316L water atomized powder/NHAP characteristic Details

Identification	SS31L PF-10F / $Ca_{10}(PO_4)_6(OH_2)$
Powder source	Epson Atomix Corp / Tilapia Fish bones
Tap density g/m^3	4.06
True pynometer density g/m^3	8.0471/ 3.156
Powder size	D10 = 2.06 μm D50 = 6.553 μm D90 = 17.712 μm

Table 2. Binder Properties

Binder	Type	Designation	Composition %	Melting Temperature ˚C	Density gcm^{-3}
Palm Stearin	Primary	PS	60	70	0.891
Low Density Polyethelene	Secondary	LDPE	40	111.5	0.92

Feedstock Mixture

Theoretically, the combination of powder and binder should be homogeneous and injectable [25, 26] Concerning to this facts, feedstock was produced by mixing the combination of NHAP/SS316L powder and binder system with suitable composition thru plastograph brabinder. Palm Stearin (PS) was acted as a primary binder where as LDPE was secondary binder system. The mixing temperature used is 150˚C by referring the melting point of LDPE which is occurs at 111.5˚C. Mixing temperature chooses based on melting point of the secondary binder system temperature. The mixing temperature were set up higher compare to highest melting point and lower than the lowest degradation temperature [27].The composition ratio between NHAP powder and SS316L is 90 % SS316L and 10% NHAP at 63 Vol % powder loading. Moreover, the composition for binder systems is 60 Vol% Palm Stearin and 40 Vol% LDPE. The optimum powder loading usually in range of 2% -5% lower than critical powder volume concentration (CPVC) [1, 28.] The value for CPVC for NHAP/SS316L feedstock was found to be 68.189 Vol %. In consequence, this paper are concentrate on investigate the feedstock with 63 Vol % powder loading consider within optimal powder loading range.

Taguchi Method

Parameter optimization of Natural Hydroxyapatite/SS316L via Injection molding process is an important process in order to produce the higher strength and great quality green part. The injection parameters are nominated based on the most significant parameter via screening trial by using classical analysis of variance (ANOVA). From ANOVA results the whole control factors are orthogonal, hence interactions effects are neglected [29, 30] the preferred injection parameters are injection temperature, mold temperature, pressure and speed.

The optimization process are conducted by using $L_9\ (3)^4$ Orthogonal Array (OA) which is proposed of three level designs of experiment with 4 selected parameter in 9 trial. Orthogonal arrays can be develops as plans of multiple factor of experiments. The columns denote to the factors, though, the entries in the represents to the test levels of the factors and the rows resemble to the test runs [31].Table 3 demonstrates the three level of injection parameter design

Table 3. Three level of injection parameter design

indicator	Parameter	0	1	2
A	Injection Temperature (°C)	165	170	175
B	Mold Temperature (°C)	40	45	50
C	Injection Pressure (%)	55	60	65
D	Speed (%)	55	60	65

Green Strength Determination (MPIF 15)

This testing are beneficial in determining the green strength of unsintered compacted powder metallurgy specimens by subjecting them to a uniformly increasing transverse loading under controlled conditions. Green strength is representing the stress required to break the powder metallurgy compact as a simple beam. The specimen is supported near the ends, and broken by applying the force midway between the fixed centres of the supports [32].

The procedures of strength test begin with position the specimen in the transverse rupture test fixture so that it centrally located and perpendicular to the supporting rods, with the top surface facing upward after that, place the loaded fixture between the platens of the compression testing machine and apply a load to the nearest 0.4N with the cross-head velocity of the compression testing machine has been set at approximately 2.5 mm/ minutes and lastly Record the result of the breaking load. [33] Figure 3 exhibit the Universal Tensile Machine

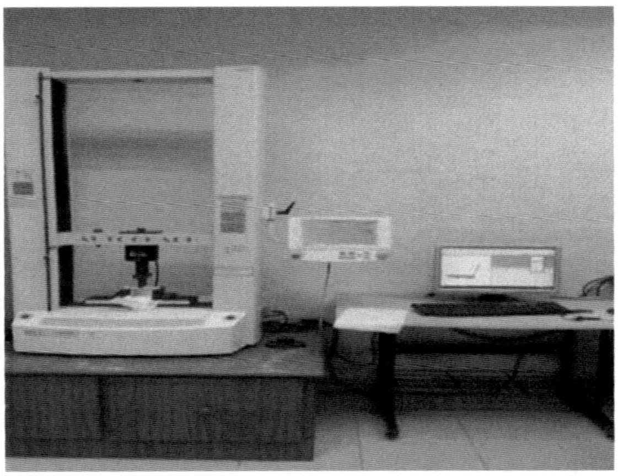

Figure 1. Universal Tensile Machine

The calculation for green strength is referred the following equation 1[6]:

$$S = 3PL/2T^2W \tag{1}$$

Where:
S = green strength (MPa)
P = force (N) required to rupture
L = length of specimen span of fixture in inches (mm)
t = thickness of specimen in inches (mm), and
w = width of specimen in inches (mm)

Results and Discussion

Green strength optimization in MIM is significance process and it focused on investigate the performance and achieves the superiority green part. Signal refer to desirables value (mean) for the output characteristic where as Noise can be observed as the undesirables value (S.D) of output characteristic [6, 34]. S/N ratios approach are applied to measure the quality characteristic diverge from the desired value [32].

Additionally, S/N ratios compromise as an alternative approach to transform the experimental outcomes into a value for the calculation characteristic in the optimal parameter analysis [34, 35]. S/N ratios can be divided into three characteristics which is larger is better, nominal is better and smaller is better [17, 36]. However, in this paper, larger the better signal type is selected as a green strength measurement characteristics. The data is interprets in S/N ratio emphasizes to obtain the optimal performance of green part [6]. S/N ratios Larger is better could be defined as equation 2

$$\frac{S}{N} = -10\log\left(\frac{1}{n}\sum_{j=1}^{n}\frac{1}{{y_{ij}}^2}\right) \tag{2}$$

Where Yij = the amount of score for the green density
N = the total number of shots for each trial

As mention before, optimization injection parameter for the strength of the green part were using Taguchi method. Optimization performance processes by Taguchi methods are very popular among researcher especially in engineering analysis and manufacturing process. Fundamentally, Taguchi method optimizes the performance characteristics over the setting of design parameters. Taguchi method is also robust tool for the design of high quality systems. A model based on L_9 orthogonal array of Taguchi method was created by employing the S/N ratio optimization process [24]. Table 4 exhibit the Taguchi's $L_9(3)^4$ orthogonal arrays and demonstrates the value of experimental trials (strength) and quality characteristic

Table 4. Taguchi's L_9 $(3)^4$ orthogonal arrays demonstrate the value of experimental trials (strength) and quality characteristic

	Factors				S/N ratio : Larger is Better	
Trial	A	B	C	D	Average	S/N ratios
1	0	0	0	0	4.996	13.973
2	0	1	1	1	4.893	13.792
3	0	2	2	2	5.045	14.058
4	1	0	1	2	5.465	14.752
5	1	1	2	0	5.374	14.605
6	1	2	0	1	5.983	15.538
7	2	0	2	1	5.372	14.603
8	2	1	0	2	5.155	14.245
9	2	2	1	0	5.146	14.229
					$\sum$	129.795
					$\bar{T}$	14.422

The highest green strength for Natural hydroxyapatite /SS316L feedstock was obtained by using the optimum configuration of injection molding. By referring figure 2, the highest peak for each parameter A_1, B_2, C_0 and D_1 presented the optimum configuration. In simplest explanation it reflect to Injection Temperature 170 °C, Mold Temperature 50 °C, Pressure 55 % and Speed 60 % .The main effects plot is developed from Table 4 above by using the mean of S/N ratio. The mean S/N value will be calculated from calculation below:

$$13.973+13.792+14.058+14.752+14.605+15.538+14.603+14.245+14.229/9$$

$$= 14.422$$

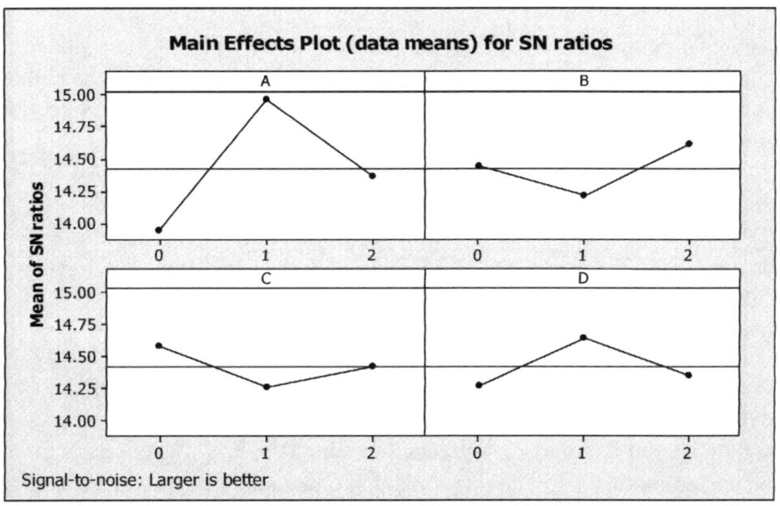

Figure 2. Main effects plot (data means) for S/N ratio

Table 5 expressed the detailed response for green strength S/N values and it can further be noted from the rank of the parameters that variation in the value of S/N ratio. The maximum parameter is injection pressure and least for pressure. The optimum of injection parameter was measured by calculates the variances sensitivity between highest and lowest value of S/N ratio for each factor [30]. The most contribute factor to produce strong green part is injection temperature with 1.02 dB delta values. The injection temperature used here is 170 °C where as to be found at moderate level. Conceivably, too higher injection temperature could lead possibilities of injection defect such as powder binder separation and other defects related with injection molding and produce a brittle green part. [6, 34, 37]. The following rank is mold temperature recorded at 0.39 dB for delta value. Based on figure 2, the optimum mold temperature are at 50 °C which is placed at higher level. The higher mold temperature could avoid possibilities of feedstock to stop flow into the die cavity, rapid freezing on the mold walls and caused cracking green part [38]. Besides, at higher mold temperature, the shear rate between the feedstock surface and cavity surface area will be decrease [39]. Consequently, heating the mold is compulsory to prevent the defects [1]. The third rank is speed with 0.38 dB delta values and the fourth rank is pressure with 0.33 dB value. The moderate speed and pressure usually used in order to produce a quality green part [1, 6].

Table 5. Response Table for S/N Values

Larger Is Better				
Level	Injection Temperature	Mold Temperature	Pressure	Speed
0	13.94	14.44	14.59	14.27
1	14.96	14.21	14.26	14.64
2	14.36	14.61	14.42	14.35
Delta	1.02	0.39	0.33	0.38
Rank	1	2	4	3

Conclusions

This paper summarizes experimental investigations carried out on optimization injection process via L_9 (3^4) Taguchi Method for SS316l/NHAP composite as a feedstock in MIM. The results shows the optimum parameter for highest green strength is A_1, B_2, C_0 and D_1 represents to Injection Temperature 170 °C, Mold Temperature 50 °C, Pressure 55 % and Speed 60.

Acknowledgments

The researchers wish to thanks to Advanced Manufacturing and Material Center (AMMC), Universiti Tun Hussein Onn Malaysia and Ministry of Higher Education Learning (Mybrain15, My PhD) for financial support in this research

References

[1] German.R and Bose, Injection Molding of Metals and Ceramics. 1997 *Powder Industries FederationU.S.A*

[2] Shivashankar TS, Enneti RK, Park SJ, German RM, Atre SV, 2013 The effects of material attributes on powder–binder separation phenomena in powder injection molding, *Powder Technology* **243** 79–84

[3] Zivic F, Babic M, Grujovic N, Mitrovic S, Adamovic D, 2013 Influence of loose PMMA bone cement particles on the corrosion assisted wear of the orthopedic AISI 316LVM stainless steel during reciprocating sliding, *Wear* **300** 65–77

[4] Ramli MI, Sulong AB, Arifin A, Andanastuti M, Muhammad N. 2012. Powder Injection Molding of SS316L / HA Composite : Rheological Properties and Mechanical Properties of the Green Part, *Journal of Applied Sciences Research*, **8(11)**: 5317-5321

[5] Kipphut, CM and German RM (1991), Powder selection for shape retention in Powder injection Molding, International *Journal of Powder Metallurgy* **27(2)**, 117-124

[6] Ibrahim MHI, Muhamad N, Sulong AB, Jamaludin KR, Ahmad S and Nor NHM (2010),Optimization of Micro Metal Injection Molding For Highest Green Strength by Using Taguchi Method *International Journal of Mechanical and Materials Engineering (IJMME)*, Vol. **5** No.2, 282-289

[7] Jamaludin KR, Muhammad.NR , Murthadhahadi MNA, Ahmad S, Ibrahim.MHI, Nor NHM (2010), Rheological Investigation of water atomized Metal Injection Molding (MIM) feedstock for processability prediction *Advance Material Research*, Vol. **83**-86 945-952.

[8] Lombardi M, Palmero P, Haberko K, Pyda W, Montanaro L 2011 Processing of a natural hydroxyapatite powder: From powder optimization to porous bodies *European Ceramic Society* **31** 2513–2518

[9] Sobczak –Kupiec A, Wzorek B , 2012 The influence of calcination parameters on free calcium oxide content in natural hydroxyapatite *Ceramics International* **38** 641–647

[10] Boutinguiza M, Pou J, Comesana R, Lusquinos F , Carlos AD, Leon B , 2012. Biological hydroxyapatite obtained from fish bones, *Materials Science and Engineering* C **32**, 478–486

[11] Ye H, Liu XY, Hong H, 2008 Fabrication of metal matrix composites by metal injection molding-A review *Journal of materials processing technology* **200** pp 12-24

[12] Arifin A, Sulong AB, Muhamad N, Syarif J, Ramli MI, 2015 Powder injection molding of HA/Ti6Al4V. Composite using palm stearin as based binder for implant material, *Materials and Design* **65** 1028–1034 Contents.

[13] Thian, ES, Loh, NH, Khor KA, Tor SB, 2002. Ti-6A1-4V / HA composite feedstock for injection molding *Materials Letters* **56**, 522–53

[14] Muhammad RR, Sulong AB, Muhammad N, Akhtar MN , Rajabi J, 2015 Effects of binder system and processing parameters on formability of porous Ti/HA composite through powder injection molding *Materials and Design* **87** 386–392

[15] Pachauri P and Hamiuddin M, 2015 Optimization of Injection Moulding Process Parameters in MIM for Impact Toughness of Sintered Parts *International Journal of Advanced Materials and Metallurgical Engineering* , Volume **1**, Issue 1, pp. 1-11

[16] Akhyar CH. Che Haron CHA, Ghani JA, 2008 Application of Taguchi Method in the Optimization of Turning Parameters for Surface Roughness *International Journal of Science Engineering and Technology* Vol. **1**

[17] Roy RK, 1990 A primer on The Taguchi Method Society of Manufacturing Engineers, Michigan

[18] Amin A, Ibrahim MHI, Asmawi R, Mustafa N, Hashim MY, 2015 Green Density Optimisation With Sustainable Sewage Fat As Binder Components In Ss316l Feedstock Of Metal Injection Moulding Process (Mim) By Taguchi Method, *Applied Mechanics and Materials* Vols **773-774** pp 173-177

[19] Muthuramalingam B & Mohan B, 2013 Taguchi-grey relational based multi response optimization of electrical process parameters in electrical discharge machining *Indian Journal of Engineering & Materials Science* Vol. **20**, pp. 471-475

[20] Kamaruddin S, Zahid AK, Wan K, 2004 The Use Of The Taguchi Method In Determining The Optimum Plastic Injection Moulding Parameters For The Production Of A Consumer Product, *Jurnal Mekanikal Disember*, Bil.**18** , 98 – 110

[21] Yang, Tarng YS, 1997 Design Optimization of Cutting parameters for Turning Operations based on Taguchi Methods *Jurnal of Material Processing Technology* **84** P 122-129

[22] Ibrahim MHI, Muhamad N, Sulong AB, Jamaludin KR , Nor NHM, Ahmad S, 2011Optimization of Micro Metal Injection Molding with Multiple Performance Characteristics using Grey Relational Grade *Chiang Mai J. Sci*; **38(2)** : 231-241

[23] Berginc B, Kampuš Z , Šuštaršič B ,2006 The use of the Taguchi approach to determine the influence of injection-moulding parameters on the properties of green parts, *Journal of Achievements in Materials and Manufacturing Engineering*, Volume **15**

[24] Nor NHM, Muhamad N, Mohd Ihsan AKA, Jamaludin KR ,2013 Sintering Parameter Optimization of Ti-6Al-4V Metal Injection Molding for Highest Strength Using Palm Stearin Binder, *Procedia Engineering* **68** pp359 – 364

[25] Liu ZY , Loh NH , Tor SB , Khor KA , Murakoshi Y , Maeda R, 2001 Binder system for micro powder injection molding, *Materials Letters* **48** , 31–38

[26] Asmawi R, Ibrahim MHI, Amin AM, 2014 Mixing and characterisation of stainless steel 316L feedstock for waste polystyrene binder system in metal injection molding (MIM), *Applied Mechanics and Materials* Vol. **607** pp 83-86

[27] Amin AM, Ibrahim MHI, Asmawi R, 2014 Mixing Homogeneity And Rheological Characterization For Optimal Binder Formulation For Metal Injection Moulding, *Applied Mechanics and Materials* Vol. **607** pp 181-184

[28] Ibrahim MHI, Mustafa N, Amin AM, Asmawi R, Processability 2015 Study of Natural Hydroxyapatite and SS316L Via Metal Injection Molding, *International Symposium on Technology Management and Emerging Technologies (ISTMET)*, p 335-33

[29] Phadke MS. 2008. Quality Engineering Using Robust Design. AT & T Bell Laboratories. Pearson Education

[30] Amin SYM, Muhamad N, Jamaludin KR, 2013 Optimization of Injection Molding Parameters for WC-Co Feedstocks, Jurnal Teknologi (Sciences & Engineering) 63, 51–54

[31] Kacker RN, Lagergren ES, and Filliben JJ , 1996 Taguchi's Orthogonal Arrays Are Classical Designs of Experiments, *Journal of Research of the National Institute of Standards and Technology*, **577** ,

[32] Nor NHM, Muhammad N, Ibrahim MHI, Jamaludin KR, 2011. Optimization of injection molding parameter of TI-6AL-4V powder mix with Palm Stearin and Polyethylene for the highest green strength by using taguchi method, *International Journal of Mechanical and Materials Engineering (IJMME)*, Vol.**6** ,No.1, 126-132

[33] Metal Powder Industries Federation Standard 15 Method for Determination of Green strength of unsintered compacted powder metullugt materials

[34] Jamaludin, KR, Muhamad , Rahman ABM, Amin, SYM, Murtadhahadi, Ismail MH. 2008 Injection molding parameter optimization using the taguchi method for highest green strength for bimodal powder mixture with SS316L in PEG and PMMA. *World Congress on Powder Metallurgy & Particulate Materials,*

[35] Oktem H, Erzurumlu T, Uzman I,2007 "Application of Taguchi optimization technique in determining plastic injection molding process parameters for a thin-shell part", Materials and Design, vol. 28, pp 1271-1278

[36] Philip JR 1996 Taguchi Techniques for Quality Engineering *McGraw- Hill* **2nd** edition

[37] Amin AM, Ibrahim MHI , Asmawi R, Mustafa N, 2015 Influences of Restaurant Waste Fats and Oils from Grease Trap As Binder on Rheological and Solvent Extraction Behavior In SS316L Metal Injection Moulding, *Metals 2015*, **5**

[38] Liu ZY , Loh NH , Tor SB , Khor KA, 2003 ,Characteristic of powder injection molding feedstock, *Materials Characterization*, 49, 313-320

[39] Huang MC and Tai CC, 2001 The effective factors in the warpage problem of an injection mold part with a shell, *A journal of Material Processing Technology* 110, 1-9

IOP Conf. Series: Materials Science and Engineering **165** (2017) 012016 doi:10.1088/1757-899X/165/1/012016

Effect of Immerse Temperature and Time on Solvent Debinding Process of Stainless Steel 316L Metal Injection Molding

Nur Hafizah Kamarudin[1,*], Mohd Halim Irwan Ibrahim[2]
[1,2] Advanced Materials and Manufacturing Center (AMMC),
Faculty of Mechanical and Manufacturing Engineering,
Universiti Tun Hussein Onn Malaysia, Johor, Malaysia
[2] Department of Material and Design Engineering,
Faculty of Mechanical and Manufacturing Engineering,
Universiti Tun Hussein Onn Malaysia, Johor, Malaysia

*Corresponding email: hd140008@siswa.uthm.edu.my

Abstract. This study was carried out to investigate solvent debinding by conducting the extraction process at temperature ranging from 40 to 80 °C within 2 to 8 hours, while keeping the heptane solvent and 12:1 of solvent to feed ratio as constant. The palm kernel loss was evaluated as an indicator of the process's performance. It was also supported by the pore evolution that was observed by Field Emission Scanning Electron Micrograph (FESEM). Results show that both parameter give large effect on the solvent debinding performance. The best immerse temperature and time for extracting maximum palm kernel in heptane solution at S/F ratio of 12:1, without sacrificing the ability of producing free defect metal part were given by 80 °C and 6 hours, respectively.

Introduction

Metal injection molding (MIM) process is drawing greatly attention for producing small and intricate part in high volume mass production [1-4]. Also, it was a versatile technology used for a variety of alloys and ferrous metals such as carbonyl iron, iron-nickel, various stainless steels, titanium, tungsten and tungsten heavy alloys, as well as, intermetallic compounds [5-7]. It becomes as a promising technology as it was producing the part with outstanding mechanical property such as strength, hardness and such on [8]. Generally, there are four process of this technology. First, the fine metal powder particles mixed with waxes and/or thermoplastic polymers to form a feedstock that can be molded. Next, the crushed feedstock is given desired shape using an injection molding machine. Then, the binders must be removed from the molded part without sacrificing the ability of retaining shape of the part, so called by debinding process. Lastly, the part is sintered at high temperatures, often to near theoretical densities [9].

However, among these processing steps, debinding process was presented as a critical stage which indicates the success of whole processes [8,10]. German & Bose 1997 reported the failure to remove

[1] Mohd Halim Irwan Ibrahim, [2] Nur Hafizah Kamarudin

most of the binders before sintering can result in part distortion such as cracking, swelling, warping and such on [5,11]. Also, Seeran et al. (2014) stated that the insufficient binder removal resulting in contamination from polymer decomposition during sintering [12]. Precisely, the effectiveness of solvent debinding process commonly based on the optimal debinding rate without compromising the part's quality [13]. Multiple steps of debinding process help to remove the binder without sacrificing shape retention of the metal compact was considered as a delicate step [8]. Often, debinding process conducted via solvent extraction and followed by thermal pyrolysis. This combination able to shorten the overall debinding cycle [14-16]. The open pore created during the solvent debinding causing the rapid loss of minor binder during thermal pyrolysis. Consequent with this, authors realize the important of critical study on the solvent debinding process need to be conducted, in purpose to deliberate the optimum performance of whole debinding process, followed by sintering.

The overall idea of solvent debinding was by taking the advantage of the high solubility of low molecular weight of binder (major binder constituent – 60 wt. % palm kernel) components in the organic solvent [8]. The green parts are immersed in a suitable organic fluid, which dissolves the binder partially. Many researchers studied different parameters that affecting the solvent debinding process. German and Bose (1997) reported that high powder loading required a long time of solvent debinding process [10]. It was due to the difficulties of the binder diffusion through the smaller pores. While, in another work, Zaky (2004) investigated the effects of different organic solvents (n-hexane, n-heptane, and isooctane) at different extraction temperatures (30, 40, 50, and 60 °C) by varying the solvent to feed (S/F) ratios (ranging from 7:1 to 15:1 by weight) within extraction times (ranging from 1 to 5 hours) on the amount of binder extracted, diffusion coefficient of paraffin wax and the shape maintenance of green parts [10]. They stated that the dissolution of binders in suitable organic solvent is regularly used for accelerating the removal of the major constituents of the binder system while leaving the minor constituents to support the structure.

Thus, the objective of this present study was to investigate the effects of immersing temperature and time when extraction process conducted in heptane solution using 12:1 of S/F ratio on the palm kernel loss and pore channel built in the green parts. To be highlighted here, the both selections as a constant was made based on previous work by Zaky (2004) and supported by

Materials and methods

The metal powder used in this study was water atomized Stainless Steel 316L (purchased from Epson Atmix Corp) having irregular shape with mean size of d_{50} = 6μm. According to Ibrahim et al (2009), the critical powder volume concentration (CPVC) for this metal powder found to be approximately 64.80 % [6] . Relating to German & Bose (1997) statement, the optimum powder loading should be kept approximately 2-5% lower than critical loading [5]. Consequent with this, the appropriate powder loading of 60 vol. % was applied for this MIM study. Details characteristic of the powder and binder system are tabulated in table 1 and 2.

Table 1. SS316L powder characteristic [6].

Characteristic	Details
Identification	SS316L, PF-10F
Tap density, g/cm^3	4.06
True Pcynometer density, g/cm^3	8.0471
Powder size	d_{10}=2.87μm
Identification	d_{50}=5.96μm
	d_{90}=10.65μm

a) **Table 2. Summary of the binder's characterization [17].**

Binders	Palm kernel (PK)	Waste plastic bag (WPB)
Binder composition (wt. %)	60	40
Degradation temperature (°C)	417.4365	473.94
Melting temperature (°C)	35.125	124.25
Density (g/cm^3)	0.8852	1.0548
Source	Sime Darby	Recycle binder

The MIM process is performed following the diagram of figure 1. In this experiment, the MIM feedstock was 60 vol. % of water atomized Stainless Steel 316L (SS316L) powder mixed with binder system made up of; 60 wt. % of palm kernel (PK) and 40 wt. % of waste plastic bag (WPB). Interestingly, the waste plastic bag as a recycle binder was applied as recycle binder in order to replace the commercial binder of High Density Polyethylene which are quite expensive. This effort was so meant in reducing the disposal problem and encourage the "Green MIM" [18]. In another works of Ibrahim et al. (2016), this feedstock formulation was proven giving a good homogeneity of feedstock [17]. The materials are mixed by Plastograph Brabender mixing machine at 135 °C for 80 minutes. In injection molding, the feedstock are molded into rectangular bar as shown in Figure 2 (a) using Horizontal screw injection molding machine. The injection molding was conducted at temperature of 180 °C, mold temperature of 50 °C, injection pressure of 40 % and injection speed of 50 %. The morphology of the green part was observed as presented in Figure 2 (b). It was used as a guidelines to observe the morphology's part, before and after solvent debinding. Figure shows the metal particles are well coated by palm kernel due to the homogenous feedstock used to produce the green compact [19].

The green parts were then subjected to a solvent extraction where the soluble binder of PK is leach out by immersing the part in heptane solution that ratio by 12:1 (S/F) at leaching temperatures of 40 to 80 °C within 2 to 8 hours. Then, debound parts were dried at 50 °C for 1 hour before the weight of the immerse part was taken. It was for evaporating the solvent from the pores. The debinding ratio (W_d) of PK were calculated based on weight percentage removed, and calculated by (1). Additionally, it was supported by the evolution of pore structure that was analyzed by Field Emission Scanning Electron Micrograph (FESEM).

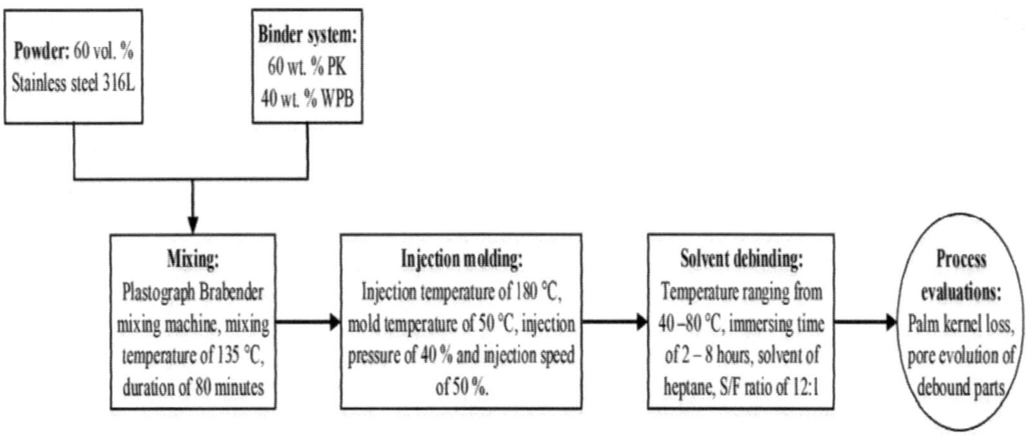

Figure 1. Diagram of the experimental procedures.

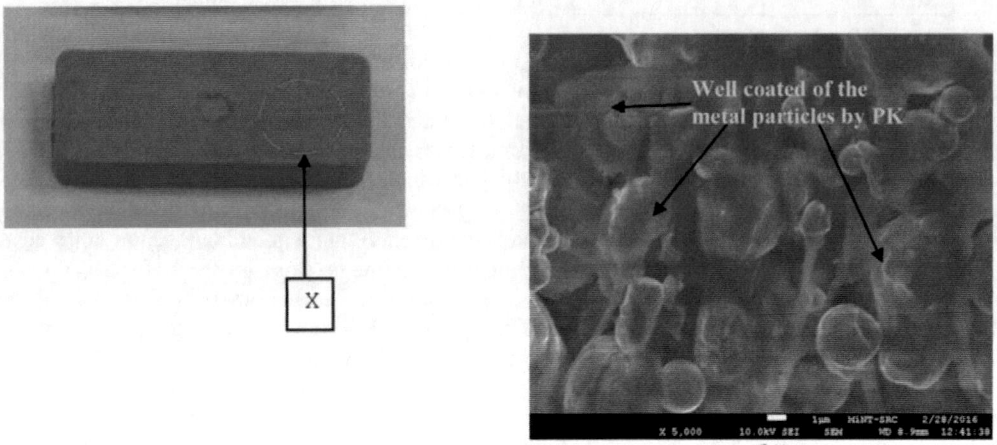

(a) (b)

Figure 2. (a) Rectangular bar shape of green part produced, (b) Morphology of region X.

$$W_d(\%) = \frac{(W_i - W)}{W_i} \times 100 \quad (1)$$

Where W_i is the initial weight of compressed bodies and W is the weight after solvent debinding. The amount of binder extracted was calculated by dividing W_d by total binder content (wt. %) in the feedstock.

Result and discussion

Among many steps in the MIM, the solvent debinding process is considered the most important step. The suitable organic solvent is important for dissolving the binders and speed up the removal of the major constituents (PK) of the binder system while leaving the minor constituents (WPB) to support the structure. This minor constituents is removed later by a thermal debinding. The porosity network created by the solvent extraction step, make the removal of minor constituents of the binder system performed rapidly.

In this study, the solvent debinding process was carried out by extracting the green part in heptane solution at range temperature of 40 to 80 °C during 2 to 8 hours. Previous study by Zaky (2004) found that a very slight increase in the amount of binder extracted when applying an increase of solvent to feed (S/F) ratio [10]. Moreover, the cost of an extraction process is related directly to the quantity of solvent required for a given feedstock. Since that, by considering the economic point of view, a medium ratio of 12:1 was applied in this study, and acts a constant debinding variable.

Trend shows in Figure 3 was built in purpose to analyse the effects of debinding parameters on the solvent debinding process. The palm kernel loss during the process was measured by applying Equation (1) and it was act as an indicator for measuring the solvent debinding performance. Additionally, it supported by the pore evolution that was observed by Field Emission Scanning

Electron Micrograph. There are several important points that can be emphasized based on this built trend.

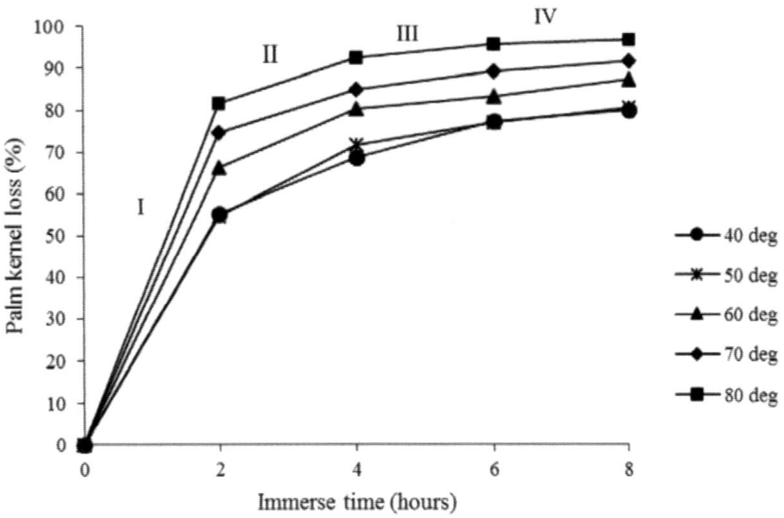

Figure 3. Trend of palm kernel loss at extraction temperature of 40, 50, 60, 70 and 80 °C when immersed in heptane solution for 2, 4, 6 and 8 hours of extraction as the 12:1 of S/F ratio used.

Firstly, the trend shows that the PK loss was increased as the extraction temperature increased. This was proven by the increase of binder loss when giving addition of extraction temperature by 10 °C. By referring to Figure 3, the binder loss for 8 hours of extraction at 40 °C (achieved palm kernel loss by 79.90 %), 50 °C (achieved palm kernel loss by 80.69 %), 60 °C (achieved palm kernel loss by 87.43 %), 70 °C (achieved palm kernel loss by 91.63 %) and 80 °C (achieved palm kernel loss by 96.71 %). So, it was about 16.81 % of PK loss obtained when increase the temperature by 40 °C. Meaning that, the average PK loss was valued by 0.42 % by adding up 1 °C of immerse temperature.

Other than that's, this fact was strongly supported by observing the pore evolution of the debound part produced when immersed in heptane solution at extraction temperature of 50 °C, 60 °C, 70 °C and 80 °C. Figure 4 shows the formation of the open pore channels as the palm kernel are extracted from the green part at different extraction temperature during 8 hours of extraction in heptane solution. The morphology structure for the solvent debinding focused on the voids created in an area dominated by a large group of binders. The existence of these pore channels within the particles facilitate the diffusion of minor constituent (WPB) of binder system during thermal debinding [12,20].

As can be seen in Figure 4 (a), (b), (c), (d) and Figure 4 (d), the higher extraction temperature create a larger void between the powder and binder particles since the greater value of binder loss recorded as the molecular chain moves quickly at the higher extraction temperature [20,21]. Therefore, the higher is the temperature, dissolution of the palm kernel from the part was increased. However, it must be remembered that a too fast removal of major constituents of binder can cause distortion [21].

It was proven that the higher PK loss was obtained as the higher temperature increased and created the larger pore channel. This result shows a good argument with Omar et al. (2003), Vielma et al. (2008), Harun et al. (2012), Chua et al. (2013), Seeran et al. (2014), Arifin et al. (2014), Ibrahim et al. (2016) and Suleiman et al. (2016) which were found that the higher extraction temperature was able to produce the larger pore channels and accelerates the binder loss [11-13,22-25].

Meanwhile, Chua et al. (2013) and Seeran et al. (2014) reported defects such as swelling, cracks, and etc. occurred at extraction temperature over than 60 °C [12,25]. This was due to the large void that created between the powder and binder particles when the major constituent of binder system diffuse out from the green part. At the same time, a large amount of water entered through the existed void created. This phenomenon was affecting the bonding between the metal powders. Also, this phenomenon leads the part to swell. But, this was in contrast with present study, which was performing the solvent debinding over the extraction temperature of 60 °C and produced free defect metal parts.

Also, by referring Figure 3, it was realized that the stage I give the rapid loss of palm kernel and it require explanation. Details of the stages were discussed as below [12-13,24]:

1) Stage I (0-2 hours): A dramatically increase of binder loss was occurred. It was due to the existing layer of a binder on the surface of the green compact. This layer was created as the metal particles are sheared during injection molding process. In this case, the surface layer of particle pack more densely and displace some binder to the surface. The layer eases the flow ability of the feedstock into the mold cavity. During solvent extraction, this surface layer can be hydrated and dissolved rapidly. In another words, the loss of the binder is very fast because the binder on the surface of the component is in direct contact with the solvent (slopes relatively very high).
2) Stage II (2-4 hours): As the debinding process continues, the pore channels extend to the inner region of the parts. The slope shows that the slow debinding rate was obtained in this stage. It was because the difficulties for the solvent to penetrate deeply into the part for dissolving and extracting the binder to the surface. The diffusion of dissolved binder to the surface of the component is facilitated by capillary forces which involve solvent extraction.
3) Stage III (4-6 hours): This stage still recorded an increase of PK loss. It was due to sufficient channels created within the metal particles which allows the diffusion of the dissolved PK to the surface.
4) Stages IV (6-8 hours): The debinding rate slows down and the mass loss remains constant (See the slopes). At this stage, a saturation point between the binder and solvent was reached.

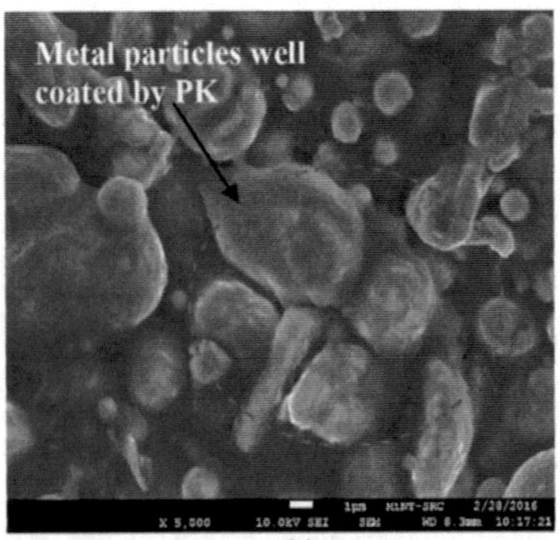

(a)

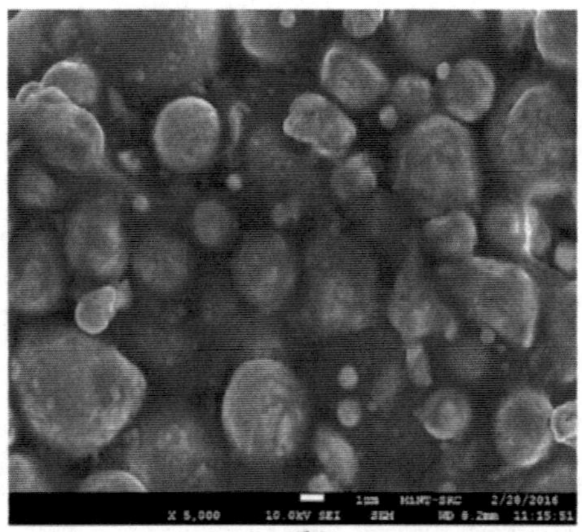
(b)

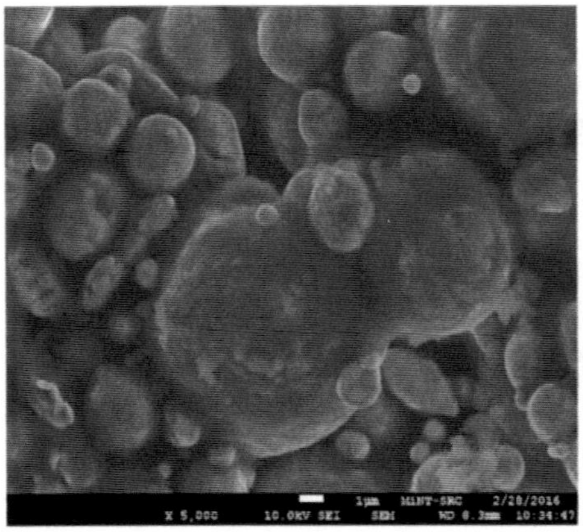

(c)

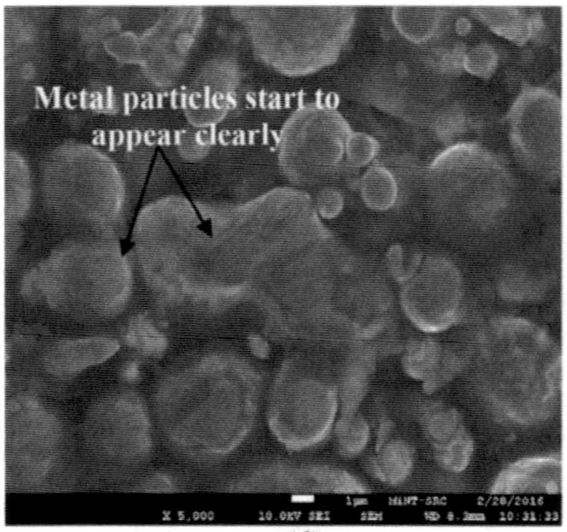

(d)

Figure 4. Open-pore evolution (magnification of x5000) of green part when immerse in heptane solution within 8 hours of extraction time at temperature of; (a) 40 °C, (b) 50 °C, (c) 60 °C, and (d) 70 °C.

German and Bose (1997) considered the time as important aspect in MIM [5]. The more time allocated for the process, the more cost required to be spent by the production line. Consequential of this, authors designed this present study in purpose to remove the major constituent of binder system in a short time with the minimum impact on the part [10]. Then, this investigation was continued by determining the best immerse time of the green part when extracted in heptane solution at 80 °C by also using 12:1 of S/F ratio.

Figure 5 presented development of the open pore channels after immersed in the heptane solution at 80 °C with given different extraction time which are 2, 4, 6 and 8 hours. The FESEM examination carried out on the solvent extracted parts indicated that all extractable organic portion of the binder was approximately removed after 6 to 8 hours of extraction. Noteworthy, the pore channels have formed in the green part after 2 hours of extraction. Figure 5 (a) clearly shows the large amount of palm kernel that coated all surface of metal particles when only immersed in 2 hours and it was supported by less value of binder loss, valued by 81.68 %. Also, it appears that the most of the waste plastic bag that links metal particles together are embedded within the palm kernel, in the necks of residual binder bridging the particles. As the extraction progresses, the removal of palm kernel in these regions reveals and exposes a network of strands waste plastic bag between the metal particles. It was clearly shown at the Figure 5 (d), the pore channels existed was largest compared to void created at the immersing time of 2, 4 and 6 hours. However, it can be noticed that there was only slightly increase of binder loss for 6 and 8 hours of immersion, valued by 95.77 % and 96.71 %, respectively. In other ways, only difference about 0.94 % for adding up the 2 hours of extraction, which is not worth to prolong the extraction duration.

Thus, by also considering the time limitation in manufacturing line [12], authors believed that the 6 hours of extraction time was the optimum time for conducting this solvent debinding process. This was strongly supported by the morphology of green part after 6 hours of extraction, as displayed in Figure 5 (c). It was shown that almost all of the palm kernel has been removed. A network of WPB ligaments remained, binding and holding the metal particles together to provide the MIM compact with sufficient brown strength to be handled [24]. Keeping in mind that the higher binder loss indicates by the larger size of pore channels formed. This was strongly supported by Omar et al. (2003), Vielma et al. (2008), Harun et al. (2012), Seeran et al. (2014) and Ibrahim et al. (2016) which stated that as extraction time increased, the weight loss of binder increased and the pore channels enlarged [8,12-13,22,24].

IOP Conf. Series: Materials Science and Engineering **165** (2017) 012016 doi:10.1088/1757-899X/165/1/012016

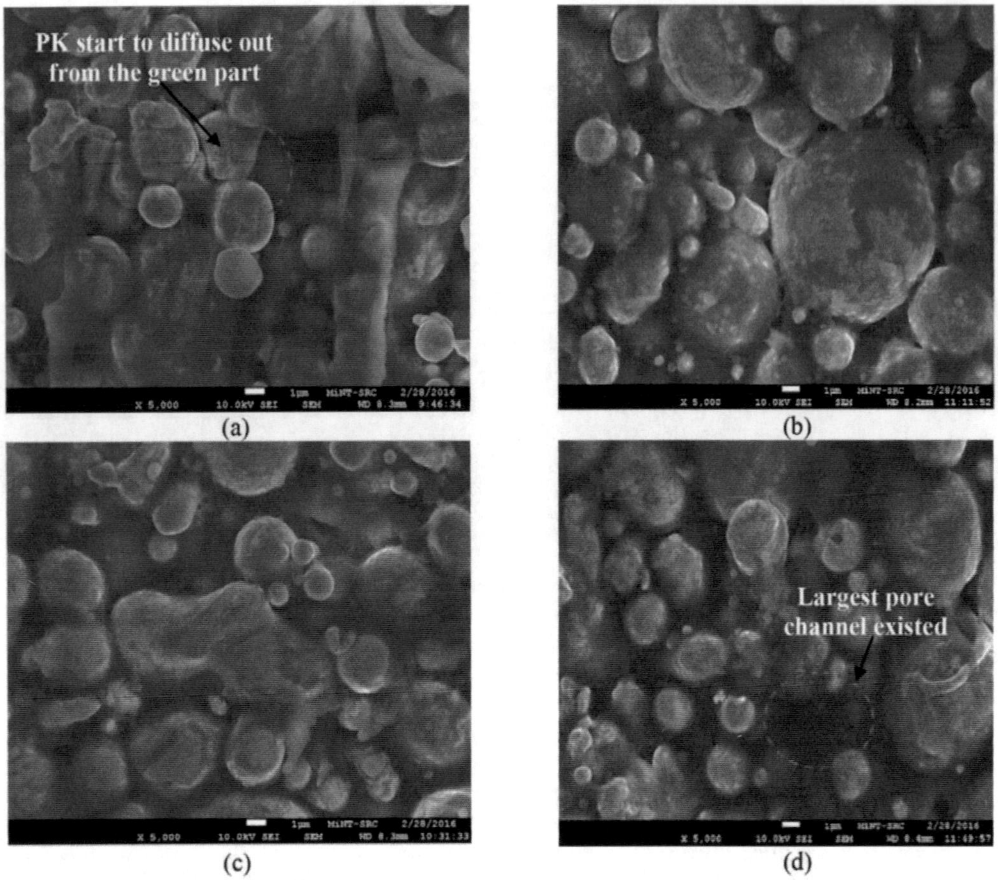

Figure 5. Open-pore evolution as a function of 80 °C of extraction temperature when immersed in heptane solution for; (a) 2 hours, (b) 4 hours, (c) 6 hours, (d) 8 hours.

During the solvent debinding process, all the specimens maintained their shape with no visible distortion and had adequate strength for handling. The remaining binder (WPB) is sufficient to hold particles in their positions. Thus, the remaining binder in both cases is sufficient to hold particles in their positions. Keeping in mind that, the residual binder can be removed rapidly during the thermal debinding process through the large pore channels existed.

Conclusion

The effect of immerse temperature and time on the solvent debinding process of palm kernel were investigated by measuring the percentage of binder loss, which supported by pore evolution observation. The findings are summarized as follows:

1) The higher is the extraction temperature, the higher PK loss obtained.
2) The longer is the immerse time, the higher PK loss recorded. But, the time limitation of manufacturing line need also to be considered.
3) The more binder loss, the larger pore channel existed in the debound part.

4) The optimum condition for the solvent debinding was conducted by immersing the green part in the heptane solution with 12:1 of S/F ratio at extraction temperature of 80 °C within 6 hours.

References

[1] Jamaludin K R, Muhamad N, Rahman A, Amin S Y M, Ibrahim M H I and Nor N H M 2009 Moulding parameter optimisation for the best sintered density *Proceedings of the World Congress on Engineering* **1** 1-5

[2] Asmawi R, Ibrahim M H I and Amin A M 2015 Mixing and characterisation of stainless steel 316L feedstock for waste polystyrene binder system in metal injection molding *Applied Mechanics and Materials* **607** 83-86

[3] Jamaludin K R, Muhamad N, Mohd M N, Ahmad S, Ibrahim M H I and Nor N H M 2009 Optimizing the injection parameter of water atomised SS316L powder with design of experiment method for best sintered density *Chiang Mai Journal of Science* **36** 349-358

[4] Mohd Amin M.A, Ibrahim M H I and Asmawi R 2014 The influence of sewage fat composition on rheological behavior of metal injection molding *Applied Mechanicals and Materials* **660** 38-42

[5] German.R and Bose 1997 *Injection Molding of Metals and Ceramics* (U.S.A: Powder Industries Federation)

[6] Ibrahim M H I, Muhamad N and Sulong A B 2009 Rheological investigation of water atomised stainless steel *International Journal of Mechanical and Materials Engineering* **4** 1-8

[7] Ibrahim M H I, Manaff M H A, Othman M H, Mustafa N, Masrol S R and Rafai N H 2014 Optimisation of processing condition using taguchi method on warpage for HDPE-clay composite *Applied Mechanics and Materials* **660** 28-32

[8] Ibrahim M H I, Mohd Amin A, Asmawi R and Mustafa N 2016 Influences of restaurant waste fats and oils (RWFO) from grease trap as binder on rheological and solvent extraction behavior in SS316L metal injection molding *Metals* **6** 19

[9] Ibrahim M H I, Muhamad N, Sulong A B and Jamaludin K R 2011 Optimization of micro metal injection molding with multiple performance characteristics using grey relational grade *Journal Science* **2** 231-241

[10] Zaky M T 2004 Effect of solvent debinding variables on the shape maintenance of green molded bodies *Journal of Materials Science* **39** 3397-3402

[11] Arifin A, Sulong A B, Gunawan and Mohruni A S 2014 Palm stearin as alternative binder for MIM: A review *Journal of Ocean, Mechanical and Aerospace* **7** 18-20

[12] Seerane M, Chikwanda H, Focke W and Machaka R 2014 A study of solvent debinding variables on Ti6Al4V green bodies *Advanced Materials Research* **1019** 204-209

[13] Harun M R, Muhamad N, Sulong A B, Mohamad N H, Jamaludin K R and Ibrahim M H I 2012 Solvent debinding process for ZK60 magnesium alloy MIM compact *Jurnal Teknologi* **59** 159-168

[14] Enneti R K, Shivashankar T S, Park S J, German R M and Atre S V 2012 Master debinding curves for solvent extraction of binders in powder injection molding *Powder Technology* **228** 14-17

[15] Amin S Y and Muhamad N 2005 Application of RSM technique to optimize the solvent debinding process for MIM compacts *International Conference on Tourism Development*

[16] Li S, Huang B, Li Y, Qu X, Liu S and Fan J 2003 A new type of binder for metal injection molding *Journal of Materials Processing Technology* **137** 70-73

[17] Ibrahim M H I and Kamarudin N H 2016 Characterization of waste plastic bag as a novel binder system and homogeneity test for stainless steel 316L metal injection molding *International Journal of Engineering and Technology*

[18] Asmawi R, Ibrahim M H I and Mohd Amin M A 2014 Mixing and characterization of stainless steel 316L feedstock for waste polystyrene binder system in metal injection molding (MIM) *Applied Mechanicals and Materials* **607** 83-86

[19] Mohd Amin M A and Ibrahim M H I 2014 Mixing homogeneity and rheological characterization for optimal binder formulation for metal injection moulding *Applied Mechanicals*

and Materials **607** 181-184
[20] Sharmin K and Schoegl I 2014 Optimization of binder removal for ceramic microfabrication via polymer co-extrusion *Ceramics International* **40** 3939-3946
[21] Abajo C and Jimenez-Morales A 2015 New processing route for ZrSiO4 by powder injection moulding using an eco-friendly binder system *Boletin de la Sociedad Espanola de Ceramica y Vidrio* **54** 93-100
[22] Vielma P T, Cervera A, Levenfeld B and Varez A 2008 Production of alumina parts by powder injection molding with a binder system based on high density polyethylene *Journal of the European Ceramic Society* **28** 763-771
[23] Suleiman M J, Johari N, Ahmad M A, Ibrahim R, Talib A R and Harmin M Y 2016 Solvent debinding of inconel 718 fabricated via metal injection molding *Advanced Materials Research* **1133** 275-279
[24] Omar M A, Ibrahim R, Sidik M I, Mustapha M and Mohamad M 2003 Rapid debinding of 316L stainless steel injection moulded component *Journal of Materials Processing Technology* **140** 397-400
[25] Chua M I H, Sulong A B, Abdullah M F and Muhamad N 2013 Optimization of injection molding and solvent debinding parameters of stainless steel powder (SS316L) based feedstock for metal injection molding *Sains Malaysiana* **42** 1743-1750

Acknowledgement
Our deepest appreciation to Advanced Manufacturing and Material Center (AMMC), Universiti Tun Hussein Onn Malaysia for research facilities and support

Review on Sintering Process of WC-Co Cemented Carbide in Metal Injection Molding Technology

M Prathabrao[1, a*], Sri Yulis M Amin[1, b] and M H I Ibrahim[1, c]

[1]Faculty of Mechanical and Manufacturing, University Tun Hussein Onn Malaysia, 86400 Parit Raja, Batu Pahat, Johor, Malaysia

Email: prathabrao16@yahoo.com[a*], yulis@uthm.edu.my[b], mdhalim@uthm.edu.my[c]

Abstract. The objective of this paper is to give an overview on sintering process of WC-Co cemented carbides in metal injection molding technology. Metal injection molding is an advanced and promising technology in producing cemented nanostructured carbides. Cemented tungsten carbide (WC-Co) hard metal is known for its high hardness and wear resistance in various applications. Moreover, areas include fine grained materials, alternative binders, and alternative sintering techniques has been discussed in this paper.

Keywords: Metal Injection Molding (MIM), WC-Co, Nano grain size cemented carbide, Microwave sintering

1. Introduction

WC-Co cemented carbides, also referred to as hard metal, exhibit excellent combination of high strength and wear resistance in making the hard metal as preferred material for wear parts and cutting tools in machining since decades ago. In the production of small and complex part in bulk quantity, Metal injection Molding (MIM) has proven to be a promising technology that can produces cemented nanostructured carbides via MIM, with the application of finer grain size of particles below 1 μm, tailored with grain growth inhibitors. Considering the elevated temperature working application, grades cubic carbides such as vanadium carbide, VC, chromium carbide, Cr_3C_2, tantalum carbide, TaC, titanium carbide, TiC and Niobium Carbide, NbC are proposed by authors in journals. The main challenge in sintering stage, whereby to retain small average grain size in the sintered product is difficult because of very high sintering activity of nanopowders. The microstructure of sintered part plays important role in determining mechanical properties such as transverse rupture strength (TRS), hardness and fracture toughness of the cemented carbide. Moreover, the study on the mechanism of different sintering method, liquid phase sintering (LPS), hot isostatic pressing sintering (sinter HIP) and microwave sintering (MW) has been studied in this paper.

2. Metal Injection Molding (MIM)

Metal Injection Molding (MIM) is a newly developed technology to form metals and alloys into desired shape. MIM is a combination of conventional plastic injection molding and powder metallurgy. This process consist of four main steps which is mixing, injection molding, debinding and sintering as shown in Figure 1. During the mixing process, the metal powder is mixed with a binder at a selected volume ratio to form a homogenous feedstock. Binder is the key component in MIM that supplies the metal powders flow ability and formability necessary for molding [1-2]. The attained feedstock from mixing step is molded to produce a "green" compact and the binders hold particles together. During debinding stage, "green" compact is processed by partially removing the binder component to produce "brown" compact. Finally, sintering process is performed to give required mechanical properties for the sintered product also known as sintered body. Thus, the development and improvement of binders results in faster debinding procedures, cost reduction and less environmental defects [2].

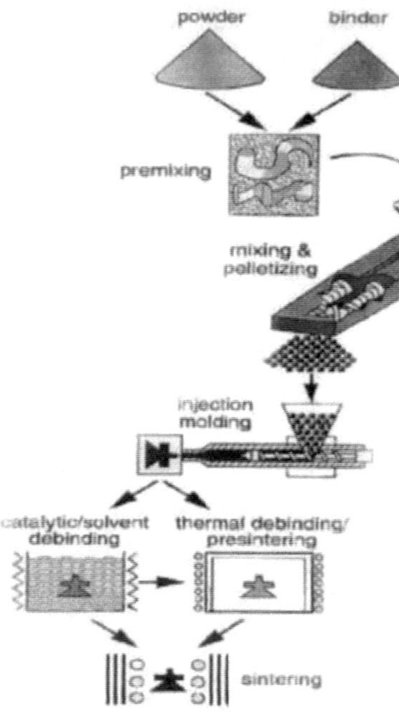

Figure 1: Schematic diagram of metal injection molding, showing the conceptual flow from powder to final structure [2]

Binders in MIM should have good flow characteristics, good debinding characteristics and favorable interactions with the metal powder used. Previous works by Prakash [3] have shown that cemented carbides with certain Fe-Ni-Co binder compositions can be of interest to the hard metal industry, in particular for applications where improved fatigue strength and toughness are required [4]. One main drawback in Fe-Ni-Co binder cemented carbides is also the more difficult control of the carbon balance, in order to achieve defect free microstructures; as well as the tendency of Fe-alloys to form martensite and react with the Fe-based workpieces during machining operations. Nickel is an interesting alternative binder, not only because of its good wettability to WC but also for a much better performance of WC-Ni cemented carbides in oxidation and corrosion conditions than that of WC-Co cemented carbides. However the mechanical properties (hardness and strength) of WC-Ni cemented carbides are relatively inferior to those of WC-Co. The major drawback of Ni-based cemented carbides is their reduced mechanical strength [5].

Ni-Co-Cr, Ni-Si and Fe-Ni binders were also investigated. Aristizabal et al. [6] reported that despite the lower hardness of cemented carbides with Ni-Co-Cr binders, they presented the same wear performances as Co based cemented carbides in wear applications. They argue that this effects is related to the higher oxidation resistance of Ni-Co-Cr compared to Co binders and the formation of continuous oxide tribofilms which reduced friction coefficients for grade with higher binder content. Correa et al. [7] investigated the effects of adding Si to Ni binders in cemented carbides for wear parts. They concluded that there is a solution strengthening of the nickel binder by silicon, which improved significantly the mechanical properties of the Ni-Si based cemented carbides, giving flexural strength and fracture toughness higher than those observed in similar WC-Co.

Good rheological properties of binder and feedstock are one of the keys to get green parts with uniform density and no defect, besides obtaining a successful debinding and sintering process and high quality products [8]. Even though the binder acts as a temporary vehicle to support metal powder especially during mixing and injection molding, but its role is very important. Its behavior especially flow properties during molding, is the most important criterion in developing the new binder system to make sure no defect occurs during molding. It was reported that palm stearin (PS) had a good attribute as a binder system together with polyethylene (PE) [9-11]. The melting point of the PS binder was originally 61°C, whereas that of PE was 127 °C. The mixing and molding temperatures should be set above the melting point of the highest melting component of the binder to ensure that all the binders will melt and that the mold will be homogenously filled with the feedstock. Mold temperature should be kept below the melting point of the minor binder to prevent the molded part from sticking into the mold cavity. The temperature should not be raised too quickly to prevent defects, such as bubbles and cracks [12]. Besides that, a wax based multi-component binder system that contained paraffin wax (PW) as a major component to improve its rheological properties and wettability, a low-density polyethylene (LDPE) as a backbone polymer to increase the strength of green parts [13-16].

3. Nanostructured WC-Co

Nanostructured cemented carbides are the most researched powder metallurgy materials but their potential applications have not yet been defined. Nanostructured cemented carbides are characterized by a unique combination of very fine grained homogenous microstructure and good mechanical properties. Mechanical properties are directly dependent on the developed microstructure in the sintered parts, which is governed by several factors which is Tungsten Carbide (WC), crystallite size, and mean free path of the binding phase and the contiguity of WC grains [14-16]. Improvement of hardness and toughness of cemented carbides can be achieved with a decrease in WC grain size to nanoscale. Thus, considerable efforts have been made over the past years to research nanostructured cemented carbides to draw certain conclusions about hardness and toughness behavior. Nanostructured cemented carbides are produced from near nano- and nano sized WC starting powders. Biggest weakness of sintering near nano- and nano scaled powders is the retaining of a small average grain size in the sintered product [17-19].

4. Metal Injection Molding of WC-Co

Cemented tungsten carbides (WC-Co) consists of tungsten carbides grains embedded in a metal binder phase. WC-Co have been widely used as machining, cutting, mining and drilling tools, as well as wear parts and chip-less forming tools, due to their extremely high hardness, excellent wear resistance and good toughness [20]. In addition, the application of the metal binder phase provides an economical and viable method to produce cemented tungsten carbide components – liquid phase sintering.

However, it was known for long time that cemented tungsten carbides with high cobalt contents had high toughness and low hardness and those with low cobalt content had high hardness and low toughness [21, 25-26]. Therefore, the fracture toughness is improved at the expense of hardness, and vice versa. It has been shown that the ultrafine grained WC-Co materials with the submicron or Nano-WC grain size have enhanced mechanical properties such as hardness, strength and wear resistance, compared to the traditional coarse grained WC-Co materials.

Although the ultrafine cemented carbides (0.1 – 0.6 μm average size) has the higher hardness and wear resistance, the fracture toughness of ultrafine hard metals is inferior to that of coarse grained cemented carbides, which can impact the application on shock environment [27-29]. With the design of WC grain size and distribution in ultrafine hard metals can improve the fracture toughness of ultrafine hard metals obviously, and achieve the combination of proper hardness and fracture toughness in ultrafine WC-(micron WC-Co) systems [30-32]. In principle, many factors will affect the microstructures and mechanical properties of WC-Co materials during the fabrication process, including the shape and granularity of the raw WC particles, composition of the WC-Co powders, selection and content of the grain growth inhibitors, milling conditions and sintering conditions [33-36]. Chen et al. [37] studied that the effect of the matching of coarse and fine powders on the mechanical properties and microstructures of WC-10Co cemented carbides, take to obtain different granularity levels of WC powders through the air classification method, mixed with the Co powder in appropriate proportion, and achieve the double grain sized WC cemented carbides, the alloy obtained a good performance in the drilling experiments.

Li et al. [37] studied that the influence of WC particle size on WC-Co cemented carbides fracture toughness, analyses the mechanism of particle size effect on fracture toughness of cemented carbides. However, the microstructure and fracture mechanism of the ultrafine WC-(micron WC-Co) hard metals have not been investigated systematically. Palm stearin suitability is based on the rheological behavior of the feedstock after being mixed with PS-PE binder system at the powder loading of 59, 61 and 63% vol. Based on the rheological properties, it was concluded that all the MIM feedstocks shows a pseudo plastic behavior and suitable for MIM, except some feedstocks from the high powder loading 63% at high temperature, that shows dilatants behavior [11]. The thermal and physical analyses of feedstock discussed that the critical powder loading spotted for the WC-Co powder is 65%. Thus, the rheological properties of the optimal feedstock with powder loading of 61% show good pseudo plastic behavior, which is suitable for injected molding [12].

5. Grain Growth Inhibitor (GGI)

One of the biggest problems of sintering near nano- and nano scaled powders is the retaining of a small average grain size in the sintered product [24]. Many attempts to achieve nanostructured cemented carbides failed because of very high sintering activity of WC Nano powders. For that reason the addition of grain growth inhibitors, GGI, is required. Small amounts of GGI's are added to starting powders [38]. The most common ones are vanadium carbide, VC, chromium carbide, Cr_3C_2, tantalum carbide, TaC, titanium carbide, TiC and niobium carbide, NbC. Their primary effect is to retain the particle size of starting powders in the sintered product, meanwhile, not only influencing the mechanical properties; increasing the value of hardness at room temperature, but also affecting the toughness, hardness and creep resistance at elevated temperatures [39].

6. Sintering Process

Since the first WC-Co hard metals have been invented and produced more than 70 years ago, activity in research and development laboratories has been directed to the improvement of the characteristics of these alloys and to optimize them for the ever increasing utilization possibilities. Particularly in the field of machining – a main utilization field of hard metals – during the further development of the materials to be processed, new hard metal alloys were continuously developed, which were characterized by an increase in not only the wear resistance of the cutting bodies, but also their strength [40]. The need for hard metals with improved properties, particularly increased hardness and strength coupled to increased ductility and toughness, has focused attention on the development of grades with finer and finer – grained powders and cemented carbides.

Nowadays, submicron and ultrafine grades of WC – Co, with a WC grain size of 0.4 – 1.0 μm, still dominate the market for cemented carbides. Essentially, they are desired to be sintered from WC – Co superfine powder with a WC grain size up to 0.3 μm or nanocomposite powder with a grain size below to 0.1 μm (100nm) [41-42]. Sintering of cemented carbides is conducted most commonly in vacuum [43-45]. But one of the crucial aspects in fine-grained hard metal sintering is the strong tendency of the very fine WC grains to coarsen, due to their high interface energies as well as differences in individual grain sizes, constituting the driving force for the growth process.

Many researchers developed novel approaches to control rapid growth, such as adding grain growth inhibitors (VC, Cr_3C_2, TaC or their combinations), conducting the sintering at lower temperature with the aid of plasma pressure compaction (P^2C) [46], spark plasma system (SPS) [47] plasma activated sintering (PAS) [48], pulse current process [43], microwave process [43] and hot isostatic pressing (Sinter + HIP or sinter hip / SIP) [43]. Packed particles heated near their melting temperature bond together by sintering. As diffusion accelerates at higher temperatures, sintering is manifested by bonding between contacting particles. Sintering occurs over a range of temperatures, but is accelerated as the particles approach their melting range. For solid – state sintering, it is appropriate to think of sintering with respect to the melting temperature. Snow sinters to form ice at temperatures near – 15 °C, while hard metals requires temperatures in excess of 1000 °C [49].

6.1. Liquid Phase Sintering (LPS)

Liquid phase sintering (LPS) is applied to alloys and composites that melt over a range of temperatures. In the typical situation, the solid grains are soluble in liquid. This solubility causes the liquid to wet the solid, providing a capillary force that pulls the grains together. At the same time, the high temperature softens the solid, further assisting densification. High – diffusion rates are associated with liquids, giving fast sintering or lower sintering temperatures. Important technical advances in LPS came in the 1930s with the development of several materials; cemented carbides (WC–Co), porous bronze (Cu–Sn), tungsten heavy alloys (W-Ni-Cu), copper steels (Fe-Cu-C), and cermet's (TiC-Fe) [49].

6.1.1. The mechanism of LPS

A conceptual view of the events taking place, as sketched in Figure 2 for the case of two mixing powders. The solid grains undergo solid-state sintering during heating. Depending on the solid-liquid solubility relations, different microstructure evolution pathways are possible. The newly formed liquid penetrates between the solid grains, dissolves the sinter bonds, and induces grain rearrangement. Further, because of solid solubility in the liquid, the liquid improves transport rates responsible for grain coarsening and densification. The surface energy associated with pores leads to their annihilation, while there is progressive microstructure coarsening and bonding to increase rigidity. The LPS are ideal for densifying hard materials that cannot be fabricated using other manufacturing approaches. The WC-Co system is a prime example, where the eutectic at 1310 °C enables the bonding of micro meter size WC grains into a dense component, such as drill or cutting insert. However, the common form of LPS is persistent LPS, where at the sintering temperature the solid is soluble in the liquid. On cooling, the liquid solidifies to produce a composite microstructure with tailored properties [50-53].

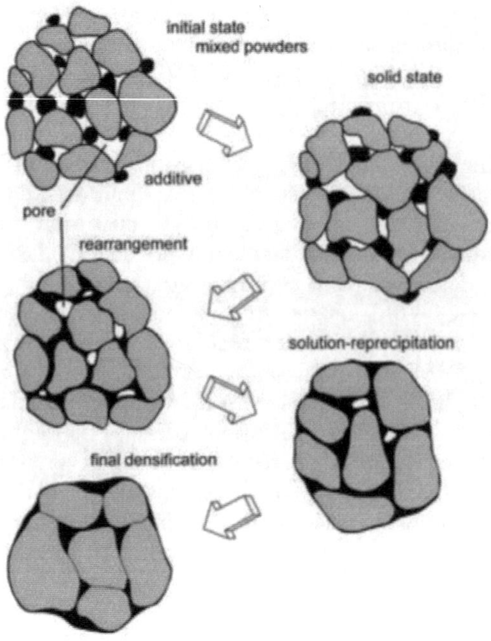

Figure 2: A schematic of the microstructure changes during LPS, starting with mixed powders and pores between the particles. During heating the particles sinter, but when a melt forms and spreads the solid grains rearrange. Subsequent densification is accompanied by coarsening. For many products there is pore annihilation as diffusion in the liquid accelerates grain shape changes that facilitates pore removal [49].

After LPS, the microstructure consists of the solid grain with a solidified liquid network, and possibly residual pores are retained for lubrication, frangibility, or filtration attributes. Thus, liquid phase sintered microstructures exist in several variants with differences in the amount, size, shape and distribution of the phases. Accordingly, substantial performance differences result, especially in properties such as hardness, strength, and elastic modulus. This is especially true for the WC – Co cemented carbides. The study of LPS focuses on linking composition, processing, and properties, with recent attention to improved dimensional precision. The glue between these factors is in the microstructure. A homogenous green structure greatly improves the LPS response. Most effective is placement of the liquid phase on the interface between the solid grains [54-55].

6.2. Microwave Sintering (MS)

Microwave sintering (MS), as a novel technology, has been employed in powder metallurgy for the last few years [56]. It has several advantages, including volumetric heating, non-thermal effect, selective heating over the conventional method [57]. These typical characteristics are beneficial to prepare materials such as accelerated heating rate, shortened processing cycle, high energy efficiency, and being environmentally friendly [58-59]. The microstructures can be improved greatly in terms of fine grain size, uniform cobalt distribution for WC-Co alloys, which can enhance the mechanical properties [60]. It is reported that the microwave sintered sample shows almost no grain growth and the cobalt phase dissolved nearly no tungsten and the sample always showed enhanced mechanical properties and better resistance towards both erosion and corrosion [61]. It is well known that carbon content plays an important role in the morphologies and microstructures.

In conventional sintering processing, the vacuum was employed to avoid the atmosphere effect on carbon content in WC – Co alloy. Nevertheless, vacuum condition is difficult to achieve in microwave furnace because of arc discharge at higher temperature (≥ 1400 °C) [62]. Therefore, the atmosphere such as Nitrogen, N_2, Argon, Ar and Hydrogen, H_2 are widely employed in microwave sintering [63]. It can be seen that the sintering atmosphere has an important effect on the microstructures. The carbon content in cemented carbide plays an important role to the morphologies and microstructures. Moreover, the carbon activity in the microwave furnace chamber is expected to be extremely sensitive because of many factors such as moisture level, purity of the protective atmosphere, oxygen content in raw materials, etc [64].

6.3. Comparison of Microwave Sintering (MS) and Conventional Sintering

Microwave sintering of cemented carbides like WC – Co also has been investigated since 1991 after pioneering work of Cheng [65-67] and thereafter by Porada [68-70]. Breval [67] et al. investigated on the microwave sintering of 0.1 – 1μm sized WC particles with cobalt as binder and compared the results with conventional sintering of the same powders. They reported that the microwave sintered sample hardly exhibits any growth and the cobalt phase does not reveal any dissolution of tungsten whereas in the conventionally sintered one. Porada and here group showed that the microwave reaction sintering of W, C, and Co powders yielded sintered WC-6Co compacts with fine and uniform microstructure (with an average grain size of 0.6μm) which exhibited a 10% increase in hardness values in comparison to tools made by conventional route [68-70].

As is well known, microwave sintering greatly reduces the time as well as the energy expended when compared to conventional sintering due to the inherent different in the heating mechanism of the two methods and the possibility of achieving very fast heating rates in the former. A typical comparison of the sintering schedules as depicted in Figure 3 shows that there is considerable reduction in the sintering time when microwaves are employed for heating the sample; 165 min for microwave sintering as compared to 330 min for conventional sintering. Figure 4 shows the photographs of the compacts before and after microwave sintering.

Initially, the microwave sintering in case of microcrystalline powder was carried out at two different temperatures, 1450 and 1550 °C, to find out at what temperature, a higher density could be achieved. It was found that at only a temperature set point of 1550 °C, a considerably higher density was obtained, though the sample densified well at 1450 °C when sintering was carried out conventionally. In case of nanocrystalline powder, different temperatures from 1100 °C up to 1550 °C were employed to verify if a well sintered sample could be obtained at a much lower temperature than samples sintered conventionally. However, as seen from Figure 5, microwave sintering of nanocrystalline powders also did not yield samples of higher density unless the temperature set point was 1550 °C, and this was true whether the sample was a compact made from microcrystalline powders or nanocrystalline powders. This is due to the expectation to observe a microwave effect, due to which sintering could be expected to take place at a lower temperature and which could be more pronounced especially in case of nanocrystalline powders [71].

IOP Conf. Series: Materials Science and Engineering **165** (2017) 012017 doi:10.1088/1757-899X/165/1/012017

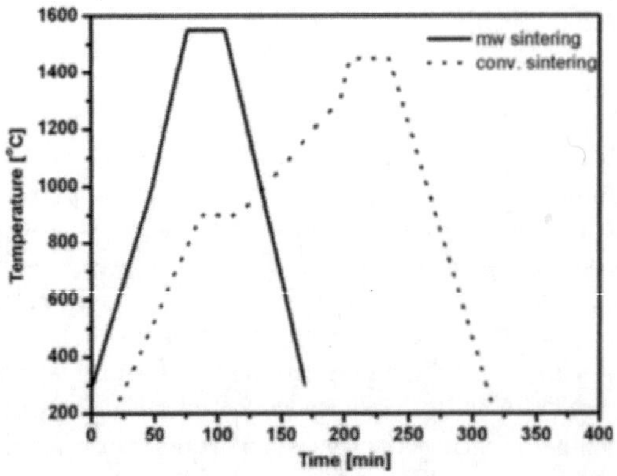

Figure 3: Comparison of sintering schedules employed for conventional and microwave processing [71]

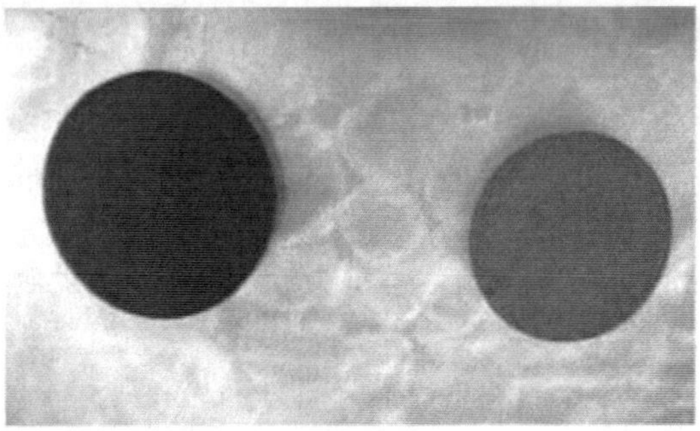

Figure 4: Photographs of the WC–12Co compacts derived from microcrystalline powders after microwave sintering in a 90N2–10H2 atmosphere, before (left hand side) and after (right hand side) sintering [71].

In all the microwave sintered samples, the periphery was seen to be less sintered when compared to the center of the sample. This aspect in the first instance was observed as a porosity difference between the periphery and center. Microwave processing is well-known to cause a volumetric heating where the heating takes place uniformly within the sample since the process is an energy conversion (microwave to heat energy) unlike a conventional heating, where heating takes place by energy transfer from the heating elements to the sample [72].

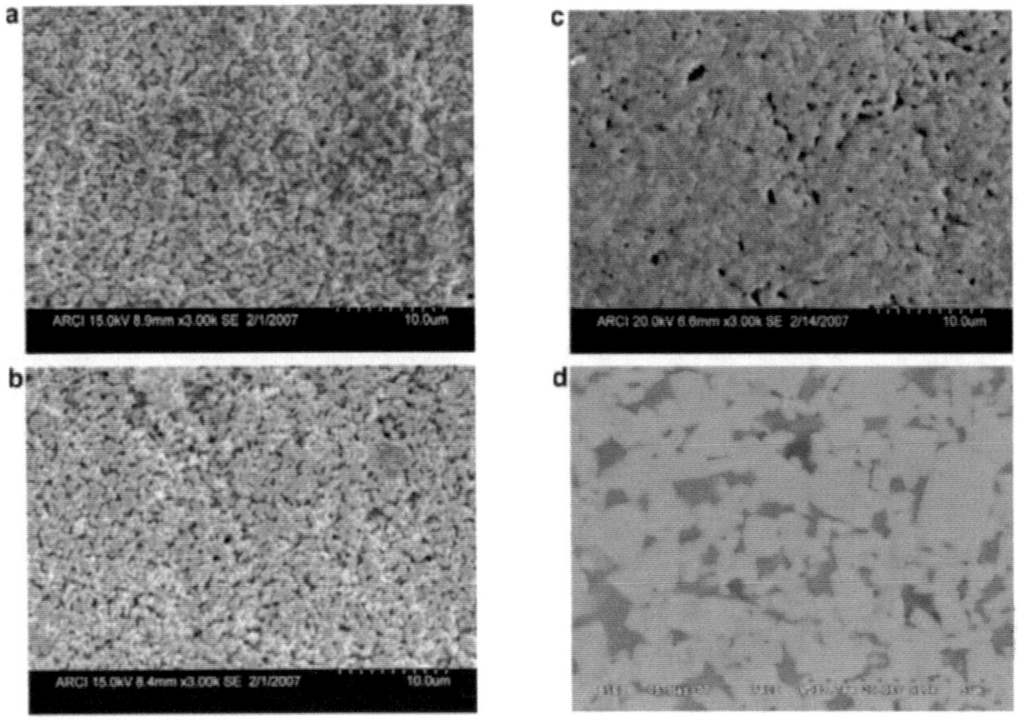

Figure 5: SEM image of microwave sintered nanocrystalline powder processed at different temperature set points; (a) 1250 °C. (b) 1350 °C (c) 1450 °C and (d) 1550 °C [72]

7. Conclusion

Based on this study it can be concluded that, metal injection molding (MIM) is newly developed technology to form metals and alloys into desired shape. Since this MIM technology is growing rapidly in Malaysia, hence it will give a good prospect for Malaysian Industries to get involved in this technology and share the benefits. Besides that, this technology helps through characterization of metal powders and binder components. The addition of grain growth inhibitor (GGI) to the starting powder also been reviewed and the role of GGI is to increase the value of hardness, fracture toughness and creep resistance at certain temperature. The use of microwave for sintering of green compacts during powder metallurgy process is emerging as a novel and innovative technology with many advantages over conventional sintering. The comparison of microwave sintering over conventional sintering has been discussed.

8. Acknowledgement

Deepest appreciation goes to Fundamental Research Grant Scheme, FRGS (UTHM) 1590, and Ministry of Higher Education Malaysia that sponsors this project. Special thanks also to University Tun Hussein Onn Malaysia for funding master research.

References

[1] Abolhasani H, Muhamad N. A new starch-based binder for metal injection molding. Journal of Materials Processing Technology. 2010 Apr 1; 210(6):961-8.

[2] German RM, Bose A. Injection Molding of Metals and Ceramics Metal Powder Industries Federation. Princeton, NJ. 1997.

[3] Prakash L. A review of the properties of tungsten carbide hardmetals with alternative binder systems. In13 th International Plansee Seminar'93. 1993 May (Vol. 2, pp. 80-109).

[4] Gille G, Bredthauer J, Gries B, Mende B, Heinrich W. Advanced and new grades of WC and binder powder–their properties and application. International Journal of Refractory Metals and Hard Materials. 2000 Mar 1;18(2):87-102.

[5] Norgren S, García J, Blomqvist A, Yin L. Trends in the P/M hard metal industry. International Journal of Refractory Metals and Hard Materials. 2015 Jan 31;48:31-45.

[6] Aristizabal M, Ardila LC, Veiga F, Arizmendi M, Fernandez J, Sánchez JM. Comparison of the friction and wear behaviour of WC–Ni–Co–Cr and WC–Co hardmetals in contact with steel at high temperatures. Wear. 2012 Mar 20;280:15-21.

[7] Correa EO, Santos JN, Klein AN. Microstructure and mechanical properties of WC Ni–Si based cemented carbides developed by powder metallurgy. International Journal of Refractory Metals and Hard Materials. 2010 Sep 30;28(5):572-5.

[8] Rhee BO, Jung YC, Lee JH. The rheological characterization of PIM feedstocks at low shear rates. Powder Injection Molding Technologies. 1998;23:79-91.

[9] Subuki I. Injection moulding of 316l stainless steel powder using palm stearin based binder system (Doctoral dissertation, Ph. D. Thesis. Universiti Teknologi Mara, Malaysia).

[10] S.Y. M.Amin., N. Muhamad, & K.R.Jamaludin, 5th Powder Metallurgy Symposium and Exhibition, Kuala Lumpur (2011).

[11] Amin SY, Muhamad N, Jamaludin KR, Fayyaza A, Yunn HS. Ball milling of WC-Co powder as injection molding feedstock. InApplied Mechanics and Materials 2012 (Vol. 110, pp. 1425-1430). Trans Tech Publications.

[12] Amin SY, Muhamad N, Jamaludin KR, Fayyaz A, Yunn HS. Characterization of the feedstock properties of metal injection-molded WC-Co with palm stearin binder system. Sains Malaysiana. 2014 Jan 1;43(1):123-8.

[13] Fayyaz A, Muhamad N, Sulong AB, Yunn HS, Amin SY, Rajabi J. Micro-powder injection molding of cemented tungsten carbide: feedstock preparation and properties. Ceramics International. 2015 Apr 30;41(3):3605-12.

[14] Liu ZY, Loh NH, Tor SB, Khor KA, Murakoshi Y, Maeda R. Binder system for micropowder injection molding. Materials Letters. 2001 Mar 31;48(1):31-8.

[15] Loebbecke B, Knitter R, Haußelt J. Rheological properties of alumina feedstocks for the low-pressure injection moulding process. Journal of the European Ceramic Society. 2009 Jun 30;29(9):1595-602.

[16] Loh NH, Tor SB, Tay BY, Murakoshi Y, Maeda R. Fabrication of micro gear by micro powder injection molding. Microsystem Technologies. 2008 Jan 1;14(1):43-50.

[17] Fabijanić TA, Alar Ž, Ćorić D. Influence of consolidation process and sintering temperature on microstructure and mechanical properties of near nano-and nano-structured WC-Co cemented carbides. International Journal of Refractory Metals and Hard Materials. 2016 Jan 31;54:82-9.

[18] Fabijanić TA, Alar Ž, Pötschke J. Potentials of nanostructured WC–Co hardmetal as reference material for Vickers hardness. International Journal of Refractory Metals and Hard Materials. 2015 May 31;50:126-32.

[19] Al-Aqeeli N, Saheb N, Laoui T, Mohammad K. The synthesis of nanostructured wc-based hardmetals using mechanical alloying and their direct consolidation. Journal of Nanomaterials. 2014 Jan 1;2014:18.

[20] Jia K, Fischer TE, Gallois B. Microstructure, hardness and toughness of nanostructured and

conventional WC-Co composites. Nanostructured Materials. 1998 Jul 31;10(5):875-91.

[21] Fang ZZ, Wang X, Ryu T, Hwang KS, Sohn HY. Synthesis, sintering, and mechanical properties of nanocrystalline cemented tungsten carbide–a review. International Journal of Refractory Metals and Hard Materials. 2009 Mar 31;27(2):288-99.

[22] Richter V, Poetschke J, Holke R, Michaelis A. Nanoscaled Hardmetals–Fiction or Reality. InProceedings of the 18th International Plansee Seminar 2013 Jun (pp. 1257-1273).

[23] Li T, Li Q, Fuh JY, Yu PC, Lu L. Two-material powder injection molding of functionally graded WC–Co components. International Journal of Refractory Metals and Hard Materials. 2009 Jan 31;27(1):95-100.

[24] Chang SH, Chen SL. Characterization and properties of sintered WC–Co and WC–Ni–Fe hard metal alloys. Journal of Alloys and Compounds. 2014 Feb 5;585:407-13.

[25] Liu C, Lin N, He Y, Wu C, Jiang Y. The effects of micron WC contents on the microstructure and mechanical properties of ultrafine WC–(micron WC–Co) cemented carbides. Journal of Alloys and Compounds. 2014 May 5;594:76-81.

[26] Lin N, He Y, Wu C, Zhang Q, Zou J, Zhao Z. Fabrication of tungsten carbide–vanadium carbide core–shell structure powders and their application as an inhibitor for the sintering of cemented carbides. Scripta Materialia. 2012 Nov 30;67(10):826-9.

[27] Zhao S, Song X, Wei C, Zhang L, Liu X, Zhang J. Effects of WC particle size on densification and properties of spark plasma sintered WC–Co cermet. International Journal of Refractory Metals and Hard Materials. 2009 Nov 30;27(6):1014-8.

[28] Sun L, Jia C, Cao R, Lin C. Effects of Cr 3 C 2 additions on the densification, grain growth and properties of ultrafine WC–11Co composites by spark plasma sintering. International Journal of Refractory Metals and Hard Materials. 2008 Jul 31;26(4):357-61.

[29] Kim HC, Shon IJ, Yoon JK, Doh JM. Consolidation of ultra fine WC and WC–Co hard materials by pulsed current activated sintering and its mechanical properties. International Journal of Refractory Metals and Hard Materials. 2007 Jan 31;25(1):46-52.

[30] Eroglu S, Baykara T. Effects of powder mixing technique and tungsten powder size on the properties of tungsten heavy alloys. Journal of Materials Processing Technology. 2000 Jun 15;103(2):288-92.

[31] Armstrong RW, Cazacu O. Indentation fracture mechanics toughness dependence on grain size and crack size: application to alumina and WC–Co. International Journal of Refractory Metals and Hard Materials. 2006 Mar 31;24(1):129-34.

[32] Deng X, Patterson BR, Chawla KK, Koopman MC, Fang Z, Lockwood G, Griffo A. Mechanical properties of a hybrid cemented carbide composite. International Journal of Refractory Metals and Hard Materials. 2001 Nov 30;19(4):547-52.

[33] Wang X, Fang ZZ, Sohn HY. Grain growth during the early stage of sintering of nanosized WC–Co powder. International Journal of Refractory Metals and Hard Materials. 2008 May 31;26(3):232-41.

[34] Gille G, Bredthauer J, Gries B, Mende B, Heinrich W. Advanced and new grades of WC and binder powder–their properties and application. International Journal of Refractory Metals and Hard Materials. 2000 Mar 1;18(2):87-102.

[35] Zhang FL, Wang CY, Zhu M. Nanostructured WC/Co composite powder prepared by high energy ball milling. Scripta materialia. 2003 Dec 31;49(11):1123-8.

[36] Carroll DF. Sintering and microstructural development in WC/Co-based alloys made with superfine WC powder. International Journal of Refractory Metals and Hard Materials. 1999 May 31;17(1):123-32.

[37] Liu C, Lin N, He Y, Wu C, Jiang Y. The effects of micron WC contents on the microstructure and mechanical properties of ultrafine WC–(micron WC–Co) cemented carbides. Journal of Alloys and Compounds. 2014 May 5;594:76-81.

[38] Weidow J, Norgren S, Andrén HO. Effect of V, Cr and Mn additions on the microstructure of WC–Co. International Journal of Refractory Metals and Hard Materials. 2009 Sep

30;27(5):817-22.
[39] Gille G, Szesny B, Dreyer K, Van Den Berg H, Schmidt J, Gestrich T, Leitner G. Submicron and ultrafine grained hardmetals for microdrills and metal cutting inserts. International Journal of Refractory Metals and Hard Materials. 2002 Jan 31;20(1):3-22.
[40] Rödiger K, Dreyer K, Willert-Porada M, Gerdes T, inventors; Widia GmbH, assignee. Hard metal or cermet sintered body and method for the production thereof. United States patent US 6,293,986. 2001 Sep 25.
[41] Lassner E, Schubert WD. Properties, Chemistry, Technology of the Element, Alloys, and Chemical Compounds. Vienna University of Technology, Vienna, Austria, Kluwer. 1999:124-5.
[42] Shao GQ, Duan XL, Xie JR, Yu XH, Zhang WF, Yuan RZ. Sintering of nanocrystalline WC-Co composite powder. Reviews on Advanced Materials Science. 2003 Dec;5(4):281-6.
[43] Kneringer G. 14th International Plansee Seminar'97: Late papers. Plansee AG; 1997.
[44] Cha SI, Hong SH, Ha GH, Kim BK. Microstructure and mechanical properties of nanocrystalline WC-10Co cemented carbides. Scripta Materialia. 2001 May 18;44(8):1535-9.
[45] Cha SI, Hong SH, Ha GH, Kim BK. Mechanical properties of WC–10Co cemented carbides sintered from nanocrystalline spray conversion processed powders. International Journal of Refractory Metals and Hard Materials. 2001 Nov 30;19(4):397-403.
[46] Sampath A, Stiglich JJ, Sudarshan TS, Singh R, Choi KS. Fast consolidation of WC–Co. Powder metallurgy. 2002 Feb 1;45(1):25-7.
[47] Yamamoto G, Omori M, Hashida T, Kimura H. A novel structure for carbon nanotube reinforced alumina composites with improved mechanical properties. Nanotechnology. 2008 Jun 24;19(31):315708.
[48] Groza JR, Dowding RJ. Nanoparticulate materials densification. Nanostructured materials. 1996 Oct 31;7(7):749-68.
[49] German RM, Suri P, Park SJ. Review: liquid phase sintering. Journal of Materials Science. 2009 Jan 1;44(1):1-39.
[50] Lenel FV. Sintering in the Presence of a Liquid Phase. AIME TRANS. 1948 Jun;15(4):1-9.
[51] Kingston WE. The physics of powder metallurgy. McGraw-Hill; 1951.
[52] Gurland J, Norton JT. Role of the binder phase in cemented tungsten carbide-cobalt alloys. Trans. AIME. 1952 Oct;194(10):1051-6.
[53] Parikh NM, Humenik M. Cermets: II, Wettability and microstructure studies in liquid-phase sintering. Journal of the American Ceramic Society. 1957 Sep 1;40(9):315-20.
[54] Riedel R. Handbook of ceramic hard materials. Wiley-VCH Verlag GmbH; 2000.
[55] Luyckx S, Love A. The dependence of the contiguity of WC on Co content and its independence from WC grain size in WC–Co alloys. International Journal of Refractory Metals and Hard Materials. 2006 Mar 31;24(1):75-9.
[56] Bykov YV, Rybakov KI, Semenov VE. High-temperature microwave processing of materials. Journal of Physics D: Applied Physics. 2001 Jul 7;34(13):R55.
[57] Demirskyi D, Ragulya A, Agrawal D. Initial stage sintering of binderless tungsten carbide powder under microwave radiation. Ceramics International. 2011 Mar 31;37(2):505-12.
[58] Demirskyi D, Borodianska H, Agrawal D, Ragulya A, Sakka Y, Vasylkiv O. Peculiarities of the neck growth process during initial stage of spark-plasma, microwave and conventional sintering of WC spheres. Journal of Alloys and Compounds. 2012 May 15;523:1-0.
[59] Guo Y, Yi J, Luo S, Zhou C, Chen L, Peng Y. Fabrication of W–Cu composites by microwave infiltration. Journal of Alloys and Compounds. 2010 Mar 4;492(1):L75-8.
[60] Bao R, Yi J, Zhang H, Peng Y. A research on WC–8Co preparation by microwave sintering. International Journal of Refractory Metals and Hard Materials. 2012 May 31;32:16-20.
[61] Breval E, Cheng JP, Agrawal DK, Gigl P, Dennis M, Roy R, Papworth AJ. Comparison between microwave and conventional sintering of WC/Co composites. Materials Science and

Engineering: A. 2005 Jan 25;391(1):285-95.

[62] Tompsett GA, Conner WC, Yngvesson KS. Microwave synthesis of nanoporous materials. ChemPhysChem. 2006 Feb 13;7(2):296-319.

[63] Zhou C, Yi J, Luo S, Peng Y, Li L, Chen G. Effect of heating rate on the microwave sintered W–Ni–Fe heavy alloys. Journal of Alloys and Compounds. 2009 Aug 12;482(1):L6-8.

[64] Bao R, Yi J. Effect of sintering atmosphere on microwave prepared WC–8wt.% Co cemented carbide. International Journal of Refractory Metals and Hard Materials. 2013 Nov 30;41:315-21.

[65] Cheng J. Study on microwave sintering technique of ceramics materials. PhD ThesisChina Wuhan University of Technology, Wuhan, China. 1991.

[66] Cheng JP, Agrawal DK, Komarneni S, Mathis M, Roy R. Microwave processing of WC-Co composities and ferroic titanates. Material Research Innovations. 1997 Jun 1;1(1):44-52.

[67] Breval E, Cheng JP, Agrawal DK, Gigl P, Dennis M, Roy R, Papworth AJ. Comparison between microwave and conventional sintering of WC/Co composites. Materials Science and Engineering: A. 2005 Jan 25;391(1):285-95.

[68] Gerdes T, Whxert-Porada M. Microwave sintering of metal-ceramic and ceramic-ceramic composites. InMRS Proceedings 1994 (Vol. 347, p. 531). Cambridge University Press.

[69] Ro K, Dreyer K, Gerdes T, Willert-Porada M. Microwave sintering of hardmetals. International Journal of Refractory Metals and Hard Materials. 1998 Dec 31;16(4):409-16.

[70] Kolaska H, Willert-Porada M, Rodiger K, Gerdes T, inventors; Widia GmbH, assignee. Composite and process for the production thereof. United States patent US 6,124,040. 2000 Sep 26.

[71] Sunil BR, Sivaprahasam D, Subasri R. Microwave sintering of nanocrystalline WC–12Co: Challenges and perspectives. International Journal of Refractory Metals and Hard Materials. 2010 Mar 31;28(2):180-6.

[72] Holcombe CE, Dykes NL. Importance of "casketing" for microwave sintering of materials. Journal of materials science letters. 1990 Apr 1;9(4):425-8.

IOP Conf. Series: Materials Science and Engineering **165** (2017) 012018 doi:10.1088/1757-899X/165/1/012018

Wave velocity characteristic for Kenaf natural fibre under impact damage

M Zaleha[1,*], S Mahzan [1], Muhamad Fitri[1], K A Kamarudin[1], Eliza Y[1], A L Mohd Tobi[1]

[1] Faculty of Mechanical and Manufacturing Engineering, Universiti Tun Hussein Onn Malaysia, 86400 Parit Raja, Batu Pahat, Johor, Malaysia

*Email: zaleha@uthm.edu.my

Abstract. This paper aims to determining the wave velocity characteristics for kenaf fibre reinforced composite (KFC) and it includes both experimental and simulation results. Lead zirconate titanate (PZT) sensor were proposed to be positioned to corresponding locations on the panel. In order to demonstrate the wave velocity, an impacts was introduced onto the panel. It is based on a classical sensor triangulation methodology, combines with experimental strain wave velocity analysis. Then the simulation was designed to replicate panel used in the experimental impacts test. This simulation was carried out using ABAQUS. It was shown that the wave velocity propagates faster in the finite element simulation. Although the experimental strain wave velocity and finite element simulation results do not match exactly, the shape of both waves is similar.

1. Introduction

In recent years, there has been an environmental awareness focused the attention to use the natural fibres as reinforcements for the polymeric matrix due to the environmental advantages. Global warming have initiated a considerable interest in using natural materials to produce green products and reduce carbon dioxide emissions. So far studies on the properties of natural fibres based composites have been the subject of a large number of papers and reviews, especially during the last decade [1-3]. The use of composite materials in structural components has increased significantly in recent years by offering potential benefits to the aerospace, automotive and marine industry. Despite the potential of natural fibres, a special consideration is required on impact problems of natural fibre composites since these material are brittle under dynamic loading particularly impact loading. The presence of this sort of problem can severely degrade the mechanical properties of composite structures, and if it is not detected in the incipient stage, it may result in a catastrophic failure of the structure. Therefore, numerous experiments and simulation have been developed to better understand the mechanisms and mechanics of impact damage in composite materials [4-8].

Recent investigation has demonstrated the use of sensor technologies in impact damage detection. Here, sensors are employed to monitor the impact strain data. The energy of impacts is estimated, which could provide substantial information related to impact severity. When these natural excitations are used instead of active excitations, the complexity of a deployed system decreases substantially by using passive only sensors and data systems that listening to the structure during operation. The data collected can be the processed using a number of methods.

Wildy et al. [9] investigated the the development of a new passive technique of on-line damage detection based on the most fundamental concept which is called strain compatibility. The development of this technique are implemented for a few typical situations of the development of crack damage in plates such as notch, edge crack and point force. It was found that, the techniques can determine a violation of these equations for a localized area, indicating the presence of cracks, voids and other types of damage in the vicinity of the cluster.

Nemat et al. [10], performed an experimental and numerical studied to characterize the acoustic wave propagation in thin glass/epoxy composites plates. Signal processing algorithms and a passive damage diagnosis system based on AE techniques were proposed for continuously monitoring and assessing the structural health of composite laminates. It can be concluded that the measured velocity agrees well with the theoretical calculation as well as the experimental measurements.

Staszewski et al. [11] and Mahzan et al. [12], have demonstrated the feasibility of manipulating the stress wave for passive damage detection approach. The results indicated that different materials produced different wave propagations depending on the configuration of the composites. Tippman et al. [13] demonstrated the ability of damage detection and localization using passively reconstructed impulse response function. The results from experiments conducted on an aluminium plate and wind turbine blade were validated with simulated damage results. It was found that the results show a promising method that can detect damage by monitoring the reciprocity of the impulse response functions.

Although a lot of researcher have investigated about the natural fibre and damage detection, however, very limited findings have been reported available for information and data dealing with the impact damage for natural fibre reinforced composite. Hence the aim of this paper is to investigate the behaviour of wave velocity characteristic in kenaf fibre composite during impact events. The impact events on the KFC eventually initiate the strain data, substantial for wave velocity characteristic. Then the simulation is developed and implemented into finite element (FE) package ABAQUS/Explicit. The impact process is simulated by FE and the wave velocity characteristic results are compared with the experiments.

2. Wave Velocity Characteristics

Theoretically, during an impact event, the structure is deflected and produced strain waves that propagate outwards in all possible directions. For an isotropic material, it is expected that the wave propagation is identical in all possible direction. However, for a natural fibre reinforced composite, that categorized as an anisotropic material, the wave propagation is unknown. However, the wave velocity characteristics can be estimated a priori for monitored composite structures using experimental analysis for all possible angles of wave propagation.

Figure 1 demonstrates the schematic arrangement for modified multilateration procedure, as reviewed in Staszewski et al. [11] and Mahzan et al. [12]. Three different sensors, e.g. P_1, P_2 and P_3 were used to 'sense' the strain wave resulted from an impact event. Three different angles, e.g. α_1, α_2 and α_3 have been randomly selected for wave propagation directions. For every transducer P_i and assumed wave propagation angle, the distance d_i between the transducer and impact position can be calculated as

$$d_i = v_i t_i \quad (i=1,2,3) \tag{1}$$

Where t_i and v_i are arrival times and velocities of the propagating strain waves, respectively. The arrival times can be estimated from the experimental strain data for all transducers.

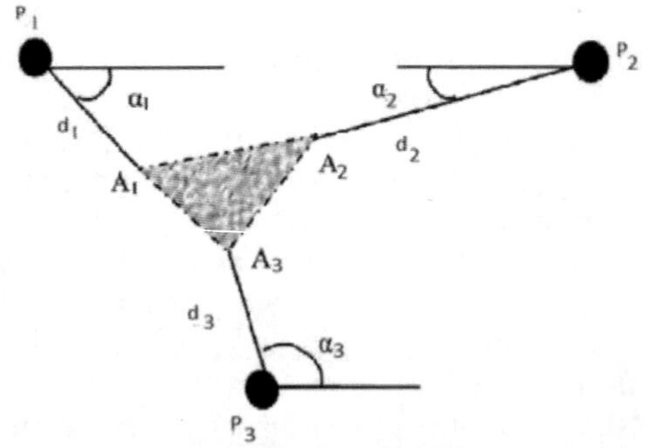

Figure 1. Schematic sensor arrangement of modified triangulation for wave propagation direction procedure [11]

3. Methodology

3.1 Sample preparation

The investigation employs the chopped kenaf fibres as the natural fibre and the epoxy as the resin matrix. The dimensions of the KFC panel were 300 mm (L) ×300 mm (W) and 3 mm thickness. The composites with fibre loading 10% of volume fraction were fabricated using compression technique. The internal surfaces of the mould were sprayed by a release agent in order to facilitate easy removal from the mould. Initially, epoxy resin and hardener were mixed together with ratio 2:1 to form a matrix. Then the chopped kenaf fibres and matrix was mixed together using a mixer for 10-20 minute to disperse fibres in the matrix. The mixture was poured into the mould and closed before manual compression took place. The sample was left to cure for about 24 hours at room temperature. Finally the panel was taken out of the mould and post-cured in the air for another 24 hours.

3.2 Experimental set up

The test performed was the experimental analysis of passive impact strain waves in the KFC panel. The objective was to obtain wave velocity characteristics based on the modified triangulation procedure, prescribed in Figure 2. An impact hammer, as used for modal testing, was applied to produce impacts on the KFC plate. The experiments were conducted on a laboratory, where the plate was positioned on foam without any mechanical constraints.

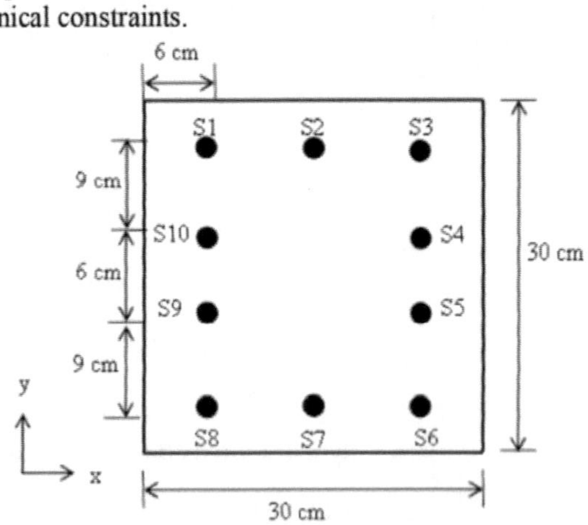

Figure 2. Schematic diagram for sensor placement on the KFC panel

4. Finite element model and analysis

The finite element model was built in ABAQUS/Explicit. The model is develop for a simple three-dimensional model of an impactor impacting a KFC panel. In this model the impactor is dropped onto the panel, as illustrated in Figure 3. The objective of this investigation is to model/replicate simulation from the experimental work. The material is modeled using certain assumptions such as the material property for all the constituents are attributed as isotropic material for both volumes. The assumptions used for this work have been gathered by the literature [14-16].

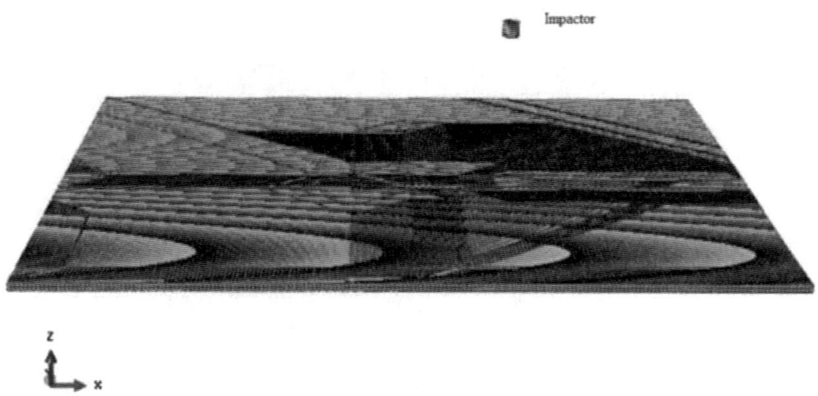

Figure 3. FE impact modelling of the KFC panel

4.1 Material and Element Selection

The impactor is a cyclindrical of 17.4 mm diameter and the target is a rectangular KFC panel of 300 mm × 300 mm × 3 mm. The material properties for KFC panel are given in Table 1. The composite panel was fabricated with kenaf fibres reinforcing an epoxy resin matrix. Due to undeformable of impactor during experimental testing, the impactor projectiles mechanical properties for finite element simulation is not determined and are considered as rigid body. All the properties of the materials for simulation were obtained from the experimental data.

Table 1. Material properties of the KFC panel

Physical Properties	KFC panel
Young Modulus (GPa)	E=2.46
Yield stress (MPa)	28.32
Poisson's ratio	0.25
Density (kg/m^3)	1400

The KFC panel was used as a deformable solid using elements C3D8R with 'reduce integration', while the projectile was modelled as an analytical rigid body using four node bilinear quadrilateral elements (R3D4). This is due to the undeformable of projectile structure during experimental testing. The KFC panel was uniformly meshed with rectangular elements.

4.2 Boundary and Loading Conditions

The square KFC panel, was simply supported at each corner, as illustrated in Figure 4. The displacements of these four corner nodes were restricted only in the z-direction, in order to create a similar condition to a free boundary condition. The projectile nose tip was assigned with a reference point (RP) moving only in the z-direction and assumed to have no rotation during impact. The initial position of the projectile was such that the projectile would only travel to impact the KFC panel. Overall the boundary conditions are also kept the same throughout all the experiments. The impact of a projectile with a mass of 2.9g on a target was then simulated. The initial impact velocity was chosen from the velocity value of which is close to the experiments work.

The thickness of the elements was equal to the thickness of the panel 3 mm. An adaptive mesh which is consists the most refined mesh was applied to the models in the impact areas. Using adaptive mesh will help to avoid error termination of the program due to excessive distortion occurred within the elements.

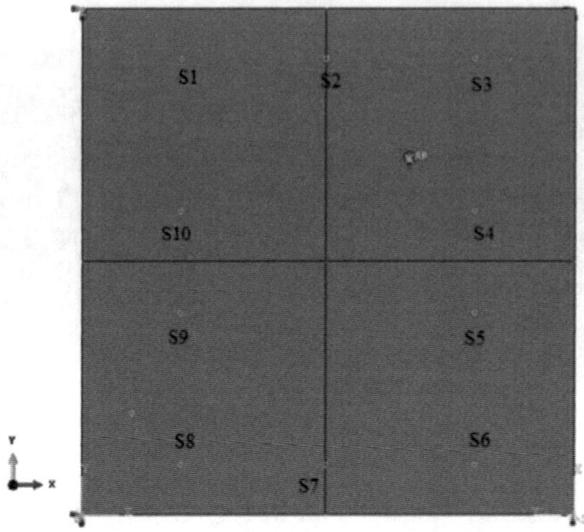

Figure 4. Boundary condition

5. Results and Discussion

A low-energy impacts were performed at the position of the panel (X=200 mm, Y=210.5). The amplitude and its arrival time, *ta,* were analysed for each impact events. All the related data such as strain data, energy, velocity, displacement were recorded each time for further analysis. Figure 5 depicts an example of the strain wave recorded for sensor 1.

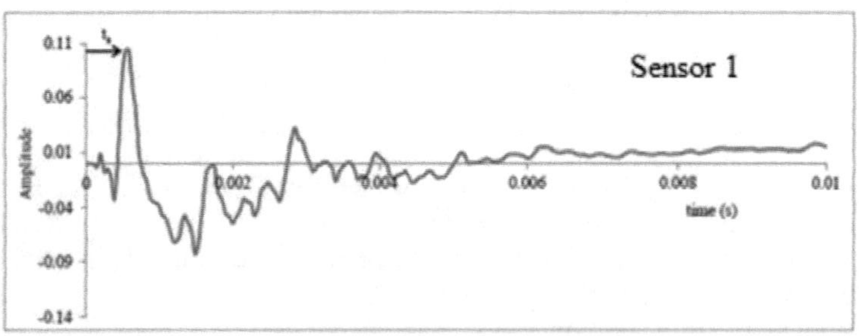

Figure 5. An example of a strain wave measurement obtained from FE model

Figure 6 shows the arrival time as a function of the distance between impact position and sensor location. The arrival time increase linearly with the impact position. The time of arrival is almost found similar for all the sensors but the maximum amplitudes are different. The near the sensor with the impact location the fastest arrival time and the higher amplitude. This is due to the fact that the energy of a strain wave decays as it propagates away from the source of impact. So it will take a longer time for the far impact position.

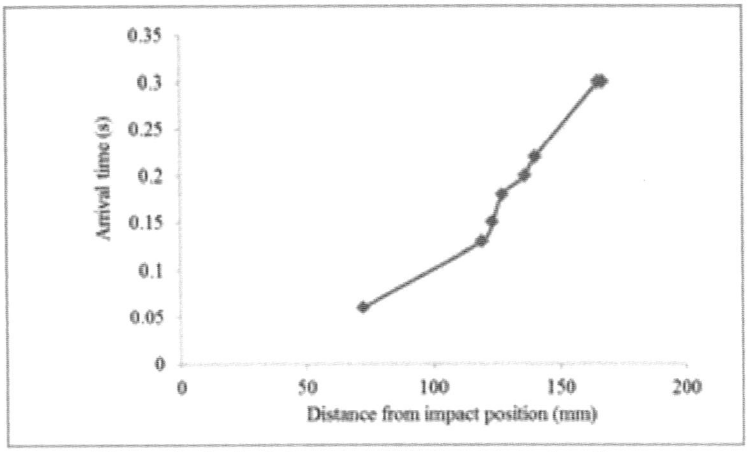

Figure 6. Arrival time versus wave propagation distance

For known distances between impact point and sensor positions, the wave velocity can be calculated, provided the arrival time was estimated. The time of arrival was estimated at the time which first maximum amplitude was obtained. The strain wave velocities were also analysed for various angles of wave propagation in FE. The results, presented in Figure 7, demonstrate that the wave velocity changes linearly from approximately 0.3 m/s for 0° to 45° and decreased nonlinearly to the value 0.04 for 180°.

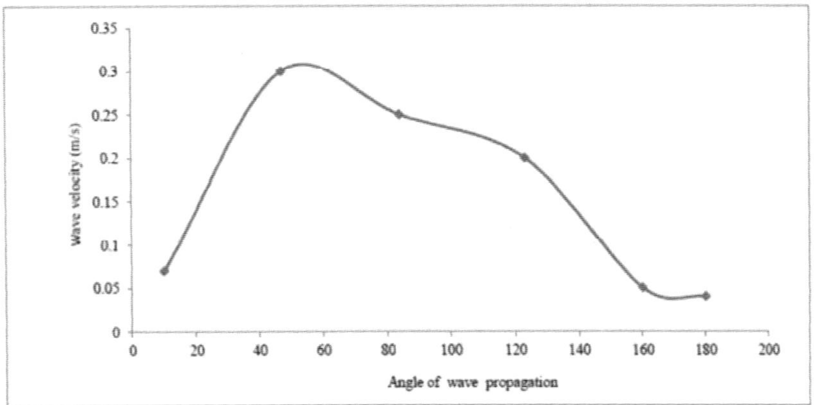

Figure 7. Wave velocity versus angle of wave propagation

Then, the amplitude normalisation was performed to match FE displacement amplitudes with experimental voltage levels of the analysed strain waves. A zero-mean type of normalisation was used in this measurement. By using this type of normalisation, the mean of the transformed set of data points is reduced to zero. The calculation of data normalisation is as follow

$$\hat{x}_i = \frac{x_i - \mu_i}{\sigma_i} \tag{2}$$

Where $\hat{x}_i$ is the normalised data, x_i is the i-th component of the original data, μ_i and σ_i are the mean and standard deviation of the original data respectively.

Figure 8 show the wave velocity characteristics comparison between experimental and FE for KFC. Although the experimental strain waves and FE modeling results do not match exactly, the shape of both waves is similar. The FE strain waves propagate faster than the experimentally measured strain waves. This is acceptable for this research work due to the boundary conditions applied to the KFC panel in the FE model could be different from those actually occurring in the experimental case. The difference between the experimental and simulation wave velocity characteristics was calculated as Mean Square Error (MSE) function. The MSE was defined as

$$MSE(u) = \frac{100}{N\sigma_\upsilon^2}\sum_{i=1}^{N}(\upsilon_i - \hat{\upsilon}_i)^2 \tag{3}$$

Where υ_i is estimated velocity value from experimental, $\hat{\upsilon}_i$ is the simulated velocity value, σ_υ^2 is the standard deviation of the estimated velocity data from experimental and N is the number of experimental velocity estimated. The MSE value for the result is 10.17 %.

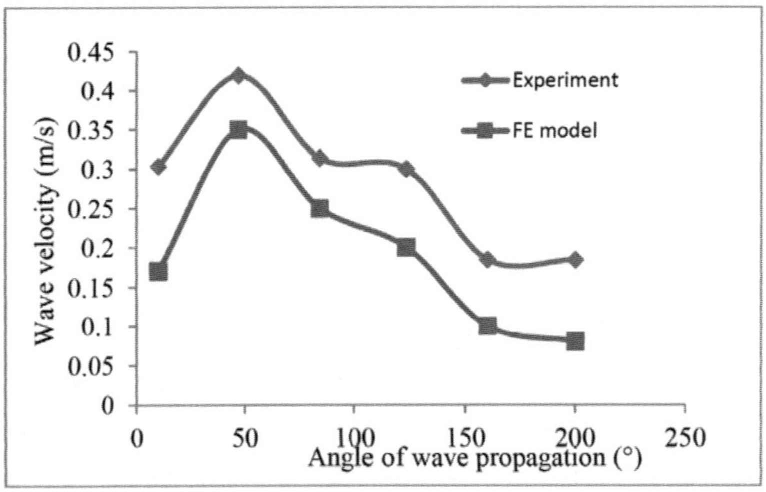

Figure 8. Strain wave velocity characteristics – comparison between the FE and experimental results

While Figure 9 (a) depicts the example of damage image from experimental work and Figure 9 (b) shows FE simulation for impact force 600N. The pattern of the damage almost same. Since the size of damage cannot be evaluated in the FE, so the comparison is just through the pattern of the damage only.

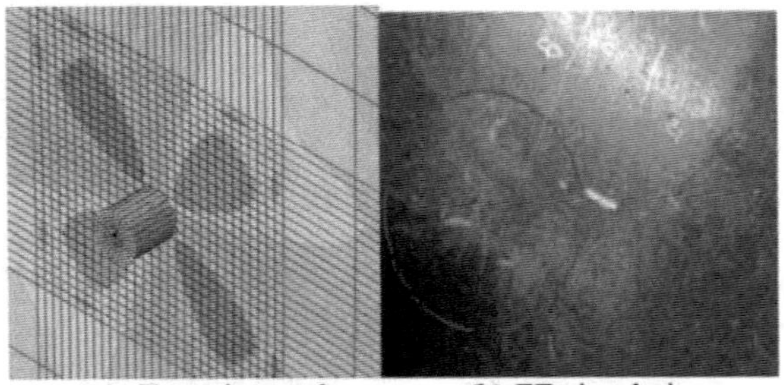

(a) Experimental (b) FE simulation

Figure 9. Example of damage image obtained from experimental and FE simulation

CONCLUSIONS

The wave velocity characteristics for a KFC has been developed using experimental impact strain data and ABAQUS simulated data. The pattern of the wave propagation of experimental and simulation agreed well with the different less than 11%. Although a classical sensor triangulation method offer a good solution but different natural fibre may produce different wave propagation depending on the configuration of the fibres. Therefore, further investigation is needed such as using different type of fibre and different type of fibre orientation in order to establish these methods. To date, only an experimental study of damage severity for chopped fibre has currently been conducted.

Acknowledgements

The author gratefully acknowledges the financial support provided by Universiti Tun Hussein Onn Malaysia (UTHM), Johor, Malaysia

References

[1] Karnani R, Krishnan. M and Ramani N 1997 Biofiber-Reinforced Polypropylene Composites. *Polymer Engineering and Science,* **37(2)** 475

[2] Faruk. O, Andrzej K.B,Fink H.P and Sain M 2012 Biocomposites reinforced with natural fibers: 2000–2010. *Progress in Polymer Science.* **37** 1552

[3] Saba N, Jawaid M, Hakeem K R , Paridah M T , Khalina A and Alothman O Y 2015 Potential of bioenergy production from industrial kenaf (Hibiscus cannabinus L.) based on Malaysian perspective, *Renewable and Sustainable Energy Reviews*. **42** 446

[4] Chang F H, Choi Y, Stanford L and Jeng S T 1990 Study on Impact Damage in Laminated Composites, *Mechanics of Materials.* **10** 83

[5] Sung D, Kim C and C. Hong 2002 Monitoring of Impact Damages in Composite Laminates using Wavelet Transform, *Science.* **33** 35

[6] Kim I G 2005 Impact Damage Detection in Composite Laminates Using PVDF and PZT Sensor Signals, *Journal of Intelligent Material Systems and Structures.* **16** 1007

[7] Aktas M, Atas C, Icten B and Karakuzu R 2009 An Experimental Investigation of the Impact Response of Composite Laminates, *Composite Structures*. **87** 307

[8] Aymerich F and Staszewski W J 2010 Impact Damage Detection in Composite Laminates using Nonlinear Acoustics, *Composites Part A: Applied Science and Manufacturing,* **41** 1084

[9] Wildy S J, Kotousov A G and Codrington J D 2008 A new passive defect detection technique based on the principle of strain compatibility, *Smart Mater. Struct.***17** 1

[10] Nemat N S, Huang Y, Fabrizia G and Trick P R 2008 2008 Passive Damage Detection in Composite Laminates with Integrated Sensing Networks *Health Monitoring of Structural and Biological Systems* Proc. of SPIE **6935** 0277

[11] Staszewski, W J, Mahzan S and Traynor R. 2009 Health monitoring of aerospace composite structures – Active and passive approach. *Composites Science and Technology*, **69** 1678

[12] Mahzan S, Staszewski W J and Worden K 2010 Experimental studies on impact damage location in composite aerospace structures using genetic algorithms and neural networks *Smart Structures and Systems*, *6*(**2**) 147

[13] Tippmann J D, X Zhu and Lanza F di Scalea. 2015 Application of damage detection methods using passive reconstruction of impulse response functions. *Philosophical Transactions*,

[14] Kebir H and Ayad R. 2014 A specific finite element procedure for the analysis of elastic behaviour of short fibre reinforced composites. The projected fibre approach. *Composite Structures*, **118** 580

[15] Bayat M. and Aghdam M 2012 A micromechanics based analysis of hollow fiber composites using dqem. *Composites Part B: Engineering*, **43(8)** 2921

[16] Behzad T and Sain M. 2007 Finite element modeling of polymer curing in natural fiber reinforced composites. *Composites Science and Technology*, **67(8)** 1666

Toenail as Non-invasive Biomarker in Metal Toxicity Measurement of Welding Fumes Exposure – A Review

S F Z Bakri[1, 2], A Hariri[1], N F Ma'arop[2] and N S A W Hussin[2]

[1]Universiti Tun Hussein Onn Malaysia, Johor, Malaysia

[2]DRB-HICOM University of Automotive Malaysia, Pahang, Malaysia

E-mail: farhana@dhu.edu.my

Abstract. Workers are exposed to a variety of heavy metal pollutants that are released into the environment as a consequence of workplace activities. This chemical pollutants are incorporated into the human by varies of routes entry and can then be stored and distributed in different tissues, consequently have a potential to lead an adverse health effects and/or diseases. As to minimize the impact, a control measures should be taken to avoid these effects and human biological marker is a very effective tool in the assessment of occupational exposure and potential related risk as the results is normally accurate and reproducible. Toenail is the ideal matrix for most common heavy metals due to its reliability and practicality compared to other biological samples as well as it is a non-invasive and this appears as a huge advantage of toenail as a biomarker. This paper reviews studies that measure the heavy metals concentration in toenail as non-invasive matrix which later may adapt in the investigation of metal fume emitted from welding process. The development of new methodology and modern analytical techniques has allowed the use of toenail as non-invasive approach. The presence of a heavy metal in this matrix reflects an exposure but the correlations between heavy metal levels in the toenail must be established to ensure that these levels are related to the total body burden. These findings suggest that further studies on interactions of these heavy metals in metal fumes utilizing toenail biomarker endpoints are highly warranted especially among welders.

1. Introduction

Biological marker or biomarker is widely known as a one of the alternative method used to measure and observed on the outcome of the disease and medical state as the results is accurate and reproducible [1]. Mayeux, (2004) had categorised biomarker into two major groups: biomarker of exposure which are used in risk prediction, and biomarker of disease, used to screen, diagnose and monitor disease progression [2]. The markers of human tissue and excreta can be obtained by invasive and non-invasive methods where non-invasive is much preferred due to greater acceptability [3]. The applications of biomarker in assessing welder's health have been conducted in a few researches especially in investigating the impact of workplace setting to health. In recent years, a few authors have begun to measure heavy metals concentration emits from welding fumes by using a biomarker. However, the scope of biomarker used is limited to a specific method. Previous study has reported that the exposure to welding fumes especially manganese has significantly higher than controls and led to altered status of serum ferritin and transferrin receptor, TfR [4]. The findings had shown that serum ferritin and TfR have a similar adequate sensitivity and were associated with age-related changes [5]. However, the results is biased as serum ferritin is an acute-phase reactant and this lead to gender differences where the results normally lower in women and this make ferritin less ideal [6], [7]. Furthermore, the issues of timing, persistence, dose and storage site need to be considered for this banked serum [2]. From the current research, Baker et al., (2015) found out that manganese influences the magnetic resonance properties in surrounding tissues since manganese is part of a paramagnetic element [8]. Despite the fact that MRI is highly sophisticated and may presented high accuracy data, but this approach also require a proficient and knowledgeable operator to operate machine, time consuming and require a special software prior to automated analysis [9].

Data from several sources have identified the used of urine as biomarker in tracing toxic metals [10]–[12]. Along the same line, there is a research revealed that the metal concentration exposed from metal fumes appeared positively in urine samples among welders with higher readings of manganese and chromium in female welders [13]. This view is supported by Golbabaei et al., (2012) who writes that the concentration of metals in welder's urine is greater for chromium, cadmium and nickel respectively [14]. The proteins and polypeptides are quite stable in urine samples anyhow exosomes in urine represent a less stability due to changes of the temperature on sample storage [15], [16].

An assessment of occupational exposure and potential related risk then had been widely explored including implementation of blood as biomarker. In 1999, Bader et al. published a paper in which they described that blood is better than urine and/or hair in identifying manganese since blood is a highly complex tissue [17],[18]. In contrast to Bader, a number of studies argue that blood is not the recommended biomarker due to the different toxicokinetic profile where bloods represent recent exposures [19], [20]. Likewise, from the experiment of assessing metal fumes exposure among welders by using blood, urine and toenail; the researchers holds a view that toenail is preferred as a biomarker for long-term exposure assessment than blood and urine [21], [22].

Studies of biomarkers show the importance of invasive and non-invasive approach for a particular clinical study [23]. This approach offers some important insights into the observation of end points in a continuum of effects prominent from exposure of environmental agents to diseases; due to sensitivity, it is able to identify risk factors for the disease outcome and provide vast value to an environmental carcinogen; and non-invasive methods are fast becoming a key instrument in assessing metal toxicity compared to invasive methods because of greater acceptability [3]. This approaches provides immense mean values in identify risk factors for a disease due to its sensitivity and particularly beneficial in measuring the progressive diseases based on manifest symptoms due to prolong exposure to welding fumes. Therefore, the ability of biomarkers to detect medical state and heavy metals constituents in welders is promising. Biomarkers used to assist in extrapolation of available data to obtain the distinct data. Biomarkers may be classified into five categories: molecular lesions, metabolized exogenous agents, endogenously produced biomolecules, cellular/tissue changes and unchanged exogenous agents [3].

Since non-invasive is preferable than invasive method, thus, the study of metals due to welding fumes exposure by using biomarkers had undergo an extensive research. In view of all that has been mentioned so far, these studies provide important insight into the exploration of toenail as an assessment tools in evaluation of long-term occupational health impact due to metal fumes exposure towards welder.

2. Objective

This paper attempts to review on crucial sources of information on application of toenail as biomarker to identify heavy metal concentration in which later may adapt in the investigation of metal fume emitted from welding process. The reliability, validity, collection and sampling of metals in toenail due to welding activities were also assessed in this study. The purpose of this paper is to review research on toenail used as a biomarker to be incorporated in welding-metal fume studies. An overall brush up of welding fume biomarkers is crucially needed to give a clear picture and direction of future innovative and exploration of welding fume studies with potential health risks among welders.

3. Methods

This review of the world literature on metal concentration and toenail as biomarkers due to exposure to welding included a search of the online electronic databases published in English language in PubMed, Science Direct, and Scopus. The relevant bibliographies in identified paper were also taken into considerations and these efforts were not limited to put into use of Google and reference manager programme as a further confirmatory search tools. The literatures searched in the reference manager were published from 2010 to 2016 and these were focused on toenail as biomarker of heavy metal only. Each database was searched through April to June 2016. A general key words used to make the full review more focused in the search are: toenail, nail, heavy metal, welding fume, metal fume, exposure to metal, health risk, welder, metal toxicity and bio-monitoring. In obtaining a specific focused on reliability and validity of toenail, the search have been made through search engine of Science Direct, PubMed and Scopus by using term "toenail AND heavy metal AND air" within year 2010 to 2016 with the search details are ("nails"[MeSH Terms] OR "nails"[All Fields] OR "toenail"[All Fields]) AND ("metals, heavy"[MeSH Terms] OR ("metals"[All Fields] AND "heavy"[All Fields]) OR "heavy metals"[All Fields] OR ("heavy"[All Fields] AND "metal"[All Fields]) OR "heavy metal"[All Fields]) AND ("air"[MeSH Terms] OR "air"[All Fields]) AND ("2010/01/01"[PubDate] : "2016/12/31"[PubDate]). Science Direct had listed 20 articles journal and only 3 articles were significant with the research scope. Whilst, PubMed had listed 180 articles with only 6 articles were found relevant where one of the articles is the same articles as listed in the Scopus search engine.

4. Results

4.1. Toenail: Biomonitoring Human Exposure to Metals

There is a growing interest in the health assessment of heavy metal exposure as these elements may initiate an adverse health effect and have a potential to induce multiple organ damage, even at a lower level of exposure since it is considered as systemic toxicants [24]. Air, surface water and soil are the medium of heavy metals emissions to the environment and the existence is derived from a wide range of processes and pathways [25]. The toxicity of this heavy metals is depends on several factors including dose, route of exposure, and chemical species, as well as the age, gender, genetics, and nutritional status of exposed individuals [26]. The evaluation of metal exposure and prediction of human health risks is important and needed for decision maker as this have potential to minimize exposure risks [27]. Apart of metal exposure along any route of entry, it has long been recognized that, in occupational settings, a significant exposure to metal either in form of dust or fume can occur among welder through inhalation and may lead to adverse health effects.

Therefore, in assessing the subsequent biological effect and risk of metal toxicity in health of workers, the biomarkers is one of the best approaches to provide important information [3]. The variety of biomarkers for heavy metals is not focusing into one method only but the availability is largely different. Blood, hair, nails, and bone or urine specimens were used previously to assess heavy metals [28]. Heavy metal in welding fumes studies which have incorporated biomarkers to examine the heavy metal exposure via inhalation of welding metal fumes were carried out in Malaysia, Germany, France, Denmark, Saudi Arabia, and the United States [29]–[38]. In the literature on biomarker of heavy metals, the relative importance of toenail has been subject to considerable decision. This is by reason of toenail samples are non-invasive yet convenient to collect and store, grow more slowly compared to hair, not susceptible from external contaminants, and represent longer-term exposures than blood or urine [39]. This research has been supported by a previous study on the use of toenail as an attractive diagnostic tool in assessing heavy metals as it is an economical method, protected from infections and contaminations [40]. In investigating a concentration of heavy metals among workers, toenail is used to provide a better insight into biological marker of welding fumes. It is belief that a diversified heavy metals constitute in fine particulate matter were arise from welding fumes and the level of these metals in toenail is corresponded with concurrent environmental exposure [41]. Therefore, all research adopted toenail as a specimen used as metal biomarker were included in this reviewed paper. A non-invasive approach of toenail concentrations is preferred and suggested as a useful biomarkers because it provide a time-integrated measure of intake in body and represent a long exposure time frame due to relatively slow growth rate [28], [42], [43]. A finding on the presences of heavy metals in the toenail from a different medium of exposure in a various sample subjects worldwide has been presented in Table 1. The results shows that toenail have a capacity to be a reliable biomarker for heavy metal tracing and this would suggest that it may be a promise tools in determine the existence of metal fumes in welders.

Table 1. Application of toenail as biomonitoring tools in assessing heavy metal concentration.

Study Area	Medium of Exposure	Subject	Heavy Metals	Findings	References
South Korea	Food: consumption of fish and seafood in dietary intakes.	Middle-aged adults	Hg	Higher toenail levels of mercury were associated with a higher risk of dysregulation of lipids.	[44]
Pakistan	Air and food: Dust exposure, drinking water and food.	Adults and children	Cr, Mn, Co, Ni, Cu, Zn, Cd, Pb.	The bioaccumulation ratios (BAF) shows that nail is an effective tools to fingerprint long term trace metals exposure.	[45]
Cambodia	Water: drinking water from groundwater.	Residential community.	Mo, Co, Pb, Ni, Cr, As, Cu, Ba, Mn, Zn, Fe.	The levels of metals accumulated in toenail were similar for hair and corresponded with the levels of metals in the groundwater.	[46]
Zambia	Air: inhalation and exposure at mining industry.	Exposed and non-exposed community adults.	Cd, Co, Cu, Pb, Ni, Se, Zn.	Metal concentrations were found higher in toenail.	[47]
Qatar	Air, ingestion of food and water contamination.	Immigrant farm workers	As, Ba, Cd, Cu, Mn, Mo, Pb, Se, U	Toenail likewise accumulates both toxic and essential elements in farm workers but there is no association of metals was found between toenail and urine.	[48]
United States	Air: metal fumes exposure	Welder	Pb, Mn, Cd, Ni, As	All toenail metal concentrations were significantly correlated with one another.	[29]
Iraq	Severe contamination of water, soil, and air due to bombardment.	Adults and children (Fallujah families).	Al, V, Cr, Mn, Fe, Co, Ni, Cu, Zn, As, Se, Mo, Cd, Pb, Th, U	The metal concentration level has been represented in toenails but hair contained significant higher amount than toenails.	[49]
United States	Air: metal fumes exposure	Welder	Mn	Manganese in the toenail appeared to have well correlated with 7 – 12 months exposure than blood or urine.	[22]

United Kingdom	Food: ingestion of contaminated soil and dust.	Residential community of the former mine site.	As	Total As concentrations in toenails from the exposed group were significantly elevated and lead to chronic exposure to high environmental contamination.	[50]
Kenya	Air: industrial and vehicle air pollution sources.	School age children	Pb, Cd, Zn, Fe, Ca	The levels of heavy metal in the nail were significantly influenced by both environmental and nutritional factors.	[40]
Northern Japan	Food: fish	Adults (women)	Hg	Toenail was significantly correlated with the estimated daily mercury intake.	[51]
Italy	Food: diet and other sources	Hospitalize male patients	Cd	Suggest that nails and toenails are a biomarker of environmental cadmium exposure.	[52]
Finland, Germany, Israel, Netherland, Norway, Russia, UK, Switzerland, Spain	Air: inhalation of atmospheric soil dust.	Adults (men)	Ce	A metal concentration in toenail level was associated with an increased risk of acute myocardial infarction (AMI).	[53]

4.2. Reliability and Validity of Toenails as Biomarker of Metal Fumes

Park et al., (2012) suggested that nail has a potential to act as a biomarker and might be a favourable tools to monitor the xenobiotic exposure and adverse health effects to human's anatomy [54]. Similar to welding fumes, tobacco smoke contained toxic materials and considerable amount of heavy metals such as cadmium, chromium, lead, nickel, aluminium, arsenic, copper, manganese, selenium and zinc [55]–[57]. Along these lines, based on research of tobacco smoke, it was decided that toenail is the best non-invasive biomarker to adopt for heavy metal investigation due to few factors: rapid growth, less exposure to external contamination, adequate sample availability and fusion of elements in the tissue [58], [59].

Finding from Sanders, Miller, Nguyen, Kotch, & Fry (2014) shows a positive results that seem to be consistent with other research which revealed a strong significant relation via Spearman's correlation of metal in toenail notably manganese, chromium, lead, cadmium and arsenic among Vietnamese children's lived in smelting craft village [60]. Although hair also is classified as non-invasive biomarkers alike toenails, by contrast, the exposure towards external contamination such as shampoo, dust, bleaching, dyeing and permanent waving may influenced the results and therefore, this is the main limitation of this biomarkers [23], [61]–[65]. Since, there is a limited study on the research of heavy metal concentration outlined in welders toenail, therefore, this review is focused on the correlation of heavy metal found by the utilization of toenail as non-invasive biomarker. The inter-comparison or the consideration of similar researches with regards to the heavy metal concentration traced in toenails via inhalation are discussed and has been presented in Table 2 and is intended to provide a clarification regarding the ability and capability of toenails as an effective tool to analyse heavy metal content in the human body. Indirectly, this comparison can reinforce a research and serve as a benchmark for measuring the use of toenails as an effective tool in the study and details the content of heavy metal in humans.

Apart of contaminated food and water, dust exposure may contribute to the burden of heavy metals in human body and approximately 50% of trace metal that exceed Agency for Toxic Substances and Disease Registry (ATSDR) is found in the Punjab's industrial areas, thus confirming on the health threat to human [45]. The studies presented thus far provide evidence that the accumulation of heavy metals found in the toenail samples is higher compared to other biomarkers and this are associated with industrial sites that had an elevated level of metals in dust which may be affected by demographic characteristics, sources contamination and a difference of exposure routes [66]. In a research of toenail concentration in mining industry, Ndilila et al., (2014) had mentioned that the metal concentration of cadmium, cobalt, copper, lead, nickel, selenium and zinc were greater where the toenail copper concentration marked 85% of the validity; and admitting the small sample size and one form of biomarker used are the limitation in this study, yet authors have been quantified the significance of toenail as long-term metal exposure [47]. Similarly, Sanders et al., (2014) found that six types of metal concentrations of cadmium, manganese, lead, chromium, mercury and arsenic had detected in the toenail of every participant and has suggested that heavy metal levels in the blood have reflect recent exposures on the order of days, whereas in the toenail the levels represent longer-term exposures till a month onwards [60].

To better understand the mechanisms of toenail and its validity, Grashow et al., (2013) had analysed the metal level in the toenail among construction workers that was exposed to metal fumes in welding hours [29]. Notably, the lead, manganese, cadmium, nickel and arsenic were detectable in the toenail clippings and there were a high correlation between toenail metals in each participant especially between manganese and cadmium. Realizing that the growth rate or clipping length of toenail is varies across individuals and may considered as a constraint in this study, this study had proven that variation would not affect the validity of the study results as it is not correlated with hours welded [29]. Thus, this outcome had supported the ability of toenail as biomarker since this biological sample may capture an internal exposure over specific time intervals and have a capability to reflect a longer term exposure. This view is supported a findings of Mordukhovich et al., (2012) who writes on the potency of toenail as biomarker in a study of a cohort of elderly men and a positive association between toenail arsenic with the blood pressure levels have been discovered [39]. Once confirmed, a little evidence of association of toenail cadmium, mercury and lead with blood pressure may give an important impact in the United States since there are no investigations have assessed relations between blood pressure with arsenic, cadmium, mercury, manganese, or lead by using toenails as biomarkers. In the same year, Amaral et al., (2012) had revealed the high relation of exocrine pancreatic cancer (EPC) risk and toenail concentrations of lead, selenium, nickel, cadmium and arsenic since this is first epidemiologic study conducted [67]. Despite the study design is a retrospective with a small and difference of the recruitment period of sample size and appears as a weakness of the study, the authors also highlight on the matching on residence area; equal age distributions; the simultaneous quantification of the trace elements in the same laboratory and under the same quality control procedures as the important strength in the research and this had present toenail as a reliable approaches in assessing past exposures.

In another major study, Al-Sabbak et al., (2012) found that most metals of aluminium, manganese, cobalt, copper, zinc, molybdenum, lead, thorium and uranium had contained a significant amount in the toenail [7]. The studies presented thus far provide evidence that toenail is able to detect a level of metal concentrations in human body with a longer exposure window. A research executed by Laohaudomchok et al., (2011) in the population recruited from welding school observed that toenail is more practical than fingernail and hair as both of samples are more frequently open to the external environment thus this subject to contamination of samples [8].

The wide range of process had influence the occurrence of heavy metals in environment for a long time and currently, due to the rapid growth of modernisation and technology, the exposure of heavy metals is increasing in some areas in the workplace. Air, water, soil and food are the pathway of environmental exposure and this normally require in-situ monitoring to evaluate the existence of heavy metals in environment. The common pathway for heavy metals in occupational exposure is via

IOP Conf. Series: Materials Science and Engineering **165** (2017) 012019 doi:10.1088/1757-899X/165/1/012019

inhalation. Therefore, in assessing of heavy metals among worker due to occupational exposure, the implementation of toenail as biomarker requires a personal sampling at work area to ensure the data gain thru biomarker is conclusive.

Table 2. The inter-comparison of trace heavy metal concentrations (μg/g) reported worldwide in toenails for year 2010 – 2016.

Study Design	Element (μg/g)										Instruments	Authors
	Cd	Cu	Fe	Mn	Ni	Zn	Pb	Cr	As	Co		
Population based study	0.005 – 1.00	0.20 – 88.0	–	0.20 – 24.0	0.10 – 19.0	8.00 – 382.0	0.30 – 34.0	0.02 – 2.00	–	0.00 – 0.3	ICP - MS	[45]
Population based study	–	–	–	–	–	–	–	–	1.68	–	ICP – MS	[66]
Cross-sectional study	0.37 – 35.5	32.5 – 2225	–	–	0.37 – 33.8	62.0 – 599.0	0.8 – 158.0	–	<DL – 1.00	0.40 – 11.5	ICP – AES	[47]
Cross-sectional pilot study	1.07	–	–	15.6	–	–	957	2.09	1.09	–	ICP – MS	[60]
Retrospective study	0.02	–	–	0.81	2.19	–	0.35	–	0.17	–	DRC – ICP – MS	[29]
Cross-sectional study	1.97	–	–	8.82	–	–	14.70	–	1.68	–	ICP – MS	[39]
Retrospective study	–	–	–	0.05 – 10.41	–	–	–	–	–	–	ICP – MS	[22]
Case-control study	0.008 – 0.029	–	–	–	0.186 – 0.885	–	0.253 – 0.975	–	0.052 – 0.106	–	ICP – MS	[67]
Epidemiological case study	0.32	3.86	63.54	1.04	2.03	116.4	0.598	1.299	0.102	0.043	ICP – MS	[49]

ICP-MS = Inductively Coupled Plasma Mass Spectrometry, DRC-ICP-MS = Dynamic Reaction Cell-Inductively Coupled Plasma Mass Spectrometry, ICP-AES = Inductively Coupled Plasma Atomic Emission Spectrometry.

4.3. Strength and Limitation of Toenail

From the past study, it has conclusively been shown that dietary selenium from food intake did not influence the concentration of selenium in the toenail [68]. In the same way, the correlation research was consistent and supported by other study on traced manganese from toenail [22], [69]. On the other hand, there is a correlation observed between intakes of arsenic in dietary with arsenic concentration in toenail among subjects in Bangladesh [70]. The most likely causes of this connection are due to presence of high concentration of arsenic in staples food in Bangladesh diet: rice and vegetables. Regardless of Kile et al., (2007) findings, a few researchers had acknowledged that in order to monitor the occupational exposure, trace elements or heavy metals in nails to be widely accepted and analysed [40], [71], [72]. In the 1970s, Hopps pointed to some of the ways in which he concluded the trace element analysis from nail samples is precise, accurate and reproducible as well as the specimen is easy to obtain and store [73].

A considerable amount of literature has been published on the adoption of both fingernails and toenails as biomarkers to trace specific metal concentrations in human body and each represent particular strength and limitation [74]–[77]. There are a number of important differences between fingernail and toenail. Compare to toenail, fingernail growth faster which this conditions is accessible for a frequent sampling among subject [78]. Nonetheless, fingernail is commonly exposed to exogenous chemicals (i.e. medication and nail polish) therefore this might leads to contamination and may affect results such as overestimation of endogenous chemical amount in body [22], [28]. In spite of this constraint, such circumstance may turn into an opportunity to assess the impact of the toxic chemicals handling.

However, unlike fingernail, toenail is more prevalent and preserved from any exogenous exposure and this provide a better option to measure any relevant metal concentration from industrial setting. This technique is scrutinize as a reliable and convenient tool in assessment of exposure because less exposure to water than other non-invasive biomarker particularly fingernail, skin and hair [79]. Moreover, toenails are easier to collect and store and this indicate a practical advantage on this approach instead of blood and urine [22], [43]. The reliability of toenail as a long-term biomarker for metal exposure has been discussed and considered and provide a more stable measure for trace elements [20], [80]–[82]. The analysis of toenail expresses a long-term exposure which reflects 2 – 12 months before collection of specimen whereas urine and blood reflects current exposure of 3 – 4 days and 2 – 3 days respectively [65], [79], [83]. Hence, the toenail clipping for heavy metal assessment in target population is required to be done at least 2 months before so that the reading of heavy metal is meticulous [75]. According to Hinwood et al., (2003), the metabolic activities in the body will be isolated from the time when nails detached from skin which causes the accumulation of heavy metals concentration in toenail is reliable [84].

Sriram et al., (2012) stated that the slow growth rate of toenail compared to fingernail is the advantage of this approach over any biological fluid i.e. blood or serum because it's allow for continuous and chronological bio-monitoring of adverse exposure and may prove an advantageous in monitoring of long-term exposure and reflects any burden in human body and/or target organ [85]. Yet, the application of toenail as biomarker have been the subject of questioned by Angerer, Ewers, & Wilhelm (2007) which claimed to be hampered due to no standard operating procedure (SOPs) have been evaluated and published. The absence of SOPs for this approach is considered as limitation for this method [86].

4.4. Toenail as Non-Invasive Method: Collecting and Sampling

Toenail is the main non-invasive method used to determine a metal concentration in human body. The toenail from all toes of each subject were clipped and stored in small envelopes together with indication of toenail clipping date. This compound was prepared by adapting the procedure used by Laohaudomchok et al., (2011) [22]. The collection method of nail specimen is one of the more practical ways of retrieve and analyses of metals concentrations in epidemiological studies even the specimen has been stored for a months or years [87], [88]. According to Kile et al., (2005) and Sanders et al., (2014), a sufficient mass of toenail from all toes were collected in pooled by a registered nurse using a clean clipper and it is best if the metal concentration in the toenail can be analysed in authorized inter-laboratory nail testing to ensure an adequate quality assessment as the certified references material is insufficient [60], [70], [77]. Briefly, a modified version of Method 3050B, USEPA was used to extract the toenail [60]. Then, each sample was spiked with 10 μL of 1000 ppm gold stock solution in order to retain mercury for measurement. Next, 0.5 mL nitric acid (HNO_3), 0.1 mL hydrochloric acid (HCl), and 1 mL deionized water were added to the samples. The samples were heated at 95°C loosely capped to allow reflux for one hour. The samples were cooled, 1 mL of 30% hydrogen peroxide was added, and the samples were heated at 95°C for 15 min. The samples were cooled, 0.4 mL of HCl was added, and the samples were heated at 95°C for an additional 15 min. The samples were brought to a final volume of 10 mL using deionized water and vortex mixed. Samples were analysed using a Thermo X-Series II ICP- MS equipped with a dual gas collision cell. The limit of detection (LOD) among toenail metals varied for each individual because the available mass differed by sample, and a higher sample mass resulted in a lower LOD.

In order to remove external contamination from nails, Kile et al., (2005) stated the sonicating samples in a 1% Triton X-100 solution is used for 20 minutes [70]. After that, toenails were then rinsed repeatedly in Milli-Q water, dried, weighed, and digested in nitric acid at room temperature [22], [39], [60], [65]. Inductively coupled plasma–mass spectrometry (ICP-MS) were used in analysed heavy metals in specimens as described by Goulle et al., (2009) after the resultant solution was diluted to 8% HNO_3 [77]. In the same vein, other past researchers however suggested that apart of accessing protein from nails for diagnosis and molecular epidemiological research, DNA in toenail also can be analysed in this biomarkers and the quality of DNA remained constant even though it has been stored for a long period [89], [90]. These statement has been proven by Nakashima et al., (2008) that found out the total called rate for DNA in old nails was 95% or higher which mean the preserved nail clipping for a prolong time were equivalent with fresh clipping nails [91].

5. Conclusions

This systematic review of a few published studies highlights the importance of biological monitoring of heavy metals by using toenail. Based on the reviewed done, the findings of this research provide insights for application of toenail as a useful tool of metal toxicity measurement and this can be administered to welder in order to minimize the exposure of heavy metals and adverse health effect. According to the studies that had been done, the results suggest that toenail may be a useful alternative matrix and attractive biomarker for the detection of heavy metal in human body.

Conflict of Interest

The authors firmly declare that there are no conflicts of interest related to this work.

Acknowledgements

No funding was received.

References

[1] K. Strimbu and J. A. Tavel, 'What are biomarkers?', *Curr. Opin.* HIV AIDS, vol. 5, no. **6**, pp. 463–466, Nov. 2010.

[2] R. Mayeux, 'Biomarkers: Potential uses and limitations', *NeuroRX*, vol. 1, no. **2**, pp. 182–188, Apr. 2004.

[3] P. Kakkar and F. N. Jaffery, 'Biological markers for metal toxicity', vol. 19, no. January 2004, pp. 335–349, 2005.

[4] L. Lu, L. Zhang, G. J. Li, W. Guo, W. Liang, and W. Zheng, 'Alteration of Serum Concentrations of Manganese, Iron, Ferritin, and Transferrin Receptor Following Exposure to Welding Fumes Among Career Welders', *Neurotoxicology*, vol. 26, no. **2**, pp. 257–265, Mar. 2005.

[5] M. Olivares, T. Walter, J. D. Cook, E. Hertrampf, and F. Pizarro, 'Usefulness of serum transferrin receptor and serum ferritin in diagnosis of iron deficiency in infancy.', *Am. J. Clin. Nutr.*, vol. 72, no. **5**, pp. 1191–5, Nov. 2000.

[6] J. B. Wish, 'Assessing Iron Status: Beyond Serum Ferritin and Transferrin Saturation', *Clin. J. Am. Soc. Nephrol.*, vol. 1, no. Supplement 1, pp. S4–S8, Sep. 2006.

[7] WHO, 'Serum ferritin concentrations for the assessment of iron status and iron deficiency in populations. Vitamin and Mineral Nutrition Information System.', Who, pp. 1–5, 2011.

[8] M. G. Baker, S. R. Criswell, B. A. Racette, C. D. Simpson, L. Sheppard, H. Checkoway, and L.

Sheppard, 'Neurological outcomes associated with low-level manganese exposure in an inception cohort of asymptomatic welding trainees', *Scand. J. Work. Environ. Heal.*, vol. 41, no. **1**, pp. 94–101, 2015.

[9] P. Mathuranath and P. Wattamwar, 'An overview of biomarkers in Alzheimer's disease', *Ann. Indian Acad. Neurol.*, vol. 13, no. **6**, p. 116, 2010.

[10] K. E. Pharr and B. T. Jones, 'Extraction of Cadmium from Urine: A Brief Review', *Appl. Spectrosc. Rev.*, vol. 42, no. **6**, pp. 563–572, Nov. 2007.

[11] G. F. Nordberg, 'Biomarkers of exposure, effects and susceptibility in humans and their application in studies of interactions among metals in China', *Toxicol. Lett.*, vol. 192, no. **1**, pp. 45–49, Jan. 2010.

[12] M. Tellez-Plaza, E. Guallar, B. V Howard, J. G. Umans, K. a Francesconi, W. Goessler, E. K. Silbergeld, R. B. Devereux, and A. Navas-Acien, 'Cadmium Exposure and Incident Cardiovascular Disease', *Epidemiology*, vol. 24, no. **3**, pp. 421–429, May 2013.

[13] V. H. Arrandale, J. Beach, G. S. Cembrowski, and N. M. Cherry, 'Urinary Metal Concentrations Among Female Welders', *Ann. Occup. Hyg.*, vol. 59, no. **1**, pp. 52–61, Jan. 2015.

[14] F. Golbabaei, M. Seyedsomea, A. Ghahri, H. Shirkhanloo, M. Khadem, H. Hassani, N. Sadeghi, and B. Dinari, 'Assessment of welders exposure to carcinogen metals from manual metal arc welding in gas transmission pipelines, Iran.', *Iran. J. Public Health*, vol. 41, no. **8**, pp. 61–70, 2012.

[15] J. Wu, Y. Chen, and W. Gu, 'Urinary proteomics as a novel tool for biomarker discovery in kidney diseases', *J. Zhejiang Univ. Sci. B*, vol. 11, no. **4**, pp. 227–237, Apr. 2010.

[16] H. Zhou, P. S. T. Yuen, T. Pisitkun, P. A. Gonzales, H. Yasuda, J. W. Dear, P. Gross, M. A. Knepper, and R. A. Star, 'Collection, storage, preservation, and normalization of human urinary exosomes for biomarker discovery', *Kidney Int.*, vol. 69, no. **8**, pp. 1471–1476, Apr. 2006.

[17] M. Bader, M. C. Dietz, A. Ihrig, and G. Triebig, 'Biomonitoring of manganese in blood, urine and axillary hair following low-dose exposure during the manufacture of dry cell batteries.', *Int. Arch. Occup. Environ. Health*, vol. 72, no. **8**, pp. 521–7, Nov. 1999.

[18] M. Thambisetty and S. Lovestone, 'Blood-based biomarkers of Alzheimer's disease: challenging but feasible', *Biomark. Med.*, vol. 4, no. **1**, pp. 65–79, Feb. 2010.

[19] P. Olmedo, A. Pla, A. F. Hernández, O. López-Guarnido, L. Rodrigo, and F. Gil, 'Validation of a method to quantify chromium, cadmium, manganese, nickel and lead in human whole blood, urine, saliva and hair samples by electrothermal atomic absorption spectrometry', *Anal. Chim. Acta,* vol. 659, no. **1–2**, pp. 60–67, Feb. 2010.

[20] S. L. O'Neal and W. Zheng, 'Manganese Toxicity Upon Overexposure: a Decade in Review', *Curr. Environ. Heal.* Reports, vol. 2, no. **3**, pp. 315–328, Sep. 2015.

[21] H. Hassani, F. Golbabaei, H. Shirkhanloo, and M. Tehrani-doust, 'Relations of biomarkers of manganese exposure and neuropsychological effects among welders and ferroalloy smelters', 2016.

[22] W. Laohaudomchok, X. Lin, R. F. Herrick, S. C. Fang, J. M. Cavallari, D. C. Christiani, and M. G. Weisskopf, 'Toenail, Blood, and Urine as Biomarkers of Manganese Exposure', *J. Occup. Environ. Med.,* vol. 53, no. **5**, pp. 506–510, May 2011.

[23] C. Jurado, P. Kintz, M. Menéndez, and M. Repetto, 'Influence of the cosmetic treatment of hair on drug testing', *Int. J. Legal Med.*, vol. 110, no. **3**, pp. 159–163, Apr. 1997.

[24] J.-L. C. M. Dorne, G. E. N. Kass, L. R. Bordajandi, B. Amzal, U. Bertelsen, A. F. Castoldi, C. Heppner, M. Eskola, S. Fabiansson, P. Ferrari, E. Scaravelli, E. Dogliotti, P. Fuerst, A. R. Boobis, and P. Verger, 'Human risk assessment of heavy metals: principles and applications.', *Met. Ions Life Sci.*, vol. 8, pp. 27–60, 2011.

[25] L. Järup, 'Hazards of heavy metal contamination.', *Br. Med. Bull.*, vol. 68, pp. 167–82, 2003.

[26] P. B. Tchounwou, C. G. Yedjou, A. K. Patlolla, and D. J. Sutton, Heavy Metal Toxicity and the Environment, vol. 101. 2012.

[27] B. Yousaf, Amina, G. Liu, R. Wang, M. Imtiaz, M. S. Rizwan, M. Zia-ur-Rehman, A. Qadir, and Y. Si, 'The importance of evaluating metal exposure and predicting human health risks in urban–periurban environments influenced by emerging industry', *Chemosphere*, vol. 150, pp. 79–89, May 2016.

[28] K. He, 'Trace elements in nails as biomarkers in clinical research', *European Journal of Clinical Investigation*, vol. 41, no. **1**. pp. 98–102, 2011.

[29] R. Grashow, J. Zhang, S. C. Fang, M. G. Weisskopf, D. C. Christiani, and J. M. Cavallari, 'Toenail metal concentration as a biomarker of occupational welding fume exposure', *J. Occup. Environ. Hyg.*, vol. 11, no. **6**, p. 131227104636009, Dec. 2013.

[30] T. Weiss, B. Pesch, A. Lotz, E. Gutwinski, R. Van Gelder, E. Punkenburg, B. Kendzia, K. Gawrych, M. Lehnert, E. Heinze, A. Hartwig, H. U. Käfferlein, J.-U. Hahn, and T. Brüning, 'Levels and predictors of airborne and internal exposure to chromium and nickel among welders—Results of the WELDOX study', *Int. J. Hyg. Environ. Health*, vol. 216, no. **2**, pp. 175–183, Mar. 2013.

[31] M. G. Baker, C. D. Simpson, B. Stover, L. Sheppard, H. Checkoway, B. A. Racette, and N. S. Seixas, 'Blood Manganese as an Exposure Biomarker: State of the Evidence', *J. Occup. Environ. Hyg.*, vol. 11, no. **4**, pp. 210–217, Apr. 2014.

[32] A. Erdely, R. Salmen-Muniz, A. Liston, T. Hulderman, P. C. Zeidler-Erdely, J. M. Antonini, and P. P. Simeonova, 'Relationship between pulmonary and systemic markers of exposure to multiple types of welding particulate matter', *Toxicology*, vol. 287, no. **1–3**, pp. 153–159, 2011.

[33] S. Hulo, N. Chérot-Kornobis, M. Howsam, S. Crucq, V. de Broucker, A. Sobaszek, and J. L. Edme, 'Manganese in exhaled breath condensate: A new marker of exposure to welding fumes', *Toxicol. Lett.*, vol. 226, no. **1**, pp. 63–69, 2014.

[34] J. P. Bond, K. S. Hansen, and R. J. Levine, 'Fertility among Danish male welders', *Scand. J. Work. Environ. Heal.*, vol. 16, no. **5**, pp. 315–322, 1990.

[35] A. R. Sørensen, A. M. Thulstrup, J. Hansen, C. H. Ramlau-Hansen, A. Meersohn, A. Skytthe, and J. P. Bonde, 'Risk of lung cancer according to mild steel and stainless steel welding', *Scand. J. Work. Environ. Health*, vol. 33, no. **5**, pp. 379–386, Oct. 2007.

[36] N. Abdull, N. Wahida, M. Hassan, and A. R. Ismail, 'Heavy metal emitting from welding fumes in automotive industry', *Int. J. Curr. Res. Acad. Rev.*, vol. 2, no. **2**, pp. 148–156, 2015.

[37] M. A. Balkhyour and M. K. Goknil, 'Total fume and metal concentrations during welding in selected factories in Jeddah, Saudi Arabia', *Int. J. Environ. Res. Public Health*, vol. 7, no. **7**, pp. 2978–2987, 2010.

[38] R. Persoons, D. Arnoux, T. Monssu, O. Culi, G. Roche, B. Duffaud, D. Chalaye, and A. Maitre, 'Determinants of occupational exposure to metals by gas metal arc welding and risk management measures: A biomonitoring study', *Toxicol. Lett.*, vol. 231, no. **2**, pp. 135–141, 2014.

[39] I. Mordukhovich, R. O. Wright, H. Hu, C. Amarasiriwardena, A. Baccarelli, A. Litonjua, D. Sparrow, P. Vokonas, and J. Schwartz, 'Associations of toenail arsenic, cadmium, mercury, manganese, and lead with blood pressure in the Normative aging study', *Environ. Health Perspect.*, vol. 120, no. **1**, pp. 98–104, 2012.

[40] F. Hussein Were, W. Njue, J. Murungi, and R. Wanjau, 'Use of human nails as bio-indicators of heavy metals environmental exposure among school age children in Kenya', *Sci. Total Environ.*, vol. 393, no. **2–3**, pp. 376–384, Apr. 2008.

[41] J. Y. Y. Wong, S. C. Fang, R. Grashow, T. Fan, and D. C. Christiani, 'The Relationship Between Occupational Metal Exposure and Arterial Compliance', *J. Occup. Environ. Med.*, vol. 57, no. **4**, pp. 355–360, Apr. 2015.

[42] M. Garland, J. S. Morris, B. A. Rosner, M. J. Stampfer, V. L. Spate, C. J. Baskett, W. C. Willett, and D. J. Hunter, 'Toenail trace element levels as biomarkers: reproducibility over a 6-year period.', *Cancer Epidemiol. Biomarkers Prev.*, vol. 2, no. **5**, pp. 493–7, 1993.

[43] F. I. Abdulrahman, J. C. Akan, Z. M. Chellube, and M. Waziri, 'Levels of Heavy Metals in Human Hair and Nail Samples from Maiduguri Metropolis, Borno State, Nigeria', *World Environ.*, vol. 2, no. **4**, pp. 81–89, Aug. 2012.

[44] K. Park and E. Seo, 'Toenail mercury and dyslipidemia: Interaction with selenium', *J. Trace Elem. Med. Biol.*, vol. 39, pp. 43–49, Jan. 2017.

[45] J. Mohmand, S. A. M. A. S. Eqani, M. Fasola, A. Alamdar, I. Mustafa, N. Ali, L. Liu, S. Peng, and H. Shen, 'Human exposure to toxic metals via contaminated dust: Bio-accumulation trends and their potential risk estimation', *Chemosphere*, vol. 132, pp. 142–151, Aug. 2015.

[46] P. Chanpiwat, S. Himeno, and S. Sthiannopkao, 'Arsenic and Other Metals' Presence in Biomarkers of Cambodians in Arsenic Contaminated Areas', *Int. J. Environ. Res. Public Health*, vol. 12, no. **11**, pp. 14285–14300, Nov. 2015.

[47] W. Ndilila, A. C. Callan, L. A. McGregor, R. M. Kalin, and A. L. Hinwood, 'Environmental and toenail metals concentrations in copper mining and non mining communities in Zambia', *Int. J. Hyg. Environ. Health*, vol. 217, no. **1**, pp. 62–69, Jan. 2014.

[48] N. Kuiper, C. Rowell, J. Nriagu, and B. Shomar, 'What do the trace metal contents of urine and toenail samples from Qatar's farm workers bioindicate?', *Environ. Res.*, vol. 131, pp. 86–94, May 2014.

[49] M. Al-Sabbak, S. S. Ali, O. Savabi, G. Savabi, S. Dastgiri, and M. Savabieasfahani, 'Metal contamination and the epidemic of congenital birth defects in Iraqi cities', *Bull. Environ. Contam. Toxicol.*, vol. 89, no. **5**, pp. 937–944, 2012.

[50] M. Button, G. R. T. Jenkin, C. F. Harrington, and M. J. Watts, 'Human toenails as a biomarker of exposure to elevated environmental arsenic', *J. Environ. Monit.*, vol. 11, no. **3**, p. 610, 2009.

[51] T. Ohno, M. Sakamoto, T. Kurosawa, M. Dakeishi, T. Iwata, and K. Murata, 'Total mercury levels in hair, toenail, and urine among women free from occupational exposure and their relations to renal tubular function', *Environ. Res.*, vol. 103, no. **2**, pp. 191–197, Feb. 2007.

[52] M. Vinceti, M. Venturelli, C. Sighinolfi, P. Trerotoli, F. Bonvicini, A. Ferrari, G. Bianchi, G. Serio, M. Bergomi, and G. Vivoli, 'Case-control study of toenail cadmium and prostate cancer risk in Italy', *Sci. Total Environ.*, vol. 373, no. **1**, pp. 77–81, Feb. 2007.

[53] J. Gómez-Aracena, R. A. Riemersma, M. Gutiérrez-Bedmar, P. Bode, J. D. Kark, A. Garcia-Rodríguez, L. Gorgojo, P. van't Veer, J. Fernández-Crehuet, F. J. Kok, and J. M. Martin-Moreno, 'Toenail cerium levels and risk of a first acute myocardial infarction: The EURAMIC and heavy metals study', *Chemosphere*, vol. 64, no. **1**, pp. 112–120, Jun. 2006.

[54] J. Park, D. Liang, J. W. Kim, Y. Luo, T. Huang, S. Y. Kim, and S. S. Chang, 'Nail DNA and possible biomarkers: A pilot study', *J. Prev. Med. Public Heal.*, vol. 45, no. **4**, pp. 235–243, 2012.

[55] M. Chiba and R. Masironi, 'Toxic and trace elements in tobacco and tobacco smoke.', *Bull. World Health Organ.*, vol. 70, no. **2**, pp. 269–275, 1992.

[56] M. W. Ashraf, 'Levels of heavy metals in popular cigarette brands and exposure to these metals via smoking.', *ScientificWorldJournal.*, vol. 2012, p. 729430, 2012.

[57] R. Caruso, R. O'Connor, W. Stephens, K. Cummings, and G. Fong, 'Toxic Metal Concentrations in Cigarettes Obtained from U.S. Smokers in 2009: Results from the International Tobacco Control (ITC) United States Survey Cohort', *Int. J. Environ. Res. Public Health*, vol. 11, no. **1**, pp. 202–217, Dec. 2013.

[58] W. K. Al-Delaimy and W. C. Willett, 'Toenail nicotine level as a novel biomarker for lung cancer risk', *Am. J. Epidemiol.*, vol. 173, no. **7**, pp. 822–828, 2011.

[59] J. A. Satia, I. B. King, J. S. Morris, K. Stratton, and E. White, 'Toenail and plasma levels as biomarkers of selenium exposure', *Ann. Epidemiol.*, vol. 16, no. **1**, pp. 53–58, 2006.

[60] A. P. Sanders, S. K. Miller, V. Nguyen, J. B. Kotch, and R. C. Fry, 'Toxic metal levels in children residing in a smelting craft village in Vietnam: a pilot biomonitoring study.', *BMC Public Health*, vol. 14, no. **1**, p. 114, 2014.

[61] R. R. Eastman, T. P. Jursa, C. Benedetti, R. G. Lucchini, and D. R. Smith, 'Hair as a biomarker of environmental manganese exposure', *Environ. Sci. Technol.*, vol. 47, no. **3**, pp. 1629–1637, 2013.
[62] T. Wang, J. Fu, Y. Wang, C. Liao, Y. Tao, and G. Jiang, 'Use of scalp hair as indicator of human exposure to heavy metals in an electronic waste recycling area', *Environ. Pollut.*, vol. 157, no. **8–9**, pp. 2445–2451, 2009.
[63] V. Bencko, 'Use of human hair as a biomarker in the assessment of exposure to pollutants in occupational and environmental settings', *Toxicology*, vol. 101, no. **1–2**, pp. 29–39, 1995.
[64] D. K. Harkins and A. S. Susten, 'Hair analysis: Exploring the state of the science', in *Environmental Health Perspectives*, 2003, vol. 111, no. **4**, pp. 576–578.
[65] G. Samanta, R. Sharma, T. Roychowdhury, and D. Chakraborti, 'Arsenic and other elements in hair, nails, and skin-scales of arsenic victims in West Bengal, India', *Sci. Total Environ.*, vol. 326, no. **1–3**, pp. 33–47, Jun. 2004.
[66] M. Subhani, I. Mustafa, A. Alamdar, I. A. Katsoyiannis, N. Ali, Q. Huang, S. Peng, H. Shen, and S. A. M. A. S. Eqani, 'Arsenic levels from different land-use settings in Pakistan: Bio-accumulation and estimation of potential human health risk via dust exposure', *Ecotoxicol. Environ. Saf.*, vol. 115, pp. 187–194, May 2015.
[67] A. F. S. Amaral, M. Porta, D. T. Silverman, R. L. Milne, M. Kogevinas, N. Rothman, K. P. Cantor, B. P. Jackson, J. A. Pumarega, T. López, A. Carrato, L. Guarner, F. X. Real, and N. Malats, 'Pancreatic cancer risk and levels of trace elements', *Gut*, vol. 61, no. **1110**, pp. 1583–1588, 2012.
[68] M. P. Longnecker, M. J. Stampfer, J. Stephen Morris, V. Spate, C. Baskett, M. Mason, and W. C. Willett, 'A 1-y trial of the effect of high-selenium bread on selenium concentrations in blood and toenails', *Am. J. Clin. Nutr.*, vol. 57, no. **3**, pp. 408–413, 1993.
[69] J. M. Guthrie, J. D. Brockman, J. S. Morris, and J. D. Robertson, 'The "one source" cohort - Evaluating the suitability of the human toenail as a manganese biomonitor', in *Journal of Radioanalytical and Nuclear Chemistry*, 2008, vol. 276, no. **1**, pp. 41–47.
[70] M. L. Kile, E. A. Houseman, E. Rodrigues, T. J. Smith, Q. Quamruzzaman, M. Rahman, G. Mahiuddin, L. Su, and D. C. Christiani, 'Toenail arsenic concentrations, GSTT1 gene polymorphisms, and arsenic exposure from drinking water', *Cancer Epidemiol. Biomarkers Prev.*, vol. 14, no. **10**, pp. 2419–2426, 2005.
[71] M. L. Kile, E. A. Houseman, C. V Breton, Q. Quamruzzaman, M. Rahman, G. Mahiuddin, and D. C. Christiani, 'Association between total ingested arsenic and toenail arsenic concentrations.', *J. Environ. Sci. Health. A. Tox. Hazard. Subst. Environ. Eng.*, vol. 42, no. **12**, pp. 1827–34, 2007.
[72] B. Nowak and H. Kozłowski, 'Heavy metals in human hair and teeth', *Biol. Trace Elem. Res.*, vol. 62, no. **3**, pp. 213–228, Jun. 1998.
[73] H. C. Hopps, 'The biologic bases for using hair and nail for analyses of trace elements', *Sci. Total Environ.*, vol. 7, no. **1**, pp. 71–89, Jan. 1977.
[74] M. Sakamoto, H. M. Chan, J. L. Domingo, R. B. Oliveira, S. Kawakami, and K. Murata, 'Significance of fingernail and toenail mercury concentrations as biomarkers for prenatal methylmercury exposure in relation to segmental hair mercury concentrations', *Environ. Res.*, vol. 136, pp. 289–294, 2015.
[75] M. J. Slotnick and J. O. Nriagu, 'Validity of human nails as a biomarker of arsenic and selenium exposure: A review', *Environ. Res.*, vol. 102, no. **1**, pp. 125–139, Sep. 2006.
[76] M. A. Davis, Z. Li, D. Gilbert-Diamond, T. A. Mackenzie, K. L. Cottingham, B. P. Jackson, J. S. Lee, E. R. Baker, C. J. Marsit, and M. R. Karagas, 'Infant toenails as a biomarker of in utero arsenic exposure', *J Expo Sci Env. Epidemiol*, vol. 24, no. **5**, pp. 467–473, 2014.
[77] J. P. Goulle, E. Saussereau, L. Mahieu, D. Bouige, S. Groenwont, M. Guerbet, and C. Lacroix, 'Application of Inductively Coupled Plasma Mass Spectrometry Multielement Analysis in Fingernail and Toenail as a Biomarker of Metal Exposure', *J. Anal. Toxicol.*, vol. 33, no. **2**,

pp. 92–98, Mar. 2009.

[78] S. Yaemsiri, N. Hou, M. M. Slining, and K. He, 'Growth rate of human fingernails and toenails in healthy American young adults', *J. Eur. Acad. Dermatology Venereol.*, vol. 24, no. **4**, pp. 420–423, 2010.

[79] B. M. Adair, E. E. Hudgens, M. T. Schmitt, R. L. Calderon, and D. J. Thomas, 'Total arsenic concentrations in toenails quantified by two techniques provide a useful biomarker of chronic arsenic exposure in drinking water', *Environ. Res.*, vol. 101, no. **2**, pp. 213–220, 2006.

[80] M. R. Karagas, 'Measurement of Low Levels of Arsenic Exposure: A Comparison of Water and Toenail Concentrations', *Am. J. Epidemiol.*, vol. 152, no. **1**, pp. 84–90, Jul. 2000.

[81] M. Hashemian, H. Poustchi, A. Pourshams, M. Khoshnia, J. D. Brockman, A. Hekmatdoost, C. C. Abnet, and R. Malekzadeh, 'The Nail as a Biomonitor of Trace Element Status in Golestan Cohort Study', *Middle East J. Dig. Dis.*, vol. 8, no. **1**, pp. 19–23, Sep. 2015.

[82] M. G. O'Doherty, C. C. Abnet, L. J. Murray, J. V. Woodside, L. A. Anderson, J. D. Brockman, and M. M. Cantwell, 'Iron intake and markers of iron status and risk of Barrett's esophagus and esophageal adenocarcinoma', *Cancer Causes Control*, vol. 21, no. **12**, pp. 2269–2279, Dec. 2010.

[83] K. Phan, S. Sthiannopkao, and K.-W. Kim, 'Surveillance on chronic arsenic exposure in the Mekong River basin of Cambodia using different biomarkers', *Int. J. Hyg. Environ. Health*, vol. 215, no. **1**, pp. 51–58, Dec. 2011.

[84] A. L. Hinwood, M. R. Sim, D. Jolley, N. de Klerk, E. B. Bastone, J. Gerostamoulos, and O. H. Drummer, 'Hair and toenail arsenic concentrations of residents living in areas with high environmental arsenic concentrations', *Environ. Health Perspect.*, vol. 111, no. **2**, pp. 187–193, 2003.

[85] K. Sriram, G. X. Lin, A. M. Jefferson, J. R. Roberts, R. N. Andrews, M. L. Kashon, and J. M. Antonini, 'Manganese accumulation in nail clippings as a biomarker of welding fume exposure and neurotoxicity', *Toxicology*, vol. 291, no. **1–3**, pp. 73–82, Jan. 2012.

[86] J. Angerer, U. Ewers, and M. Wilhelm, 'Human biomonitoring: State of the art', *Int. J. Hyg. Environ. Health*, vol. 210, no. **3–4**, pp. 201–228, May 2007.

[87] P. Xun, K. Liu, J. S. Morris, M. L. Daviglus, and K. He, 'Longitudinal association between toenail selenium levels and measures of subclinical atherosclerosis: The CARDIA trace element study', *Atherosclerosis*, vol. 210, no. **2**, pp. 662–667, Jun. 2010.

[88] P. Xun, K. Liu, J. Steven Morris, M. L. Daviglus, J. Stevens, D. R. Jacobs, and K. He, 'Associations of toenail selenium levels with inflammatory biomarkers of fibrinogen, high-sensitivity C-reactive protein, and interleukin-6', *Am. J. Epidemiol.*, vol. 171, no. **7**, pp. 793–800, 2010.

[89] T. D. Anderson, J. P. Ross, R. K. Roby, D. A. Lee, and M. M. Holland, 'A validation study for the extraction and analysis of DNA from human nail material and its application to forensic casework', *J. Forensic Sci.*, vol. 44, no. **5**, pp. 1053–1056, 1999.

[90] S. G. Van Breda, J. G. Hogervorst, L. J. Schouten, A. M. Knaapen, J. H. van Delft, R. A. Goldbohm, F. J. van Schooten, and P. A. van den Brandt, 'Toenails: An Easily Accessible and Long-Term Stable Source of DNA for Genetic Analyses in Large-Scale Epidemiological Studies', *Clin. Chem.*, vol. 53, no. **6**, pp. 1168–1170, Apr. 2007.

[91] M. Nakashima, M. Tsuda, A. Kinoshita, T. Kishino, S. Kondo, O. Shimokawa, N. Niikawa, and K. -i. Yoshiura, 'Precision of High-Throughput Single-Nucleotide Polymorphism Genotyping with Fingernail DNA: Comparison with Blood DNA', *Clin. Chem.*, vol. 54, no. **10**, pp. 1746–1748, Aug. 2008.

Abrasive Wear Failure Analysis of Tungsten Carbide Hard facing on Carbon Steel Blade

A. L. Mohd Tobi[1], Z. Kamdi[1], M. I. Ismail[*,1], M. Nagentrau[1], L. N. H. Roslan[1], Z. Mohamad[1], A. S. Omar[1], N. Abdul Latif[1]

[1]Faculty of Mechanical and Manufacturing Engineering University Tun Hussein onn Malaysia (UTHM), 86400, Batu Pahat, Johor, Malaysia.

***E-mail: muhdiqzaham@yahoo.com**

Abstract. This study investigate the abrasive wear failure of tungsten carbide hardfacing on continuous digester (CD) blade (carbon steel) in an environment of sulphuric acid and ilmenite ore mixture. Comparison being made on the hardness, thickness and microstructural of the hardfacing between unworn and 3 months old worn blade on few locations around the blade. The cross sections of the blade revealed non-uniform coverage of the hardfacing on the blade for both worn and unworn blade. The edge of the blade has the least amount of hardfacing thickness which with time acts as the point of failure during the wear process. The hardness obtained from both the unworn and worn samples are around 25% lower from the hardfacing electrode manufacturer's hardness specification. Microstructural micrograph analysis of the hardfacing revealed non uniform size carbide with non-uniform distributed of carbide in the hardfacing layer.

1. Introduction

Wear is the degradation material from friction or when two surface slide together that could lead to a failure point of the component. Wear of machine part and component is one of the major problem faced by the industry in our world. Furthermore, the material to reduce the wear rate and increase the life of the component is being investigated and needed by all of the industry. Moreover, Wear component also could lead in increasing the maintenance cost of the component to the company.

Hardfacing is a process of depositing a material on a component by using certain method such Welding, thermal spraying and cladding. Hardfacing process is one of the method that will add a coating layer that act as a protective layer that will reduce the wear rate and increase the component life. Furthermore, the hardfacing alloys show clear superiority in abrasive wear resistance in contrast to unhardfaced alloy [1]. But the effect of hardfacing on component life and performance will depend upon the hardfacing material and the application process [2]. The high hardness that protective layer poses is one of the factors that help improve in abrasive wear of material that reduce the wear rate [3-4].

Shielded metal arc welding (SMAW), or also known as manual metal arc welding (MMA or MMAW), flux shielded arc welding or informally as stick welding, is a manual arc welding process that uses a consumable electrode covered with afflux to lay the weld. SMAW is much cheaper than other type of fusion welding such as Gas–metal arc welding, Plasma arc welding, Electro-slag welding and many more which require more specific expensive tools and equipment. SMAW is one of the way of hardfacing. Furthermore, hardfacing SMAW applications gives more importance to the economical and metallurgical aspects, looking for a balance between quality and efficiency [5].

Tungsten carbide (WC) had been wide used by industry and still now it is being investigated. Tungsten carbide is one of the hard material that have a good properties to act as a protective coating layer on a component to increase wear resistance. Furthermore, Tungsten Carbide (WC) have a high hardness properties with low wear rate, it has become one of the material that are used in hardfacing. The wear resistance of any alloy will increase when WC particles existed at the coating [6]. Furthermore, an addition of WC in coating can increases the micro hardness of the coating [7]. Higher carbide contain will have a higher hardness thus will increase the wear resistance. Moreover, the size of tungsten carbide particle and the distribution of tungsten carbide through the coating layer is one of the effecting properties that effect the hardness. Better distribution or finer grain of carbide will give a better hardness. Moreover, the microstructure characteristic such as the type, amount, size, and the distribution of carbides, has a great influenced and effect on the wear property. [8-9]

The major purpose of the continuous digester blade are to mix the ilmenite and sulphuric acid (H2SO4) to produce titanium dioxide. The acidic and abrasive properties of the mixture causes the blade to wear. The CD blade are hardface by Tungsten Carbide (WC) electrode using SMAW method to increase the wear resistance of the blade. Figure 1 shows the condition of the blade after 6 month of process. The red circle shows the major abrasion occurs on the blade. Moreover, the other part also shown an abrasion wear occurs but the wear are just minor wear. In this study, the blade will be investigated on coating thickness, hardness, and microstructure.

Figure 1: Blade Condition after 6 Month of Process.

2. Methodology

2.1 Material and hardfacing welding

The substrate or blade material used in CD blade are carbon steel. Figure 2 shows the illustration of the blade before the after hardfacing process. The size of the blade is approximately 370 mm times 955 mm and the thickness are range between 25-30 mm. Even though the blade is already hard, the hardfacing still need to be done cause by the acidic and abrasive factors occurs during the digestion process.

The hardfacing process of the CD blade was done continuously until one whole blade is complete hardfaced. Before the hardfacing process, the electrode (WC) will be dry in the oven for 0.5 to 1.0

hours at the temperature 80-120 °C. Table 1 shows the chemical composition of the electrode used for hardfacing. The hardfacing process will be conducted by the company welder using 300-320 DC current range with direction flow of weld from top to down, with the average slag size of 30 to 40 mm produce. The blade are weld part by part until all the hardfacing is done, refer Figure 4 for complete hardfacing process.

In this study, two blade with different time process are investigated. New unworn blade and worn blade that had been used for 3 month. New unworn blade is a blade that not being used yet for the digestion process, while the 3 month worn blade is a blade that had been used for three month in the digestion process. Both of the blade was hardfaced with the same parameter and same welding method. The blade are hardfaced by the company welder.

Table 1: Chemical Composition of Electrode

Chemical composition				
C	Si	Mn	W	Fe
3.2	0.4	1.5	54.2	bal

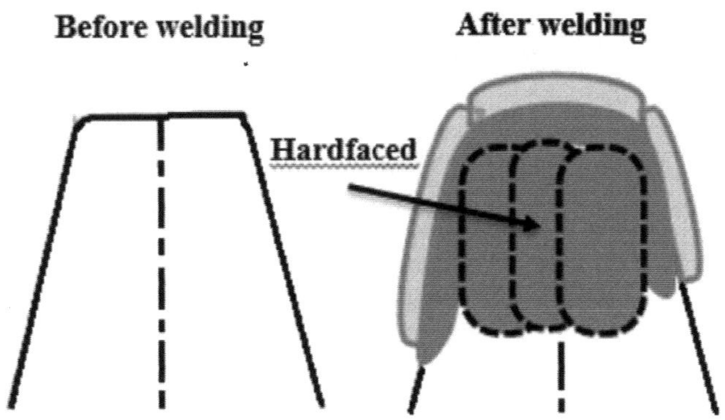

Figure 2: Blade illustration Condition before Hardfacing and after Hardfacing Process.

2.1 Sample Preparation

The hardfaced blade was section into 7 section 1, 2, 3, 4, 5, 6 and 7. All the seven section are cut by Electrical Discharge Machining (EDM) wire cut. Before cutting process, the blade need to undergo cleaning process by removing the slag obtain by welding process. The seven section of the sample are being selected throughout the blade as shown in Figure 3.

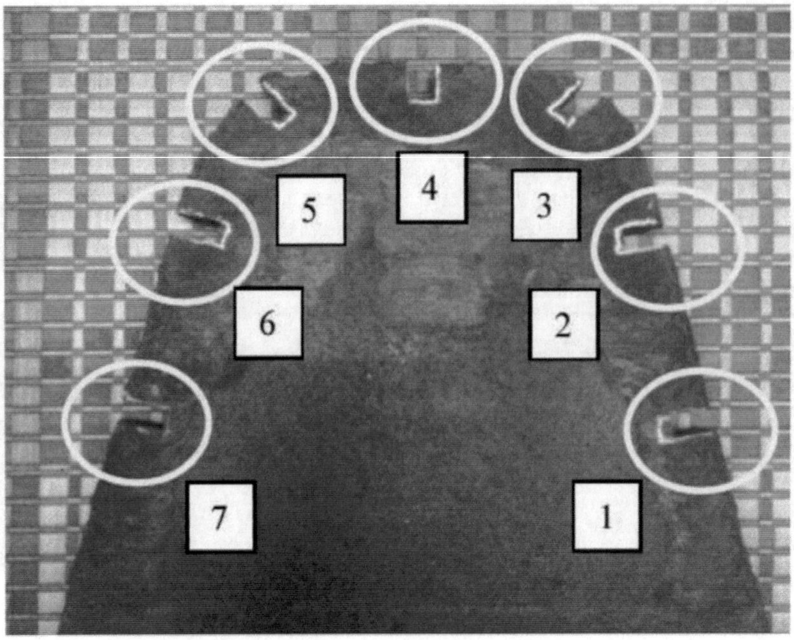

Figure 3: Blade sectioning and numbering specimen

2.2 Coating thickness, Hardness, and Microstructure analysis

All sample are grind using sand paper from grit 200, 400, 600, 800, 1200, and 1500 and are polish using polisher with additional of alumina for microstructure analysis. The microstructure are observe using Scanning Electron Machine (SEM) and Optical Microscope (OM). While the Coating thickness are measured using image J software (commercial image software). The sample are capture with a measurement instrument (ruler) and the measurement instrument will be the indicator of the measurement. The micro hardness of substrate and WC are being tested by Vickers Hardness using load 0.5HV and the dwelling time of 15 s. 3 readings are taken for each specimen and an average reading are calculated as the final hardness.

Each of the specimen the coating thickness are measured at 5 different points A, B, C, D, and E using image J. Figure 4 shows the 5 points of coating thickness measured. The red colour region are the coating while the blue colour region are the substrate.

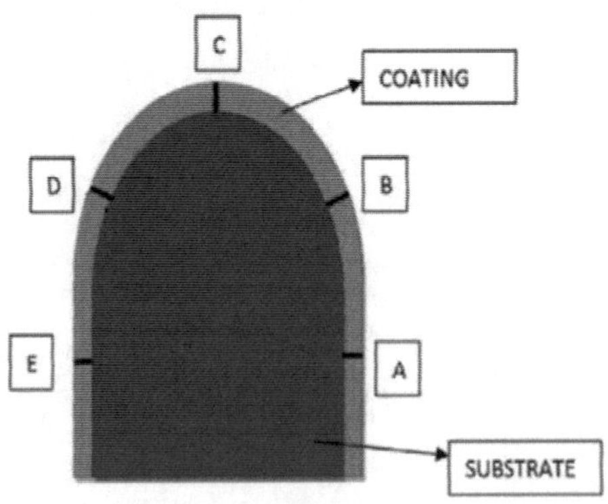

Figure 4: The Positon of the Coating Thickness Measure

3. Result and Discussion

3.1 Coating thickness

Each of the specimen the coating thickness are measured at 5 different points A, B, C, D, and E using image J. Figure 4 shows the 5 points of coating thickness measured. The red colour region are the coating while the blue colour region are the substrate.

From unworn coating thickness result Figure 5, it shows a W shape pattern of thickness. From all the 5 points, it shows that points B and D shows the lowest coating thickness compare to points A, C, and E. The W shape pattern shows at sample 1, 2, 6, and 7. While Sample 3, 4, and 5 shows almost a straight line pattern where it have a same coating thickness through all the points. Furthermore, only sample 4 that have not coating at all at point C.

From worn coating thickness result Figure 6, it also show a W shape pattern of thickness but the coating thickness are less than coating thickness of unworn blade. Furthermore, at point B and D the coating thickness are almost zero.

Furthermore, effect from the low coating thickness, Point B, and D have the possibilities to become the point of failure of the blade and cause early damage after the coating loss, but coating thickness will not affect much on coating hardness [10]. Figure 7 shows the illustration of the hardfaced blade, the red arrow point at the edge of the blade that have low coating thickness are the place where points B and D are situated on the blade.

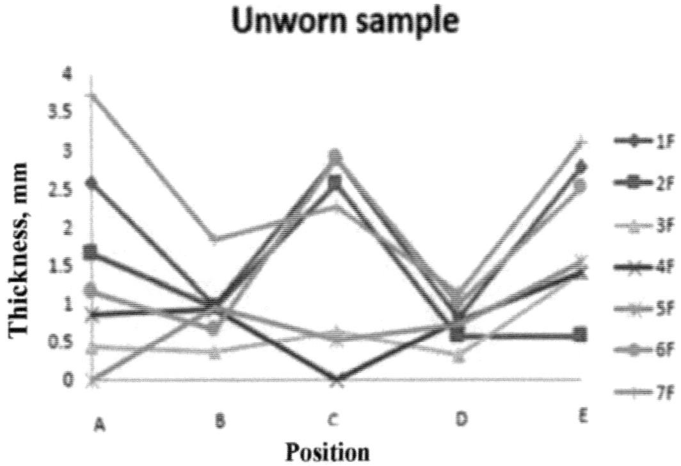

Figure 5: Coating Thickness Result on Unworn Blade

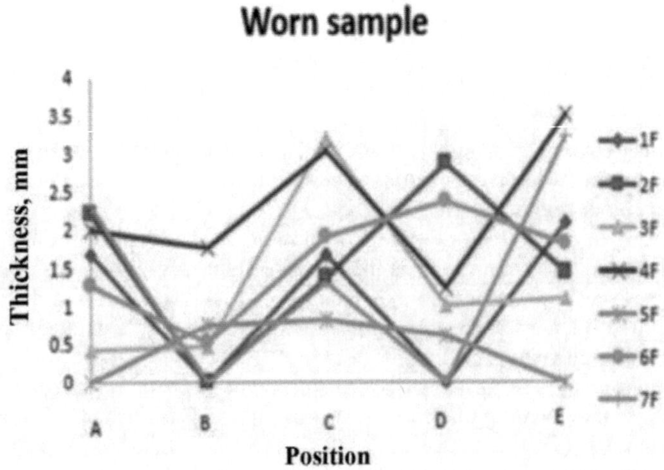

Figure 6: Coating Thickness Result on Worn Blade

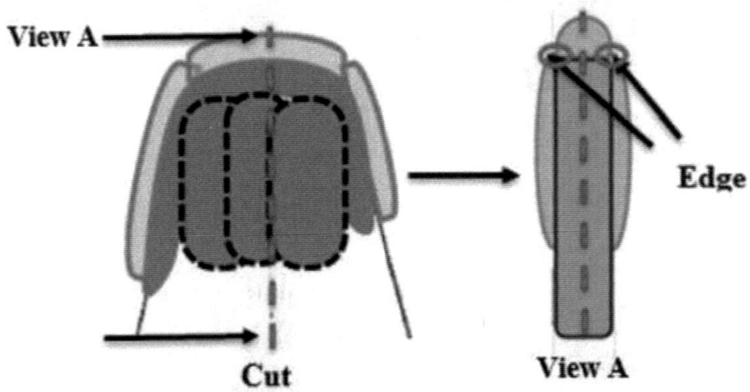

Figure 7: The Observation of the Early Damage of Abrasion Wear

3.2. Hardness

The hardness test was done on two condition, at the substrate and the coating of the blade. Figure 8 shows the result for substrate on both of the blade and Figure 9 shows the result for coating for both blade also. The substrate hardness for worn and unworn blade is approximately at 230 HV. The substrate hardness pattern shows a slightly increase in hardness for all sample for worn blade. Furthermore, The coating hardness for worn and unworn blade are approximately at 750 HV, but at sample 2 and 4 the hardness value for coating are approximately at 1200 HV. Moreover, the coating hardness pattern shows that the unworn blade has a slightly higher hardness than worn blade.

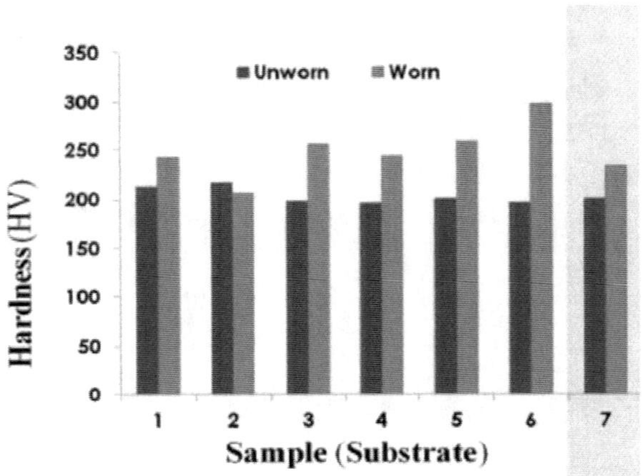

Figure 8: Hardness at Substrate for Worn and Unworn Blade

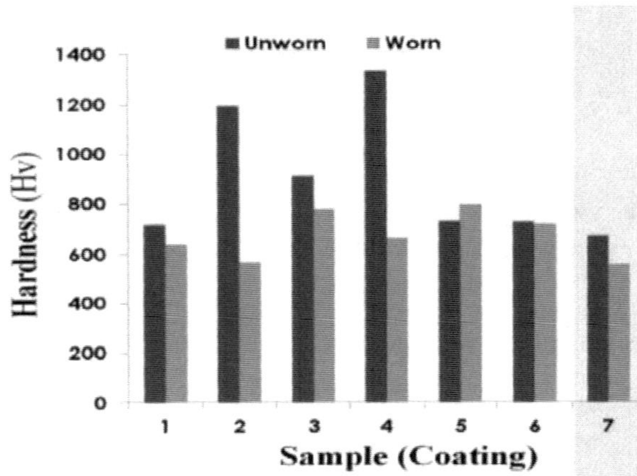

Figure 9: Hardness at Coating Surface for Worn and Unworn Blade

3.3. Microstructure

The micrograph of the coating of the blade are shown on Figure 10. There are three types of tungsten carbide microstructure that are present at the coating, large carbide, small carbide and Non carbide Region. The hardness of the coating are constant except sample 2 and 4, it shows that the hardness test conducted might land on the large carbide size that have a higher hardness. Furthermore, to achieve best wear resistance, a high content of tungsten carbides which are uniformly distributed within the microstructure is necessary [11-12]. Higher contain of carbide will give higher hardness value. Furthermore, higher carbide contain with finer grain of microstructure will produce a higher hardness value and increase wear resistance. The result shows that the value of hardness obtain except sample at position 2 and 4 decrease about 25% from the hardfacing electrode manufacturer's hardness specification. New hardfacing process need to be introduce to overcome this problem.

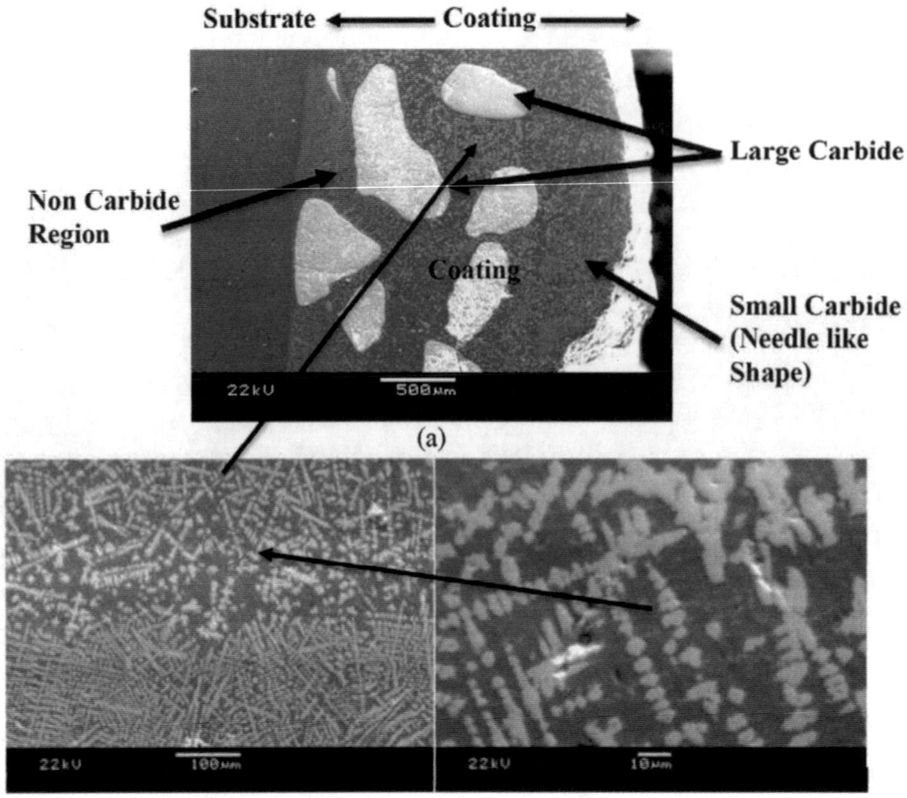

(a)

(b) (c)

Figure 11: Interface microstructure of substrate and coating (WC) at point C (a) Needle like of WC microstructure (b), (c)

4. Conclusion

In this study, the continuous digester blade that was used in the digestion process of titanium dioxide has been analyzed. The digester blade was made up from the carbon steel and coated with tungsten carbide (WC) based materials. The main purpose of this study was to characterized and compared the used with the new digester blade. The following conclusion can be drawn from this study:-

- Inconsistent thickness of coating around the blade along position A until E, due the inconsistent welding of the coating.
- Both worn and unworn blade has minimum thickness located at edges B and D of the cross sections due to improper weld at edge location. This could lead to insignificant factors during operation to edges B and D that will be exposed to abrasive environment and the coating might wear, thus unable to give protection to the blade.
- Edge B and D will act as the point of failure upon further wear.
- The carbide distribution along the coating area was not fully participated at the coating area. This will give impact to the original hardness that should be obtain by the blade to reduce wear rate.

- The welding parameter used by the industry does not follow the actual parameter according to the actual specs, this could be also one of the reason and factors that cause the major reduce in the hardness value for the unworn blade.

References

[1] Amardeep S K, Gurmeet S C, Shivali S, 2014, *Wear Behavior of Hardfacings on Rotary Tiller Blades,* Procedia Engineering **94**, 1442-1451.

[2] Vineet S, Vikas C, 2013, *Enhancement in Wear Resistance by Hardfacing: A* Review, ISSN **2**, 2320-2491.

[3] Jun L, Shuai Y, Weisheng X, Xin J and Chibin G, 2016, *Microstructure and Wear Resistance Performance of Cu-Ni-Mn Alloy Based Hardfacing Coating Reinforced by WC Particles*, Journal of Alloy and Compound **654**, 63-70.

[4] Xin L, Liang L, Xiaoqiang L, Yuanyuan L, (2015) *Abrasion Wear Behaviour of Wc–10ni3al Cermet with Plate-Like Triangular Prismatic Wc Grains*, ceramic international **41**, 5147–5158.

[5] Amado C C, Americo S, Manuel R P, 2008, *Operational Behavior Assesment of Coated Tubular Electrodes for SMAW Hardfacing*, 265-273.

[6] P. Wu, H.M. Du, X.L. Chen, Z.Q. Li, H.L. Bai, E.Y. Jiang, 2004, *Influence of WC particle behavior on the wear resistance properties of Ni–WC composite coatings*, **257** Wear, 142–147.

[7] S. Harsha, D.K. Dwivedi and A. Agrawal, 2007, *Influence of WC addition in Co–Cr–W–Ni–C flame sprayed coatings on microstructure, microhardness and* wear, behaviour Surface & Coatings Technology **201**, 5766 – 5775

[8] V.M. Desai, C.M. Rao, T.H. Kosel, N.F. Fiore, 1984, *Wear* **94**, 89.

[9] H.R. Shetty, T.H. Kosel, N.F. Fiore, 1982, *Wear* **80**, 347.

[10] Hongxia Deng, Huiji Shi, Seiji Tsuruoka, 2010, *Influence of coating thickness and temperature on mechanical properties of steel deposited with Co-based alloy hardfacing coating*, Surface & Coatings Technology **204**, 3927–3934.

[11] K. Van Acker, D. Vanhoyweghen, R. Persoons, J. Vangrunderbeek, 2005, *Influence of tungsten carbide particle size and distribution on the wear resistance of laser clad WC/Ni coatings*, Wear **258**, 194–202.

[12] E. Badisch, M. Kirchgaßner, 2008, *Influence of welding parameters on microstructure and wear behaviour of a typical NiCrBSi hardfacing alloy reinforced with tungsten carbide*, Surface & Coatings Technology **202**, 6016-6022.

UV radiation effect towards mechanical properties of Natural Fibre Reinforced Composite material: A Review

Shahruddin Mahzan[1,*], Muhamad Fitri[2] and M. Zaleha[3]
[1,2,3] Faculty of Mechanical and Manufacturing Engineering, Universiti Tun Hussein Onn Malaysia, Parit Raja 86400 Batu Pahat, Johor, Malaysia

Email:sharudin@uthm.edu.my or gd140052@siswa.uthm.edu.my

Abstract. The use of natural fibres as reinforcement material have become common in human applications. Many of them are used in composite materials especially in the polymer matrix composites. The use of natural fibres as reinforcement also provide alternative solution of usage instead of being a waste materials. In some applications, these natural reinforced polymer composites were used as the outer layer, making them exposed to ultra violet exposure, hence prone to UV radiation. This paper reviews the effect of UV radiation towards the mechanical properties of natural fibre reinforced polymer matrix composite material. The effect of chemical treatment towards the natural fibre is also investigated. One of the important features that was critically explored was the degradation of the composite materials. The influence of UV radiation on the degradation rate involve several parameters such as wavelength, intensity and exposure time. This review highlights the influence of these parameters in order to provide better solution for polymer matrix composite's development.

1. Introduction

Polymer materials have been extensively used in many applications ranging from automotive parts, furniture, housewares and many others. There are various forms of polymer materials such as plastics, rubber and fibres [1-3]. In some applications, pure polymer was used. Nonetheless, the advancement in composite materials have seen polymers are used as matrices of composite material. There are many research works have been conducted in exploring the polymer fibre as matrix reinforcement in composite materials. Two types of polymer fibres have been used namely the natural fibres and the synthetic fibres. The conventional fibres for reinforcement of polymer matrix composites usually use glass fibre, carbon and organic fibre (such as aramid) [1].

Recent years have seen the utilization of natural fibres as reinforcement in composite materials. The choice of kenaf, oil palm, jute, banana and many other natural fibres have been introduced to the polymeric matrix composite materials [4-7]. Besides reducing the cost (i.e. cost of synthetic fibres are comparatively expensive), the natural fibres offer environment safe and recycling benefits. For an example, for a 1 hectare of oil palm plantation, an oil palm industry produced about 55 tons dry matters in the form of fibrous biomass annually [8-9]. This fibrous biomass consists of lignocellulosic fibre which was extracted from trunk, fruit mesocarps and empty fruit bunch (EFB). EFB contributes about 73% of waste of which 1.1 tons of EFB has to be disposed annually from oil palm industry [10]. Instead of being unused waste, this EFB was subsequently recycled and exploited to be a potential natural fibre, hence turning it from waste to wealth as well as environment safe.

Polymers have unique responses to mechanical loads. The properties of polymer (mechanical, electrical and optical) are time dependent, and cannot be treated mathematically by the laws of either solid or fluids [11]. One of the polymers that are usually used is polypropylene (PP). Polypropylene is the simplest chemical polymer which is classified as thermoplastic. Polypropylenes are tough, flexible and water repellent but having low strength. Polypropylene pipe preferably used for potable water which some of them used above ground [12] and exposed to sunlight. It is also used for electrical insulation, cold water pipes, containers and packing sheets [12]. Although polypropylene is very similar to high density polyethylene (HDPE), it has comparatively lower density (0.9 g/cm^3) and a higher softening point enable to withstand to boiling water and much steam sterilizing operation. Polypropylene has been used at large in transportation, particularly in automotive and battery cases. It has secured of about 90% in this market as a result of the drive in automotive industry to reduce weight and cost. In addition to being lightweight, polypropylene is also provides outstanding resistance to creep and fatigue, high temperature rigidity, impact strength and resistance to corrosion. In general, 30% of polypropylene are used as fibre products, 15% as housewares and toys, 15% for automotive parts and 5% for appliance parts 1 [13].

There are three types of fibres dominated as reinforcement, e.g. fiberglass, carbon or graphite fibres, and organic fibres (the most important is generally called aramid). The choice of which fibre were used in particular applications depend on its cost and performance. Fiberglass is rather expensive fibre, carbon fibre is the stiffest of the three major fibres and aramid fibres are the toughest [1]. Natural fibres are exploited as replacement of conventional fibre such as glass, aramid and carbon, due to its low cost and their properties. The properties of natural fibres are fairly good mechanical properties, high specific strength, and non-abrasive, eco-friendly and biodegradability characteristics [14]. Table 1 shows the properties of selected natural fibres and man-made fibres.

Table 1 Properties of selected natural and manmade fibres [14]

Fibre	Density (g/cm^3)	Elongation (%)	Tensile Strength (MPa)	Elastic Modulus (GPa)
Jute	1.3	1.5-1.8	393-773	26.5
Flax	1.5	2.7-3.2	500-1500	27.6
Hemp	1.47	2-4'	600	70
Kenaf	1.45	1.6	930	53
Ramie	N/A	3.6-3.8	400-938	61.4-128
Sisal	1.5	2.0-2.5	511-635	9.4-22
Coir	1.2	30	593	4.0-6.0
Softwood Kraft pulp	1.5	4.4	1000	40
E-glass	2.5	0.5	2000-3500	70
S-glass	2.5	2.8	4570	86
Aramid (std)	1.4	3.3-3.7	3000-3150	63.0-67.0
Carbon (Std. PAN-based)	1.4	1.4-1.8	4000	230-240

Most of natural fibres are used as reinforcement of polymer in a polymer matrix composite material which is designed to achieve combination properties that will be better than properties of fibre itself or properties of the polymer. Compared to fibreglass, natural fibres are cost saving and reduce in density. However, the disadvantages of using natural fibres as reinforcement of polymer composites include the incompatibility between fibres and polymer matrices, poor moisture resistance, inferior fire resistance, tendency to form aggregates during processing, limited processing temperatures, lower durability, variation in quality and price, and difficulty in using established manufacturing process [15]. The fibre reinforced composites efficiency also depends on the manufacturing process that able to transfer stress from the matrix to fibre [7]. The fibre–matrix adhesion in composite can be improved by doing some treatment to the fibre. The fibre treatment can be chemical treatment by using Methanol [16], or NaOH and acid [17], or hot water and NaOH [18], or NaOH and Anthraquinone solution [19], or hot water, NaOH and Ultrasonic [20]. Therefore some processes needed to overcome these drawbacks have incurred additional cost and needed to be optimised [15].

The addition of natural fibres into some of polymer is expected not only to increase the properties of the composite material, but also to retain the degradation rate. Natural fibre reinforced polymer composites which are used for outdoor environment are prone to be degraded with influence of various factors such as moisture, temperature, microorganism activities and ultraviolet radiation [15]. Natural fibre composites which are used outdoor with direct sunlight are subjected to radiation and thus break the covalent bonds in organic polymer causing colour fading, weight loss, surface roughening, mechanical property deterioration and embrittlement with more reduction in wetter condition [15]. However, it was observed that different reinforcement materials produced different surface responses under UV accelerated degradation conditions [21].

Polymer degradation is a change in the properties of a polymer or polymer based product under the influence of one or more environmental factor such as heat, light or chemical. The changes in the properties can be the changes of tensile strength, discoloration or change of shape [22]. Ultraviolet (UV) radiation is one of environmental factors that cause the polymer degraded. Ultraviolet light or ultraviolet radiation is an invisible electromagnetic radiation of the same nature as visible light, but having higher energies and shorter wavelengths [23]. Conventionally, there are three regions categorised of UV radiation band based on the wavelength, e.g. UV-A for wavelength 315 – 400 nm, UV-B for wavelength 280 – 315 nm, and UV-C for wavelength 100 – 280 nm. UV-C rays are totally absorbed by the atmospheric ozone. Only UV-A and about 10% of UV-B reached the surface of the earth [23]. Quantities of UV radiation are expressed using radiometric terminology as shown in Table 2 [24]. Radiant energy and radiant flux are the terms relating to the radiation passing through space. Radiant intensity and radiance are the terms relating to sources of radiation. Irradiance is the term which is commonly used for photobiology and radiant exposure is strictly the time integral of the irradiance, but sometimes expressed as exposure dose.

Table 2 Radiometric term and unit [24]

Terminology	Unit	Symbol
Wavelength	nm	λ
Radiant Energy	J	Q
Radiant Flux	W	Φ
Radiant Intensity	$W\ sr^{-1}$	I
Radiance	$W\ m^{-2}\ sR^{-1}$	L
Irradiance	$W\ m^{-2}$	E
Radiant exposure	Jm^{-2}	H

Fig. 1 presented a schematic diagram of the degradation of natural fibre/polymer composite due to UV exposure [15]. The polymer, which absorbs the ultraviolet radiation, modifies the chemical structure, providing molecular chain scission and/or crosslinking [15]. All scales of polymer dimension, including the monomer unit, the chain (crosslinking or chain scission), and the morphology (breakdown of tie molecules and crystal) are changed due to photo-degradation. Due to the degradation of fibres and matrix after weathering periods, the polymer composite lost its tensile strength [15]. Therefore, it is necessary to highlight some of the research evolution of UV radiation effects towards composite material and open further areas of study related to UV degradation towards properties of polymer matrix composites.

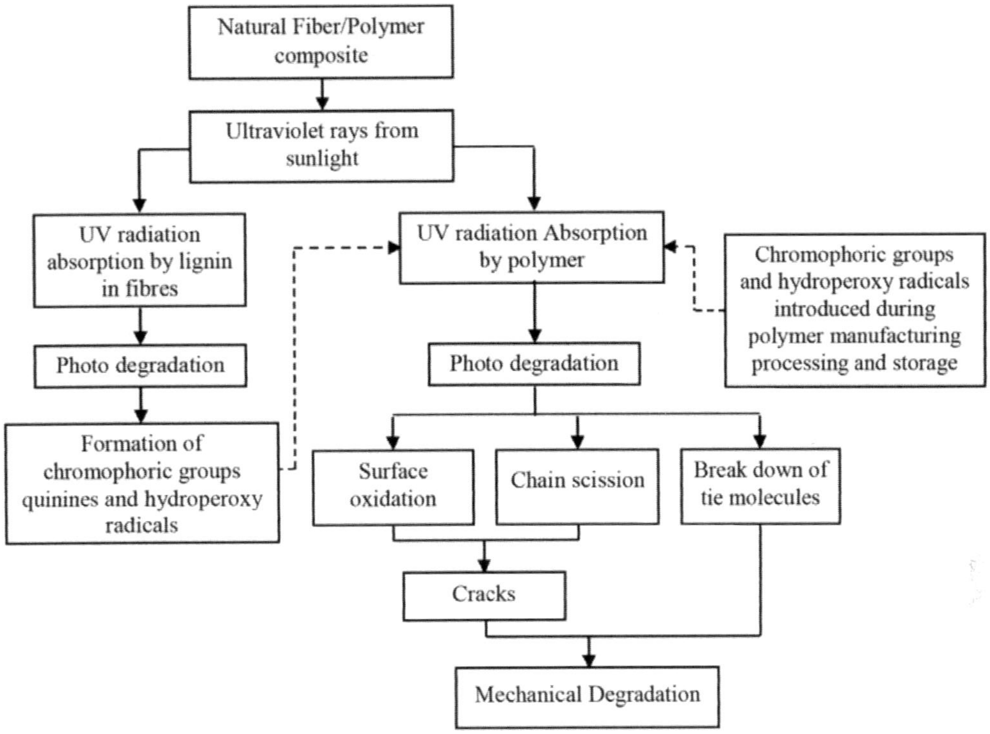

Figure 1. The schematic diagram representing the mechanical degradation processes due to UV exposure [15].

2. Previous Research

2.1 The radiation exposure and its effect

Nowadays, there are so many applications of polymer matrix composite materials which are UV exposed, including the sporting goods, aircraft components, automotive parts, consumer goods and many other applications [1, 25 – 28]. A polymer material may be degraded by chemical changes due to reaction with component in the environment. Basically most pure organic synthetic polymers (polypropylene, polyvinyl chloride, and polystyrene) do not absorb wavelengths longer than 300nm owing to their ideal structure, and hence should not be affected by sunlight [15]. However, polymer is also undergone the photo-oxidation, the oxidation which is induced by light. The following statement briefly introduced on the oxidation process. Polymer reacts so slowly with oxygen at normal temperature, thus the oxidation is only become apparent after a long time. But if UV light irradiates the polymer at the same condition for 12 days, the strong band in the spectrum are appeared. In this case, the polymer free radicals are formed by the synergetic action of factors like electromagnetic radiation and thermal energy on the oxidation. This condition is observed to occur in Polypropylene, Polystyrene and also natural rubber [15].

The effect of UV radiation and associated elevated temperature on mechanical performance of steel / CFRP (Carbon Fibre reinforced Polymer) double strap joints have been investigated. Epoxy adhesive samples were fabricated by pouring two parts of Araldite 420 epoxy into the mould. These adhesives are suitable for bonding both steel and polymeric composite [29]. The wet lay-up method was used for preparing the CFRP specimen. According to the Australian natural weather condition

[28], the total outdoor UV dose for a year in Australia ranges from 250MJ/m^2 to 385 MJ/m^2. This data were used to simulate outdoor the UV exposure. Radiation power was measured as 340 nm following the standard practice according to ASTM G155. The result shows that the ultimate strain in tension of the epoxy adhesive specimens (Araldite 420) has decreased under UV exposure and associated temperature. UV exposure for 372 x 2h does not influence the tensile elastic modulus and tensile strength of Carbon fibre and CFRP. The effective bond length of steel / CFRP adhesively bonded joints has increased with the duration of UV exposure [29].

2.2 Weather condition effects of degradation

The effect of weather condition on degradation has been investigated on jute / phenolic composites [15]. It shows that for 2 years UV exposure on jute / phenolic composites has decreased the tensile strength of about 50%. It is also observed that resin cracking, bulging, fibrillation and black spots have existed on the composites. An exposure to natural weather for 9 months on palm fibre / polypropylene also has showed that the composite experienced only a little drop in its strength as compared to pure polypropylene which experienced 50% strength reduction. Moreover, it was also found that increment in exposure time has decreased the fibre strength [15]. It can be assumed that the tensile strength and modulus have decreased due to embrittlement of the matrix [15]. Similar characteristics were observed for sisal reinforced polypropylene composites [15]. After 3 months of UV radiation exposure, it shows that the tensile strength was inversely behaved to the fibre loading. For a neat PP, the tensile strength was decreased by 92.5%. When 10% sisal fibre loading was introduced to the composite, the strength has reduced to 58%. For fibre loading of 20% and 30%, the tensile strength of composites has decreased to 37% and 23%, respectively.

It shows that by adding polypropylene into the composite materials help to alleviate crack propagation to some extent. The extent of retention in tensile properties has increased with the increased in fibre loading [15]. The accelerated weather conditions testing are preferred in investigating weather exposure as this method reduced the duration required for UV exposure. However, accelerated weather test did not exactly correlate to what happen in real condition due to the irregularity of the cycle, duration, intensity and exposure condition [15].

2.3 Fibre Treatment

The fibre treatment is conducted in order to improve the adhesion between the fibres and the matrices. There are several of fibre treatments available such as alkali treatment [30] and fibre coating [31]. For instance, fibre treatment may consist of mercerization, latex coating, γ radiation treatment, silane treatment, acetylation treatment, or peroxide treatment. Effect of fibre treatments on both morphological and single fibre tensile strength of oil palm empty fruit bunches (OPEFB) has been evaluated [18]. The result showed that alkali treatment on OPEFB fibre enhanced the tensile strength and thermal stability of the fibre samples. The effect of oil palm shell (OPS) powder on the mechanical performance and thermal stability of polyester composite has been investigated [16]. The influence of washing OPS powder in methanol to remove surface impurities was investigated with tensile and flexural strengths. The result showed that the composite tensile and the flexural strength generally increased with decreasing powder size with the strength of the composite containing 75 – 150 μm powder being similar to that of the pure matrix. However, the tensile and flexural moduli of the composite were found to be essentially independent to powder size. Thermogravimetry analysis (TGA) in flowing oxygen indicated that the addition of OPS powder shifted the thermal degradation peak of the bio composite from 370 °C to 480 °C.

The effect of fibre surface modification on water sorption characteristic of oil palm fibre has been investigated [17]. It was found that the mechanical properties of fibres have decreased upon sorption and regains on desorption [15]. The effect of coating on oil palm empty fruit bunches (OPEFB) fibres with acrylonitrile butadiene styrene (ABS) has been evaluated [31]. The result shows that the sheer strength parameter of the fibre reinforced soil was improved significantly. It was found that the Young's modulus of the coated fibre was improved with ABS coating. However, the tensile strength of the pure fibre indicated less increased in comparison to untreated fibre. The ABS coated fibre was found to be more durable against physical and chemical destruction compared to uncoated fibre.

3. Defining Analysis

3.1 Effect of UV radiation to nanocomposite coating on Metal

The effect of UV radiation to the tribology performance of the nanocomposite coating to steel has been investigated [32]. The UV radiation was evaluated by using weather tester equipment for simulating sunlight with a light source of fluorescent UVA-340. The samples were exposed to UV radiation for 300 h at a temperature 60°C without the water condensation or spraying cycle, with intensity 0.71W/m^{2}nm (according to ASTM G154-06). The result shows that exposure to UV for 300h at normal load of 4 N and a rotational speed of 2000 rpm (linear velocity = 0.41 m/s) didn't affect the tribology performance of the nanocomposite coating until 240,000 cycles [32].

3.2 Effect of UV light Exposure to Composite Material

Lopez et.al. [4] have investigated the effect of UV radiation to wood plastic composite. Here, a fluorescent lamp with peak emissions of 365 nm wavelength (UVA) and 313 nm wavelength (UVB) were used to expose the test coupons (20 XZ 160 X 3 mm^3) of composite with UV radiation in the environment chamber. The fluorescent lamp was set up to illuminate the samples for two periods, e.g. 2000 and 4000 hours of UV exposure. The result indicated that the UV exposure provides the same effect to the Modulus of Rupture (MOR) and the Modulus of Elasticity (MOE) of the samples. It was demonstrated that the UV exposure has increased both the MOR and MOE of the samples at high temperature (40°C) with low relative humidity (34%) or low temperature (23°C) with high relative humidity (93%) [4]. Vice versa, the UV exposure has decreased MOR and MOE of the samples at high temperature (40°C) with high relative humidity(93%) or low temperature (23°C) with low relative humidity (34%) [4]. This indicated that the relative humidity also influenced to the strength of the polymer composites.

Zhang et.al. [33] have investigated the effect of UV irradiation to magnetism in $SrTiO_3$. The single crystalline $SrTiO_3$ samples were prepared from $SrCO_3$ and TiO_2 by the modified composite-hydroxide-mediated approach. The results showed that the ferromagnetism of $SrTiO_3$ has enhanced after UV irradiation. It indicated that new diluted magnetic material of oxygen vacancy rich $SrTiO_3$ can be obtained by UV Irradiation [31]. The presence of compatibilising agent in polymer composite has also been studied by Kruzskowska et.al [34]. The degradation of polylactide based film composites exposed to UV-C radiation at 254 nm wavelength was evaluated. It was observed that the amount of unmodified montmorillonide as nanofillers and a compatibilising agent influenced the rate of photo-oxidative degradation of composite film [34]. Incorporating nanoparticles into epoxy polymer composites and delay onset of deleterious effect of UV radiation by partially curing the samples has been investigated by Narteh et.al [35]. It shows that the addition of nanoclay has increased the viscoelastic properties of the polymer composites. It was also observed that the storage modulus and activation energy of decomposition has evolved over exposure time period. Samples with partially cured has found delayed the UV radiation degradation effects [35].

A comparative evaluation on actual and artificial UV radiation was also investigated. Sil and Chakrabarti [36] have compared both tropical UV sunlight and artificial UV radiation on PVC-ZnO composite film for its photocatalytic degradation. The degradation was measured by weight loss and followed a pseudo first order rate equation. It was observed that after 90 minutes of radiation, an artificial UV radiation decomposed about 6.2% of its original weight, whereas tropical UV sunlight decomposed about 14.2% of its original PVC-ZnO weight [36].

Yakimets et.al. [37] have investigated the effect of UV photo oxidation cracks on behavior of thick polypropylene samples. The ageing degradation tests were done using the Suntest XLS+ containing a filtered xenon lamp, with total intensity of the lamp being about 600W/m^2. The samples were exposed continuously without day-night cycle. The samples were divided into four well defined periods. The first period was the incubation period (3-5 days). During this period, the effect of photodegradation of polypropylene was invisible. The spontaneous cracks were formed on the sample surfaces in the period of forming surface cracks (5 days - 3 weeks). These cracks caused the embrittlement of the material. In the third period (3 – 5 weeks) the sample surfaces with many cracks offer the optimal conditions for the development of chemy-crystallization which produces small independent crystallities, therefore the material lost its mechanical resistance. The sample surfaces totally affected by chemical crystallisation in the stabilisation period (5-7 weeks) which was in the last period of the research. There wasn't any more cracks formed at this stage due to the exposed surfaces were completely photo oxidized [37].

Davis et al. [38] have investigated the effect of photo-oxidative degradation on fracture of the ABS (Acrylonitrile-butadiene-styrene) pipe resins. The analysis of DN10 Class 15 ABS pipe predicted a transition to brittle fracture failure occurs after 137 years. However, to accurately analyse, the effect of pipe wall thickness should also be quantified [38].

The environmental effects on the degradation behavior of sisal fibre reinforced polypropylene have been investigated by Joseph et al. [39]. There are some chemical treatments were given to sisal fibre in this research, i.e. MA-PP (Maleic anhydride polypropylene), TDI-PPG (Toluene 2,4 diisocyanate) - (Polypropylene glycol), PMPPIC (Poly[methylene poly (phenyl isocyanate)]. After exposure to UV radiation in a weatherometer for some period of time, i.e. 2, 4, 6,8 and 12 weeks, the tensile properties of both untreated and chemically treated composites were observed to have decreased due to chain scission as a result of photooxidation. After being exposured to UV radiation for 12 weeks, the untreated fibres composite still has the highest retention in tensile strength e.g. 63.0 MPa while MA-PP treated fibres composite has the lowest retention in tensile strength, e.g. 50.4 MPa [39]. The MA-PP treated fibres composite has the highest Young's modulus even for unirradiated samples (1685 MPa) or 2,4,6,8 or 12 weeks irradiated samples (i.e. Young's modulus for 12 week irradiated sample was 1063 MPa).

Thermal degradation of kenaf fibre/epoxy composite using thermo gravimetric analysis have been investigated by Azwa et al. [40]. There were 2 types of fibre used in this research, i.e. untreated and the treated fibre. For the treated fibre, during samples preparation, the kenaf fibres were treated by solution of 6% sodium hydroxide for 24 hours at room temperature. The degradation of the samples analyzed by thermo gravimetric analysis (TGA). The study concluded that the treated fibre has lowered the decomposition temperature and produced lesser char than untreated composite, caused by removal of lignin [40].

Hybrid flax reinforced supersap composites have been investigated by Zhu et al.[41]. There are five kinds of conditions of flax fibres used in these experiments, i.e. Untreated, NaOH Treatment (5% NaOH), BTCA Treatment (Alkali + BTCA (Butanetetracarboxylic acid)), APS Treatment (Alkali + APS (Amiopropyltriethoxysilane)) and Lacasse Doga (LD) Treatment. The UV aging conditioning was done in QUV accelerated weathering tester and freezer by using UVB-313 in the temperature range of -15 to 25°C, cycles for 552h (23 days). Similarly samples were kept in the Q-Sun Xe-3-HS and Q-SUN Xenon test Chamber XE-3-H/HSB/HS for being exposed to artificial weathering for 500 h with xenon arc light. Another condition of samples was the normal condition samples, which were samples without UV radiation and xenon light. The results obtained demonstrated that the untreated fibres composite has the highest tensile strength compared to others for normal condition and UV radiation test (i.e.: 185.4±8.5 MPa and 185.4±7.8 MPa respectively). But in xenon aging test, APS treatment samples have the highest tensile strength (i.e 175.1 ± 6.2 MPa) [41]. The untreated fibres composite also has highest tensile modulus compared to others, for normal condition and UV radiation test (i.e.: 13.9±0.4 GPa and 14.0±0.3 GPa respectively). But in xenon aging test, LD treatment samples has the highest tensile modulus (i.e 11.9 ± 0.3 MPa) [41].

4. Conclusion

This paper has reviewed the effect of UV radiations on composite materials; in particular the polymer matrix reinforced natural fibre composites. It has been noted that some investigations have been conducted through some natural fibres reinforcing composite. From the review, it is well described that UV has played significant role in degrading the material strength. The fibre tensile strength was found to decrease with the duration of UV exposure time. UV radiation has been linked to the degradation of composite materials with the degradation rate depends on some key parameters such as UV wavelength, exposure time and UV intensity. The influence of using accelerated UV equipment also contributing to the degradation rate, where it degrades much faster than the normal UV exposure. Therefore this review could provide some significant areas for further development of UV exposure analysis particularly towards natural fibre reinforced polymer matrix composite materials.

Acknowledgements

The authors would like to thank the Ministry of Higher Education Malaysia and Universiti Tun Hussein Onn Malaysia (UTHM) for providing the financial support through FRGS Grant Vot 1541. Also, special thanks to Office for Research, Innovation, Commercialization and Consultancy Management (ORICC) and Faculty of Mechanical and Manufacturing Engineering for providing the facilities and assistance throughout the project.

References

[1] Strong A B 2000 *Plastics: Materials and Processing* (New Jersey: Prentice Hall)
[2] Gay D 2015 *Composite Materials: Design and Applications* (UK: CRC Press)
[3] McCrum N G, Buckley C P and Bucknail C B 2005 *Principles of Polymer Engineering* (UK: Oxford Science Publication)
[4] Lopez J L, Sain M and Cooper P 2006 *Appl. Polymer Sci.* **99** pp 2570.
[5] Karina M, Syampurwadi A and Oggo H 2007 *J. Biol. Sci.* **7(2)** pp 393.
[6] Zaleha M, Mahzan S and Maizlinda Izwana I 2012 Applied Mech. Mtrls **229-231** 276.
[7] Samuel O D, Agbo S and Adekanye T A 2012 *J Mineral and Mtrls Charact Eng.* **11** pp 780.
[8] Hasamudin W and Soom R M 2002 *Malaysian Palm Oil Board*, Kuala Lumpur Malaysia.
[9] Shinoj S, Visvanathan R, Panigrahi S and Kochubabu M 2011 *Industrial Crops and Products.* **33** pp 7.
[10] Kakou C A, Arrakhiz F Z, Trokourey A, Bouhfid R, Qaiss A and Rodrigue D. 2014 *Mater Des.* **63** pp 641.
[11] Brinson H F and Brinson L C, 2008 *Polymer Engineering Science and Viscoelasticity: An Introduction.* (New York: Springer).
[12] Horath L. 2001 *Fundamentals of Material Science for Technologist*, (USA: New Jersey: Prentice Hall).
[13] Guneri A 2005 *Polymers in Construction.* (USA: Smithers Rapra Technology).
[14] Ku H, Wang H, Pattarachaiyakoop N and Trada M 2011 *Composites Part B* **42** pp 856.
[15] Azwa Z N, Yousif B F, Manalo A C and Karunasena W, 2013 *Mater Des.* **47** pp 424.
[16] Nabinejad O, Sujan D, Rahman M E, Ian J D 2015 *Mater Des* **65** pp 823.
[17] Sreekala M S and Thomas S 2003 *Comp Scie Tech* **63** pp 861.
[18] Norul Izani M A, Paridah M T, Anwar U M K, Mohd Nor M Y and H'ng P S, 2013 *Composites Part B* **45** pp 1251.

IOP Conf. Series: Materials Science and Engineering **165** (2017) 012021 doi:10.1088/1757-899X/165/1/012021

[19] Khalil A H P S, Marliana M M, Issam A M, Bakare I O 2011 *Mater Des* **32 pp** 2604.
[20] Moshiul Alam A K M, Reddy Prasad D M, Khan M R and Mina F 2012 *Composites Part A* **43 (11)** pp 1921.
[21] Awaja F and Pigram P J, 2009 *Polym Degrad Stab.* **94** pp 651.
[22] Van Krevelen D W, Nijenhuis K T, 2009 *Properties of Polymer*. (Netherland Amsterdam: Elsevier).
[23] Stempien Z, Dominiak J and Sulerzyeka-Bil M, 2013 *Eastern Europe* **21** pp 96.
[24] Diffey, BL. 2002 *Methods* **28** 4.
[25] Witold B 2000 *Performance of Plastics*. (USA: Hanser Publisher Ohio).
[26] Schwartz M M. 1997 *Composite Materials, Volume II: Processing, Fabrication, and Applications*. (USA New Jersey: Prentice Hall).
[27] Dutton S and Baker A A 2004 *Composite Materials for Aircraft Structures. AIAA education series*, (USA Virginia: Blacksburg).
[28] Sanjay K M 2002 *Composites Manufacturing, Materials, Product, and Process Engineering*. (USA Florida: CRC Press, Boca Raton).
[29] Nguyen T C, Bai Y, Zhao X L and Al Mahidi R, 2012 *Composite Structures* **94** pp 3563.
[30] Hashim M Y, Roslan M N, Mahzan S and Ariffin S 2014 *App. Mech. And Mater.* **660** pp. 285.
[31] Farshid B, Fauziah A, Ahmad Shukri Y and Mastura A, 2011 *Constr Build Mat.* **25 pp** 1824.
[32] Samad M A and Sinha S K 2011 *Wear* **271** pp 2759.
[33] Zhang S, Guo D, Wang M, Javed M S and Hu C 2015 *Appl. Surface Scie.* **335 pp** 115.
[34] Kruszkowska E O, Koter I, Wisniewska S J and Richert, 2015 *J. Photochem. and Photobio. A: Chemistry* **311** pp 144.
[35] Narteh A T, Hosur M, Triggs E, Owuor P and Jelaani S, 2014 *Polymer Degrad and Stability* **101** pp 81.
[36] Sil D and Chakrabarti S, 2010 *Solar Energy* **84** pp 476.
[37] Yakimets I, Lai D and Guigon M, 2004 *Polymer Degrad and Stability* **86** pp 59.
[38] Davis P, Tiganis B E and Burn L S, 2004 *Polymer Degrad and Stability* **84** pp 233.
[39] Joseph P V, Rabello M S, Mattoso L H C, Joseph K, Thomas S, 2002 *Comp. Sci. and Tech.* **62** pp 1357.
[40] Azwa Z N and Yousif B F, 2013 *3rd Malaysian Postgraduates conference* **16** pp 256.
[41] Zhu J, Immonen K, Avril C, Brighton J, Zhu H, and Abhyankar H, 2015 *Fibers* **3** pp 76.

Characterization of WC-10Ni HVOF Coating for Carbon Steel Blade

N.A.Ahmad[1,*], Z. Kamdi[1], Z. Mohamad[1], A. S. Omar[1], N. Abdul Latif[1], A. L. Mohd Tobi[1]

[1]Faculty of Mechanical and Manufacturing Engineering University Tun Hussein onn Malaysia (UTHM), 86400, Batu Pahat, Johor, Malaysia.

*E-mail: aqilahahmad09@yahoo.com

Abstract. High Velocity Oxy-Fuel, HVOF is a depositing methods of a material layer over a base metal or substrate with characteristics of high flame velocity and moderate temperature. Where, tungsten carbide, WC cermet HVOF coatings is widely used to protect machine components from wear and corrosion. The main purpose of this present paper is to characterize the WC-10Ni coating deposited by HVOF thermal spray onto a carbon steel blade. The morphology and chemical composition of the coating were characterized by Scanning Electron Microstructure (SEM), electron dispersive spectrometer (EDS), and X-ray diffraction (XRD). The hardness test was carried out by using Vickers micro-hardness tester with loads of 490.3 mN (HV0.05). From XRD results, no sharp nickel peak was identified and has been replaced by a hump which indicate the amorphous Ni. The major crystalline phases were compounds WC, W_2C and metallic phase of W. The WC-10Ni coating shows high hardness with low porosity distribution.

1. Introduction

HVOF is one of the thermal spraying technique for WC feedstock powder in order to obtain good bond strength and higher density. It is widely used to deposit carbide-based cements, metals and alloys coatings such as WC-Co, WC-Ni and stainless steel material [1].

HVOF has special features of higher particle velocity and lower particle temperature compared to other spray processes. According to Oksa *et al.* [2], this combination leads to adherent coatings with lower porosity, less extent decarburization and higher wear resistance. A large number of researchers have reported that decarburization of WC has a detrimental effect on the abrasive wear resistance of coating. This is due to the increasing brittleness and decrease in hard particle content. Thus, the drawback of decarburization can be avoided by the cold spray technique, a recently emerging technique operating at temperatures much lower than the melting point of spray materials [3].

WC has been proven their superiority in great number of other tooling and engineering applications due to their unique properties. Over the years, many studies showed the fabricated WC-based coatings exhibited higher hardness, fracture toughness and correspondingly improve wear resistance compared to the other carbide [4]. According to Van Acker [5], WC were reinforced by the binders such as Cobalt (Co) and Nickel (Ni) in order to increase wear and corrosion resistance. For cemented WC, the properties of the components are superimposed where the carbide phase WC provides hardness and wear resistance. While, the ductile binder contributes to toughness and strength [6]. However, the microstructure of WC cracks, large pores and poor bonding occurred as the percentage of binder's content increase [5-7]. From the previous researches, due to the large pores, the coating facing the abrasion and erosion behaviour, where it leads to reduction of wear resistance.

Co has been the dominating in metal-ceramic composites known as hard metals since the emergence of the first WC-Co cemented carbide in 1923 [8]. This is due to the chemical bonding between WC and Co resulting in a very low interfacial energy, nearly perfect wetting and a very good adhesion in solid state [9]. However, the high price Co metal together with the need for improving the performance of cemented carbides under severe working condition such as corrosion, alternative binder's Ni have been promoted [10]. Ni is a metal that has similarities as Co in structure and properties, besides it is a good corrosion resistance.

In this study, the Blade was used to stir the mixture of sulphuric acid (H_2SO_4) and ilmenite in order to maximizing Titanium Dioxide (T_iO_2) pigments. This blade exposed to deterioration, wear and corrosion due to the corrosive (acid) and erosive (ilminite) properties of the mixture. Thus, as modern method is quite effective in improving the wear and corrosion resistance, one of the option is thermal spraying process (HVOF) with WC-based materials. As many coating enhancements have been reported by researchers, more understanding on the characteristics of the HVOF thermal spraying is needed, so that this technique could be commercialize for various industrial applications in the next future. In this paper, the characterization of WC-10Ni HVOF coated on the carbon steel blade was analysed.

2. Methodology

2.1. Coating Deposition

The powder employed in the present work are known as WOKA 3302. This powder is mainly composed of 90% WC and 10% Ni, where it prepared by agglomeration and sintering. Table 1 summarizes the deposition condition, gun characteristic and particle size for the powder, which correspond to those typically applied industrially. The substrate (blade) that used in this study was a mild carbon steel (Cast steel BS 3100 GR A3) supplied by vendor (Huntsman Tioxide). The chemical composition of both substrate and powder materials shown in Table 2.

Before the deposition of the coating, the substrate surface was thoroughly cleaned by acetone in order to remove mill and rust scale, dirt and foreign materials. Grit blasting was carried out by using an Empire 350 abrasive equipment at a pressure of 2 kg/cm^2 with alumina grit. The purpose of this process is to ensure the spray powder (coating) strongly bond to the substrate. The WC-10Ni coating subsequently deposited had average thicknesses in the range of 200 μm to 300 μm.

Table 1. Deposition parameters supplied by the industrial coating facility

	WOKA 3302
Thermal spray gun	The Diamond Jet (SULZER METCO)
Powder feeding rate (SCEH)	28.5
Particles size	-45+15 μm
Oxygen Pressure (psig)	150
Fuel Pressure (psig)	110
Air Pressure (psig)	100
Air flow (SCEH)	893
Fuel flux (SCEH)	432
Oxygen flux (SCEH)	635
Spraying distance (in.)	9
Spray Rate (lbs/h)	5

Table 2. Chemical composition of powder (WC-10Ni) and substrate (medium carbon steel)

Wt. [%]	W	C	Ni	Fe	Si	Mn	P	S
Powder WC-10Ni	Bal	5.2-6.0	8.5-11.5	<0.2	-	-	-	-
Carbon Steel(BS 3100)	-	0.45	-	-	0.06	1.0	0.05	0.06

2.2. *Microstructure Characterization*
Samples was grind by using sand paper starting from grit 360, 400, 600, 800,1000,1200 and 1500. Metcon polisher was used to polish the samples with additional of alumina 0.05 μm. The morphological characterization of the coatings was carried out by means of optical microscope (OM) and scanning electron microscope (SEM, Jeol 2000) with energy dispersive spectroscopy (EDS). The coatings analysis was conducted both on the deposition surface (S) and cross sectional area (CS). The apparent porosity of the coating was determined by using SEM. The porosity values correspond to the analysis of a mean of 20 images taken at a magnification of 200x. X-Ray Diffraction (XRD) with Cu Kα (λ = 0.154 nm) radiation were used to identified the phase present in the coating. All the patterns were determined in the interval $20^\circ < 2\Theta < 80^\circ$ at a scanning step of 2°/ min and 0.02° with time step of 0.3s.

2.3. *Mechanical Characterization*
The coating hardness was evaluated by means of the Vickers Micro-Hardness. The measurements were carried out under indention loads of 490.3 mN (0.05HV) within 10 s with flat shape of test piece for 10 times and the average of the readings were calculated. From the other researches, at least 5 indentions were conducted at each applied load, both on the coating surface and cross section. The test was carried out randomly in zones apparently homogeneous, free of pores and other discontinuities, as far as the observation under the optical microscope is concerned. All the test was done using Micrometer (USA) micro-hardness measurement set up.

3. Result and Discussion

3.1. *Microstructure Analysis of Powder and Coating.*
The SEM micrographs (Fig. 1) shows the characteristic of an agglomerated and sintered powder. The particles size of the powder was from the range of 11 μm to 50 μm. The powder particles consist of W, C and Ni was proven by EDS in Fig 2. While, Fig.3 shows the existence of W, C, and Ni in the coating of WC-10Ni.

Fig.4 illustrate the Backscattered electron (BSE) microstructure with high magnification of the WC-10Ni coating layers. It can be seen that the coating consists of angular carbide which dissolve into the metallic binder of Ni. Furthermore, WC-10Ni shows a greater amount of rounded carbide particle which means the carbide grains of the WC is preferentially melted into the Ni phases in high temperature during thermal spray process [11].

The cross-sectional micrograph revealed a good adhesion between dense coating layer and substrate as shown in Fig 5. The high velocity impact of coating materials to the substrate accompanies some severe plastic deformation and rapid solidification. Thus, spray coating layers are likely to possess inevitable pores and cracks [12]. The thickness of the coating was within the range 130μm to 150μm.

IOP Conf. Series: Materials Science and Engineering **165** (2017) 012022 doi:10.1088/1757-899X/165/1/012022

Figure 1a. SEM micrographs of experimental WC-10Ni powder particle

Figure 1b. Particle size of WC-10Ni powder

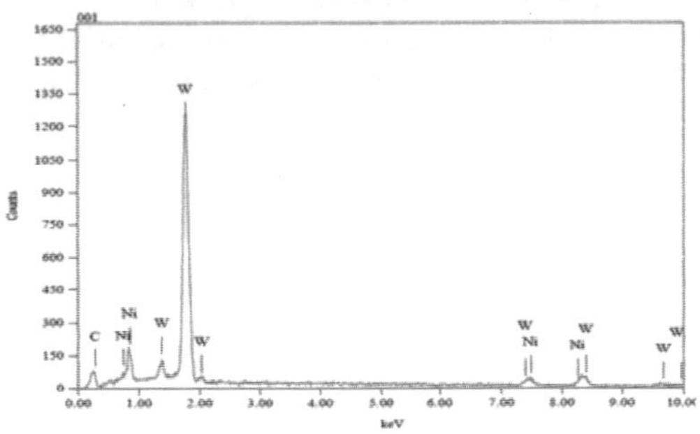

Figure 2. EDS analysis of WC-10Ni powder

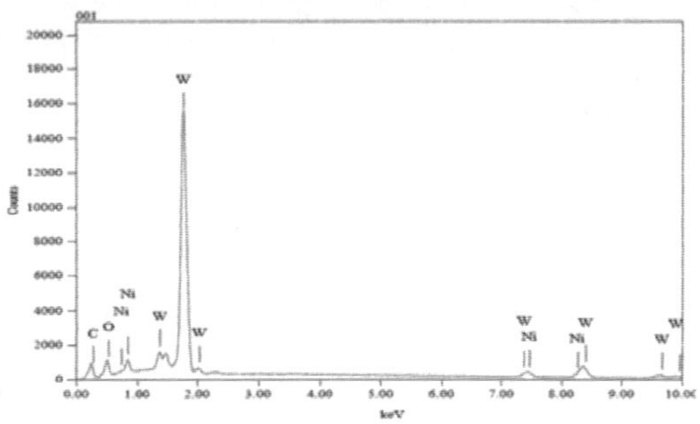

Figure.3. EDS analysis of coating

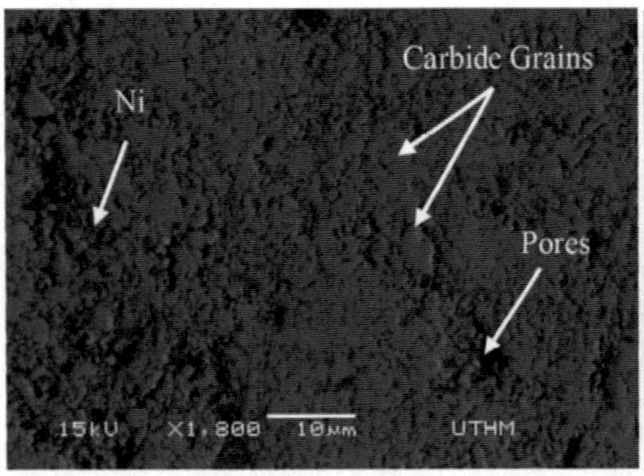

Figure 4. BSE micrograph of coating

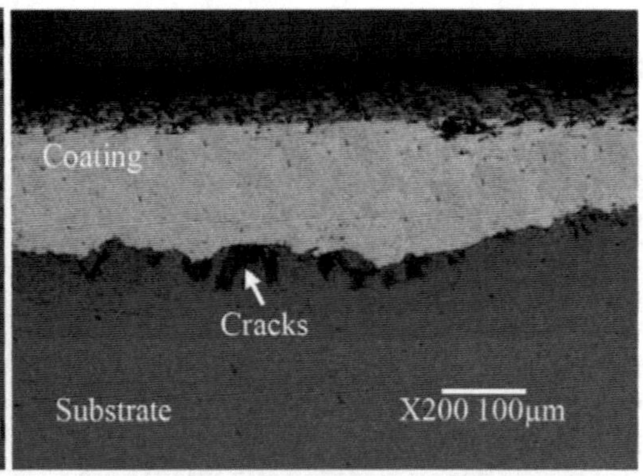

Figure 5. Cross section micrograph of coating

3.2. XRD characterization of powders and coatings.

Fig. 6 illustrate the results of the X-ray analysis conducted on the WC-10%Ni powder and corresponding HVOF sprayed coatings. The peaks of HVOF coatings are broader than those of the Scorresponding spray powders. Broader peaks are the result of a smaller crystal size and/or microstrains inside crystal which is occurred due to the thermal and kinetic process encountered by powder particles during thermal spraying. Indeed, the particles are heated and accelerated with high velocity, and might impinge on the substrate in unmelted, partially molten or fully molten condition [13].

The structure of the powder showed that only WC and Ni phases existed in the powder. There was no evidence of intermetallic phases based on Ni and WC had been formed. While, the XRD patterns for coatings showed WC, W_2C, W and Ni phases. In coating, there is a larger hump in the intensity of the background of the XRD pattern at 2θ between 40 and 46°. It indicates that some development of amorphous phases has taken place during the thermal spray process. The amorphous region was formed as a consequence of the carbide or tungsten dissolution and/or diffusion into the matrix, which leads to a supersaturation of the matrix [14]. In addition, previous researches also stated that the formation of amorphous phases is due to the combination of extremely high cooling rate of impinging molten droplets and the slow crystallization kinetics of the sprayed coating process [15].

During HVOF thermal spraying, it is possible that the secondary phases of W_2C and W are formed due to the decarburization process at high temperature in the flame. The loss of carbon from WC phase occurred during the coating degradation which leads to formation of metallic W phase [9].

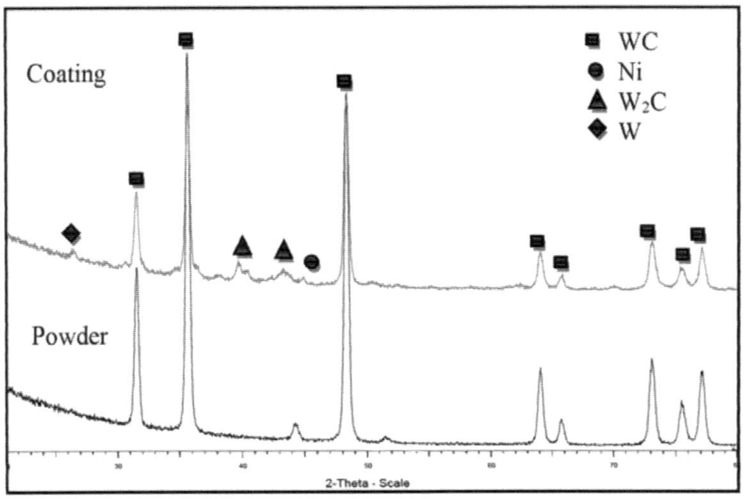

Figure 6. XRD patterns and the powders and corresponding as-sprayed HVOF coatings.

3.3. Mechanical Properties of Coatings.

The hardness distribution is illustrated in Fig 7. Mechanical analysis show that the hardness of the WC-10Ni coating ranges from 1480 Hv to 1544 Hv indented under load of 490.3 mN (0.05HV). It can be seen that the hardness of substrate is 80% lower than hardness of coating. Where, the hardness of the coating is a result of balance microstructure, phase contribution (amounts of W_2C, amorphous phase) and porosity [16-17]. It is reasonable that the hardness increases with existence of W_2C in the coating due to the phase is harder than WC phase. Additionally, the lower hardness of the coating can be explained by the high porosity content. The porosity in terms of area fraction of pores and hardness of the coating are compared in Table 3.

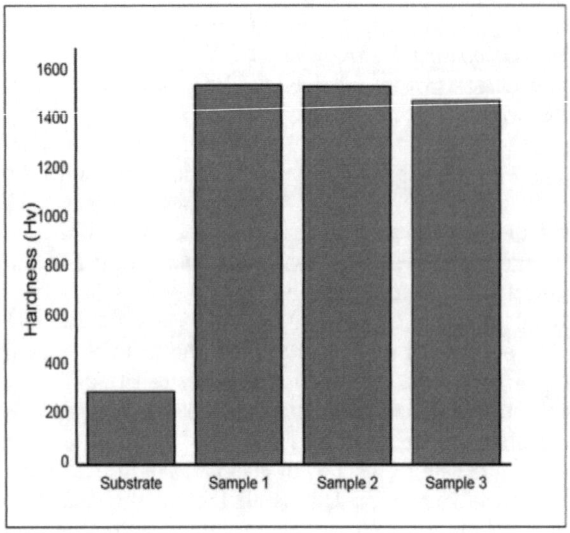

Figure 7. Distribution of Hardness of substrate and coating of WC-10Ni

Table 3. The hardness and the porosity value of the coating

	Sample 1	Sample 2	Sample 3
Hardncss (Hv)	1544	1540	1480
Porosity (%)	1.22	1.28	1.34

4. Conclusion

The blade coating with WC-10Ni has been analyzed in terms of surface morphology, particles in the coating, thickness and hardness of the coating. The conclusion that can be drawn from this study are:

- It has been shown that the carbide grains of the WC are preferentially melted into the Ni phases, where the coating has a greater amount of rounded carbide particles.
- The secondary phases of W_2C and W are formed due to the decarburization process at high temperature in the flame during HVOF thermal spraying.
- The hardness of the coating depends on the pores distribution. The hardness increases with decreasing of porosity contents, where WC-10Ni coating has high hardness with low porosity distribution.
- Thus, HVOF WC-10Ni sprayed coating has strong ability to increase the performance of blade due to results obtained in this research.

5. Acknowledgement

This research was financially supported by Knowledge Transfer Programmed (KTP) project under grand no 1483.

IOP Conf. Series: Materials Science and Engineering **165** (2017) 012022 doi:10.1088/1757-899X/165/1/012022

References

[1] Sun, B., Fukanuma, H., and Saitama, J 2000: *Study on Stainless Steel 316L Coatings Sprayed by High Pressure HVOF,* Surface and Coating Technology **239**, 58-64.

[2] Maria Oksa, Erja Turunen, Tomi Suhonen, Tommi Varis and Simo-Pekka Hannula 2011: *Optimization and Characterization of High Velocity Oxy-Fuel Sprayed Coating:* Techniques, Materials and Applications, Coatings **1**, 17-52.

[3] Wang, Q., Zhang, S., Cheng, Y., Xiang, J., Zhao, X., and Yang, G. 2013: *Wear and Corrosion Performance of WC-10CO4Cr Coatings Deposited by Different HVOF and HVAF Spraying Processes*, Surface and Coating Technologies **218**, 127-136.

[4] B. Weilage, H. Pokhmurska, A. Wank, G. Reisel, S. Steinhaeuser, M. Woezel 2003: *Influence of Thermal Spraying Method on the Properties of Tungsten Carbide Coatings*, Advance in Technology and Application.

[5] Van Acker, K., Vanhoyweghen, D., Persoons, R. & Vangrunderbeek, J. 2005: *Influence of Tungsten Carbide Particle Size and Distribution on The Wear Resistance of Laser Clad WC/Ni Coatings*, Wear **258**, 194-202.

[6] C.M. Fernandes, A.M.R. Senos, M.T. Vieira and J.M. Antunes 2008: *Mechanical characterization of composites prepared from WC powders coated with Ni rich binders*, International Journal of Refractory Metals & Hard Materials **6**, 491-498.

[7] Su, W., Sun, Y., Liu, J., Feng, J., and Ruan, J 2015: *Effects of Ni on the Microstructures and Properties of WC-6CO Cemented Carbides Fabricated by WC-6(CO, Ni) Composite Powders*, Ceramic International **41**, 3169-3177.

[8] Ettmayer P, Kolaska H, Ortner HM 2014: *History of Hardmetals*, Comprehensive Hard Materials **1**. Elsivier.

[9] J.M. Tarrago, C. Ferrari, B. Reig, D. Coureaux, L. Schneider and L. Llanes 2015: *Mechanics and Mechanisms of fatigue in a WC-Ni Hardmetal and a Comparative Study with Respect to WC-Co Hardmetals*, International Journal of Fatigue **70**, 1-150.

[10] J.E. Cho, S.Y. Hwang, and K.Y. Kim 2006: *Corrosion behavior of thermal sprayed WC cermet coatings having various metallic binders in strong acidic environment*, Surface & Coatings Technology **200**, 2653-2662.

[11] Lining Gua, Jiwu Huanga, Yunfeng Tang, Chenhui Xie, and Shuangshuang Gao 2015: *Influence of Different Post Treatments on Microstructure and Properties of WC-Co Cemented Carbides*, Journal of Alloys and Compounds **620**, 116-119.

[12] J.G. La Barbera-Sosa, Y.Y. Santana, M.H. Staia, D. Chicot, J. Lesage, J. Caro, G. Mesmacque, E.S. Puchi-Cabrera 2008: *Microstructural and mechanical characterization of Ni- base thermal spray coatings deposited by HVOF*, Surface & Coatings Technology **202**, 4552- 4559.

[13] Giovanni Bolelli, Luca Lusvarghi, and Roberto Giovanardi 2008: *A Comparison Between the Corrosion Resistances of some HVOF-Sprayed Metal Alloy Coatings*, Surface & Coatings Technology **202**, 4793-4809.

[14] M. Magnani, P.H. Suegama, N. Espallargas, S. Dosta, C.S. Fugivara, J.M. Guilemany, and A.V. Benedetti 2008: *Influence of HVOF parameters on the corrosion and wear resistance of WC-Co coatings sprayed on AA7050 T7*, Surface & Coatings Technology **202**, 4746-4757.

[15] T. Sahraoui, H.I. Feraoun, N. Fenineche, G. Montavon, H. Aourag, and C. Coddet 2004: *HVOF-Sprayed Tribaloyn-400: Microstructure and First Principle Calculations*, Materials Letters **58**, 2433-2436.

IOP Conf. Series: Materials Science and Engineering **165** (2017) 012022 doi:10.1088/1757-899X/165/1/012022

[16] Pornthep Chivavibul, Makoto Watanabe, Seiji Kuroda, and Kentaro Shinoda 2007: *Effects of Carbide Size and Co Content on the Microstructure and Mechanical Properties of HVOF-Sprayed WC–Co Coatings*, Surface & Coatings Technology **202**, 509-521.
[17] L.M. Berger, S. Saaro, T. Naumann, M. Wiener, V. Weihnacht, S. Thiele, and J. Suchánek 2008: *Microstructure and Properties of HVOF-Sprayed Chromium Alloyed WC–Co and WC–Ni Coatings*, Surface & Coatings Technology **202**,4417-4421.

IOP Conf. Series: Materials Science and Engineering **165** (2017) 012023 doi:10.1088/1757-899X/165/1/012023

Theoretical Modelling of Sound Radiation from Plate

I Zaman[1,a], S A M Rozlan[1], A Yusoff[2], M A Madlan[1], S W Chan[3]

[1]Faculty of Mechanical and Manufacturing Engineering, Universiti Tun Hussein Onn Malaysia, 86400 Batu Pahat, Johor, Malaysia
[2]Kolej Kemahiran Tinggi MARA Pasir Mas, 17000 Pasir Mas, Kelantan, Malaysia
[3]Faculty of Technology Management and Business, Universiti Tun Hussein Onn Malaysia, 86400 Batu Pahat, Johor, Malaysia

[a]Email: izzuddin@uthm.edu.my

Abstract. Recently the development of aerospace, automotive and building industries demands the use of lightweight materials such as thin plates. However, the plates can possibly add to significant vibration and sound radiation, which eventually lead to increased noise in the community. So, in this study, the fundamental concept of sound pressure radiated from a simply-supported thin plate (SSP) was analyzed using the derivation of mathematical equations and numerical simulation of ANSYS®. The solution to mathematical equations of sound radiated from a SSP was visualized using MATLAB®. The responses of sound pressure level were measured at far field as well as near field in the frequency range of 0–200 Hz. Result shows that there are four resonance frequencies; 12 Hz, 60 Hz, 106 Hz and 158 Hz were identified which represented by the total number of the peaks in the frequency response function graph. The outcome also indicates that the mathematical derivation correlated well with the simulation model of ANSYS® in which the error found is less than 10%. It can be concluded that the obtained model is reliable and can be applied for further analysis such as to reduce noise emitted from a vibrating thin plate.

1. Introduction

Formerly about 4 decades ago, the vibrations of complex engineering systems was done using rough model with only a few degrees of freedom. However, since the advent of high-speed digital and interactive computers in the early 2000s had managed to produce an approximate solution that can display thousands of degrees of freedom efficiently through matrices. The concurrent development of the finite element method allows engineers to use the digital computer to perform a detailed vibration analysis of mechanical systems and complex structure [1,2]. According to Clough [3], the finite element method as known today was introduced by Turner, Clough, Martin and Topp through their research related to analysis of aircraft structures.

In engineering and construction industry, machines or equipment are certainly tool that used to help people to accomplish the objectives as planned by the company. Nevertheless, the use of process technology produces a huge problem to vibration and noise [4,5]. For example in the event of machinery and equipment are not treated, it can emit considerable sound power. In this case, workers who operating the equipment and machines as well as its surrounded people will be directly exposed to vibration and noise affection. For the employee, it can cause the loss of efficiency and discomfort.

Moreover, it also affects the public and the environment, and at the same time can bring harm to the machine itself.

Meanwhile, plate is one of the most extensively used structures in industrial due to its lightweight and economy advantages. Some of the example application of plates can be found in aircraft, automobile, machinery and modern houses. Despite this advantage, the utilization of plate has potential to contribute substantial vibration and sound radiation which leads to excessive noise. Putra and Thompson [6] in his study state the vibration of engineering structures, particularly those consisting of thin plate-like members can be a significant source of noise. This is why the study of sound radiation of a plate become of great interest to both researchers and practicing engineers since the noise produced from plate can be meaningfully reduced. In study by Snyder and Tanaka [7] stated that acoustic radiation from plates with various boundary conditions and loading features can be minimized by altering the relative amplitudes of the vibration modes. Similar study was carried out by Park et al. [8] who observed that the sound emitted from plate can be weaken by modifying the boundary support configurations.

Therefore, this paper strive to present details derivation of a mathematical model which allows the calculation of frequency response function and sound radiation of a simply-supported plate. The outcome of this study will provide a helpful reference to future researchers who endeavor to find an equation for sound radiation of simply-supported plate and use it for further analysis in sound reduction.

2. Mathematical Modelling

This section describes the derivation of a mathematical model which allows the calculation of sound radiation of a simply-supported plate. At first, the mathematical equation for dynamic response of a simply-supported plate is derived, followed by the sound radiation from a simply-supported plate.

2.1. Simply-Supported Plate (SSP)

The equation of motion of a simply supported plate is written as:

$$EI\left(\frac{\partial^4\omega}{\partial x^4}+2\frac{\partial^2\omega}{\partial x^2\omega\partial y^2}+\frac{\partial^4\omega}{\partial y^4}\right)+\rho h\frac{\partial^2\omega}{\partial t^2}=-F(\mathrm{x,y,t}) \tag{1}$$

where E is the Young's modulus, I is the area moment of inertia, ρ is the density of plate and h is the thickness of plate. Eq. (2) defined the area moment of inertia for plate, where v is the Poisson's ratio.

$$I=\frac{h^3}{12\left(1-v^2\right)} \tag{2}$$

The solution of transverse modal displacement for a plate is given by Eq. (3) which is the summation of all individual modal amplitude responses multiplied by their mode shapes at that point.

$$\omega(x,y,t)=\sum_{m=1}^{\infty}\sum_{n=1}^{\infty}W_{mn}\cdot\psi_{mn}(x,y)e^{j\omega_n t} \tag{3}$$

where W_{mn} is the modal amplitude, $\Psi_{mn}(x,y)$ is the mode shape of plate, and m and n are modal integers.

The general mode shape of a simply-supported plate can be calculated with:

$$\psi_{mn}(x,y)=2\sin\left(\frac{m\pi x}{a}\right)\sin\left(\frac{n\pi x}{b}\right) \tag{4}$$

where a and b are the length and width of a plate, respectively.

The natural frequencies of a simply-supported plate can be calculated from:

$$\omega_n = \sqrt{\frac{EI}{\rho h}}\left[\left(\frac{m\pi}{a}\right)^2 + \left(\frac{n\pi}{b}\right)^2\right] \tag{5}$$

By neglecting the exponential time varying term, an expression of the total response if simply-supported plate incorporating the viscous damping ζ is given in Eq. (6) as derived in our previous study [9,10].

$$\omega(x,y,t) = \frac{F}{\rho hab}\sum_{m=1}^{\infty}\sum_{n=1}^{\infty}\frac{\psi_{mn}(x,y)\psi_{mn}(x_i,y_i)}{\omega_{mn}^2 - \omega^2 + j2\zeta\omega\omega_{mn}} \tag{6}$$

2.2. Sound Radiation by a Simply-Supported Plate
The simplest approach to calculate the sound field radiated by a vibrating surface that is surrounded by a rigid infinite panel is the evaluation of the Rayleigh Integral, which given as follows [11]:

$$p(x_0,y_0,t_0) = \frac{j\omega p_a \mathring{\omega}(x,y,z)\exp(j\omega t)\exp(-jkR)}{2\pi r}dS \tag{7}$$

where $\mathring{\omega}(x,y,z)$ is the component of the complex velocity normal to the surface, ρ_a is the density of the acoustic medium, ω is the frequency in rad/s, r is the distance from the observation point (x_0,y_0,z_0) to the coordinate origin and $R^2 = (x - x_0)^2+(y - y_0)^2 + (z - z_0)^2$.

The classical assumption made in order to evaluate the far-field pressure is that the value of R is approximately by [11]:

$$R \approx r - x\sin\theta\cos\phi - y\sin\theta\sin\phi \tag{8}$$

where x and y define the coordinate position on the plate and (r, θ, ϕ) are the coordinate of the field point. This assumption is valid for provided $R \geq a,b$.

A particular form of out-of-plane vibration for a simply-supported rectangular plate with the above assumptions, leads to an analytically tractable form of equation which is given by:

$$\mathring{\omega}(x,y,z) = \sum_{m=1}^{\infty}\sum_{n=1}^{\infty}\sin\left(\frac{m\pi x}{a}\right)\sin\left(\frac{n\pi y}{b}\right) \tag{9}$$

By substituting Eq. (8) into Eq. (9), the sound pressure radiated by a simply-supported plate in an infinite baffle then can be written as:

$$p(r,\theta,\phi) = \sum_{m=1}^{\infty}\sum_{n=1}^{\infty}\frac{j\omega p_a \mathring{W}_{mn}\exp(j\omega t)\exp(-jkR)}{2\pi r}\times\sin\frac{m\pi x}{a}\sin\frac{n\pi y}{b}\exp\left[j\left(\frac{\alpha x}{a}+\frac{\beta y}{b}\right)\right]dxdy \tag{10}$$

where $\alpha = ka\sin\theta\cos\phi$ and $\beta = kb\sin\theta\sin\phi$.

This integral equation has been evaluated by [12] who gives the solution:

$$p(r,\theta,\phi) = \frac{j\omega p_a \overset{\circ}{W}_{mn} \exp(j\omega t)\exp(-jkR)}{2\pi r} \cdot \frac{ab}{mn\pi^2} \times \left[\frac{(-1)^m \exp(-j\alpha)-1}{\left(\alpha/m\pi\right)^2 - 1} \right] \times \left[\frac{(-1)^n \exp(-j\beta)-1}{\left(\beta/m\pi\right)^2 - 1} \right] \tag{11}$$

3. Result and Discussion

A sound radiation from SSP in an infinite baffle subjected to a harmonic load $F(x_i,y_i,z_i)$ is shown in Figure 1. The point load, F (1N) is applied at the center of plate (0,0,0) which has a uniform thickness, h; width, a; and length, b. In this analysis, the sound pressure $p(x_o,y_o,z_o)$ radiated from vibrating plate was determined by using finite element analysis (FEA) and mathematical model. The FEA was carried out using ANSYS® and mathematical model was plotted in MATLAB®. The frequency range was set between 0 to 200 Hz with approximately step size of 2 Hz increments. The responses of sound pressure level were measured at far field coordinates (0,0,0.8) and near field (0,0,0.4). The parameters of SSP and air used in this analysis are shown in Tables 1 and 2, respectively.

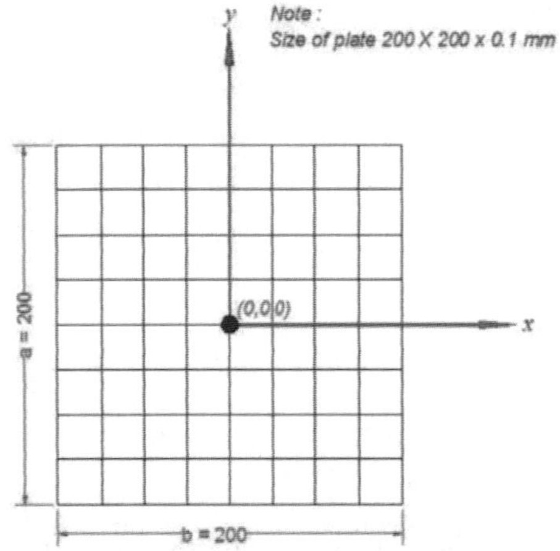

Figure 1. Starting coordinate and size of plate

Table 1. Parameters of SSP structure

Parameters	Description	Value	Units
A	Length	0.2	*m*
B	Width	0.2	*m*
H	Thickness	0.0001	*m*
I	Area moment of inertia	9.2 x 10^{-8}	m^4
E	Young's modulus	2.1 x 10^2	*GPa*
v	Poisson ratio	0.3	-
ρ	Density	7.85 x 10^3	kg/m^3
ζ	Damping ratio	1	%

Table 2. Parameters of air medium

Parameters	Description	Value	Units
ρ_a	Density	1.21	kg/m^3
C	Speed of sound	95450.0	m/s
R_a	Radius of infinite acoustic elements	1.0	m

Figures 2 and 3 show the comparison graph of sound pressure level radiated from SSP by using ANSYS® and mathematical model of MATLAB® in which the pressure measured at the respective far field and near field. Obviously, it can be seen that the pressure level increases when the sound was measured at near field [11]. Both of the graph show almost identical pattern which indicate that the mathematical model derived in Eq. (11) is in a good agreement with FEA result of ANSYS®.

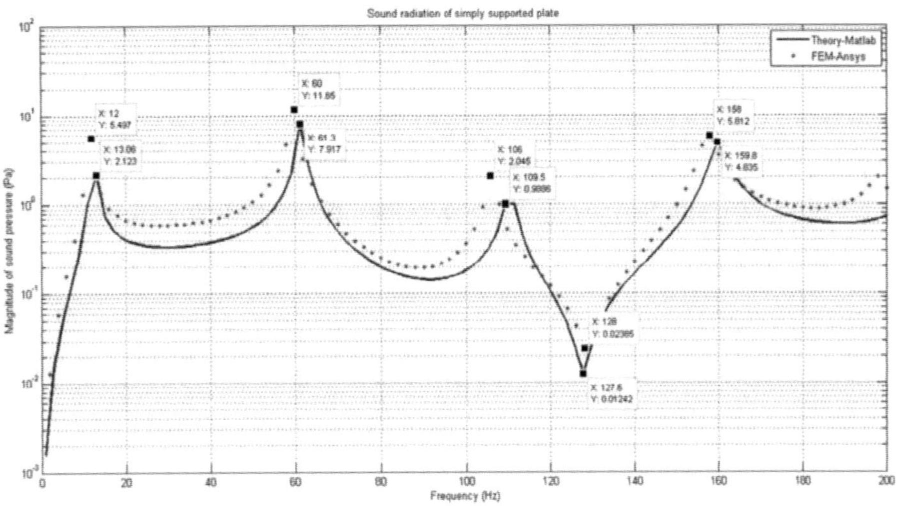

Figure 2. Graph comparison between ANSYS® and MATLAB® on point (0,0,0.8)

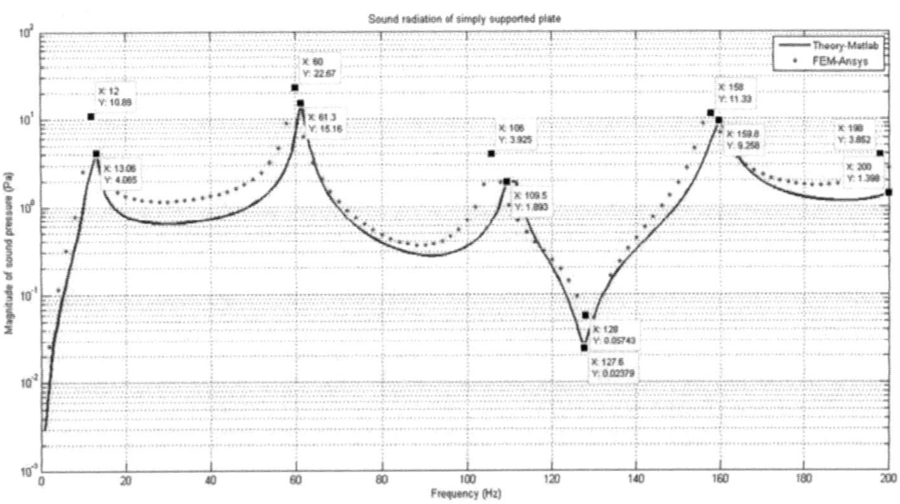

Figure 3. Graph comparison between ANSYS® and MATLAB® on point (0,0,0.4)

There are four resonance frequencies were identified in the frequency range of 0–200 Hz which represented by the total number of the peaks shown in Figures 2 and 3. Table 3 tabulates the value of the natural frequencies of sound radiation from SSP which determined directly from frequencies where the peaks are occurred. These natural frequencies are found similar for both sound pressure measurement at far field (0,0,0.8) and near field (0,0,0.4). The errors between the result obtained from ANSYS® and MATLAB® are found below 10% which again demonstrate the reliability of mathematical model derived in the study.

Table 3. Errors of sound radiation from SSP between theory and FEM at (0,0,0.8) and (0,0,0.4)

Mode n^{th}	ANSYS®	MATLAB®	Error (%)
1	12	13.06	8.83
2	60	61.3	2.17
3	106	109.5	3.30
4	158	159.8	1.14

4. Conclusion

A mathematical model was developed to enable the prediction of sound pressure level radiated from a simply-supported plate. Later, a comparison with finite element analysis of ANSYS® is performed. The outcome displays errors difference between FEA and mathematical model are less than 10% which can be negligible.

Acknowledgements

The authors thank to Office for Research, Innovation, Commercialization and Consultancy Management and Universiti Tun Hussein Onn Malaysia for the financial support under RSGS Grant Vot U105.

References

[1] Fantuzzi N, Tornabene F and Viola E 2014 *International Journal of Mechanical Sciences* **79** 216-251
[2] Salleh M M and Zaman I 2016 *ARPN Journal of Engineering and Applied Sciences* **11** 2336-2339
[3] Clough R W 2004 *International journal for numerical methods in engineering* **60** 283-287
[4] Qatu M S, Abdelhamid M K, Pang J and Sheng G 2009 *International Journal of Vehicle Noise and Vibration* **5** 135
[5] Griffin M 2004 *Occupational and Environmental Medicine* **61** 387-397
[6] Putra A and Thompson D 2010 *Applied Acoustics* **71** 1113-1125
[7] Tanaka N and Sanada T 2006 *Smart Materials and Structures* **16** 36
[8] Park J, Mongeau L and Siegmund T 2003 *Journal of Sound and Vibration* **264** 775-794
[9] Zaman I, Salleh M M, Ismon M, Manshoor B, Khalid A, Sani M S M and Araby S 2014 *MATEC Web of Conferences* **13** 03003
[10] Zaman I, Salleh M M, Manshoor B, Khalid A, Ngali M Z and Sani M S M 2015 *Applied Mechanics and Materials* **773-774** 33-37
[11] Fahy F J and Gardonio P 2007 *Sound and structural vibration: radiation, transmission and response* (UK: Academic press)
[12] Li W and Gibeling H 2000 *Journal of Sound and Vibration* **229** 1213-1233

IOP Conf. Series: Materials Science and Engineering **165** (2017) 012024 doi:10.1088/1757-899X/165/1/012024

Numerical Tension Adjustment of X-Ray Membrane to Represent Goat Skin Kompang

M Syiddiq[1], W A Siswanto[2]

[1,2]Faculty of Mechanical Engineering, Universiti Tun Hussein Onn Malaysia, 86400 Batu Pahat, Johor, Malaysia

[1]cdix91@gmail.com, [2] waluyo@uthm.edu.my

Abstract. This paper presents a numerical membrane model of traditional musical instrument *kompang* that will be used to find the parameter of membrane tension of x-ray membrane representing the classical goat-skin membrane of *kompang*. In this study, the experiment towards the *kompang* is first conducted in an acoustical anechoic enclosure and in parallel a mathematical model of the *kompang* membrane is developed to simulate the vibration of the *kompang* membrane in polar coordinate by implementing Fourier-Bessel wave function. The wave equation in polar direction in mode 0,1 is applied to provide the corresponding natural frequencies of the circular membrane. The value of initial and boundary conditions in the function is determined from experiment to allow the correct development of numerical equation. The numerical mathematical model is coded in SMath for the accurate numerical analysis as well as the plotting tool. Two *kompang* membrane cases with different membrane materials, i.e. goat skin and x-ray film membranes with fixed radius of 0.1 m are used in the experiment. An alternative of *kompang's* membrane made of x-ray film with the appropriate tension setting can be used to represent the sound of traditional goat-skin *kompang*. The tension setting of the membrane to resemble the goat-skin is 24N. An effective numerical tool has been used to help *kompang* maker to set the tension of x-ray membrane. In the future application, any traditional *kompang* with different size can be replaced by another membrane material if the tension is set to the correct tension value. The numerical tool used is useful and handy to calculate the tension of the alternative membrane material.

1. Introduction

Traditional musical instruments are unique and special which not only producing sounds but also representing the cultural value of a certain region [1]. Some research works for the traditional musical instruments have been done such as *tampani* [2], *wadaiko* [3], *angklung* [4], *kantele* [5], also *gong* and *cymbals* [6]. Those traditional musical instruments and percussions have been investigated in various aspects by researchers, some derived the mathematical models to represent the vibration characteristics. A report on related to kettledrums has been documented by [7]. For snare drums are found in [8] and [9]. There is also a report on sound reproduction of traditional musical instruments such as reproduction of *kompang* sound [10].

The *kompang* model defined in this this paper is the traditional percussion instrument found in Johor, Malaysia. There are only two interacting components: a membrane and a rigid shell for constraining the membrane (See Figure 1). The membrane of the *kompang* is fixed to the wood shell firmly. The incorrect placement of the membrane to the shell may cause a false sound characteristic

[3]. *Kompang* has various sizes, normally having a radius in between 22 cm and 35 cm. The height of the frame is in between 4 cm and 6 cm [1], while tom-toms are generally smaller [11].

A tap on *kompang* can cause displacements of several millimeters along with many modes can be formed by membranes where different modes tend to gives a different vibration attribute [12].

In this study, musical instrument *kompang* is used to produce a numerical equation. The musical instruments are generally categorized by the musical instrument type whether string, drum or flute type, with based on its physical frameworks [6]. This is supported by [13] which stated that different type of musical instrument have distinctive characteristic which becomes its trademark thus requiring different type of physical modeling analysis. By dissecting and analyzing the component of the musical instrument, more information about that particular musical instrument styles and characteristic can be obtained for better musical instrument data and information for documentation [14].

By dissecting the components, the physical framework of the musical instrument *kompang*, are found to be almost identical to the framework of the snare drum [6] and the timpani drum [15], which also build with rigid shell as main frame and attached with circular membrane. Despite exhibit the same basic framework, the sound characteristic produced will be different between these musical instruments as different musical instrument is drafted with unique properties in comparison with one another in terms of material, size and its detailed design [16]. The frequency of percussion musical instrument however, reported by [2] are mainly collaborated with the modes formed by the membrane, although other factors such as membrane density or radius may also play a role to some extent. It is also found that in percussion type of musical instrument, different modes produce different frequency where mode 1,1 tend to execute the highest membrane velocity, especially at higher membrane tension compared to any other modes.

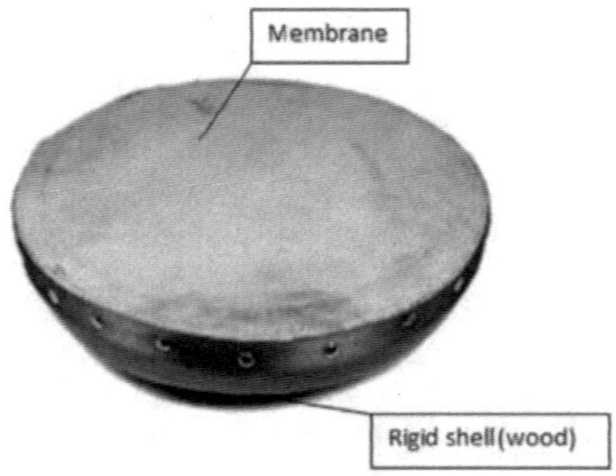

Figure 1: Musical instrument kompang with goat skin as membrane

Figure 2: Musical instrument kompang with x-ray film as membrane

The understanding of traditional musical instrument characteristic is crucial as it represent the cultural values in some region as stated by [5] which study traditional musical instrument Kantele for Finland region, while this study focus on traditional musical instrument kompang for Malaysia region. In this paper, two types of *kompang* are considered; using goat skin as membrane (see Figure 1) and x-ray film as membrane (see Figure 2). The objective of this study is to use a correct mathematical model for traditional *kompang* and using that equation, the frequency of x-ray kompang would be similar with goat skin *kompang*, by altering the variable found in the numerical equation.

2. Experimental Setup

2.1. Design of Experiment

There are several equipment involved in the experiment, for measuring the weight and the thickness of the membrane, the tension of the membrane and for capturing the vibration of the membrane.

In this study, the weight and volume of sample goat skin membrane and X-ray film is measured to determine the density values. The tension of the membrane is determined by using TAMA tension watch. The measuring tools and are illustrated in Figure 3.

The capture of the membrane's vibration is conducted by employing non-contact vibration measurement apparatus Laser Doppler Vibrometer (LDV) - Polytec PDV 100. The set-up illustration of the vibration capture is shown in Figure 4. Prior to the measurement the vibration apparatus was calibrated using Kistler Portable Shaker Accelerometer Calibrator Type 8921.

2.2. Experimental Procedure

The sample of goat skin and X-ray film is firstly carefully cut using a paper cutter with the size of 10 cm × 10 cm dimension then the thickness is measured using digital Vernier caliper. The weight is scaled by using Shimadzu AUW 220D Dual-Range Semi-Micro Balance. The density can be calculated based on the two measured parameters. The relative tension of each *kompang* is first measured by using TAMA tension watch as shown in Figure 3 and then the actual tension is determined using slotted weight approach. The recorded results are compiled in Table 1. These data collected will also be used for the input variables in the mathematical modelling.

Figure 3. Measurement tools

Table 1. *Kompang* parameters and their values.

Symbol	Kompang Membrane Parameters	Membrane Type	
		Goat Skin	X-ray Film
1	Radius, R (m)	0.1	0.1
2	Tension, T (N)	40.16	39.94
3	Density, ρ (kg/m^3)	552.905	1402.56
4	Thickness, h (V)	0.74×10^3	0.18×10^3

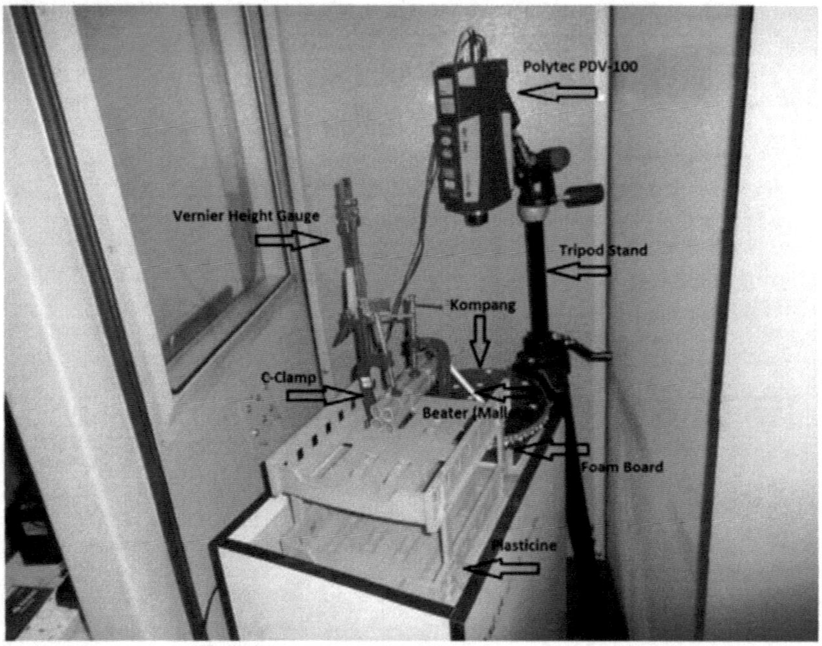

Figure 4. Experimental setup to determine the frequency of membrane

Figure 4 shows the experimental setup for the experiment to determine the frequency of membrane. The experimental procedure are:

1) Setting the experimental setup.
2) Boot up the computer and LDV.
3) Open up the laser cover and adjust the laser point to the smallest possible as it is at the most focus.
4) Calibrate the LDV using Kistler Portable Shaker Accelerometer Calibrator Type 8921.
5) Pointed out the laser from LDV toward the designed point on membrane.

6) With the help of Vernier height gauge, lift the mallet up to 25 cm and hit the center of membrane with mallet and record the response of membrane.
7) Determines the frequency reading of the points from the plotted graph of FFT domain.
8) Repeat the step 5 to step 7 for different *kompang*.
9) After finish, shut down the LDV and the result collected to be analysed.

3. Mathematical Modelling

The mathematical formula for 3D vibration simulation of circular membrane in Smath software can be found from [17]:

$$u_{mn}(r,t,\theta) = [a_{mn}\cos(ck_{mn}t) + b_{mn}sin(ck_{mn}t)]J_n(k_{mn}r)cos(n\theta) \tag{1}$$

$$c = \left(\frac{T}{\rho h}\right)^{1/2} \tag{2}$$

With Fourier series coefficients of:

$$a_{mn} = \frac{2}{R^2 J_1^2(\alpha_{mn})}\int_0^R rp(r)J_0\left(\frac{\alpha_{mn}}{R}r\right)dr \tag{3}$$

$$b_{mn} = \frac{2}{c\alpha_{mn}RJ_1^2(\alpha_{mn})}\int_0^R rq(r)J_0\left(\frac{\alpha_{mn}}{R}r\right)dr \tag{4}$$

where r is the radius of the point on membrane, T tension of the membrane per length, ρ is the membrane density, h is membrane thickness, R is radius of circular membrane, t is time and θ is the angle around the axis of the membrane. For the circular membrane equation to works, the assumption for initial displacement, $p(r)$ and initial velocity, $q(r)$ is necessary by following [18]:

$$p(r) = u(r,0) \text{ and } q(r) = \left.\frac{\partial u}{\partial t}\right|_{t=0} \tag{5}$$

The "α_{mn}" is defined as the mth positive zeros of $J_n(s)$, where m represent the nodal circle and n represent nodal line. The positive zeros can be calculate and determined by plotting the zero-th order Bessel function as shown in Table 2, and can be found from [19].

Table 2: Values for α_{mn}

	n			
m	1	2	3	4
0	2.4048	5.520	8.654	11.792
1	3.832	7.016	10.173	13.323
2	5.135	8.417	11.620	14.796

3.1. Natural Frequency

The equation for the frequency of the membrane is [20]:

$$\lambda_{mn} = \frac{\alpha_{mn}}{R}\sqrt{\frac{T}{\rho h}} \tag{6}$$

4. Results and Discussion

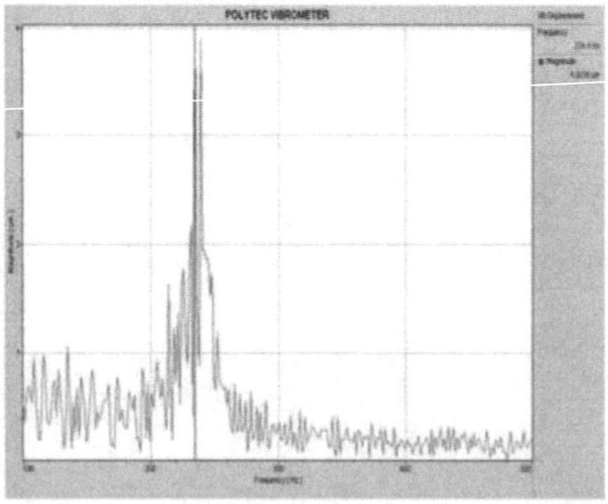

(a)

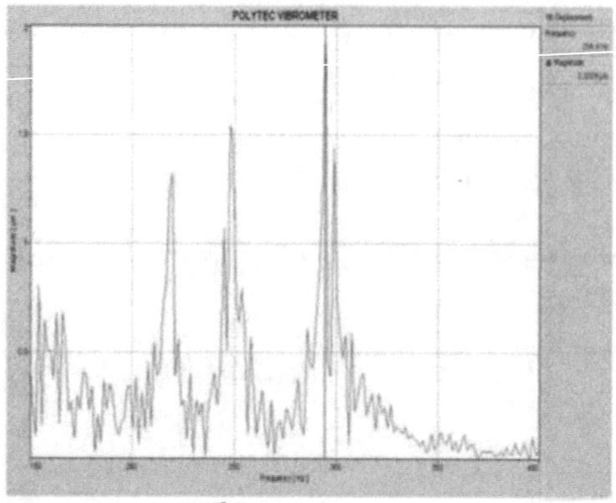

(b)

Figure 5. Experimental result (a) Goat skin *kompang*; (b) X-ray film *Kompang*

From the experiment, the frequency reading of both type of *kompang* is recorded using LDV which running the vibration analysis software Vibsoft 3.0. The graph is plotted in FFT domain to find the highest peak of displacement which shows the natural frequency of the membrane. In the Figure 5 (a), the goat skin *kompang* shows the reading of 234.4 Hz while in Figure 5 (b), the X-ray *kompang* yield an astonishing reading peak at 294.4 Hz. This is probably due to the difference in membrane density and thickness which plays a major role in this comparison as shown in equation (6).

4.1. Numerical Modelling

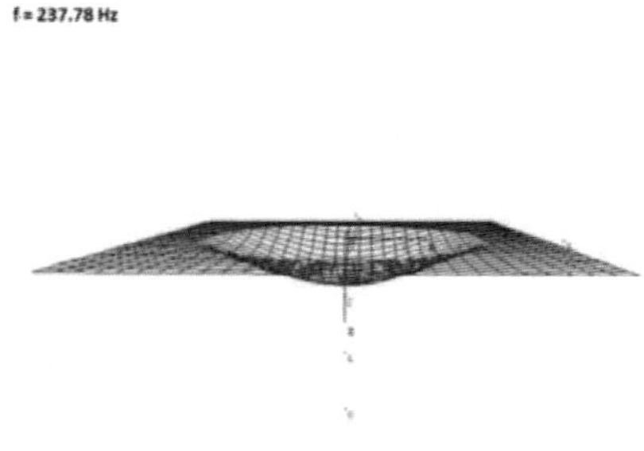

(a)

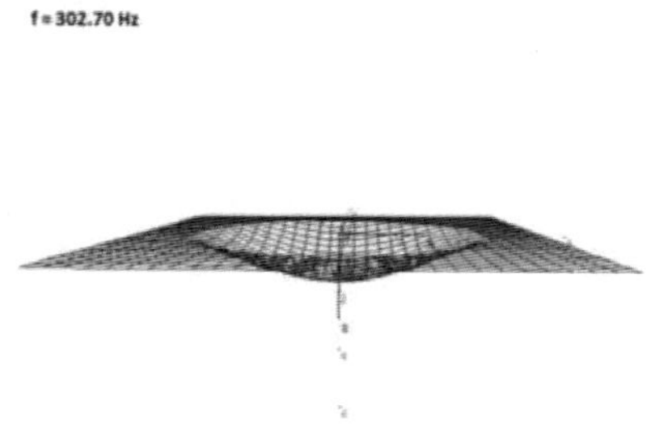

(b)

Figure 6. Numerical result (a) Goat skin *kompang*; (b) X-ray film *kompang*

Figure 6 shows the result from the numerical equation calculated in Smath software. The simulated 3D image is generated using equation (1) while the frequency is obtained from equation (6). It is calculated from the Figure 6 (a) that the goat skin membrane produce a result of 237.8 Hz while in Figure 6 (b), the result of 302.7 Hz is obtained. This result shows an almost similar frequency when being compared with the result from Figure 5.

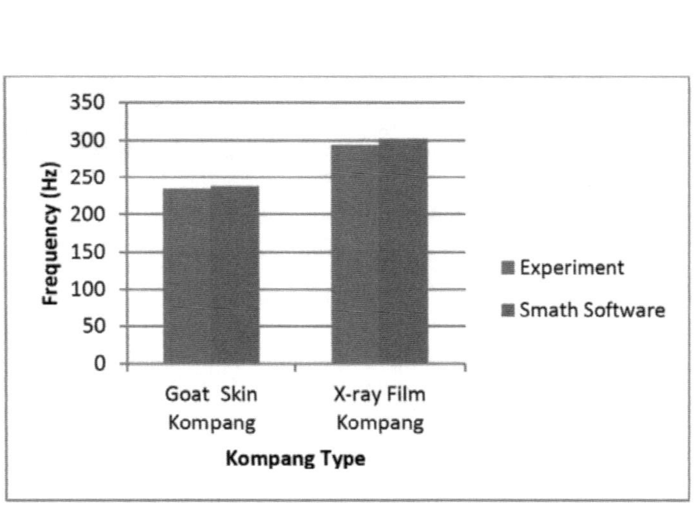

Figure 7. Comparison experiment and numerical results

The Figure 7 shows a chart as comparison between experimental and numerical calculation result. It is observed that the difference of frequency between experimental approach and calculation in Smath software is very small that from the plotted chart, the human eye can barely see the differences.

4.2. Percentage Difference

In the numerical study, it is important to know the exact error to confirm accuracy of the data collected for both from experiment and numerical calculation. The difference of the numerical simulation and the experiment results for both goat-skin and x-ray membranes are considerably very small, i.e. 1.45% and 2.82% respectively. The numerical model can be used to find the correct tension of the x-ray model to represent the goat-skin membrane.

4.3. X-Ray Membrane Tension Adjustment

In the experiment that have been conducted, the tension reading of x-ray film membrane is 39.94 N, resulting the frequency of 294.4 Hz, while the goat-skin *kompang* has a fundamental frequency of 237.8 Hz. Since the tension proportionally corresponding to the frequency, it can be seen that if the tension is lowered, a frequency similar to the goat-skin membrane can be achieved.

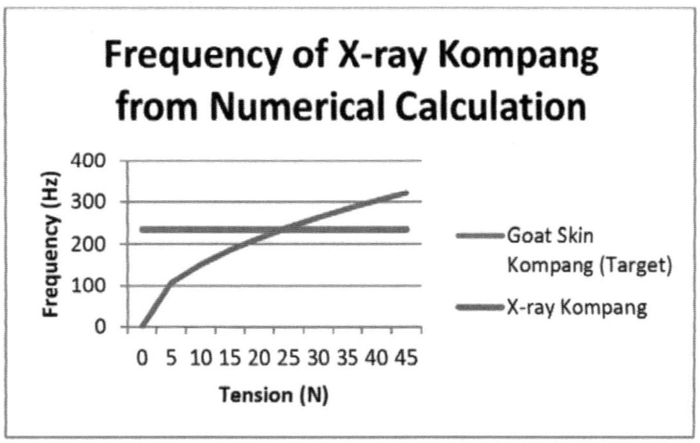

Figure 8. Fundamental frequency of x-ray *kompang* with various tension settings

IOP Conf. Series: Materials Science and Engineering **165** (2017) 012024 doi:10.1088/1757-899X/165/1/012024

Several numerical attempts are then conducted. The results of frequency of these numerical simulations for x-ray *kompang* are plotted in Figure 8. By plotting line chart as shown in Figure 8, the tension parameter value of x-ray membrane to represent the goat-skin membrane should be 23.99 N (~24 N).

With the new tension setting, the sound of the traditional goat-skin membrane can be represented by an alternative x-ray membrane.

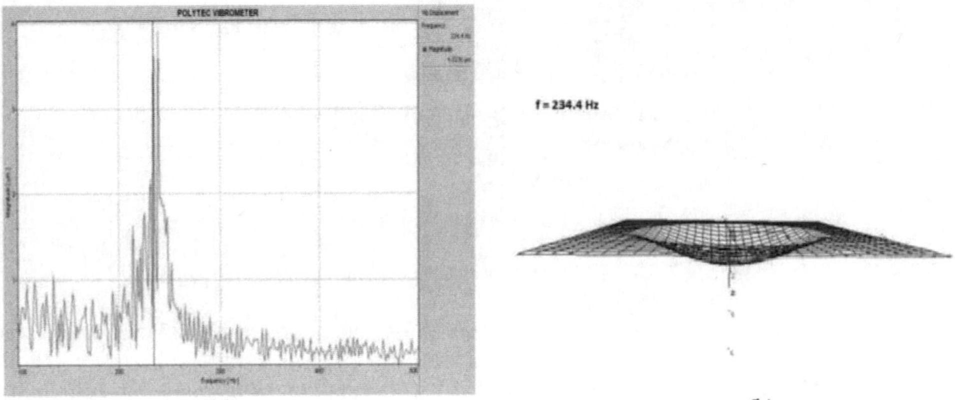

(a) (b)

Figure 9. Final result of frequency (a) goat skin *kompang* frequency during experiment; (b) X-ray *kompang* new frequency after altering the tension value.

Figure 9 shows the result of *kompang* fundamental frequency of goat skin and X-ray film membranes with the new tension. Both of them generates fundamental frequencies of 294.4 Hz.

5. Results and Discussion

The sound of traditional goat-skin membrane of *kompang* can be replaced by x-ray film membrane. The tension for the x-ray membrane is 24 N to have similar fundamental frequency of the goat-skin *kompang*.

In this research a numerical tool is used to calculate the fundamental frequency of *kompang's* membrane. The difference between the experimental and simulation is small for both x-ray and goat-skin membranes (i.e. 1.45% for goat-skin and 2.82% for x-ray film). The simulation tool is proofed to be practical and handy for the *kompang* maker in determining the tension of the membrane.

The usage of the alternative x-ray film for the *kompang* membrane has an advantage that the sound is stable not easily changed by the humidity. The sound also standard since the thickness is always the same.

Acknowledgements

The authors would like to be obliged to Universiti Tun Hussein Onn Malaysia for providing laboratory facilities and financial assistance under project Fundamental Research Grant Scheme Vot. 1206. This paper was also partly sponsored by the Centre for Graduate Studies UTHM.

References

[1] Abdullah M H 2005 Kompang: an organological and ethnomusicological study of a Malay frame drum

[2] Bertsch M 2001 Vibration patterns and sound analysis of the Viennese Timpani *Proceedings ISMA*

[3] Ono T, Takahashi I, Takasu Y, Miura Y & Watanabe U 2009 Acoustic characteristics of Wadaiko (traditional Japanese drum) with wood plastic shell *The Journal of Acoustical Science and Technology.* **30**(6) 410 – 416

[4] Siswanto W A, Tam L & Kasron M Z 2012 Sound characteristics and sound prediction of the traditional musical instrument the three-rattle angklung *International Journal of Acoustics and Vibration, Inst Acoustics & Vibration Auburn Univ, Mechanical Engineering Dept, 270 Ross Hall, Auburn, AL 36849 USA.* **17** 120-126

[5] Karjalainen M, Backman J & Polkki J 1993 Analysis, modeling, and real-time sound synthesis of the kantele, a traditional Finnish string instrument *Proceedings of the 1993 IEEE International Conference on Acoustics, Speech, and Signal Processing: Plenary, Special, Audio, Underwater Acoustics, VLSI, Neural Networks.* **1** 229 – 232

[6] Chaigne A, Touzé C & Thomas O 2005 Nonlinear vibrations and chaos in gongs and cymbals Acoustical science and technology *Acoustical Society of Japan.* **26** 403-409

[7] Rhaouti L, Chaigne A & Joly P 1999 Time-domain modeling and numerical simulation of a kettledrum *The Journal of Acoustical Society of America.* **105**(6) 3545 – 3562

[8] Rossing T D, Bork I, Zhao H & Fystrom D O 1992 Acoustics of snare drums *The Journal of Acoustical Society of America.* **92**(1) 84 – 94

[9] Avanzini F & Marogna R 2010 A modular physically based approach to the sound synthesis of membrane percussion instruments *The Journal of Audio, Speech, and Language Processing, IEEE Transactions on.* **18**(4) 891 – 902

[10] Siswanto W A, Che Wahab W M A, Yahya M N , Ismail A E & Nawi I 2014 A platform for digital reproduction sound of traditional musical instrument Kompang *Applied Mechanics and Materials.* **660** 823-827

[11] Fletcher N & Rossing T 2012 The physics of musical instruments *Springer Science & Business Media*

[12] Baricz 2010 Generalized Bessel functions of the first kind *Springer*

[13] Golzari S, Doraisamy S, Sulaiman M, Udzir N & Norowi N 2008 Artificial immune recognition system with nonlinear resource allocation method and application to traditional malay music genre classification. **5132** 132 – 141

[14] Ferguson S 2006 Learning musical instrument skills through interactive sonification *Proceedings of the 2006 conference on New interfaces for musical expression, IRCAM - Centre Pompidou.* 384 – 389

[15] Bilbao S Webb C Craig J 2013 Physical modeling of timpani drums in 3D on GPGPUs," *Acoustics and Audio Group, University of Edinburgh, Edinburgh, UK.* **61**(10) 737-748

[16] Wieczorkowska A 1999 Rough sets as a tool for audio signal classification *Proceedings of the 11th International Symposium on Foundations of Intelligent Systems. London. UK :Springer-Verlag.* **1609** 367 – 375

[17] Morse P M 1948 Vibration and sound *McGraw-Hill New York*

[18] Nguyen D M, Anton E, Allan R G & Jens G 2011 Isogeometric shape optimization of vibrating membranes *Journal of Computer Methods in Applied Mechanics and Engineering.* **200**(13) 1343 – 1353

[19] Watson G N 1995 A treatise on the theory of Bessel Functions *Cambridge University Press*

[20] Kreyszig E 2011 Advanced engineering mathematics *Wiley*

The Application of Coconut Fiber as Dissipative Silencer

M. A. Madlan[1,2*], M. I. Ghazali[1,2], I. Zaman[1,2], M. Z. Kasron[1,2] and T. C. Ying[1]
[1]Faculty of Mechanical Engineering & Manufacturing,
Universiti Tun Hussein Onn Malaysia,
86400 Batu Pahat, Johor, Malaysia
[2]Noise and Vibration Analysis Research Group,
Universiti Tun Hussein Onn Malaysia,
86400 Batu Pahat, Johor, Malaysia

*Email: amran@uthm.edu.my

Abstract. *Heat ventilation air conditioning system (HVAC) is one of the ducting systems that broadly applied in the building. There are HVAC silencers in the market, however the sound absorptive material commonly used is mineral wool. In this research study, a sound absorptive material made of coconut fiber was tested to identify its performance as a potential replacement of green material for ducting silencer. The experiment was carried out in a testing apparatus that follows the BS EN ISO 11691:2009 standard. Different configurations of sound absorptive material and contents of coconut fiber were investigated in the study. The trend of insertion loss at 1/3 octave frequency was identified where at frequency below 3000Hz, the insertion loss of dissipative silencer is observed high at certain frequency with a very narrow range. At 3000Hz, the insertion loss of 4dB to 6dB is constant until 4000Hz and drops until 5000Hz before it increases again steadily up to 13dB at 10000Hz. A similar trend was observed for different configuration of sound absorptive material. Despite the configuration different, the outcome shows that the insertion loss is increasing with higher content of coconut fiber.*

1. Introduction

Heat ventilation air conditioning system (HVAC) is a ducting system that commonly found inside the building. HVAC system serves the purpose of making the people inside the building feel comfortable by moving the air for cooling or heating which is depends on the weather of the area. However, the HVAC system at the same time will produce noise that makes people feel uncomfortable. The sources of noise are from fan, passage of air through straight ducts and impact of air flowing through components such as elbows, branches and mixing boxes [1].

This is where a dissipative silencer required to be installed in HVAC duct systems in order to reduce noise transmitted through the ducting system. The principal advantages of this device are: (1) provides good absorption at medium and high frequencies, and (2) useful for narrow and broadband noise [2]. Theoretically, silencer works by transforming partial of incident acoustic energy into heat by causing motion in silencer's sound absorbing material when the sound passing through the silencer [3]. This device typically contains fibrous or porous materials and it is commonly made from mineral wool.

On the other hand, coconut fiber is a renewable resource and it is a CO_2 neutral material. The coconut fiber is abundant in Malaysia and the price also is very cheap. In contrast to mineral wool and other polymer materials, coconut fiber is biodegradable, low in density, good mechanical properties and light weight [4,5]. The fiber's structure is determined by the dimension and arrangement of various unit cells. Lumen is a central hollow cavity that located at transverse section of the unit cell and it serves as an acoustic and thermal insulator because its presence decreases the bulk density of the fiber.

Of all natural fibers, coconut fiber has a good ability to absorb noise in which the absorption coefficient was found to be 0.42 at frequency 500 Hz [6]. The thickness and the spacing of this absorptive material are critically important as these can affect the performance of the silencer. Hence, this work is carried out to investigate the potential of using coconut fiber as a raw material for sound absorptive material in the ducting system. The performance of fabricated dissipative silencer made of coconut fiber with different configurations and fiber content are determined in the study. The outcome of the study can help in further utilize the use of natural fiber material and at the same time reduce the cost of purchasing the dissipative silencer that made of other materials which are more expensive than the natural fiber of coconut.

2. Experimental Set Up

2.1. Design of Experiment
The apparatus setup and testing was conducted based on the BS EN ISO11691:2009. The testing apparatus are made of Perspex which consists of two test duct and one silencer duct where the sound absorptive material is arranged inside the silencer duct. The cross section of the testing apparatus is 150 mm x 150 mm. The length of test duct and silencer is 875 mm and 200 mm respectively. The molds are made of mild steel with dimension of 155 mm x 205 mm x 15 mm.

In the making of the sound absorptive materials, a certain amount of coconut fiber is put inside the mold. The coconut fiber is compressed so that it takes the shape of the mold with the same dimension. The fiber is removed from the mold. The mass that used are 30grams and 60grams but the sound absorptive materials that produced are the same dimension. The sound absorptive material is tested in the impedance tube to determine its ability to absorb noise at various frequencies.

2.2. Experimental Procedure
There are various type of configuration of dissipative silencer such as lined duct, splitter silencer and bar silencer. Each of these configurations has its own advantages and disadvantages. The dissipative silencer has the ability to filter out the acoustic energy over a wide range of frequencies [7]. In this study, the sound absorptive materials are arranged in the silencer duct with different configuration (5 pieces, 4 pieces, 3 pieces, 2 pieces and 4 sides). Noise is generated inside the test tube. The TES sound analyzer is used to collect the data of noise at frequency of 1/3 octave. The data collected in the TES sound analyzer is transfer to the computer. The data taken is interpreted into visual form for better analysis.

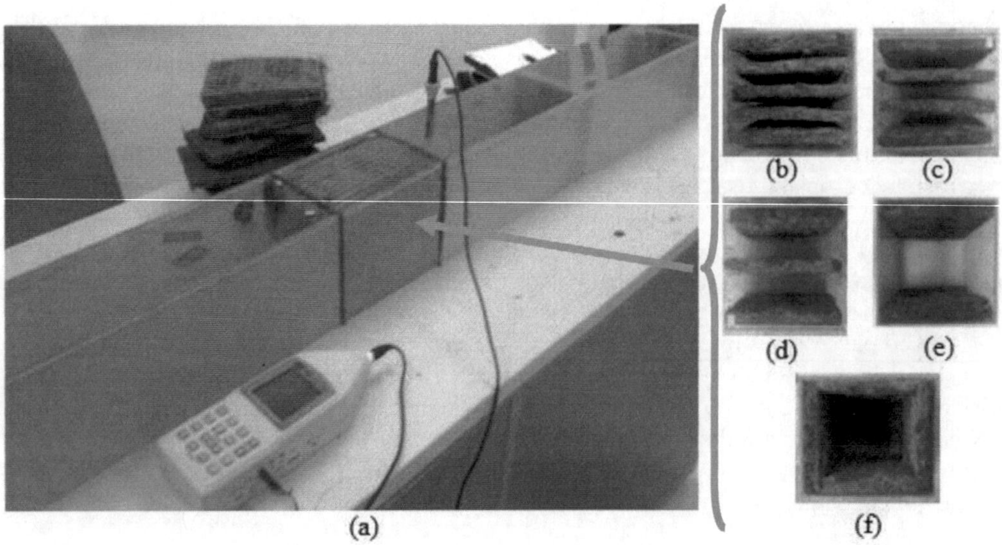

Figure 1. (a) Experimental setup; (b) 5 pieces configuration; (c) 4 pieces configuration; (d) 3 pieces configuration; (e) 2 pieces configuration; (f) 4 sides configuration

3. Results and Discussion

Basically, there are two comparison that are made. The first comparison is the insertion loss of silencer of sound absorptive material with same content of coconut fiber at different configurations. The second comparison is the insertion loss of silencer with same configuration with different content of fiber.

Figure 2 shows the result of noise inside duct before installed the silencer. The result comparison of silencer with same content of coconut fiber at different configurations, and same configuration but different fiber content can be seen in Figure 3 to Figure 7, and Figure 8 to Figure 11 respectively.

From the experiment, the trends insertion of sound absorptive material silencer with 60g fiber but with different configuration are identical. From Figure 3, we can see the insertion loss at below 3000Hz has very narrow range where the insertion loss is high at certain frequency only. Figure 4 shows that at 3000Hz, the insertion loss is constant until 4000Hz and decrease until 5000Hz. After 5000Hz, the insertion loss increase steadily until 10000Hz.

The trends of the insertion of silencer of sound absorptive material with fiber content of 30g with the configuration of 5 pieces, 4 pieces, 3 pieces, 2 pieces and 4 sides silencer is same with fiber content of 60g where the insertion loss at below 3000Hz has very narrow range where the insertion loss is high at certain frequency only. At 3000Hz, the insertion loss is constant until 4000Hz. After that, the insertion loss decrease until 5000Hz. After 5000Hz, the insertion loss increase steadily until 10000Hz.

The differences of all these configuration of 5 pieces, 4 pieces, 3 pieces and 2 pieces silencer shows that the more sound absorptive material in the silencer, the higher the insertion loss. While the trend of the silencer with configuration of 4 sides is same as other except for the insertion loss after 8000Hz where the insertion loss of the 4 sides configuration become stable and constant (Figure 11).

From the observation that made on both experiments above, we can conclude that the more sound absorptive material in the silencer, the higher the insertion loss.

IOP Conf. Series: Materials Science and Engineering **165** (2017) 012025 doi:10.1088/1757-899X/165/1/012025

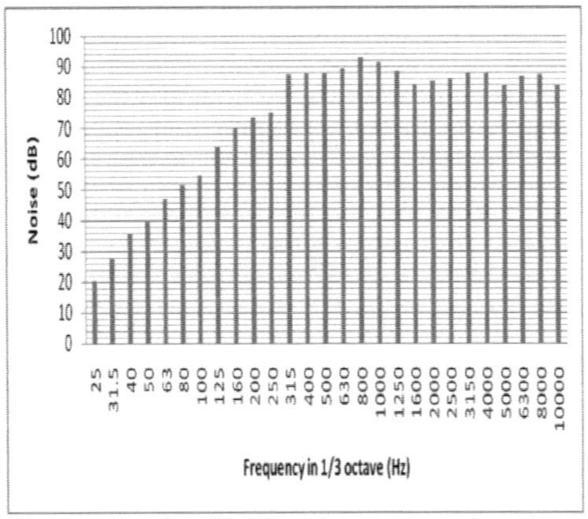

Figure 2. Graph of noise before insertion of silencer versus frequency (Hz) in 1/3 octave

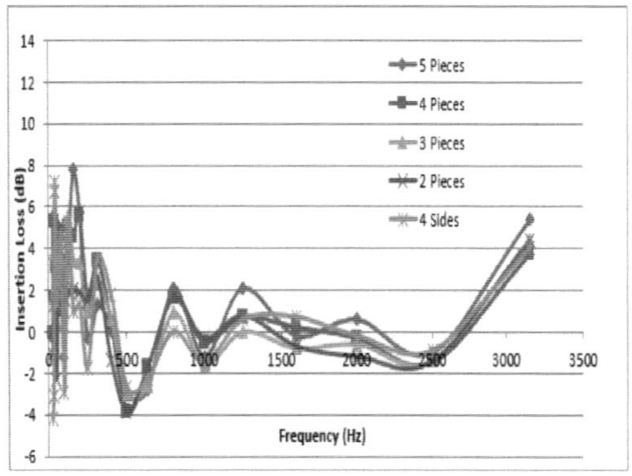

Figure 3. Comparison of Insertion Loss for all 60g Sound Absorptive Material at frequency below 3150Hz

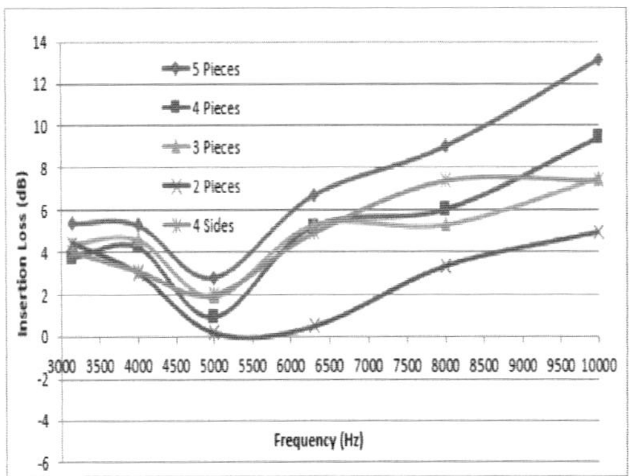

Figure 4. Comparison of Insertion Loss for all 60g Sound Absorptive Material at frequency above 3150Hz

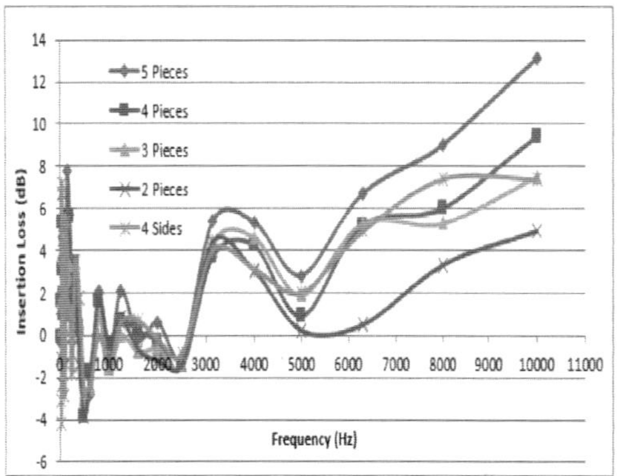

Figure 5. Comparison of Insertion Loss for all 60g Sound Absorptive Material in 5 Pieces Configuration

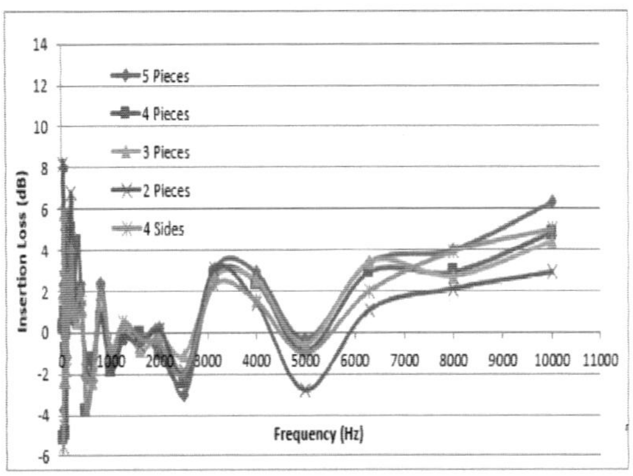

Figure 6. Comparison of Insertion Loss for all 30g Sound Absorptive Material

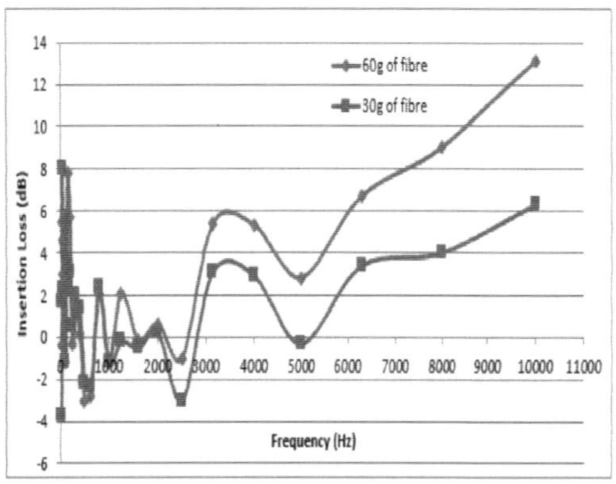

Figure 7. Comparison of Insertion Loss for 30g and 60g Sound Absorptive Material in 5 Pieces Configuration

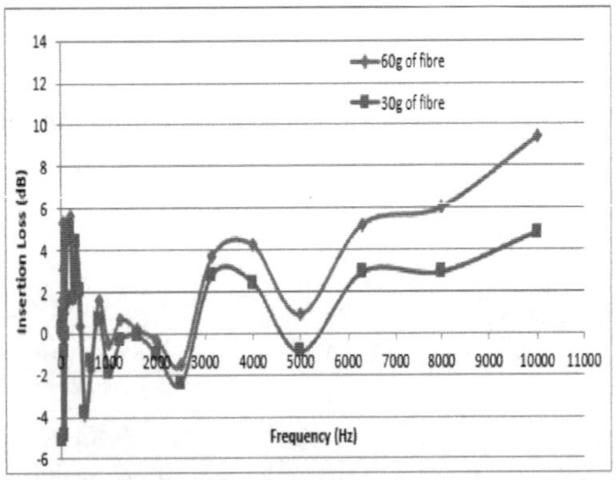

Figure 8. Comparison of Insertion Loss for 30g and 60g Sound Absorptive Material in 4 Pieces Configuration

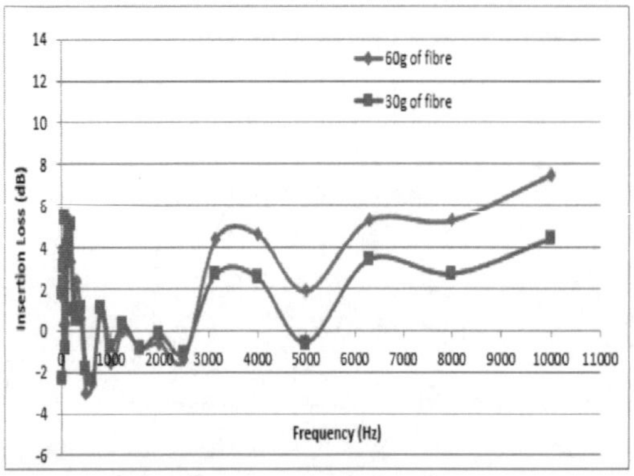

Figure 9. Comparison of Insertion Loss for 30g and 60g Sound Absorptive Material in 3 Pieces Configuration

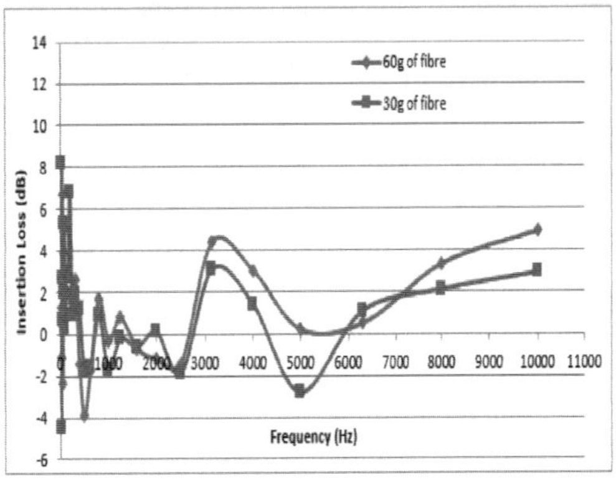

Figure 10. Comparison of Insertion Loss for 30g and 60g Sound Absorptive Material in 2 Pieces Configuration

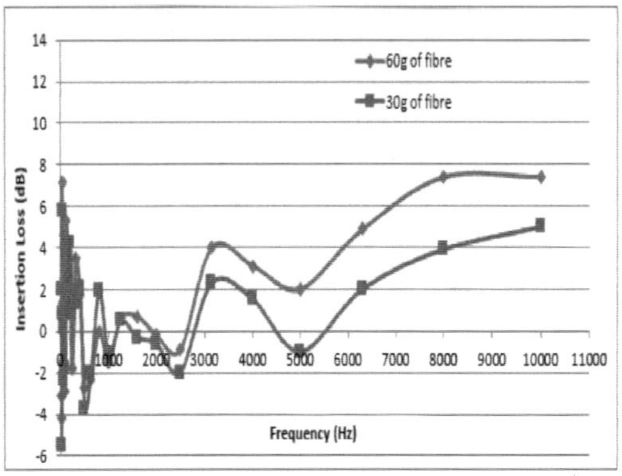

Figure 11. Comparison of Insertion Loss for 30g and 60g Sound Absorptive Material in 4 sides Configuration

4. Conclusions

In this paper, there are two types of sound absorptive materials which are 60g fiber and 30g fiber with the same dimension. The data collected are the noise at 1/3 octave frequencies before and after silencer is installed. The silencer has five different configuration testing. The analysis that made are the comparison of silencer with same mass of fiber but different configuration and same configuration but different mass of fiber. In this study, the insertion loss increase to 13dB at 10000Hz as the pieces of sound absorptive materials in the silencer increase. Similarly, the insertion loss was found increase as the mass that coconut fiber used in the sound absorptive materials increase. In addition, the insertion loss of silencer with same configuration with different content of fiber has been tested. It is found that the insertion loss of sound absorptive material with higher content of fiber has higher insertion loss at each frequency.

5. Acknowledgements

The authors would like to acknowledge Noise and Vibration Analysis Research Group (NOVIA) of Universiti Tun Hussein Onn Malaysia for providing laboratory facilities and Ministry of Higher Education Malaysia for the financial support under Fundamental Research Grant Scheme vote 1546.

References

[1] Hansen, C., Snyder, S., Qiu, X., Brooks, L., and Moreau, D., Active control of noise and vibration. 2012: CRC Press.

[2] Nennig, B., Perrey-Debain, E., and Tahar, M.B., A mode matching method for modeling dissipative silencers lined with poroelastic materials and containing mean flow. The Journal of the Acoustical Society of America, 2010. 128(6): p. 3308-3320.

[3] Rairat, V. Design and Fabrication of Plenum Chamber Silencer for Noise Attenuation. Mahidol University: Master's Thesis; 2003.

[4] Bujang, I.Z., Awang, M.K., and Ismail, A.E. Study on the dynamic characteristic of coconut fibre reinforced composites. in Regional Conference on Engineering Mathematics, Mechanics, Manufacturing & Architecture (EM3ARC). 27-28 November 2007. Putrajaya, Malaysia.

[5] Zaman, I., Ismail, A.E., and Awang, M.K., Influence of fiber volume fraction on the tensile properties and dynamic characteristics of coconut fiber reinforced composite. Journal of Science and Technology, 2009. 1(1): p. 55-71.

[6] Latif, H.A., Yahya, M.N., Zaman, I., Sambu, M., Ghazali, M.I., and Hatta, M.N.M., The influence of physical properties and different percentage of the oil palm mesocarp natural fiber. ARPN Journal of Engineering and Applied Sciences, 2016. 11(4): p. 2462-2466.

[7] Rozli, Z. and Zulkarnain, Z., Noise control using coconut coir fiber sound absorber with porous layer backing and perforated panel. American Journal of Applied Sciences, 2010. 7(2): p. 260-264.

Prevalence of Hand-transmitted Vibration Exposure among Grass-cutting Workers using Objective and Subjective Measures

N. A. Azmir[1,*], M. N. Yahya[1]

[1]Noise and Vibration Analysis Research Group,
Faculty of Mechanical and Manufacturing Engineering,
Universiti Tun Hussein Onn Malaysia, Batu Pahat, Johor, Malaysia

[*]azali@uthm.edu.my

Abstract. Extended exposure to hand-transmitted vibration from vibrating machine is associated with an increased occurrence of symptoms of occupational disease related to hand disorder. The present case study is to determine the prevalence and correlation of significant subjective as well as objective variables that induce to hand arm vibration syndrome (HAVS) among hand-held grass-cutting workers in Malaysia. Thus, recommendations are made for grass-cutting workers and grass maintenance service management based on findings. A cross sectional study using adopted subjective Hand Arm Vibration Exposure Risk Assessment (HAVERA) questionnaire from Vibration Injury Network on hand disorder signs and symptoms was distributed to a sample of one hundred and sixty eight male workers from grass and turf maintenance industry that use vibrating machine as part of their work. For objective measure, hand-transmitted vibration measurement was collected on site during operation by the following ISO 5349-1, 2001. Two groups were identified in this research comprising of high exposure group and low-moderate exposure group. Workers also gave information about their personal identification, social history, workers' health, occupational history and machine safety inspection. There was positive HAVS symptoms relationship between the low-moderate exposure group and high exposure group among hand-held grass-cutting workers. The prevalence ratio (PR) was considered high for experiencing white colour change at fingers and fingers go numb which are 3.63 (1.41 to 9.39) and 4.24 (2.18 to 8.27), respectively. The estimated daily vibration exposure, A(8) differs between 2.1 to 20.7 ms^{-2} for right hand while 2.7 to 29.1 ms^{-2} for left hand. The subjects claimed that the feel of numbness at left hand is much stronger compared to right hand. The results suggest that HAVS is diagnosed in Malaysia especially in agriculture sector. The A(8) indicates that the exposure value is more than exposure limit value which is 5 ms^{-2}. Thus, control measure such as engineering and administrative control should be implemented to reduce the severity of hand-transmitted vibration hazard.

1. Introduction

Hand-held grass-cutting machine is widely applied in various tropical countries for maintenance of grass area landscape. A typical public grass compound needs monthly grass-cutting operation either in highway or facilities location. It produces a service industry with large numbers of workers involved using the hand-held grass-cutting to maintain grass growth. The workers involved in these duties are generally contract workers with less awareness related to occupational safety and health such as hand arm vibration syndrome [1]. Statistics of occupational accidents by sector, which caused death, permanent disability and non-permanent disability, have been widely investigated and represented a significant problem to the Malaysian industry. Hand-held grass-cutting workers were investigated under the agricultural sector which reported the second highest cause to non-permanent disability behind industrial sector [2]. Information from the Social Security Organization (SOCSO), Malaysia shows increased number of cases related to occupational musculoskeletal disorder from year 2005 to 2014 [3].

Extended exposure to hand-transmitted vibration (HTV) can interfere the normal circulation as well as nervous and musculoskeletal system that causes a syndrome known as hand arm vibration (HAVS). HTV occurs when vibration enters the body through the hands, for example in various work activity where rotating or percussive power tools or vibrating machine was held by the hand or fingers [4]. HAVS developed after long exposure of hand vibrating instrument which are irreversible injury, chronic pain in the hands and arms which also lead to difficulties in the implementation of daily activities and introduce work incapacity [5].

The prevalence of HAVS ranged from 2.5% to more than 82.8% globally depending on the duration and magnitude of vibration exposure as well as climatic factors [6]. However, this percentage has increased as the working environment temperature decreased. Thus, the vibration white fingers (VWF) are renowned by low ambient temperature countries compared to warm environment countries. Its prevalence in Malaysia is not well established compared to European Union (EU) countries since there is no regulation concern about occupational HTV hazard. Recent study from Malaysia showed that the prevalence of HAVS among construction workers exposed to hand arm vibration from hand-held power tolls in warm environments was 18% [7].

The Department of Occupational Safety & Health (DOSH) Malaysia establishes the minimum health and safety requirements regarding the occupational vibration of workers involved in vibrating tools. It states that the vibration transmitted to the hand arm system is to be measured in accordance with the International Standard [8]. Furthermore, American Conference of Governmental Industrial Hygienists (ACGIH) specifies the threshold limit value and duration of exposure to be followed for the purpose of mitigating HAVS [9]. Besides that, European Directive established the regulation to protect workers from vibration hazard [10]. Although, there is several detailed health descriptions related to human response to vibration, it indicated that hand arm vibration disease reduced after implementing the Directive 2002/44/EC in Great Britain [11]. For hand arm vibration assessment, articles 3 of Directive 2002/44/EC established the following safety limits for A(8) by means, exposure limit value (ELV) equals to 5 ms^{-2} while exposure action value (EAV) equals to 2.5 ms^{-2}. The ELV is the maximum amount of hand arm vibration worker which may be exposed in any single day. It indicates a potential high risk above which workers should not be exposed. The EAV is the daily amount of hand arm vibration exposure which employers are required to take acceptable and suitable control measures.

The purpose of this research was to investigate the prevalence of HAVS among groups of hand-held grass-cutting workers and determine the correlation of significant symptoms. Besides that, the HTV exposure characteristics during onsite operation for five types of grass-cutting machine model were evaluated according to established safety standard. The result of this study would best predict the diagnosis of HAVS. These significant characteristics can be used in occupational safety and health screening as indicator to the health effect due to HTV.

2. Methodology

This is a cross-sectional study among hand-held grass-cutting workers covering highway grass maintenance subcontractor and personal contractor in central and southern region of peninsular Malaysia. Average daily working time is 8 hours starting from 8 am to 5 pm depending on target grass cutting area in a day. About hundred sixty eight hand-held grass-cutting workers were selected into the study. The workers were involved in general cutting operation location consists of flat, slope and drain area. Some of the workers were working full time as the grass cutter while others as their part time routine work in maintaining facilities landscape. In this study, high HAV risk exposure group was identified as full time operating grass-cutting machine while low and moderate HAV risk exposure group was classified under part time grass-cutting workers. Furthermore, five different types of hand-held grass-cutting machine were identified and detailed specifications of each machine type are as shown in **Table 1**.

Table 1. The technical specification of the survey of hand-held grass-cutting machine

Specification	Type A	Type B	Type C	Type D	Type E
Model	TU33	TB33 / TL33	KNC 3340	TB43	BG328
Discharge capacity (cc)	32.6	32.8	33	43	30.5-32.8
Net weight (kg)	9	9.5 – 9.8	9.7	8 – 9.4	7.8 – 10.5
Standard power (kW)	0.97	0.9	1.26	1.25 – 1.57	0.81 – 0.85
Max output (rpm)	6500	6500	6500	7000	6000 - 7500

All subjects were interviewed using a modified HAVERA questionnaire with Malay translation adopted from the Hand-Transmitted Vibration Health Surveillance – Initial Questionnaire and Clinical Assessment, published by the Research Network on Detection and Prevention of Injuries due to Occupational Vibration Exposure (Vibration Injury Network). Some of questions stated in the HAVERA questionnaire were based on the study of hand arm vibration syndrome among group of construction workers in Malaysia [7] which was found to be reliable and suitable for use in different working activity. The interview consists of questions on personal identification, social history, occupational health, occupational history, machine safety inspection and symptoms. Detailed health effect information including the experience of any finger colour changes and weakness in hand grip strength were also asked. Subjects specifically need to recall any sustained injury to body parts and the curing process to regain normal health condition.

Blaze software version 6.1.1 for Larson Davis human vibration meter (HVM 100, PCB Piezotronic, 00496) was used to analyse hand arm vibration exposures referring to ISO 5349-1. Raw, h-weighted tri-axial hand arm vibration measurements were collected using ICP accelerometer mounted on both left and right handles of hand-held grass-cutting machine, respectively. The tri-axial piezoelectric accelerometer handle mounted adapter inserted between the worker's fingers and the handle was fixed on the handle by tape [12] [13]. The right handle was used for operating the speed cutting control. Each handle was measured one time with sampling rate of 1 second period for 4 minutes. However, to produce the statistical significance of vibration data, at least 60 seconds should be enough for vibration data collection [14].

The subjective data analysis was carried out using SPSS version 19. Categorical data from different exposure groups were compared using two independent sample tests. The corresponding health symptoms among subjects with high versus low-moderate HAV exposure data were analysed using prevalence ration (PR) and Spearman correlation coefficient. The significant level for all statistical tests was set at 95% confident interval. For objective measurement, the hand arm vibration exposure depends on the magnitude of the total vibration and the duration of exposure. According to ISO 5349-1, daily vibration exposure should be stated in terms of 8 hours equivalent energy frequency h-weighted vibration total values, A(8). So, the average value of A(8) will be compared by means of the types of grass-cutting machine.

3. Results

Majority of the subjects were foreigners, with 38% from Indonesia, 26.2% from Bangladesh, 16.7% from Malaysia, 8.3% from India, 7.7% from Pakistan and the rest from Myanmar. The mean age for subjects was 32.0 (SD 9.57) years old. Most workers (51.8%) had finished at least primary academic level, while 41.1% had completed secondary level. All of them were fluent in Malay language and understood the HAVERA questionnaires. All subjects were in normal build range (mean BMI 22.65) with 97.6% were right-hand dominants. Median duration of working experience using hand-held grass-cutting machines and mean cutting duration of a day were 24.0 (range from 1 to 240) months and 5.46 (range from 2 to 6) hours per day, respectively. Sixty four per cent of the subjects were smokers with ninety one per cent non-alcoholics. None of the subjects had a hobby involving vibration tool and work experience involving neurotoxic chemicals in the past. All workers claimed that grass-cutting operation involved repeated work done with the forearms and hands with forceful and twisting movements. There were 100 (59.5%) workers in high exposure group (working at highway area) and the rest were low-moderate exposure group (working at facilities area).

Physical observation of hand disease symptoms among two group of subject indicates there is a possibility of inducing hand-transmitted vibration disorders as shows in **Table 2**. Finger colour change and numbness were experienced by 81% and 36% of all subjects, respectively. Thirteen subjects suffered weakness on left hand grip while 14% suffered weakness on right hand grip for high group exposure. Other reported symptoms related to HAVS were experiencing cold fingers (6%), difficulty in manipulating small objects (9%), difficulty in opening tight jars (10%) and experience a tingling sensation in fingers (17%). Reported information and knowledge of health hazard in vibrating machine and awareness of any health control measure taken was 39% and 14% of all subjects, respectively.

Table 2. Prevalence ratio (PR) and Spearman correlation coefficient of hand arm vibration syndrome (HAVS) symptoms among subjects with high versus low-moderate HAV exposure.

Symptoms	HAV risk exposure, n (%)		PR (95% CI)	Correlation Coefficient
	Low-moderate group (n=68)	High group (n=100)		
Experience any colour changes at fingers	62 (91.1)	74 (74)	3.63 (1.41 to 9.39)	0.215
Experienced a tingling sensation in your fingers	24 (35.3)	17 (17)	2.66 (1.30 to 5.48)	0.209
Fingers go numb	38 (55.9)	23 (23)	4.24 (2.18 to 8.27)	0.336
Suffer from weakness at left hand grip	13 (19.1)	13 (13)	1.58 (0.68 to 3.66)	0.083
Suffer from weakness at right hand grip	12 (17.6)	14 (14)	1.32 (0.59 to 3.05)	0.049
Experience cold fingers	10 (14.7)	6 (6)	2.70 (0.93 to 7.83)	0.146
Difficulty in picking up or handling small objects	6 (8.8)	9 (9)	0.98 (0.33 to 2.89)	-0.003
Difficulty in opening tight jars	6 (8.8)	10 (10)	0.87 (0.30 to 2.52)	-0.200
Information of vibratory tools is hazardous	32 (47.1)	34 (34)	1.73 (0.92 to 3.24)	0.131
Awareness of vibration hazard	13 (19.1)	11 (11)	1.91 (0.80 to 4.57)	0.114

* CI – Confident interval

Correlation test is a test to measure the strength of the linear relationship between two variables studied. There are three types of relationships that may exist between variables of linear correlation which are positive, negative or no correlation. The magnitude of the coefficient shows the strength of the linear relationship between two variables while the (+) or (-) signs indicate the direction of the relationship either positive or negative linear correlation. In the case study, fingers go numb is the highest correlation contributing to hand arm vibration syndrome with positive 0.336 value.

Statistical data for daily vibration exposure, A(8) for left and right hands by type of hand-held grass-cutting machine are presented in Table 3. Both left hand and right hand vibration measurement are exceeding the EAV value for mean A(8) for each type of machine. As seen from **Table 3**, the A(8) values are almost similar for all of the machines except for machine type C which shows less vibration magnitude for both left and right hands. However, the A(8) value exceeded the ELV for left hand and below EAV for right hand. The extreme maximum A(8) value reveals that the workers have high possibility of exposed to impulsive force due to impact between dynamic rotation of cutting blade with stone, wood etc.

Table 3. Statistical data for daily vibration exposure, A(8) for left and right hands by type of hand-held grass-cutting machine.

Type of hand held grass cutter	Mean operating hours in a day	Left hand, A(8) (m/s^2)	Right hand, A(8) (m/s^2)
		Mean (S.D)	Mean (S.D)
A	5.57	9.92 (4.92)	6.73 (3.94)
B	5.40	8.13 (5.27)	5.91 (2.70)
C	5.96	6.30 (3.34)	4.77 (2.27)
D	4.70	8.23 (4.60)	6.25 (2.84)
E	5.62	10.82 (3.94)	6.57 (1.96)

S.D, standard deviation

4. Discussion

In the early 2000s HAVS was reported in quarry miners using rock drill machine in Vietnam [15]. Subsequent reports came from construction and property management [16], together with detailed research on the quality of HAVS workers life [5]. Previous study revealed that the HAVS prevalence was between 18% to 50%. Based on the onsite observation of physical examination, the prevalence of HAVS was found to be 36% in the study. The result shown slightly corresponds with the prevalence of HAVS among construction workers in a tropical environment in Malaysia [7].

There was positive HAVS symptoms relationship between the low-moderate exposure group and high exposure group among hand-held grass-cutting workers. However, the numbness symptoms showed the highest significant value of 0.336 compared to other symptoms such as finger colour changes and tingling sensation. The prevalence ratio (PR) was considered high for experiencing any colour change at fingers and fingers go numb which are 3.63 (1.41 to 9.39) and 4.24 (2.18 to 8.27), respectively. The current subject reported numbness at left hand is much stronger compared to right hand. Hence, it supports the theory of human response to hand arm vibration stating that the effect of health depends on the vibration exposure either by duration or intensity of exposure [17].

The measurement for 5 types of hand-held grass-cutting machine with the standard power ranged in kilowatt from 8.0 to 10.5 kW and the discharge capacity ranged in cubic centimetre from 30.5 to 43.0 cc. An analysis of monitoring HAVERA measurement has found that the mean value of 168 subjects indicates estimated right and left hand daily vibration exposure, A(8) is 6.05 ± 2.74 m/s2 and 8.68 ± 4.41 m/s2, respectively. Based on the hand-held grass-cutting machine type, type A machine produces the highest A(8) value at right hand with 6.73 m/s2, while type E machine produces the highest A(8) value at left hand with 10.82 m/s2. However, only type B and type C machines produce below exposure limit value (ELV) for the right hand daily vibration exposure. The left hand daily vibration exposure exceeds the exposure action value (ELV) for all types of machine.

By referring to the guidelines on occupational vibration, hand-held grass-cutting workers should operate the machine for maximum of 2 hours per day to protect them from HAVS diseases. Thus, risks of HAVS have to be observed not only in the form of high vibration magnitude exposed but will associate to the duration of operating the machine. Control measures were not being implemented in reducing the vibration level such as handle vibration damper or anti vibration glove. So, engineering and administrative control measure will be the most suitable recommendation to reduce the HAV risk exposure and it must be associated with the management safety and health budget.

5. Conclusion

The current study has suggested that HAVS is diagnosed in Malaysia especially in agriculture sector among hand-held grass-cutting workers exposed to hand transmitted vibration in a tropical environment. These findings are indicating significant symptoms which contribute to HAVS prevalence. Employers should provide provision of such information, instruction, training and supervision to ensure, so far as is practicable, to mitigate the hand arm vibration risk. Hence, proper work schedule should be arranged to protect workers induced with HAVS. The machine type C provides the most comfortable operating condition for workers compared to types A, B, D and E. Therefore, machine type C should be recommended for grass-cutting operation. Besides that, the finger colour change and numbness showed high prevalence induced to HAVS. In association with those symptoms, further research will study on the health consequences of hand arm vibration using Vascular and Neurological Stockholm Workshop Scale and hand grip strength assessment.

References

[1] Ko, Y. H., Ean, O. L. & Ripin, Z. M. (2011). The design and development of suspended handles for reducing hand arm vibration in petrol driven grass trimmer. *International Journal of Industrial Ergonomics*, 41 (2): 459-470.

[2] Department of Occupational Safety & Health. (2013). Occupational Accident Statistics By Sector Online. http://www.dosh.gov.my/index.php?option=com_content&view=article&id=843&Itemid=545&lang=ms. Retrieved on 31st December 2013.

[3] Social Security Organization. (2015). Statistik Penyakit Muskuloskletal Pekerjaaan 2005-2014 Online. http://www.dosh.gov.my/index.php?option=com_docman&task=doc_download&gid=330&Itemid=496. Retrieved on 6th July 2015.

[4] Bovenzi, M. (2005). Health effects of mechanical vibration. *G Ital med Lav Erg*, 27 (1): 58-64.

[5] Cederlund, R. & Iwarsson, S. (2007). Quality of life in Swedish workers exposed to hand – arm vibration. *Occup. Ther. Int.*, 14: 156-169.

[6] Lin, W., Chunzhi, Z., Qiang, Z., Kai, Z. & Xiaoli, Z. (2005). The study of hand-arm vibration syndrome in China. *Ind. Health*, 43 (3): 480.

[7] Su, T. A., Hoe, V. C. W., Masilamani, R. & Awang Mahmud, A. B. (2011). Hand-arm vibration syndrome among a group of construction workers in Malaysia. *Occup. Environ. Med.*, 68 (1): 58-63.

[8] BS EN ISO 5349-1. (2001). Mechanical Vibration – Measurement and Evaluation of Human Exposure to Hand-transmitted vibration – Part 1: General Requirement. British Standard.

[9] Janicak, C. A. Preventing HAVS in the Workplace. *Professional Safety*, 49 (1): 35-40.

[10] Directive 2002/44/EC of The European Parliament and of The Council of 25th June 2002 on The Minimum Health and Safety Requirements Regarding The Exposure to Workers to The Risks Arising From Physical Agents (Vibration), *The Official Journal Of European Communities*, 45, 13-19.

[11] Brereton, P. (2011). Impact of European Directive 2002/44/EC on the risk of developing hand-arm vibration syndrome in Great Britain, *Canadian Acoustics*, 39 (2): 108-109.

[12] Cakmak, B., Saracoglu, T., Alayunt, F. N. & Ozarsla, C. (2011). Vibration and Noise Characteristics of Flap Type Olive Harvester, *Applied Ergonomics*, 42 (3): 397-402.

[13] Dewangan, K. N. & Tewari, V. K. (2009). Characteirstics of Hand-transmitted Vibration of a Hand Tractor used in Three Operational Modes, *International Journal of Industrial Ergonomic*, 39 (1): 239-245.

[14] Wasserman, J. F., Wasserman, D. E. & Wilder, D. (2005). Occupational Vibration: A concise Perspective, *Handbook of Human Factor and Ergonomics Methods*, CRC Press: 73-1.

[15] Futatsuka, M., Shono, M., Sakakibara, H. & Quan, P. Q. (2005). Hand arm vibration syndrome among quarry workers in Vietnam, *Journal Occupational Health*, 47 (2): 165-170.

[16] Coggins, M. A., Van Lente, E., McCallig, M., Paddan, G. & Moore, K. (2010). Evaluation of hand-arm and whole-body vibration in construction and property management, *Annual Occupational Hygiene*, 54 (8): 904-914.

[17] Griffin, M. J. (1997). Measurement, evaluation and assessment of occupational exposure to hand-transmitted vibration, *Occupational Environment Medicine*, 54 (2): 73-89.

Dynamic Simulation and Analysis of Human Walking Mechanism

Athirah Azahari*, W.A. Siswanto, M.Z. Ngali, S. Md. Salleh, Eliza M. Yusup

Department of Engineering Mechanics, Faculty of Mechanical and Manufacturing Engineering, Universiti Tun Hussein Onn Malaysia.

*coresponding author: athirahz_1302@yahoo.com

Abstract. Behaviour such as gait or posture may affect a person with the physiological condition during daily activities. The characteristic of human gait cycle phase is one of the important parameter which used to described the human movement whether it is in normal gait or abnormal gait. This research investigates four types of crouch walking (upright, interpolated, crouched and severe) by simulation approach. The assessment are conducting by looking the parameters of hamstring muscle joint, knee joint and ankle joint. The analysis results show that based on gait analysis approach, the crouch walking have a weak pattern of walking and postures. Short hamstring and knee joint is the most influence factor contributing to the crouch walking due to excessive hip flexion that typically accompanies knee flexion.

1. Introduction

Human gait is one of the most important movement for develop of the human being activities in life. Human movement involves a complex movement of skeleton and muscles in various combination of bending, stretching, compressing and rotating actions [1]. During movements, the human body can move in a large number of degrees of freedom. In walking movement, it is required an appropriate coupling and coordinate movement of body segment to produce smooth motion, maintain balance and minimize energy expenditure [2].

Walking seems so simple that most of us do it daily since childhood. However, for those with an orthopaedic or neurological impairment, walking is no longer automatic nor simple [3]. For them, walking may have become struggle and pose a threat to one independence and quality of life.

Normal gait is a pattern of walking human normally adopted in undisturbed situation. Gait pattern are considered to be either abnormal or impaired if they do not satisfy certain key attributes of normal gait. It has previously been postulated that normal gait is more energy efficient than abnormal or impaired gait. Crouch gait pattern is characterized by increase knee flexion through stance phase. Individual with cerebral palsy who walk in a crouch gait often receive strength training as part of their therapy. The crouch gait will increase the energy costs of walking and can lead to joint pain and degeneration when during walking. It generally considered being a negative symptom of cerebral palsy [4].

Several factor have been linked with crouch gait, including muscle weakness, spasticity, skeletal deformities and tightness [4]. A link between crouched gait postures and the capacity of muscle to extend the hip or knee joints, individual may be required to exert more muscle force to maintain a crouched posture and to avoid injuries from falling and tripping.

Posture is the orientation of body segment that relative to the gravitational vector. It is an angular measure from the vertical. Balance is the dynamic of body posture to prevent falling or in another word is a process whereby the body's state of equilibrium is controlled for a given purpose. The difficulty which equilibrium can be disturbed is by encounter frequently in activities demanding static postures. Many skills and tasks involving the moving system. This concept also can be called as stability. It is related to the inertial forces acting on the body and the inertial characteristic of body segment [5].

The objective of this study is to know the posture of walking in a proper way and the parameter contributing in crouch walking will be identified. Besides, the method used to achieve of the study is conducted in simulation approach.

2. Methodology

In this research, the musculoskeletal modelling is all conducted in OpenSim [6]. The walking motion is assessed and analysed on some walking parameters. The investigation parameters involved in this work are hamstring, range of motion for right knee and ankle plantar flexion.

2.1 Human model

The main goal of this research was to evaluate between normal walking and crouch walking in a first stage by using a simplified version of a musculoskeletal model on which applied simple motion. In this study, the model is 3DGaitModel2392 [6]. This model consists of body, lower extremities and torso. It had developed by 23 degrees of freedom mechanical linkages, whereas the lower extremities and torso has 92 muscle-tendon actuators to represent 76 muscles.

The normal walking and four types of crouch walking (crouch 1 walking, crouch 2 walking, crouch 3 walking and crouch 4 walking) are simulated in this model. Crouch 1 is upright posture. This type of posture was defined from the average of gait data. Crouch 2 is interpolated posture. Crouch 3 is crouched posture, where this type of posture defined from the person with cerebral palsy and crouch gait. Lastly is crouch 4 is severe crouch posture or extrapolated posture. This type of posture had a huge abduction of tight.

2.2.1 Normal Walking Posture

Normal walking posture is walking in a proper way with a good posture shown in figure 1. A good body posture or structure is the bones and joints in the correct alignment. Thus, it can minimizes stress on muscles, bones and joints.

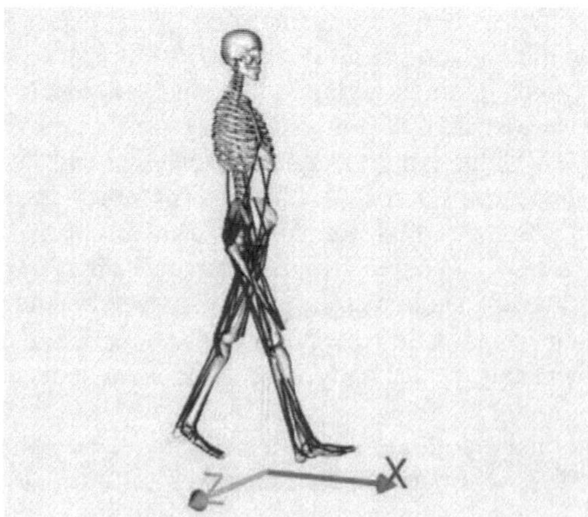

Figure 1: Normal Walking Posture

2.2.3 Comparison of Walking Posture

This stage is to compare the differentiation of walking and the parameter that contribute during this walking process. Normal walking as a reference to compared between the others walking. From there, it can be know what are the parameters that make someone cannot walk in the proper way. There were four type of comparison which is type 1 shown in figure 2, type 2 shown in figure 3, type 3 shown in figure 4 and lastly is type 4 shown in figure 5. During these walking activity, there will be flexion and extension of the legs. Flexion is the bending movement that result in decrease of the angle in a joints by bringing the bones together, while the extension is straightening movement that results in an increase of the angle in a joint by moving the bones apart. Hamstring muscle, knee and ankle is the parameter that interested to investigate which is the most factor in walking process.

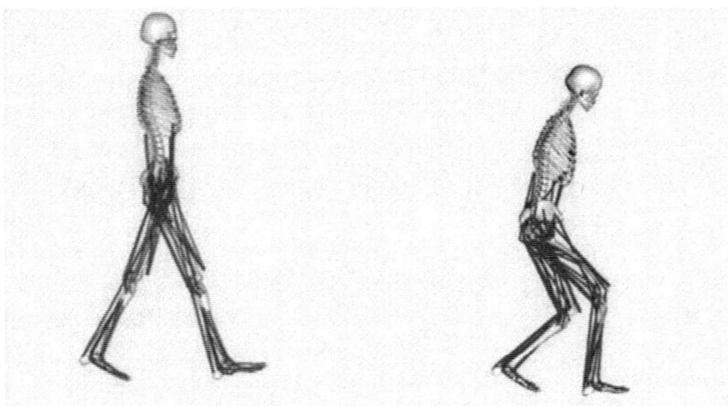

Figure 2: Normal Walking Posture and Crouch 1 Walking Posture (Type 1)

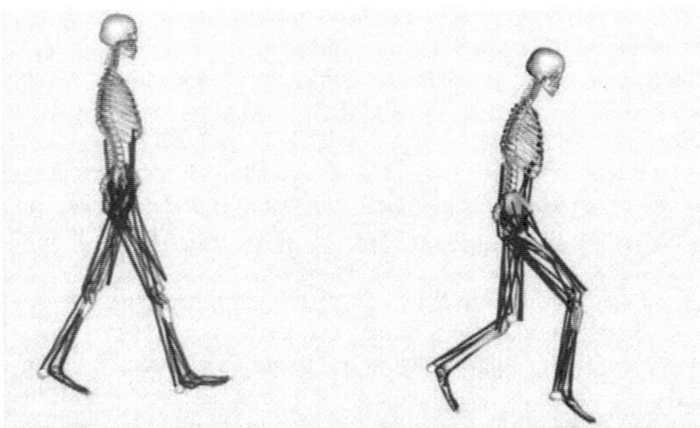

Figure 3: Normal Walking Posture and Crouch 2 Walking Posture (Type 2)

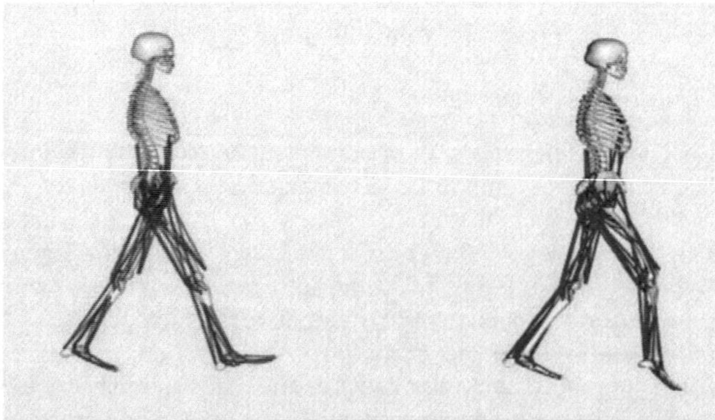

Figure 4: Normal Walking Posture and Crouch 3 Walking Posture (Type 3)

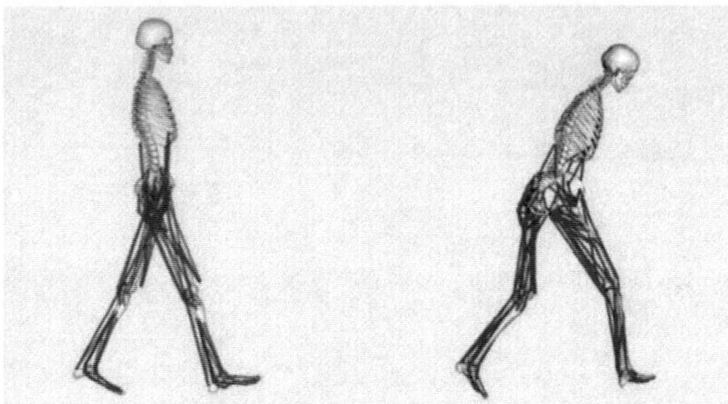

Figure 5: Normal Walking Posture and Crouch 4 Walking Posture (Type 4)

3. Results and Discussion

Investigation parameters are plot and presented to see the difference on each crouch posture.

3.1 Hamstring Muscle

In daily life activities, muscle strength is important. Force that provide from our muscle that allow us to walk, jump, climb stairs, run and perform athletic activities. Humans can increase their performance with training even for run faster, jump higher and lift more weight. The limitation performance of activities will be loss of muscle strength if it is disuse, aging and disease [7].

Walking may become impaired for human when muscle become excessively weak during daily life activities. The muscle strength is an important element for walking performance. The comparison of walking is to determine with joint rotation have the greatest influences on hamstring length during walking [8]. Data that gather from the simulation of hamstring is to see the different length of muscle. Muscle of hamstring type 1 shown in figure 6 and 7. The data shows that crouch 1 have small different hamstring length compared to normal. During initial to mid stance stage, there had some flexion of leg which is the foot is quickly change. Then, continue until initial swing which the leg had a huge of extension.

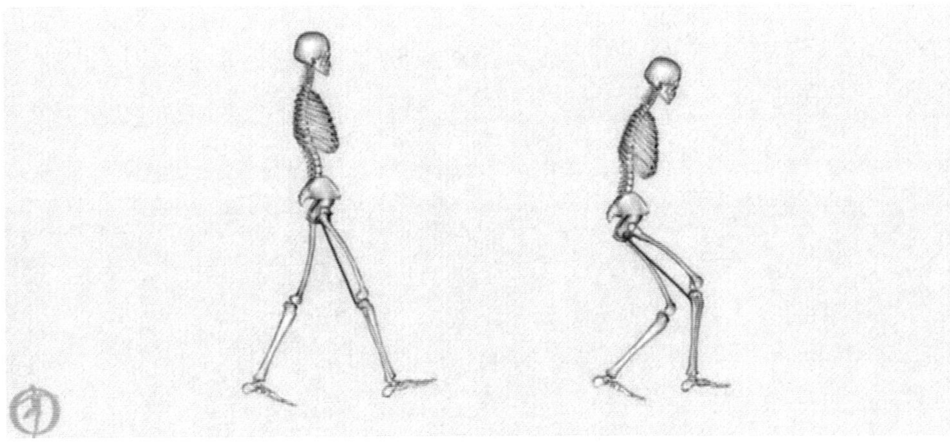

Figure 6: Muscle of hamstring in models type 1

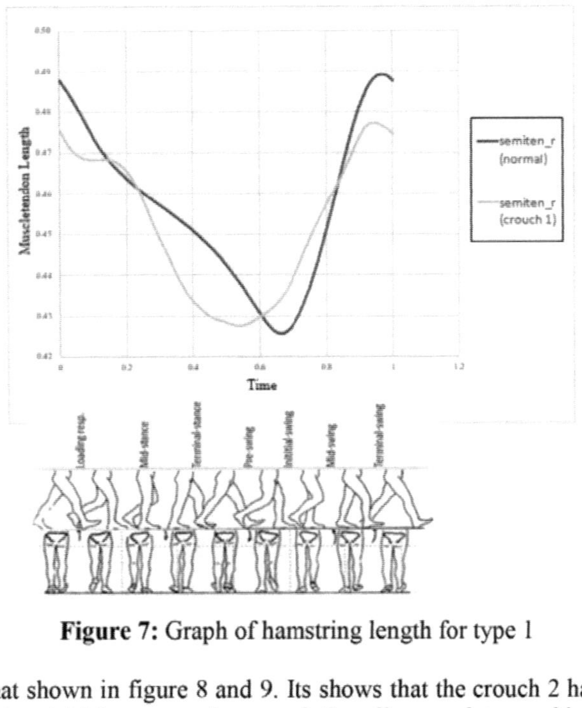

Figure 7: Graph of hamstring length for type 1

The data for type 2 that shown in figure 8 and 9. Its shows that the crouch 2 have a huge different of hamstring length. During initial contact, the crouch 2 walker need to use high of force to start the motion. Besides, during the terminal stance to swing phase, the changes of leg to fast and take a huge of flexion until the toe off. This because the length of hamstring are the one of factor that the leg stops early.

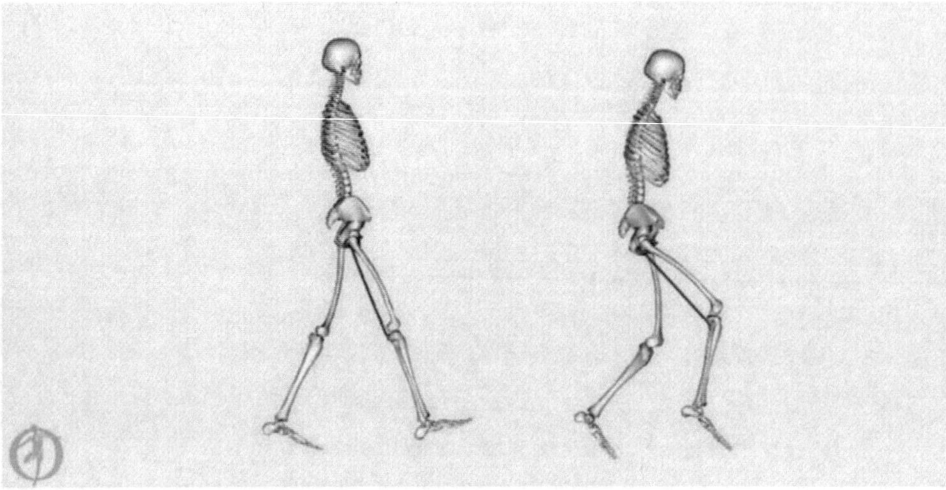

Figure 8: Muscle of hamstring in models type 2

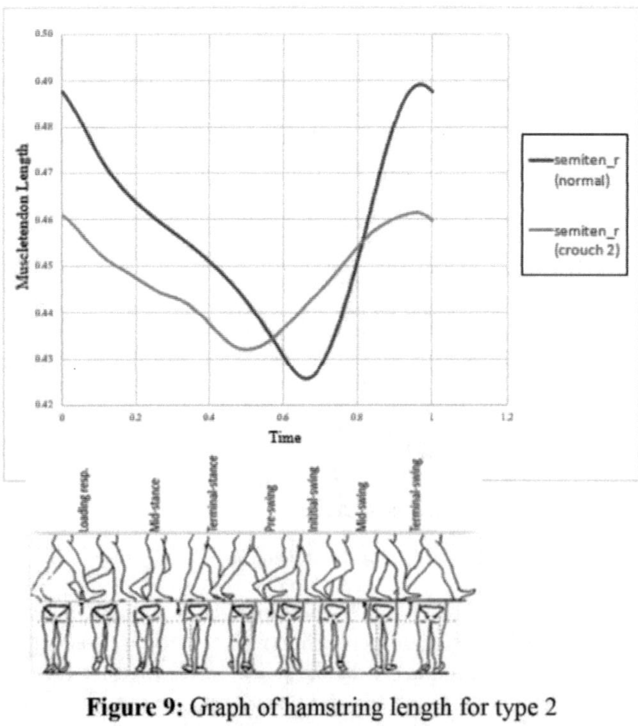

Figure 9: Graph of hamstring length for type 2

The data for type 3 shown in figure 10 and 11. It shows that the crouch 3 walker have similarity with the normal walker but it have a smaller different of hamstring length. During the initial contact the walking activity still in a right path, but when enter to mid stance stage the walking activity take quickly changes during flexion and extension. During the terminal swing which is the final stage to toe off, the crouch 3 had stop early due to smaller different of length.

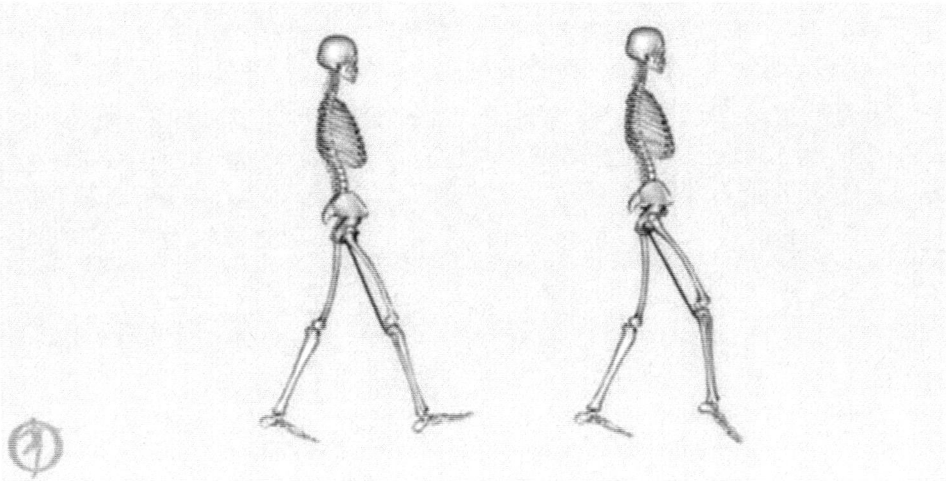

Figure 10: Muscle of hamstring in models type 3

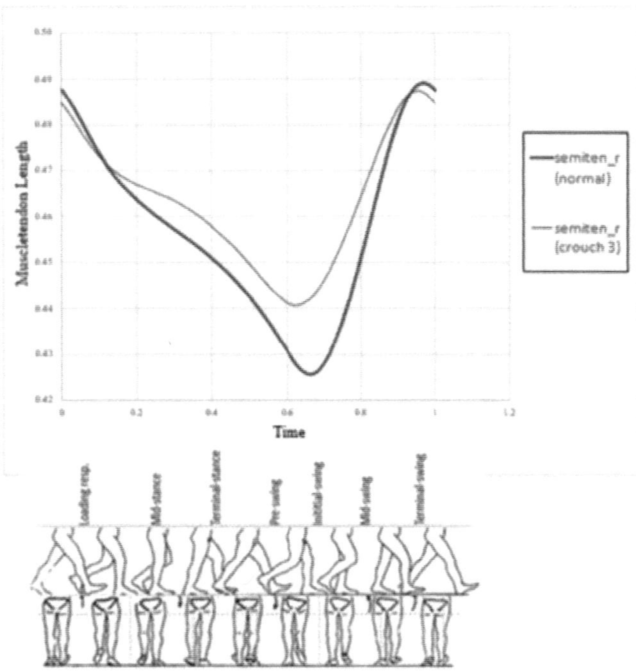

Figure 11: Graph of hamstring length for type 3

The data for type 4 shown in figure 12 and 13. It shows that the crouch 4 have a huge different of hamstring length compared to normal walking. This is because during the initial contact and process to swing phase, the changes of the leg is very quick.

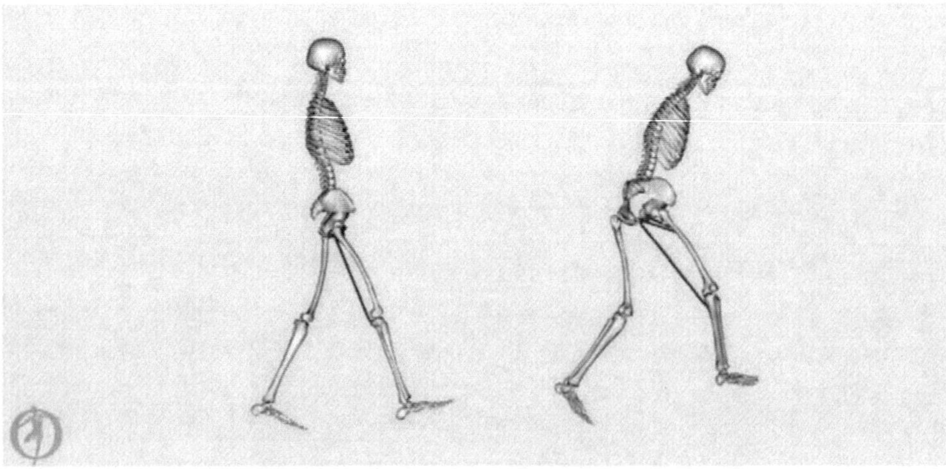

Figure 12: Muscle of hamstring in models type 4

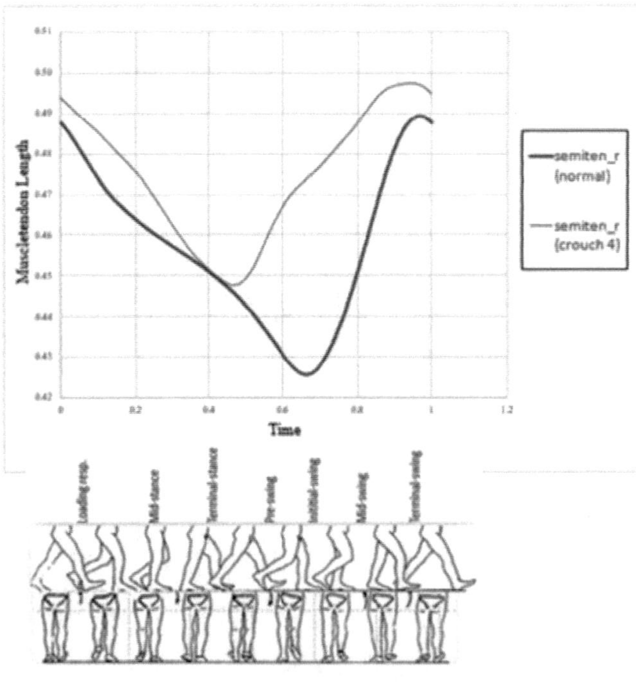

Figure 13: Graph of hamstring length for type 4

This is the overall of data hamstring length vs gait cycle for normal walking and all type of crouch walking as shown in figure 14. Hence, we can see that the most shorten hamstring length is crouch 4, next is crouch 2 continue to crouch 3.

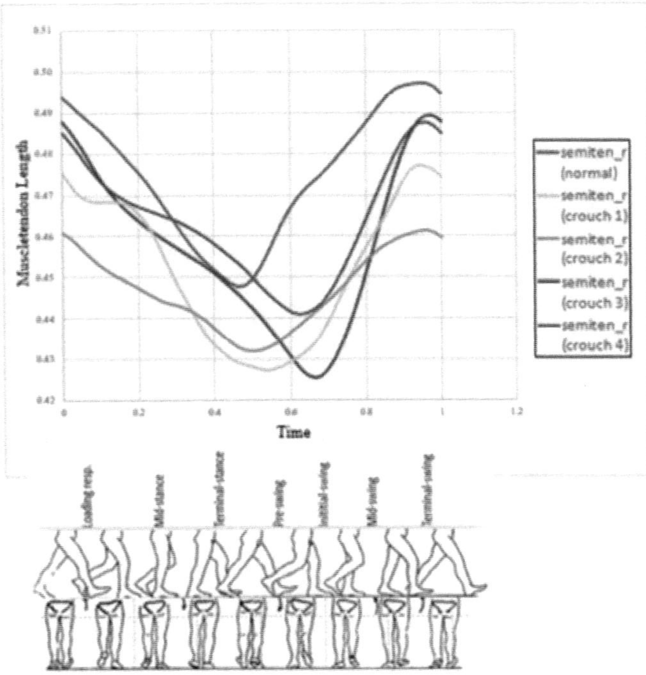

Figure 14: Graph overall of hamstring length for each model

3.2 Range Of Motion

Range of motion is the motion of the knee in vertical displacement during the walking activity. Joints and muscles are included in this motion. The data for type 1 shown in figure 15 and 16. It shows that the crouch 1 walking have a knee flexion angle. The knee angle for normal is -3.94 deg, while the crouch 1 walking is -59.26 deg. The motion for normal is smoother than crouch 1. This is because during the phase there have some extension of the hip joint then continue to knee joint and lastly to ankle joint. For crouch 1 walking have small changes that related to hip, knee and ankle joint. There is small flexion during the mid stance to terminal stance. The smaller extension occurring during mid swing stage.

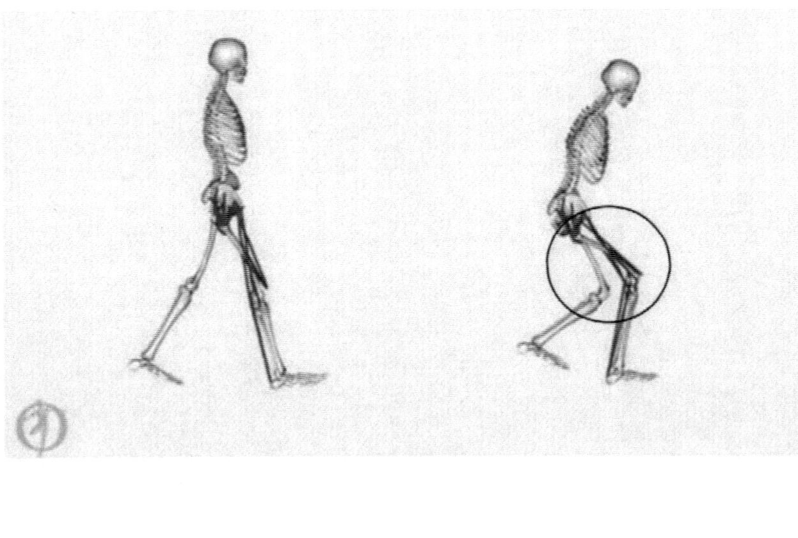

Figure 15: Knee flexion angles type 1

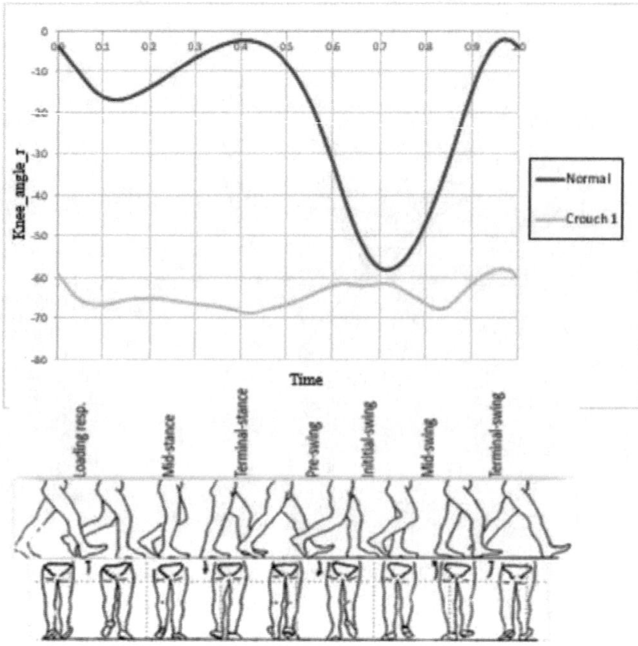

Figure 16: Graph knee flexion angles type 1

The data for type 2 is shown in figure 17 and 18. It shows that the crouch 2 walking have a knee flexion angle. The knee angle for crouch 2 walking is -54.94 deg. For type 2 the pattern of range of motion is same with type 1 but the crouch 2 pattern is smoother compared to crouch 1. During the stance phase and swing phase, the flexion and extension occur in small changes by hip joint, knee joint and ankle joint.

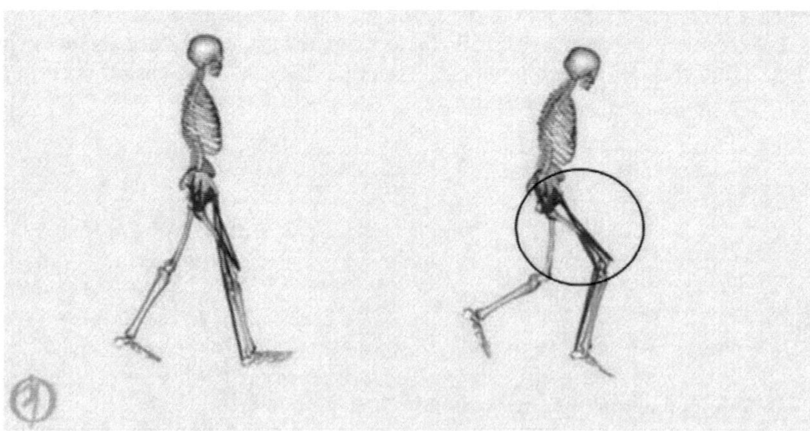

Figure 17: Knee flexion angles type 2

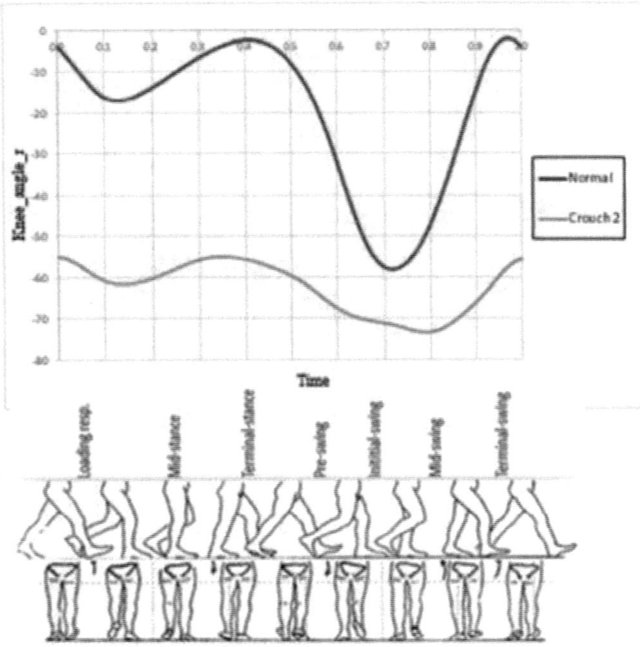

Figure 18: Graph knee flexion angles type 2

The data for type 3 shows in figure 19 and 20 It shows that the crouch 3 walking have a knee flexion angle. The knee angle for crouch 3 walking is -26.26 deg. During the initial contact, the pattern of the motion is same but the value is different with the normal walking. When enter the mild stance, there is small flexion occur. This pattern is continue until enter the initial swing. During this stage until terminal swing, there is small extension occur by hip joint, knee joint and ankle joint.

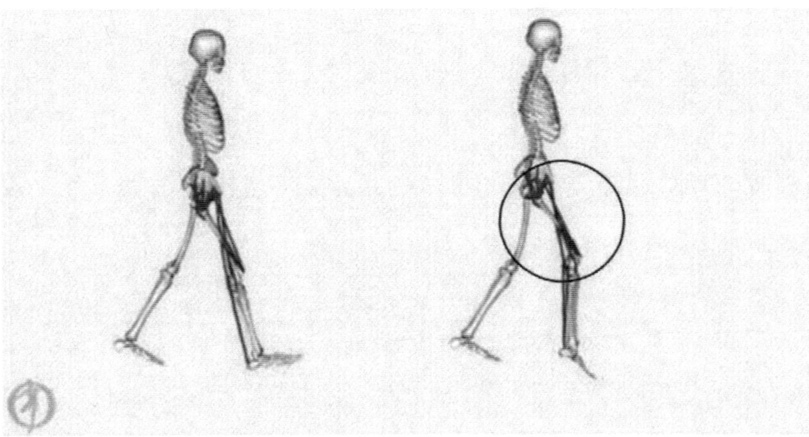

Figure 19: Knee flexion angles type 3

IOP Conf. Series: Materials Science and Engineering **165** (2017) 012027 doi:10.1088/1757-899X/165/1/012027

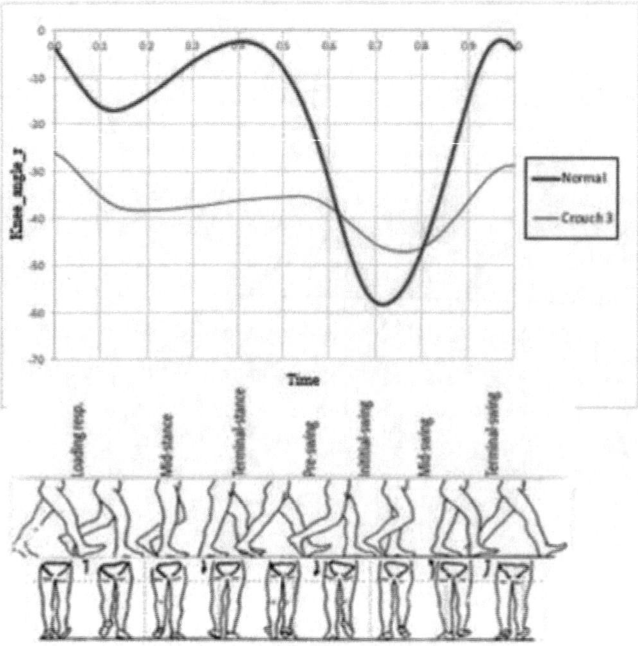

Figure 20: Graph knee flexion angles type 3

The data for type 4 shows in figure 21 and 22. It shows that the crouch 4 walking have a knee flexion angle. The knee angle for crouch 4 walking is -45.12 deg. The pattern of the graph motion for crouch 4 is same with normal walking but during the terminal stance to swing phase, the graph shows not smooth in changes because of the extension of the hip joint, knee joint and ankle joint is very quick.

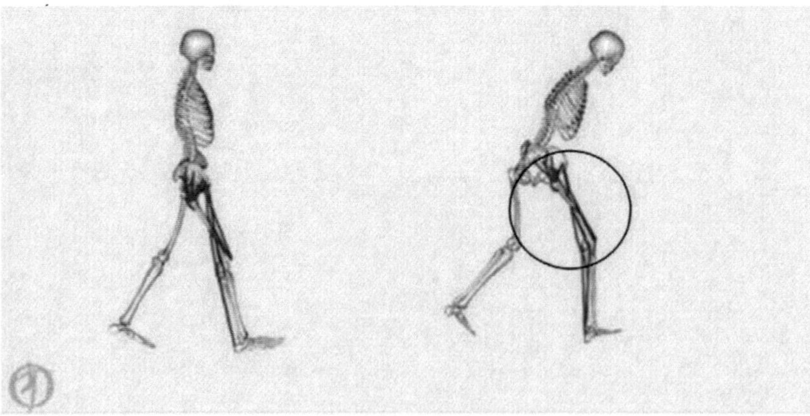

Figure 21: Knee flexion angles type 4

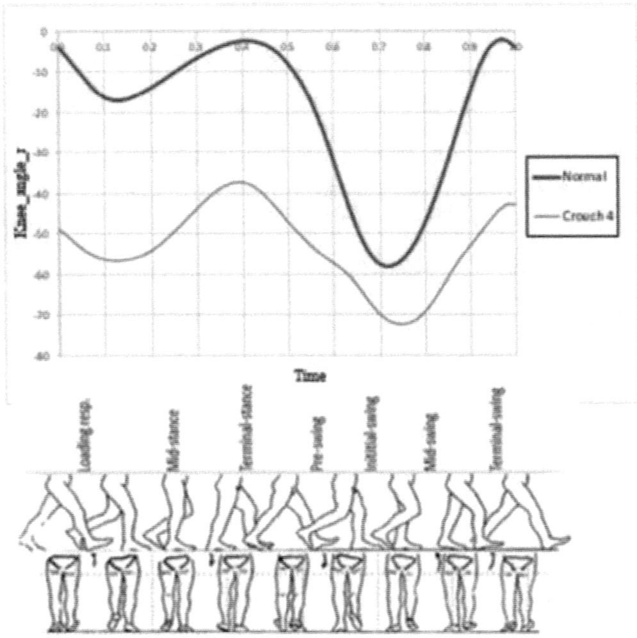

Figure 22: Graph knee flexion angles type 4

This is the overall of data right knee angle vs gait cycle for normal walking and all type of crouch walking as shown in figure 23. Hence, we can see the most bending angle for knee is crouch 1, next is crouch 3 continue to crouch 2 and lastly is crouch 4.

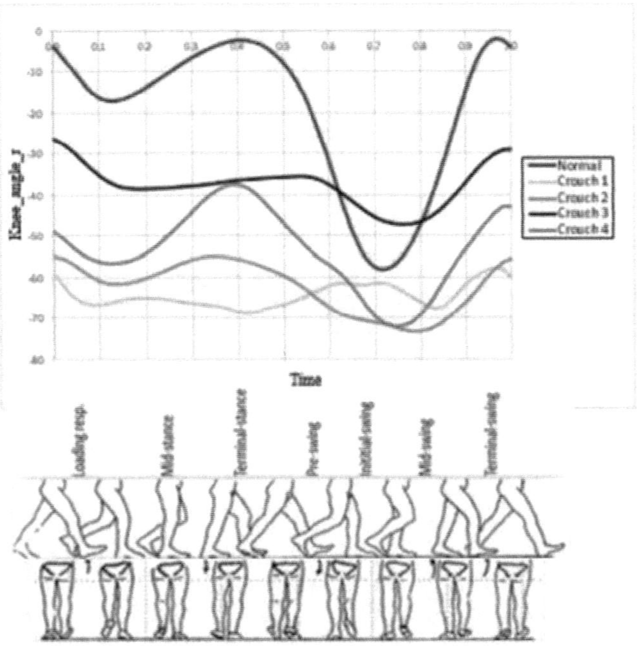

Figure 23: Graph overall of range of motion for each model

3.3 Ankle

Movement of ankle is the one of the factor to walking, running, jumping and others activity. The movement of ankle have two type which is when the foot of the top is called dorsiflexion while when the foot in the bottom is called plantar flexion. The data for type 1 shown in figure 24 and 25. It shows that, pattern of the graph for the ankle plantar flexion for crouch 1 walking is same with normal walking. The ankle take action at stance phase in loading response, mild stance and terminal stance. The ankle angle for normal is -1.7 deg while crouch 1 is 13.07 deg. It shows that the graph of crouch 1 is upper than normal walking. This is because the angle of the ankle. During initial contact, the crouch 1 had directly flexion while for normal there had some time to start flexion.

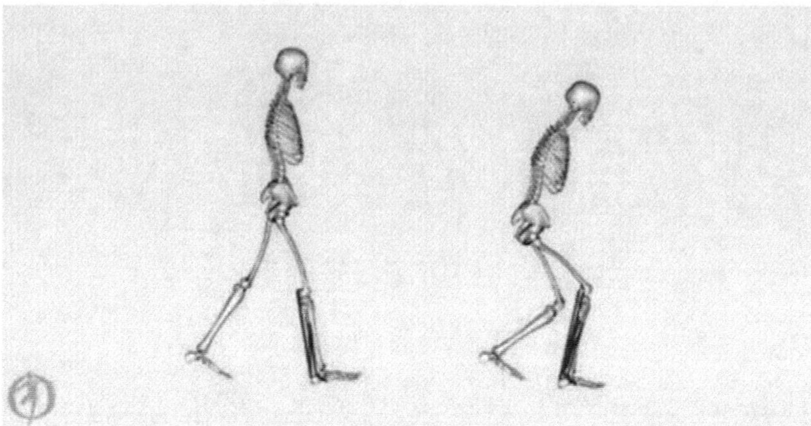

Figure 24: Ankle Plantar Flexion type 1

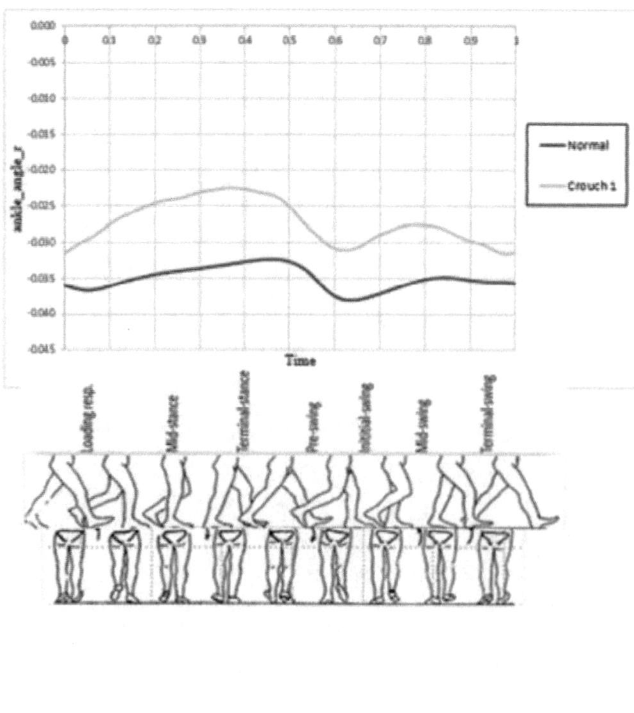

IOP Conf. Series: Materials Science and Engineering **165** (2017) 012027 doi:10.1088/1757-899X/165/1/012027

Figure 25: Graph Ankle Plantar Flexion type 1

The data for type 2 shown in figure 26 and 27. It shows that, crouch 2 walking pattern. The ankle take action at stance phase in loading response, mild stance and terminal stance. The ankle angle for crouch 2 walking is -16.17 deg. The graph shows the crouch 2 during the initial contact, the flexion is directly occur during the walking in small changes. Then, there is no extension occur There is small changes of ankle plantar flexion compared to normal walking.

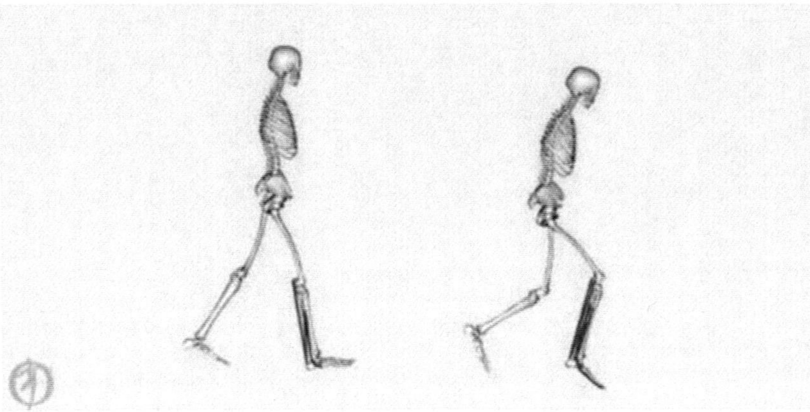

Figure 26: Ankle Plantar Flexion type 2

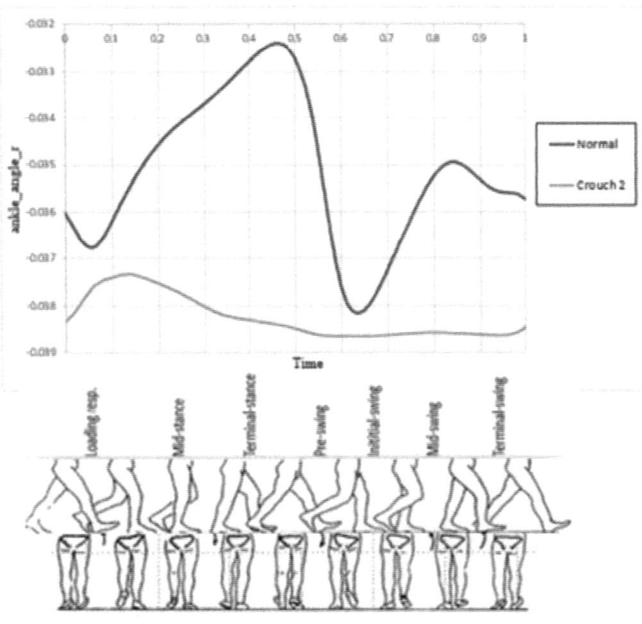

Figure 27: Graph Ankle Plantar Flexion type 2

The data for type 3 shown in figure 28 and 29. It shows that, the crouch 3 walking pattern. The ankle take action at stance phase in loading response, mild stance and terminal stance. The ankle angle for crouch 3 walking is -39.07 deg. The graph shows the crouch 3 during loading respond, there is no flexion. Then, during the extension process at pre swing until mid swing there is flexion occur with small changes during the walking and it continued until toe off.

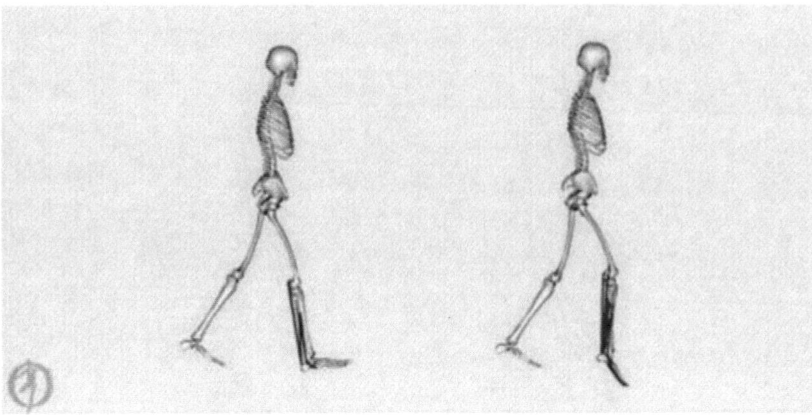

Figure 28: Ankle Plantar Flexion type 3

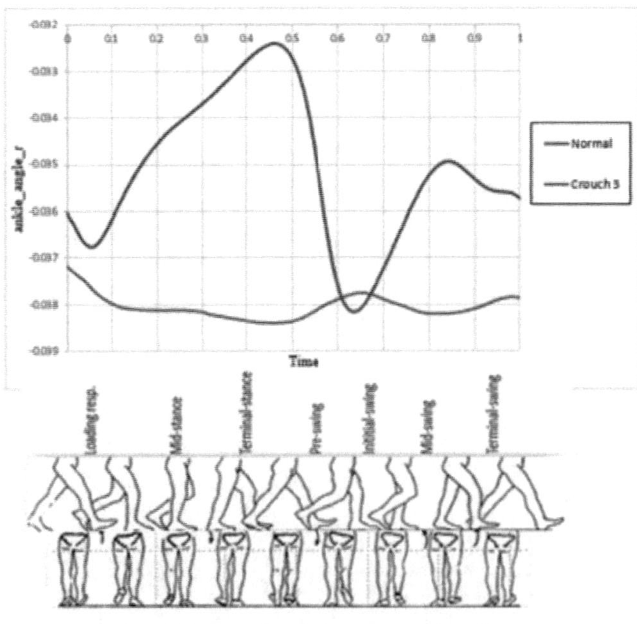

Figure 29: Graph Ankle Plantar Flexion type 3

The data for type 4 shown in figure 30 and 31. It shows that, pattern of the graph for the ankle plantar flexion for crouch 4 walking is same with normal walking. The ankle take action at stance phase in loading response, mild stance and terminal stance. The graph of crouch 4 is upper than

IOP Conf. Series: Materials Science and Engineering **165** (2017) 012027 doi:10.1088/1757-899X/165/1/012027

normal walking. This is because the angle of the ankle. The ankle angle for crouch 4 walking is 20.45 deg. The graph pattern is not smooth and during extension occur, there had smaller changes. Besides, during initial contact to terminal stance the changes is very quick so that the extension take a quickly too.

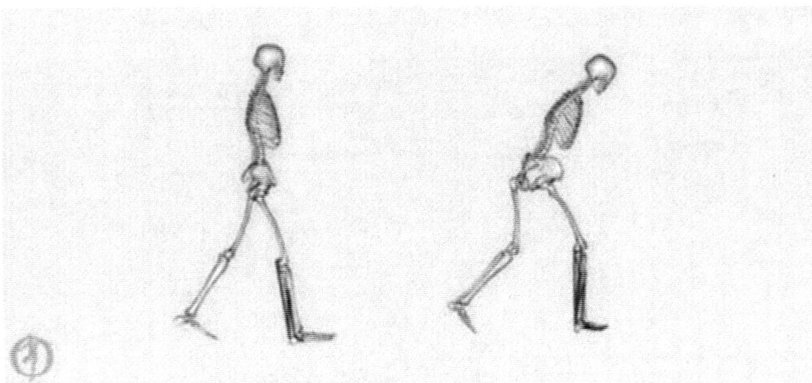

Figure 30: Ankle Plantar Flexion type 4

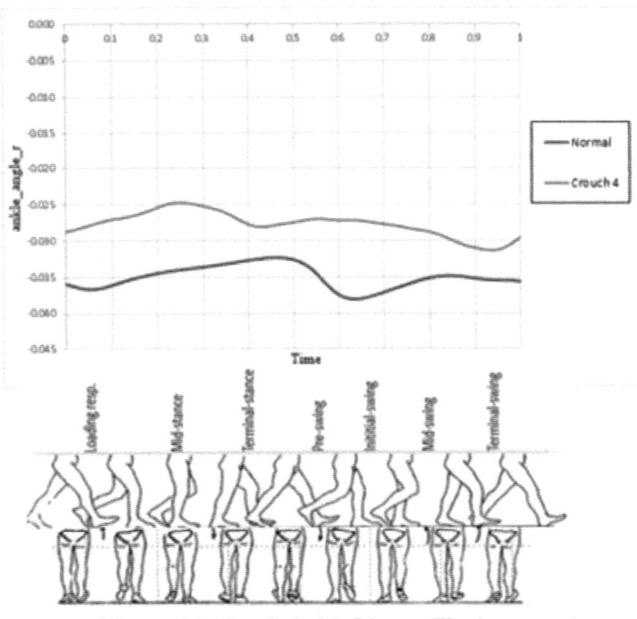

Figure 31: Graph Ankle Plantar Flexion type 4

This is the overall of data ankle plantar flexion vs gait cycle for normal walking and all type of crouch walking as shown in figure 32. Hence, we can see that, the most bending ankle angle is crouch 3, next is crouch 2 continue to normal, crouch 1 and lastly is crouch 4.

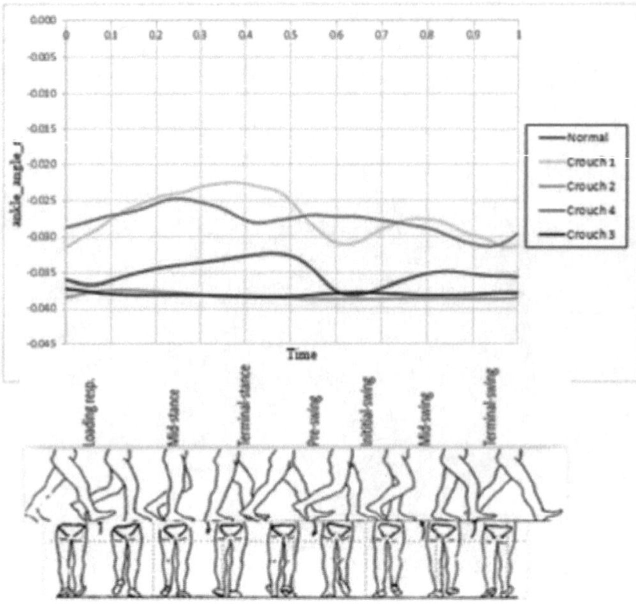

Figure 32: Graph overall of ankle plantar flexion for each models

3.4 Gait Analysis

The result indicate to investigate the posture of walking in a proper way and the parameter contributing in crouch walking. Analysis of this were performed in one type of normal walking and four type of crouch walking. Besides, by using a simulation platform to evaluate the human motion and its disorder. Other than that, the effect of human movement.

Individual who walk in crouched posture during single support may increase activation of the hamstring to accelerate the hip and knee toward extension since the capacity of the muscle group. Besides, usually the person received strength training as part of their therapy. Generally, crouch gait is considered to be disadvantages but however some an athlete adopts the crouched posture to rise their ability to move in any chosen direction [9].

During stance phase, the joint and muscle activity in heel strike stage at the beginning the hip joint is partially flexed with gluteus maximus and the hamstring. This contract immediately to initiate the hip extension. The knee joint in condition full extension or flexed about 5 degree and the quadriceps working eccentrically to control the knee flexion which follow immediately after heel strike. The ankle joint on heel strike usually to neutral position which up to 10 degree of dorsiflexion or plantar flexion. Next, in foot flat stage the hip joint is started to move into extension by concentric action of the hip. The knee has flexed further in order the effect of heel strike. This also is to reduce the vertical displacement of the center gravity that occur the body passes over the stance limb. At the ankle joint, the controlled plantar flexion to lower the foot to the ground which is undertaken by eccentric work of dorsiflexors. There is small amount of eversion to transfer body weight when the foot achieve good foot-ground contact. Then to mid stance stage, the hip extension continue and being produce by momentum and the muscle. Lastly, heel off which is the leading to push off. At the beginning of the phase, the center of gravity in stance foot, the force of gravity will increase the range of hip extension and dorsiflexion. The heel will rise off the ground and the plantar flexors will contact concentrically when full range of dorsiflexion to provide the propulsive component of push off. The walking speed from slow to moderate this contraction not usually very large as momentum as the major factor in moving the body forward. For normal walking, the hip extensor not active in this stage, but actually the hip and knee joints are usually starting to flex to prepare for swing phase.

During the swing phase, the joint and muscle activity have three stages which is acceleration, mid-swing and deceleration. In acceleration stages, the minor force is generated to push off the hip flexors and plantar flexors to accelerate the limb forwards. It is assisted by momentum and gravity. The both flexing which is hip and knee joint and the rapid movement towards dorsiflexion is to ensure that the toes do not catch on the ground. In mid swing stages, the flexion of knee and hip joint continues to keep the foot raised and avoid from toes catching the ground. This stage, the foot is lowered into slight plantar flexion. Lastly, deceleration process is the hip continues to flexion. The movement mainly produce by momentum and the hamstring act eccentrically to slow down the movement at the hip joint. The flexion to extension by the knee in the movement. The knee joint extension make the whole of the lower limb being moved forward by flexion of the hip and the resulting momentum. Towards the end of deceleration stage, extension of knee joint may have to be slow down and this is achieved by eccentric action of hamstrings. During the toe off or heel strike, the dorsiflexion contract strongly to ensure that the foot in optimum position condition.

The ground reaction force in gait is the foot contact with the ground, there will be vertical, anterior-posterior and medio-lateral force acting between the feet. At the beginning and end of stance phase, about 25% greater than body weight from the vertical ground reaction force and it is also same for anterior posterior ground reaction force [10]. The relatively small forces involved and the design of the human body that facilitates shock absorption. The cushioning effect of the knee flexion on initial foot and ground contact mean that repetitive strain injuries caused by ground reaction force are rare in walking.

The amount of energy required to walk at a comfortable pace is very small compared to other activities. Normal structure of human body or in normal pattern of walking are mostly to increase the energy expenditure significantly. Energy is conserved through the utilization of momentum, kinetic and potential. From the rhythmical, the center of gravity that produce vertical fluctuations in normal gait is similar fluctuations in kinetic energy and gravitational potential energy. In mid stance stage, when the center of gravity is highest, the gravitational potential also highest but the kinetic energy is lowest. Normal walking have a smoothness rhythmic walking compared to the crouch walking based on their posture, muscle pattern and regular stepping [11].

4. Conclusion

In conclusion, the human walking pattern is based on the human body structure or gait. For normal walking, the rhythmic of walking is very smooth based of their good body posture or structures. For the crouch walking that have problem in lower limb, they required high energy to make their body forward. Usually these crouch walker, have unstable during stance phase. The capacity of the muscle to generate extension accelerations at the knee. It also depends on the muscle activation level and physiological ability to generate force like hamstring, the bending of the knee and the ankle. The muscle must work harder to maintain their walking pattern. The moment of the hip, knee joint and ankle are related to ground reaction force. For further research, this simulation platform can be used to investigate other types of gait pattern such as running to improve the performance of athletes. The dynamic analysis of this study could also be implemented to analyze the gait kinematics and bone geometry.

5. References

[1] Potter, J.M., Evans, A.L. & Duncan. G. (1995). *Gait speed and activities of daily living function in geriatric patients*. Archives of Physical Medicine and Rehabilitation, **76(11),** pp. 997-999.

[2] Miller, R.H. (2014). *A comparison of muscle energy models for simulating human walking in three dimensions*. Journal of Biomechanics, **47(6),** PP. 1373-1381.

[3] Whittle, M.W. (1996). *Clinical gait analysis: A review. Human Movement Sciences*, **15(3),** pp. 369-387.

[4] Hicks, J.L., Schwartz, M.H., Arnold, A.S. & Delp, S.L (2008). *Crouched Postures Reduce the capacity of muscle to extended the hip and knee during the single limb phase*

stance phas gait. Journal of biomechanics, **41(5),** pp. 960-967.

[5] Winter, D. (1995). *Human balance and posture control during satnding and walking*. Gait & Posture, **3(4)**, pp. 215-232.

[6] Delp, S., Anderson, F., Arnold, A., Loan, P., Habib, A., John, C., Guendelman, E. & Thelen, D. (2007). *Opensim: Open-source software to create and analyze dynamic simulation of movement*, **54(11)**, Pp 1940-1950.

[7] Van der Krogt, M.M., Delp, S.L. & Schwartz, M.H (2012). *How robust is human gait to muscle weakness?* Gait & Posture, **36(1)**, pp.113-119.

[8] Agarwal-Harding, K.J., Schwarts, M.J. & Delp, S.L. (2010). *Variation of hamstring lengths and velocities with walking speed*. Journal of Biomechanics, **43(8)**, pp. 1522-1526.

[9] Reinbolt, J.A., Seth, A. & Delp, S.L. (2011). *Simulation of human movement : application using opensim*. Procedia IUTAM, **2**, PP. 186-198.

[10] Elftman, H. (1939). *Forces and energy changes in the leg during walking*. American Journal of Physiology – Legacy Content, **125(2)**, pp.339-1950

[11] Branch, J.S., McGurl, D., Wert, D., VanSwearingen, J.M., Perera, S., Charn, R. & Studenski, S. (2010). *Validation of a measure of smoothness of walking*. The Journals of Gerontology Series A: Biological Sciences and Medical Sciences, 66A (1), pp.136-141.

Biomechanics Analysis of Combat Sport (Silat) By Using Motion Capture System

Muhammad Zulhilmi Kaharuddin*, Siti Badriah Khairu Razak, Muhammad Ikram Kushairi, Mohamed Syawal Abd. Rahman, Wee Chang An, Z. Ngali, W.A. Siswanto, S M Salleh, E M Yusup

Department of Mechanical Engineering, Universiti Tun Hussein Onn Malaysia
Faculty of Mechanical and Manufacturing Engineering, Universiti Tun Hussein Onn Malaysia

*zulhilmikaharuddin.psrc1314@gmail.com

Abstract. 'Silat' is a Malay traditional martial art that is practiced in both amateur and in professional levels. The intensity of the motion spurs the scientific research in biomechanics. The main purpose of this abstract is to present the biomechanics method used in the study of 'silat'. By using the 3D Depth Camera motion capture system, two subjects are to perform 'Jurus Satu' in three repetitions each. One subject is set as the benchmark for the research. The videos are captured and its data is processed using the 3D Depth Camera server system in the form of 16 3D body joint coordinates which then will be transformed into displacement, velocity and acceleration components by using Microsoft excel for data calculation and Matlab software for simulation of the body. The translated data obtained serves as an input to differentiate both subjects' execution of the 'Jurus Satu'. Nine primary movements with the addition of five secondary movements are observed visually frame by frame from the simulation obtained to get the exact frame that the movement takes place. Further analysis involves the differentiation of both subjects' execution by referring to the average mean and standard deviation of joints for each parameter stated. The findings provide useful data for joints kinematic parameters as well as to improve the execution of 'Jurus Satu' and to exhibit the process of learning a movement that is relatively unknown by the use of a motion capture system.

1. Introduction

Biomechanics is an application of mechanical laws to living structure, specifically to human locomotion system [1]. Biomechanics can also be defined as the study of the structure and capacity of biological system of biological by means of the methods [2]. Mechanically, biomechanics is the study of forces acting on the object and what its impact on the movement, shape, size, and structure [3]. It can also be associated with the study of the internal and external forces acting on the human body and the effects produced by action [4]. The role of biomechanics in life is to grasp the concept of mechanical cause and effect relationships that determine the living organism's motion. Sport biomechanics is a use of biomechanics concept that underlines the comprehension of exhibition on athletic event through mechanical modeling, simulation and calculation. In sport, biomechanics helps to analyze and improve an athlete performance as well as reducing the risk of injuries during any sporting events.

"Silat" or "pencak silat" is the traditional Malay self defense art. "Pencak" or "pentjak" is a self defense training method: it consists of myriads range of regulated body movements directed to training while "silat" is the application of the training method which is the combat [5]. Based on the advancement of the science of biomechanics in sports arena, there are numerous studies and examinations were completed. Among them is to choose the distinction in kinematics between the individual contenders who have speedy and moderate acceleration in sports [6]. The study of the golf swing kinetics and kinematics (3-dimensional) also was created. From the data acquired will be separated in purpose of interest and the relationship made is firmly associated with four swing types. This study communicates that the golf swing is included with the movement of individuals. Four athletes were taken a gander at in perspective of their swing. It is revealed that the golf swing is encouraged with the movement of individuals and subjected to the assortment of the individual strokes. This study focuses on the significance of wrist in creating velocity the head of the club and the club face orientation [7].

By using Ergonometric for rowing sports, three types of studies have been finished. The first is the examination of the nerve coordination rowing cycle, the second is a study on the attributes of kinematics and dynamics paddle, while the third is to choose the maximum power of muscle limit in the midst of rowing [8]. All things considered, a vast part of the studies related to the kinematic parameters and motor coordinated together to ensure the advancement of control capacities for each sport discipline. Normally, analysis of motions in 'silat' is limited to qualitative evaluations both in amateur and professional levels.

The purpose of this project is to conduct the biomechanics analysis on the "bunga silat"' execution by several person by using Kinect motion capture system. The movement execution will be recorded and numerically analyzed in three session. The analysis is aimed to investigate and evaluate the effect of joint displacement before and after execution, to provide descriptive kinematic factors that correlates with the movement speed and acceleration during motion between experienced and new practitioners.

2. Methodology

2.1. Subject

The study involves a 14-year experienced Seni Silat Gayong practitioner as benchmark and a new practitioner with no martial art experience. The subjects are a 28 years old male and a 25 years old male, 173cm of height and weight 65kg, 165cm of height and weighs 57kg, respectively. Motions of these subjects are compared specifically for selected 'silat' movements. The "bunga silat" that will be used are from the "Seni Silat Gayong Jurus Satu" as shown in Figure 1.

Fig.1 Nine primary steps in "seni silat gayong jurus satu"

There are nine steps in "jurus satu" execution. The first step is to establish "kekuda" or footing stance whereas the left foot at the front and the right foot at the back with right fist clenched to the front. The second step is right arm parry or deflection with the "kekuda" in the middle (both foot is parallel to each other). Third step is right side hit with the right foot advances forward. The fourth step is left foot advances forward with hands in capture like posture. The fifth step is the right elbow parrying downwards with the left foot forward and in crouching position. The sixth step is right foot T-kick to the front. The seventh step is right elbow to the front. The eighth step is the right hand and leg advances forward. The last step is the right leg move the side, open palm right hand and leg rose up to the left knee.

2.2. Motion Capture System Protocol

The Kinect motion capture system is able to capture motion operating at 30 fps with a distance between the sensor and the subject is approximately 1.5m minimum and 5m maximum. The Kinect motion capture system and its sensor are able to detect 20 body joints and the data extracted from the sensor are in the form of .txt file per frame captured. The subjects were asked to repeat the execution of ' Jurus Satu' three times and as well as to teach the inexperienced subject on the 'Jurus Satu' execution. The post-processing data were compiled in a Microsoft spreadsheet after being analyzed by the Kinect server software. The kinematic parameters are calculated using these equations:

$$Displacement(m) = \sqrt{(x_f^2 - x_i^2) + (y_f^2 - y_i^2) + (z_f^2 - z_i^2)} \quad (1)$$

$$Velocity\left(\frac{m}{s}\right) = \frac{d_f - d_i}{t_f - t_i} \quad (2)$$

$$Acceleration\left(\frac{m}{s^2}\right) = \frac{v_f - v_i}{t_f - t_i} \quad (3)$$

The temporal simulation of the 3D coordinates obtained were also reconstructed to cross reference with the video obtained visually.

3. Results and Discussion

Tables 1 shows a subject data retrieved during the execution of 'Jurus Satu' by the first subject. The mean and standard deviation of the displacements of each joints, obtained by the kinematic equations were tabulated based the data obtained throughout the three sets of executions. The relatively small values of standard deviations for all joints of subject number one exhibit the consistency of movements by the 14-year experienced 'silat' practitioner. Throughout the execution of 'Jurus Satu', the subject maintained the same level of consistency hence became the benchmark for the movement by the inexperienced subject number two.

Table 1 Mean and standard deviation for joint displacements per joint of 'Jurus Satu' taken from subject 1 and set 1

	Displacement	
Type of Joint	Mean (m)	Standard deviation
Spine	0.02779473	0.04445259
Head	0.03080678	0.04644384
Shoulder left	0.02671781	0.04434403
Elbow left	0.03476577	0.04152885
Wrist left	0.04541242	0.04412944
Hand left	0.0526389	0.04297978
Shoulder Right	0.03368637	0.04709784
Elbow right	0.04389255	0.04474416
Wrist right	0.05492126	0.04423981
Hand right	0.06407865	0.04580653
Hip left	0.02931514	0.04395879
Knee left	0.02623266	0.04409291
Foot left	0.03579723	0.04819077
Hip right	0.03064572	0.04453037
Knee right	0.03654757	0.04491509
Foot right	0.0564036	0.05640314

Figure 2 shows the joint displacements versus frame number taken from the experienced, first subject. The pattern shows clearly the variations in 'silat' movements that involve both slow motions and fast motions. The huge range of displacement magnitudes also shows that 'silat' has both smooth and aggressive movements.

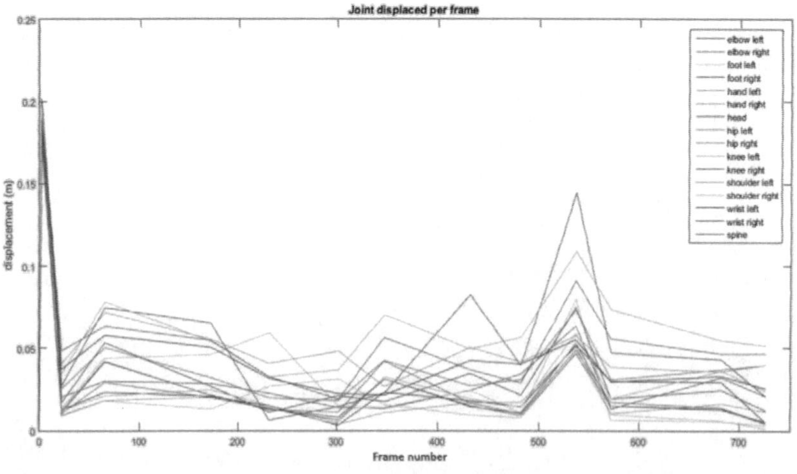

Fig.2 Joint displacement throughout execution of set 1 for subject number one

In order to differentiate the two experienced and inexperienced subjects, a table of average mean values are tabulated and the difference of mean values are calculated. Table 2 shows the average mean values of joint displacement and velocity between the two subjects and its differences:

Table 2 average mean of displacement and velocity between two subjects and its difference

	Average mean of displacement			Average mean of velocity		
	Subject 1	Subject 2	Difference	Subject 1	Subject 2	Difference
Spine	0.0246	0.0253	0.0007	0.2716	-0.0096	-0.2813
Head	0.0264	0.0299	0.0035	0.2935	-0.0075	-0.301
Shoulder left	0.0236	0.0276	0.004	0.2714	-0.0075	-0.279
Elbow left	0.0294	0.0344	0.005	0.2637	-0.0076	-0.2713
Wrist left	0.0379	0.0459	0.008	0.2953	-0.0069	-0.3022
Hand left	0.044	0.0547	0.0107	0.2983	-0.0069	-0.3052
Shoulder right	0.0291	0.0309	0.0018	0.2919	-0.0083	-0.3002
Elbow right	0.038	0.042	0.004	0.294	-0.0082	-0.3022
Wrist right	0.0502	0.0543	0.0041	0.2962	-0.007	-0.3032
Hand right	0.0601	0.0623	0.0021	0.3084	-0.0055	-0.3139
Hip left	0.0254	0.0274	0.002	0.2688	-0.0085	-0.2774
Knee left	0.0224	0.0248	0.0024	0.2674	-0.0073	-0.2747
Foot left	0.0316	0.0322	0.0006	0.2816	-0.0082	-0.2897
Hip right	0.0259	0.0276	0.0016	0.2743	-0.0089	-0.2832
Knee right	0.0313	0.036	0.0047	0.2769	-0.0095	-0.2864
Foot right	0.0488	0.0519	0.0031	0.3083	-0.0083	-0.3166

The mean and standard deviation values were used in this research due to the continuous nature of the data obtained. In addition to the large number of data obtained during capture, the mean and standard deviation helped to generalize and compare the data between two subjects much easier. The table above shows that the closer the difference between two subjects and as the values closing towards zero it can be said that the subject number one movement is relatively identical to subject number two movement. This confirm the fact that both subjects, eventhough had a different level of experience were actually executing the same 'Jurus Satu' if 'Seni Silat Gayong'.

4. Conclusion

Biomechanic analyses of 'silat' as both combat and art sports are proven to be feasible via Kinect-based Motion Capture System. Among others, initial capture of the subjects' motions showed that 'Jurus Satu' has interestingly good variety of slow and fast motions, trivial and aggressive motions throughout the execution. The results also illustrated that both subjects were doing the same 'Jurus Satu' where the mean displacements and accelerations throughout the practice were relatively identical in its patterns. The patterns of displacements and velocities during the course of the executions will never be alike if the subjects were doing dissimilar motions. In order to distinguish the quality of the motion between the two subjects, further analyses on the smoothness of the motion, timing of each movements withing the jurus, exact positioning of each joints, angles of joints and also the precision of the motions must be taken into considerations. More comprehensive analyses are expected in the future via the same motion capture system used in this work.

Acknowledgment

Author would like to thank the University of Tun Hussein Onn Malaysia for the opportunity given and provided sufficient facilities and equipment to conduct the experiment regarding on the Biomechanics analysis of Combat Sport (Silat) and this research is supported by the Fundamental Research Grant Scheme (FRGS), Grant No: FRGS Vot 1543 by Ministry of Education, Malaysia.

.

References

[1] Garner, B. (2003). Joint Anatomy and Basic Biomechanics. *Joint Antomy*, 568. BJ Schoenfeld, (2010), Squatting Kinematics and Kinetics and Their Application to Exercise Performance, Journal of Strength and Conditioning Research , Vol. 24, 12 ,p 3497-3506.

[2] Hatze, H. (2015). The meaning of the term "biomechanics." Journal of Biomechanics, 7(2), 189–190. http://doi.org/10.1016/0021-9290(74)90060-8

[3] WATKINS, J. Structure and Function of the Musculoskeletal SystemChampaign, Ill: Human Kinetics, 1999, 376 pp. ISBN 0-88011-686-2

[4] Hall, S. J. (2012). Kinematic Concepts for Analyzing Human Motion. Retrieved from http://kinesiology. boisestate.edu/wp-content/uploads/2011/05/Chapter-2.pdf

[5] Silat, P., & Draeger, D. F. (1972). Pentjak Silat Past and Present,Pentjak Silat The Indonesian Fighting Art. Hatze, H. (2015). The meaning of the term "biomechanics." *Journal of Biomechanics*, *7*(2), 189–190. http://doi.org /10.1016/0021-9290(74)90060-8

[6] Murphy AJ., Lockie RC, & Coutts AJ. (1993). Kinematics Determinats of Early Acceleration in Field Sport Athletes. *"J Sports Sci Med",* 2003 Dec 1;2(4):144-50. eCollection 2003

[7] Nesbit SM & McGinnis R (2008). Kinematic Analyses of the Golf Swing Hub Path and Its Role in Golfer/Club Kinetic Transfers. *"Journal of Sports Science & Medicine",* (2009) 8, 235-246

[8] I A Bernstein, O Webber., & R Woledge (2002). An Ergonomic Comparison of Rowing Machine Designs: Possible Implications for Safety. *"Br J Sports Med"*, 36: 108-112 doi:10.1136/bjsm.36.2. 108

Preliminary Design on Screw Press Model of Palm Oil Extraction Machine

Muhammad Firdaus[1], S M Salleh[2], I. Nawi[3], Z. Ngali[4], W.A. Siswanto[5], E M Yusup[6]
Faculty of Mechanical and Manufacturing Engineering, Universiti Tun Hussein Onn Malaysia, Parit Raja, Batu Pahat, Johor, Malaysia

E-mail: [2]saliha@uthm.edu.my, [1]firdaussaarin@gmail.com

Abstract. The concept of the screw press is to compress the fruit bunch between the main screw and travelling cones to extract the palm oil. Visual inspection, model development and simulation of screw press by using *Solidworks 2016* and calculation of design properties were performed to support the investigation. The project aims to analyse different design of screw press which improves in reducing maintenance cost and increasing lifespan. The currently existing of screw press can endure between 500 to 900 hours and requires frequent maintenance. Different configurations have been tried in determination of best design properties in screw press. The results specify that screw press with tapered inner shaft has more total lifespan (hours) compared existing screw press. The selection of the screw press with tapered inner shaft can reduce maintenance cost and increase lifespan of the screw press.

1. Introduction

A screw press is used to compress the fresh fruit bunch between the main screw and the travelling cones to extract the palm oil. It has been used in solid-liquid separation of extraction palm oil machine purposely is to separate between oil and non-oily solid (N.O.S). The effectiveness of the screw press is very critical in optimizing the oil extraction. Ineffective press will result in high oil loss and maintenance cost of the machine. There are various types of screw presses used for the extraction of palm oil. The earliest was introduced to the palm oil industry in the early 1960's. This type of machine had been developed for a multitude of application, both for extraction and extruding. The concept of the screw press is to compress the fruit mash between the main screw and the travelling cones to extract the palm oil [1]. The output from the screw press is crude palm oil and press cake. The extracted oil is mixed with fruit particles, water, sand and dirt. The press cake consists of palm fibre and nuts. Figure 1 shows the arrangement of each component. In the screw press, the stability of press cage pressure is an important factor in determining the oil loss and nut breakage in the press cake. A high pressure causes a better recovery of the oil but unfortunately produce a higher nut breakage. This study are carried out to analyze the different design of screw press which can increase effectiveness the machine to extract crude oil. Screw press with current design are originated to work between 500 hours until 900 hours. However, after this limit of time, it must be replaced with a new screw press which intermittent the production due to maintenance works.

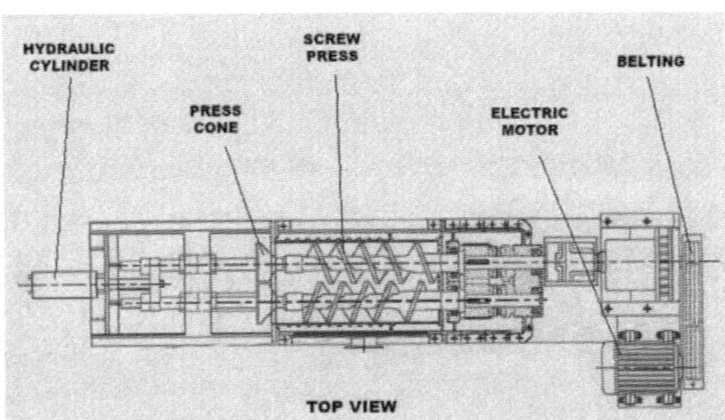

Figure 1. Main component of screw press machine

This study will include the preliminary design of screw press with altered shape of screw and shaft. Four (4) configurations of pitch distance and shaft diameter are attempted using *Solidworks2016* simulation and components design properties to estimate longer life span in screw press.

2 Literature review

Some types of screw presses machine can be obtained in [2] which have wide range of capacity to be processed. A straight screw shaft configuration is commonly used screw press and easy to produce. These following type of screw shaft configuration are used in processing seed oil industry over 50 years [3]. For first configuration of screw press, the pitch and the root diameter are constant through the screw press as shown in Figure 2. The pressure level in hydraulic presses will increase linearly. At one pressing, process flow in the screw press is continuously replaced after each stroke in a hydraulic press.

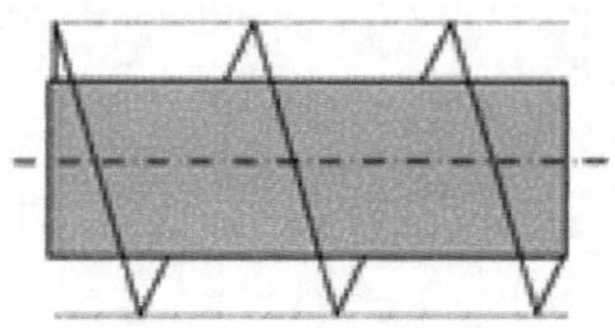

Figure 2. Constant Pitch Screw

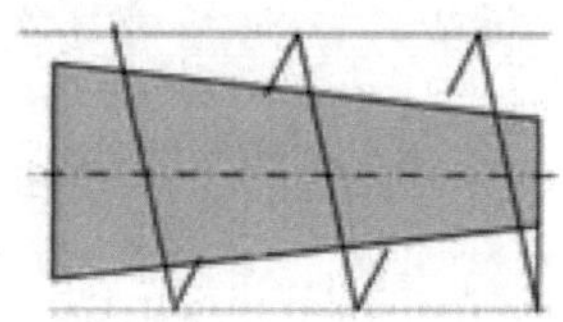

Figure 3. Tapered Shaft and Constant Pitch Screw

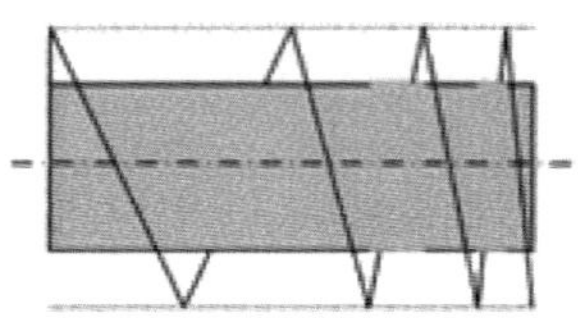

Figure 4. Variable Pitch Screw

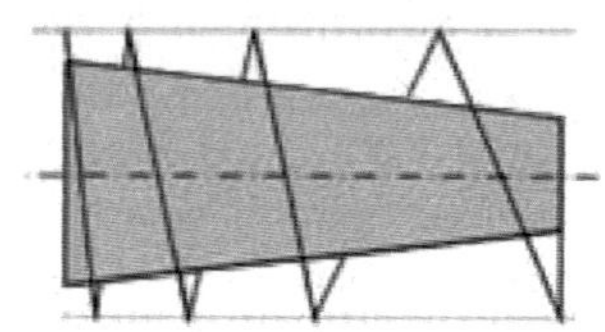

Figure 5. Tapered Shaft and Variable Pitch Screw

The second configuration of screw is tapered shaft and constant pitch. The annular area is decreasing through the length of the screw and takes its minimum value at the end of the screw which is shown in Figure 3. The volume swept by the screw thread in each turn is the multiplication of the annular area and the pitch distance. The rate of pressure increase in tapered screw press is higher than straight screw shaft type. This unusual part of tapered screw type requires a CNC machine.

The fourth type of screw is variable pitch with decreasing pitch screw type as shown in Figure 4. This unusual part of screw thread requires 5-axes CNC machine tool. Practically the total shaft is divided into several different sections with different pitch. Screws with a maximum pitch, which the palm kernel is fed, is useful in increasing in the total capacity of the kernel. Similar to the tapered shaft system, in this type of screw shaft system, volume is decrease by an amount in each turn. The main difference between them is the rate of pressure increase through each thread. In tapered screw, pressure increase linearly through the screw press shaft whereas in screw with variable pitches distance, pressure is constant through each thread and increase at the transitions [4].

The fifth configuration is a screw with tapered shaft and variable pitch which combine screw press types with variable pitch and tapered shaft as represented in Figure 5. Rate of pressure increase in this of screw is higher when compared with the other types of screw. The same pressure can be determined in a shorter time because the pressure attains its maximum value in fewer revolutions when compared with the same sized screw types [4].

In producing worm screw press particularly for palm oil mill, Cast Carbon Steel material typically used. The mechanical properties of this material shows in Table 1.

Table 1 Physical and mechanical properties for material Cast Carbon Steel, [5].

Property	Value	Units
Elastic modulus	200000	MPa
Poisson's ratio	0.32	N/A
Shear modulus	7600	MPa
Mass density	7800	kg/m^3
Tensile strength	482549000	N/m^2
Yield strength	248168000	N/m^2
Thermal expansion coefficient	$1.2x10^{-0.05}$	/K
Thermal conductivity	30	W/(m.K)
Specific heat	500	J/(kg.K)

In starting this study and research, industrial visit is required to expose the researchers with real problem. Preliminary studies was carried out in a crude palm oil processing plant where the company is owned by Kilang Kelapa Sawit Pertubuhan Peladang Negeri Johor (PPNJ), Kahang, Johor. This study focused on primary process to produce crude palm oil. Results from industrial visit has given the vital information related to the screw press machine. This include how the machine mechanism operates, dimension of machine, advantages and disadvantages using screw press machine. Figure 6 shows the new screw press and in Figure 7 shows after its lifespan, the screw press set becomes tear and wear.

Figure 6. New screw press

Figure 7. Used screw press

3. Model Development and Simulation

The results from industry visit on the operational of screw press are used to model the situation using a *Solidworks 2016* software. Figure 8 shows the dimension of straight screw press shaft which screw shaft frequently used in palm oil mill.

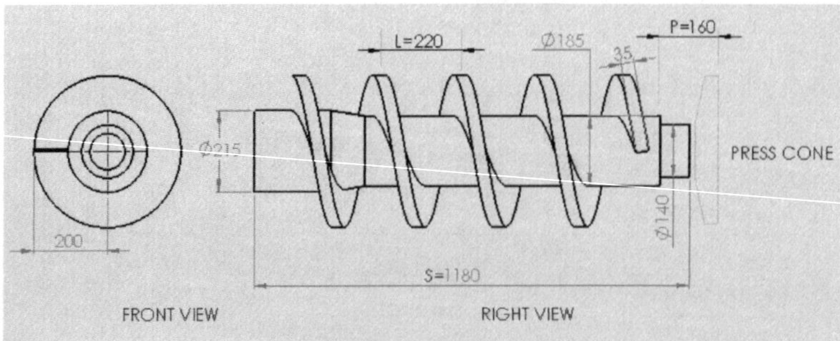

Figure 8. Full dimension of straight screw press

Based on the simulation design, *Solidworks 2016* software is used to identify the critical parts of screw press. The software displays the analysis results in contours of color to identify critical areas in screw press. The main purpose of this analysis is to evaluate the minimum and maximum value of Von Mises stress for screw press part. Figure 9-12 shows the Von Misses stress analysis on the screw press. It shows that higher yield strength are predicted in constant pitch both in straight and tapered shaft designs. The configurations of constant pitch yields higher in increase pressure than variable pitch as the variable pitch can cause different increasing pressure at each sections. Thus, it can be predicted variable pitch screw will damage earlier than constant pitch screw as lower in yield strength in variable pitch design.

Figure 9. Von Misses stress analysis of screw press part with Constant pitch

Figure 10. Von Misses stress analysis of screw press part with variable pitch

Figure 11. Von Misses stress analysis of screw press part with Tapered Shaft and Constant Pitch

Figure 12. Von Misses stress analysis screw press part with tapered shaft and variable pitch

After run the simulation on each model, the result from simulation shows critical areas of screw press where values of minimum and maximum Von Misses stress are presented in Table 2.

Table 2 Result of the maximum and minimum Von Misses Stress

Physical Properties / Screw Press Type	Max. Von Misses stress (MPa)	Min.Von Misses stress (MPa)
Constant pitch Screw	$1.071 x 10^7$	$7.882 x 10^4$
Variable Pitch Screw	$1.362 x 10^8$	$1.486 x 10^5$
Tapered Shaft and Constant Pitch Screw	$8.104 x 10^6$	$1.046 \ x 10^5$
Tapered Shaft and Variable Pitch Screw	$3.982 x 10^8$	$1.067 \ x 10^5$

4. Result and Analysis

The effect of fatigue life is affected by the manufacturing process and component configuration. Several factors such as surface condition, size, stress concentration and temperature are significantly affect fatigue life of materials which being processed. The simulation results are used to find of design properties or design parameters for screw press components. The value of Von Mises provide σ_{min} and σ_{max} from estimated simulation data using Solidworks software. These values are used to compute Stress Concentration Factor, K_t. Using interpolation in [6], q can be found and K_f can be calculated which used to find reliability factor, k_e as refer in equation (1). All factors as surface condition, size, stress concentration and temperature are applied to calculate endurance limit, S_e as summarized in equation (2).Then, total rotation of screw, N applied the ultimate strength S_f= 482.549 MPa to complete the calculation of equation (3). Finally, total life span is valued through the ratio between total rotation of screw and total rotation within one hour as written in equation (4).

$$k_e = \frac{1}{K_f} \quad (1)$$

$$Se = Ka.\ Kb.\ Kd.\ Ke.\ Se' \quad (2)$$

$$N = 10^{\frac{-c}{b}} S_f^{\frac{1}{b}} \quad (3)$$

$$L_t = \frac{N}{N_P} \quad (4)$$

The steps of calculation and detailed values are given in Table 3 for example calculation of design properties in constant pitch screw type. Similar to these calculations are applied to other types of screw press configurations.

Table 3. Example calculation of design properties in constant pitch screw type.

Reliability factor, k_e	Endurance limit, S_e (MPa)	Total rotation of screw, N	Total lifespan (hour)
Stress Concentration Factor, $K_t = \frac{\sigma_{\max(FE\ data)}}{\sigma_{\min(FE\ data)}}$ $= \frac{1.071\times10^{13}}{7.882\times10^{10}}$ = 135.9 Using interpolation, $\frac{4-q}{4-0} = \frac{100-70}{100-60}$ q = 0.8 K_f = 1 + q(K_t -1) = 1+0.8(135.9-1) = 108.92	The ultimate strength, (use Table 1) Sut = 482.549 MPa = 69987.8 psi Rotary beam test endurance limit, Se' = 0.504 Sut = 0.504 (69987.9) = 35273.8512 psi Surface modification factor, k_a = 0.766 Size modification factor, d = $\sqrt{(95^2 - 33^2)}$ = 89.08 mm $k_b = 1.189\ (89.08)^{-0.097}$ = 0.77 Temperature modification factor k_d = 1	S_f= 482.549 MPa b = $-\frac{1}{3}\log\frac{0.8\,S_{ut}}{S_e}$ = $-\frac{1}{3}\log\frac{0.8(69987.8)}{190.99}$ = - 0.822; c = $\log\frac{(0.8\,S_{ut})^2}{S_e}$ = $\log\frac{(0.8\times69987.8)^2}{190.99}$ = 7.22 $-\frac{C}{b} = -\frac{-7.22}{(-0.822)}$ =-8.78; $\frac{1}{b} = \frac{1}{(-0.822)}$ = -1.22	Distance of pitch, L = 173mm N = 12 rpm v= $\frac{173}{4.5}$ = 38.4 mm/s $t = \frac{s}{v} = \frac{1180\ mm}{38.4\ mm/s}$ = 30.73 s $t_p = \frac{P}{V} = \frac{160\ mm}{38.4\ mm/s}$ = 4.17 s Time to complete one cycle $t_{total} = t + t_p$ = 34.89 s Total rotation within 1 hr $N_p = \frac{3600}{t_{total}} = \frac{3600}{34.89}$ = 103 rotation
$k_e = \frac{1}{K_f}$ $= \frac{1}{108.92}$ = 0.00918	Se = Ka. Kb. Kd. Ke. Se' = 190.99 psi @ 1.317 MPa	$N = 10^{\frac{-c}{b}} S_f^{\frac{1}{b}}$ $= 10^{8.78}(482.549)^{-1.22}$ = 320.69 (10^3) rotation	$L_t = \frac{N}{N_P}$ $= \frac{320.69\ (10^3)}{103}$ = 3113 hours

Table 4 shows the overall results of design properties in four different types of screw press. From this table, constant pitch screw in both straight and tapered shaft give higher total life span compared with other two type in variable pitch. This result also agreed with simulation data which give higher Von Mises stress in both configurations. This comparison of different design of screw press type will be used to get a best design which can be applied to screw press machine palm oil mill.

Table 4. Result summary in each types of screw press

Types / Properties	Constant pitch Screw 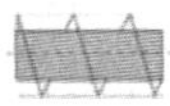	Variable Pitch Screw 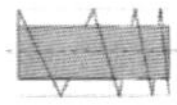	Tapered Shaft and Constant Pitch Screw 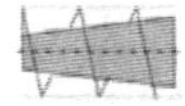	Tapered Shaft and Variable Pitch Screw 
Reliability factor, k_e	0.00918	0.001363	0.0161	0.00335
Endurance limit, S_e (MPa)	1.317	0.1955	2.303	0.4802
Rotation of screw presses	320.69 (10^3)	75.39 (10^3)	612.66 (10^3)	132 (10^3)
Total lifespan (hours)	3113	731	5948	1281

5. Conclusion

Investigation on the frequent occurrences of screw press failures in several palm oil screw press machines in a Malaysia palm oil mill company was presented. The findings indicate that screw press with tapered inner shaft has more total lifespan (hours) which is 5948 hours compared with others design of screw press. Which is screw press with tapered inner shaft has double the lifespan compared with existing of screw press. The selection of the screw press with tapered inner shaft can reduce maintenance cost, increase lifespan of the screw press and it can be decrease frequency of maintenance work to change screw press.

6. Reference

[1] R.H.V Corley, P.B. Tinker. (2003). *World Agriculture Series: The Oil Palm*. Fourth edition. Garsington Road, Oxford. Blackwell Publisher.

[2] Webpage of MHSK Engineering Sdn Bhd., *http://www.mshk.com.my/press.html*, assessed at 20th October 2016.

[3] Pelin Sari. *Preliminary Design and Construction of a Prototype Canola Seed Oil Extraction Machine*. The Degree of Master. Middle East Techincal University ; 2006

[4] Tekad Sitepu. *Analisa Umur Pemakaian Screw Press Pada Mesin Pengekstraksi Minyak Mentah Kelapa Sawit*. 2010. Vol 1, No. 7: 1-9.

[5] Harper, Charles A., Handbook of Materials fo Product Design. 3rd ed. New York: McGraw-Hill, 2001.

[6] Shigley, Joseph Edward., and Charles R.Mischke. *Mechanical Engineering Design*. 5th ed. New York: McGraw-Hill, 1989.

Computational Modelling and Movement Analysis of Hip Joint with Muscles

W.A. Siswanto*, C.C. Yoon, S. Md. Salleh, M.Z. Ngali, Eliza M. Yusup

Department of Engineering Mechanics, Faculty of Mechanical and manufacturing Engineering, Universiti Tun Hussein Onn Malaysia

*corresponding author: waluyo@uthm.edu.my

Abstract. In this study, the model of hip joint and the main muscles are modelled by finite elements. The parts included in the model are hip joint, hemi pelvis, gluteus maximus, quadratus femoris and gamellus inferior. The materials that used in these model are isotropic elastic, Mooney Rivlin and Neo-hookean. The hip resultant force of the normal gait and stair climbing are applied on the model of hip joint. The responses of displacement, stress and strain of the muscles are then recorded. FEBio non-linear solver for biomechanics is employed to conduct the simulation of the model of hip joint with muscles. The contact interfaces that used in this model are sliding contact and tied contact. From the analysis results, the gluteus maximus has the maximum displacement, stress and strain in the stair climbing. Quadratus femoris and gamellus inferior has the maximum displacement and strain in the normal gait however the maximum stress in the stair climbing. Besides that, the computational model of hip joint with muscles is produced for research and investigation platform. The model can be used as a visualization platform of hip joint.

1. Introduction

Hip joint also known as acetabulofemoral joint which is a joint that located between the femur and acetabulum of the pelvis. Hip joint brings a good mobility to human such as walking, sitting and squatting in life. There are severals of movement of the femur wih the hip joint which is abduction and adduction, flexion and extension, inward and outward rotation. These movements are constrained by ligaments, muscles and bony structure of the hip.

According to Anthony Chila [1], hip joint can require high muscle contraction forces under load conditions. When a combination of ground reaction force and the muscle contraction forces directed into the hip joint, the joint reaction forces are also high relative to the body weight. Joint reaction forces are important in consideration of stresses on the hip joint in replacement.

Muscle, acetabular cup and ligaments are provided translational hip stability. However, muscles are only responsible for rotational stability. The full function of the hip is dependent on the coordination of bones, muscles, tendons, ligaments and nerves. Hence, a hip joint modeling is needed and important to do a comprehensive simulation before it is manufactured for a real joint replacement. The model should consist of bone and muscles.

The purpose of this study was to simulate loaded condition on the model of hip joint with muscles.

2. Literature Review

An initial finding that hip adductors muscles such as adductor longus, adductor brevis, adductor magnus, and gracilis muscles are mostly suffer contusions and strains occurring in the pelvis region has been revealed by Ozkaya and his coworkers in 1991 [2].

Sir John Charnley, a British Orthopaedic surgeon, developed the fundamental principles of the artificial hip. He designed a hip prosthesis in the mid to late 1960's as recorded by Jackson [3]. After the original prosthesis is introduced, several variants of the artificial hip joint have been developed and none have proved to be superior in the clinical setting. The longevity of the Charnley prosthesis appears to be enhanced by the smaller diameter head, which reduces friction and the amount of wear debris produced and results in a lower incidence of acetabular loosening.[4,5]

Siopack [6] found that there are many early complications and late complications occur after replacing a prosthesis joint. The complications are include fracture in arthroplasties with non-cemented prostheses, nerve injury, dislocation of the femoral head component out of the acetabular socket, wound complications, infections and loosening. From previous study of Kadaba et al, [7], electromyographic studies of tensor fasciae latae and the gluteus medius showed that tendon in normal subjects during the gait cycle and in isolated abduction of the lower limb. Surfaces electrodes were used in all subject as good correlation had been obtained with fine wire electrodes in four subjects.

According to Gottschalk et al. [8], they propose a new model of the anatomy and function of the glutei, whereby the gluteus medius is a segmented muscle with each part separately innervated and with a phasic action in different directions. This anatomical configuration suggests a different function for each of the three parts rather than a total single muscle action.

For the material properties, Sivasankar and his team's research [9] suggests the general modulus of elasticity, ultimate tensile strength and tensile yield strength of a number materials used for any surgical implants. In soft tissue, it is very difficult to be considered in global models as the used to analyze prostheses and implants because the complexity to the different biological tissues and great amount of non-linearities do the convergence practically unfeasible [10].

Musculoskeletal simulation of human movement commonly use Hill muscle models to predict muscles force. Hill component models represent the active and passive properties of the musculo-tendenous unit [11]. This is consistent with a typical postprocesser display overlays colored contours representing stress levels on the model, showing a full field picture similar to that of photoelastic or moire experimental results found in [12].

Van Leeuwen [13] introduced a muscle model with the behaviour of a muscle can be separated into a passive and an active state. The input parameters for the passive properties are muscle density, Poisson's ratio and the stress-strain relation [14]. The application of boundary conditions in biomechanical FEA is based on assumptions including forces and pressures acting in the human body. The muscle forces acting at the hip joint as well as the hip resultant during normal gait and during stair climbing were extracted from telemetric measurements and corresponding calculations.[4,15-16] Kluess research [17] shows the attachment areas of the simulated muscle forces in the FE-model acting at the hip joint.

3. Methodology

In this research work, the simulation is conducted in FEBio finite element program [18]. FEBio has been verified to be reliable as reported in ref [19]. The hip model and the muscles are firstly drawn in SolidWorks then mesh in GiD [20]. The finite element meshing is then transferred to FeBIO in gmsh format. The material properties used are listed in Table 1.

Table 1: Physical and mechanical properties.

Model	Material	Young's Modulus, $E\,(GPa)$	Bulk Modulus, $K\,(GPa)$	Density, ρ (Mkg/l)	Poisson's Ratio, ν	Mooney Rivlin Constant, $a_{10}\,(GPa)$	Mooney Rivlin Constant, $a_{01}\,(GPa)$
Hip joint	Isotropic elastic	110	107.84	4.5×10^{-6}	0.33	-	-
Hemi pelvis	Neo-hookean	12.4	7.38	1.7×10^{-6}	0.22	-	-
Muscles	Mooney Rivlin	0.0012	2	1×10^{-9}	0.4999	10	0

3.1. Geometry Modelling

The finite element modelling started with designing the model of hip joint, hemi pelvis, glutues maximus, quadratus femoris and gamellus inferior in SolidWorks. Since PreView preprocessor in FEBio cannot import file directly from SolidWorks, the model need to be converted in another format (gmsh) in GiD. The hip model and the muscles are shown in Figure 1.

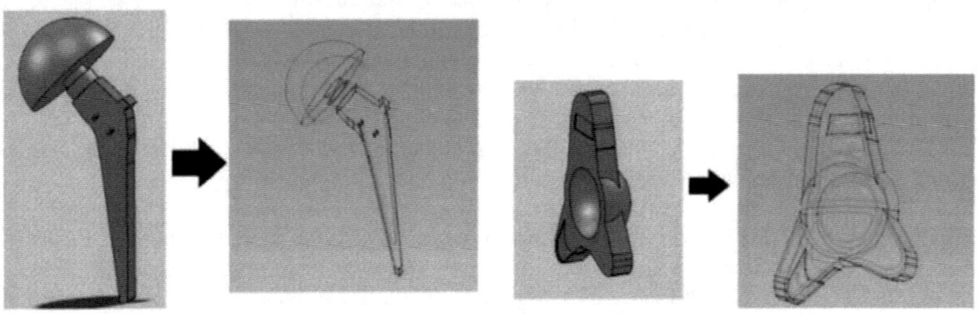

(a) Hip joint model. (b) Hemi pelvis model.

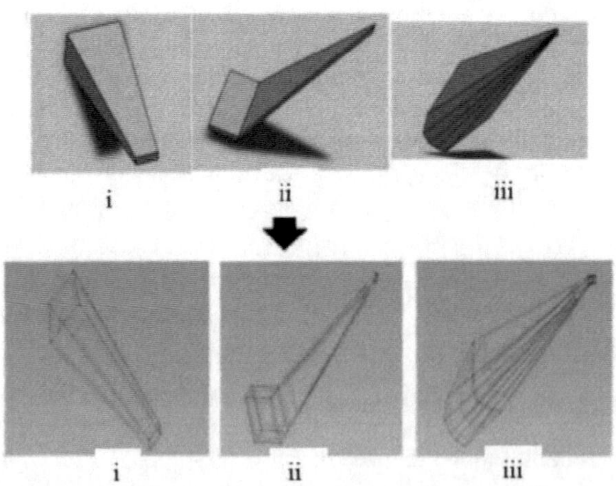

(c) i- Gluteus maximus, ii- Gamellus inferior, iii- Quadratus femoris.

Figure 1: Hip joint model with muscles.

IOP Conf. Series: Materials Science and Engineering **165** (2017) 012030 doi:10.1088/1757-899X/165/1/012030

3.2. Type of FEM Meshing

The tetrahedral meshing elements were used in the model of hip joint with muscles. The meshed FEM model developed by GiD is shown in Figure 2.

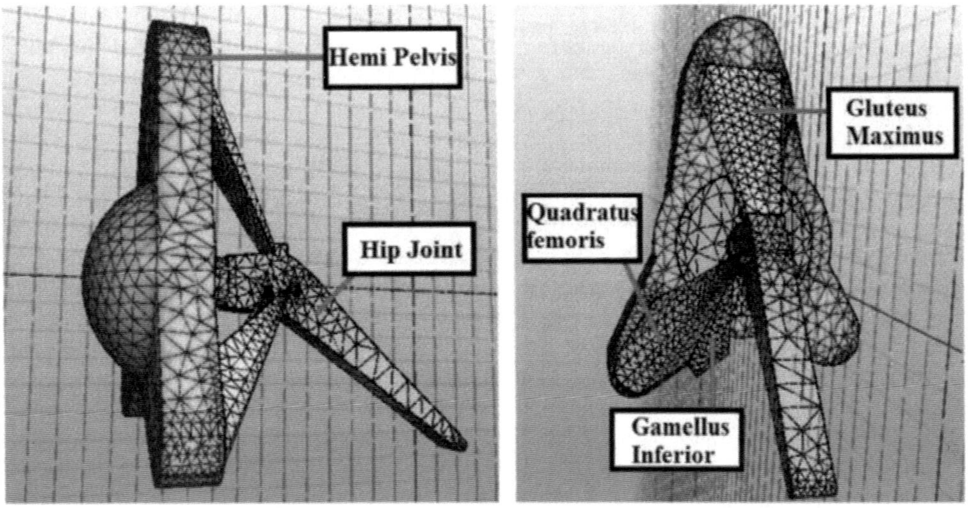

(a) Side view (b) Front view

Figure 2: FEM meshing of hip joint model with muscles.

The total nodes and tetrahedral elements for the entire model is depicted in the Table 2.

Table 2: Total nodes and tetrahedral elements of the model.

Model	Total nodes	Total tetrahedral elements
Hip joint	804	2373
Hemi pelvis	2141	6923
Gluteus maximus	244	687
Gamellus inferior	169	436
Quadratus femoris	448	1332

3.3. Type of Interface

There are two types of contact surfaces are involved i.e. sliding contact and tied contact.

Sliding contact is applied to the surface between the hip joint and pelvis and the surface between the gamellus inferior and quadratus femoris with the hip joint when the hip joint is moved in different direction. Tied contact is applied to the contact of the muscles with the hemi pelvis and hip joint model, so it will become a complete modeling hip joint with muscles.

4. Results and Discussion

The muscles are divided into three cases A, B and C, representing for Gluteus maximus, Quadratus femoris and Gamellus inferior, respectively. The summary of the simulation results are documented and compared in Table 3 and Table 4.

The results of the normal gait and stair climbing will be compared and analyzed. Table 3 shows the comparison of the total displacement between normal gait and stair climbing. From the overall graph, the displacement of the muscles for stair climbing is higher than normal gait. For case A, the maximum displacement of the gluteus maximus for normal gait and stair climbing are 9.31mm and 10.45mm. Thus, the gluteus maximus of the stair climbing is stretched more than normal gait. For case B, the maximum displacement of the quadratus femoris for normal gait and stair climbing are 6.21mm and 5.77mm.

The quadratus femoris of the normal gait is stretched more than stair climbing. For case C, the maximum displacement of the gamellus inferior for normal gait and stair climbing are 8.05mm and 7.66mm. The gamellus inferior of the normal gait is stretched more than stair climbing.

The graph of the effective stress versus time for the normal gait and stair climbing shows in Table 4. The stress of the muscle is low when there is some stretching occurs, so that the muscle can contract and expand easily. For case A, the maximum stress of the normal gait and stair climbing are 2.65GPa and 2.66GPa. For case B, the maximum stress of the normal gait and stair climbing are 6.67GPa and 6.69GPa. For case C, the maximum stress of the normal gait and stair climbing are 9.33GPa and 9.39GPa. From the three cases, the muscles of the stair climbing are more stress than normal gait.

For further improvement, the model of hip joint with muscles can improve by adding more group of muscles such as illiacus, gluteus minimus and gluteus medius to make the model become a complete model. Since hemi pelvis has many free form surfaces and the main investigation of this research is to test the simulation of the computational hip joint with muscles, thus the model of hemi pelvis for this research is using predicted dimension.

The way to improve the model of hemi pelvis looks more realistic is using 3D scanner to scan the real model of the pelvis to get an accurate results. Besides that, the computational model of hip joint with muscles is used to test the other types of gait pattern such as standing up and down stair to analyzed and get accurate results for the group of muscles.

Table 3: Total displacement versus time off the normal gait and stair climbing in different cases.

Case	Comparison charts
A	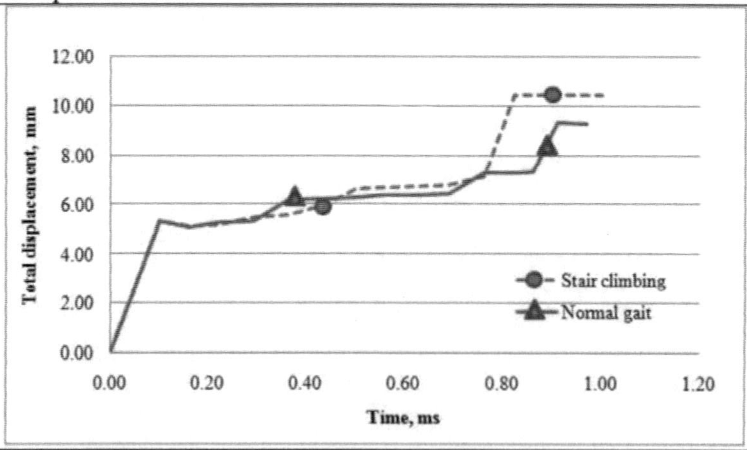

<table>
<tr><td>B</td><td>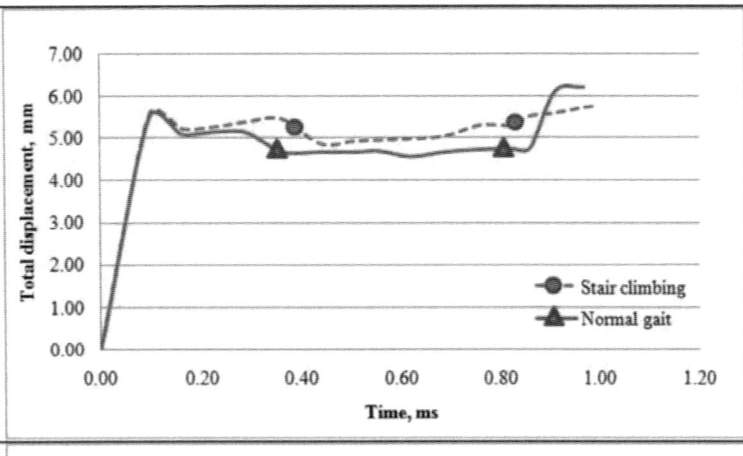
</td></tr>
<tr><td>C</td><td>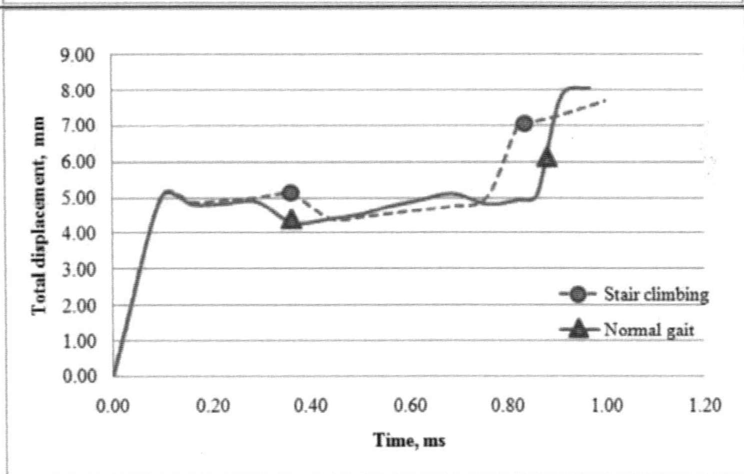
</td></tr>
</table>

Table 4: Effective stress versus time of the normal gait and stair climbing in different cases.

<table>
<tr><th>Case</th><th>Comparison charts</th></tr>
<tr><td>A</td><td>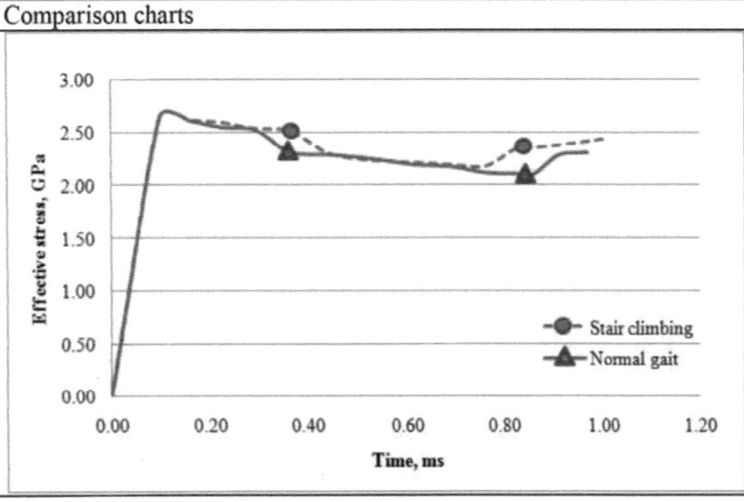
</td></tr>
</table>

B	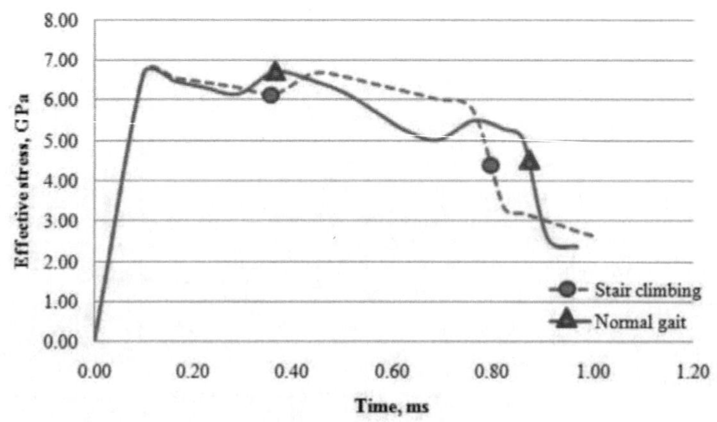
C	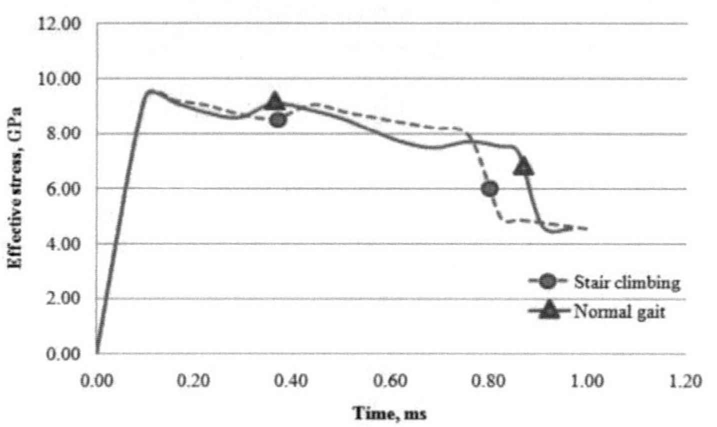

5. Conclusion

In conclusion, the computational model of hip joint with muscles is produced. The model can be used as a visualization platform of hip joint. The maximum displacement, stress and strain of the muscles that could withstand by the hip joint modeling under loaded condition are obtained.

From the analysis, gluteus maximus shows maximum displacement in the gait pattern of stair climbing. However, the quadratus femoris and gamellus inferior show the maximum displacement in normal gait. The maximum stress is happened in gluteus maximus, quadratus femoris and gamellus inferior, when the model of hip joint with muscles is simulated in the gait pattern of stair climbing. The gluteus maximus has the same maximum strain in the normal gait and stair climbing. The maximum strain of the quadratus femoris and gamellus inferior are occurred in the normal gait.

References

[1] Anthony Chila, A.O.A. (2010). Foundations of Osteopathic Medicine. Lippincott Williams & Wilkins, 2010, 3rd edition.

[2] Ozkaya, N. & Nordin, M. (1991). Fundamentals of Biomechanics Equilibrium, Motion, and Deformation. Van Nostran Reinhold.

[3] Jackson, J. (2011). Father of the modern hip replacement: Professor Sir John Charnley (1911–82), *Journal Medical Biography*, November 2011 **19**(4) pp.151-156.

[4] Hall, M. & Owings, M. (2002). 2000 national hospital discharge survey. *PubMed*, **329**, pp.1-18.

[5] Morrey, B. & Ilstrup, D. (1989). Size of the femoral head and acetabular revision in total hip replacement arthroplasty. *The Journal of bone and joint surgery*, **71**, pp. 50–55

[6] Siopack, J. & Jergesen, H. (1995). Total hip anthroplasty. *Western journal of medicine*, **162**, p. 243-249.

[7] Kadaba, M.P., Wootten, M.E., Gainey, J. & Cochran, G.V.B. (1985). Repeatability of phasic muscle activity: Performance of surface and intramuscular wire electrodes in gait analysis. *Journal of Orthopaedic Research*, **3**(3), pp. 350–359.

[8] Gottschalk, F., Kourosh, S. & Leveau, B. (1989). The functional anatomy of tensor fasciae latae and gluteus medius and minimus. *Journal of anatomy*, **166**, pp. 179–180.

[9] Sivasankar, M., Senthil, R., Dhandapani, N., S.Venkatesan, D.Balaji, Raajkumar, N., S.Iyappan, J.Yuvaraj & K.Raja (2010). Fatigue analysis of cementless hip prosthesis. *Recent Research in Science and Technology*, **3**(3), pp.129–134.

[10] Herrera, A., Gracia, L., Ibarz, E., J.Panisello, J., Cegonino, J., Mateo, J., Rodriguez-Vela, J. & Puertolas, S. (2010). Finite element simulation. Applications in Orthopaedic Surgery and Traumatology. Sciyo.

[11] Scovil, C.Y. & Ronsky, J.L. (2006). Sensitivity of a hill-based muscle model to perturbations in model parameters. *Journal of Biomechanics*, **39**(11), pp. 2055 – 2063.

[12] Roylance, D. (2001). Finite element analysis. Lecture notes-Department of Material Science and Engineering, Massachusetts Institute of Technology, Cambridge, MA, 2139.

[13] Van Leeuwen, J. (1992). Muscle function in locomotion. *Mechanics of animal locomotion*, **11**, pp. 191–249.

[14] Meier, p. & Blickhan, R. (2000). Fem simulation of skeletal muscle: The influence of inertia during activation and deactivation. Skeletal muscle mechanics: from mechanisms to function, pp. 207-223.

[15] Bergmann, G., Deuretzbacher, G.,Heller, M., Graichen, F., Rohlmann, A., Strauss, J. & Duda, G. (2001). Hip contact forces and gait patterns from routine activities. *Journal of Biomechanics*, **34**(7), pp. 859 – 871.

[16] Bergmann, G., Graichen, F. & Rohlmann, A. (1993). Hip joint loading during walking and running, measured in two patients. *Journal of Biomechanics*, **26**(8), pp. 969 – 990.

[17] Kluess, D., Wieding, J., Souffrant, R., Mittelmeier, W. & Bader, R. (2010). Finite Element Analysis in Orthopaedic Biomechanics (chapter 6). In book Finite Element Analysis Edited by David Moratal, ISBN 978-953-307-123-7, 698 pages, Publisher: Sciyo.

[18] Maas, Ellis, S., B.J., Ateshian, G.A., Weiss & J.A. (2012). Febio: Finite elements for biomechanics. *Journal of Biomechanical Engineering*, **134**(1), pp.011005.

[19] Maas, Ellis, S., Rawlins, B., Weiss, D. & J.A. (2009). A comparison of febio, abaqus, and nike3d results for a suite of verification problems. SCI Technical Report, University of Utah, UUSCI-2009-009.

[20] Melendo, A. and Coll, A. and Pasenau, M. and Escolano, E. and Monros, A. http://www.gidhome.com, 2016, Online; last accessed Jun-2016.

Programme Learning Outcomes Assessment and Continuous Quality Improvement in Faculty of Mechanical and Manufacturing, UTHM

H. Taib[1], S M Salleh[2], B A Md Zain, M.A. Azlan, S. Mahzan, Hafeez Z A, Ong, P., S. Ahmad, M. N.A Rahman, N.F. Nasir, M. Azham Azmi, H.A Rahman, Z.Ngali

Faculty of Mechanical and Manufacturing Engineering, University Tun Hussein Onn Malaysia (UTHM).

E-mail: [2]saliha@uthm.edu.my, [1]hariati@uthm.edu.my, [3]aisham@uthm.edu.my

Abstract. This paper describes the assessment and continuous quality improvement of Programme Learning Outcomes (PLOs) in the Faculty of Mechanical Engineering and Manufacturing. PLO is known as an elementary requirement in Outcome Based Education (OBE) system. All PLOs have been mapped with graduate attributes by EAC Manual 2012. Conceptual process for establishing and reviewing PLOs has been explained in the Plan-Check-Do-Act cycle. PLO assessment has been shown in different types which classified as direct and indirect methods. Continuous Quality Improvement has been extracted from a variety of assessment and has been discussed. Seven (7) CQIs are identified using different assessment methods of PLO during years 2013 to 2016 and subsequent improvement actions have been taken by the faculty within three years.

1. Introduction

The development of OBE system requires main parts of Programme Educational Outcomes (PEO), PLOs and Course Learning Outcomes (CLO). Thus, the importance of PLO is realised as the primary element and assessment report are needed to ensure OBE system has been implemented. The 13 PLOs of the programme were established according to the Engineering Programme Accreditation Manual 2012 and the Ministry of Higher Education requirements. For this academic programme, the senate and faculty has approved and decided on the thirteen (13) of PLO as shown in FKMP PLO column in Table 1. In order to promote the PLOs, some initiatives have been taken to publish them through the media such as

(i) Faculty website (*http://fkmp.uthm.edu.my/*)
(ii) Poster-placed at strategic locations including all laboratories, lecturer rooms, faculty classroom and main office
(iii) Font Cards – All PLOs are published and distributed to the Faculty staffs and student in form of font cards

Table 2 shows the corresponding PLOs mapped with EAC Manual 2012 with additional one attribute satisfying Ministry of Higher Education (MOHE) requirement.

Detailed descriptions of individual PLOs which are related to the EAC graduate attribute areas as given in the Engineering Programme Accreditation Manual 2012 are shown in Table .

Table 1. Detailed Indicators of PLOs associated to EAC Graduate Attributes

EAC Attributes	FKMP PLO	Indicators of PLO Achievement by the Students
Knowledge	PLO 1	Ability to apply knowledge of mathematics, science, engineering fundamentals and an engineering specialisation to the solution of complex engineering problems
Problem Analysis	PLO 11	Identify, formulate, research literature and analyse complex engineering problems reaching substantiated conclusions using first principles of mathematics, natural sciences and engineering sciences
Design/Development of Solution	PLO10	Design solution for complex engineering problems and design systems, components or processes that meet specified need with appropriate consideration for public health and safety, cultural, societal, and environmental considerations
Investigation	PLO 4	Create, select and apply appropriate techniques, resources, and modern engineering and IT tools, including prediction and modelling, to complex engineering activities, with an understating of the limitation
Modern Tool Usage	PLO 2	Conduct investigate into complex problems using research based knowledge and research methods including design of experiments, analysis and interpretation of data, and synthesis of information to provide valid conclusion
The Engineer and Society	PLO 13	Apply reasoning informed by contextual knowledge to assess society, health, safety, legal and cultural issues and the consequent, responsibilities relevant to professional engineering practice
Environment and Sustainability	PLO 12	Understand the impact of professional engineering solutions in societal and environmental contexts and demonstrate knowledge of and need for sustainable development
Ethics	PLO 8	Apply ethical principles and commit to professional ethics and responsibilities and norms of engineering practice
Communication	PLO 3	Communication effectively on complex engineering activities with the engineering community and with society at large, such as being able to comprehend and write effective reports and design documentation, make effective presentations, and give and receive clear instructions

Table 1. Detailed Indicators of PLOs associated to EAC Graduate Attributes (cont.)

EAC Attributes	FKMP PLO	Indicators of PLO Achievement by the Students
Individual and Team Work	PLO 5	Function effectively as an individual, and as a number or leader in diverse teams and in multi-disciplinary setting
Life Long Learning	PLO 6	Recognise the need for, and have preparation and ability to engage in independent and life-long learning in the broadest context of technological change
Project management and Finance	PLO9	Demonstrate knowledge and understanding of engineering and management principles and apply these to one's own work, as a member and leader in a team, to management projects and in multidisciplinary environments
Entrepreneurship	PLO7	Recognise the importance of entrepreneurship in mechanical engineering and its' related discipline. - As required by Ministry of Higher Education (MOHE)

Table 2. Relationship of Current PLOs with the EAC Accreditation Manual

Programme Learning Outcomes (PLOs)	Program Outcomes (EAC)											
	PO1	PO2	PO3	PO4	PO5	PO6	PO7	PO8	PO9	PO10	PO11	PO12
	Engineering Knowledge	Problem Analysis	Design/Development of Solution	Investigation	Modern Tool Usage	The Engineer and Society	Environment and Sustainability	Ethics	Communication	Individual and Team Work	Life Long Learning	Project management and Finance
PLO1	✓											
PLO2					✓							
PLO3									✓			
PLO4				✓								
PLO5										✓		
PLO6											✓	
PLO7	Mapped to Ministry of Higher Education (MOHE) Requirement											
PLO8								✓				
PLO9												✓
PLO10			✓									
PLO11		✓										
PLO12							✓					
PLO13						✓						

2. PLOs Conceptual Process

The PLOs were first established in 2004. The conceptual process for establishing and reviewing PLOs is as shown in Figure 1. Plan-Do-Check-Act cycle have been used to describe this conceptual process which contribute to establishment of PLOs.

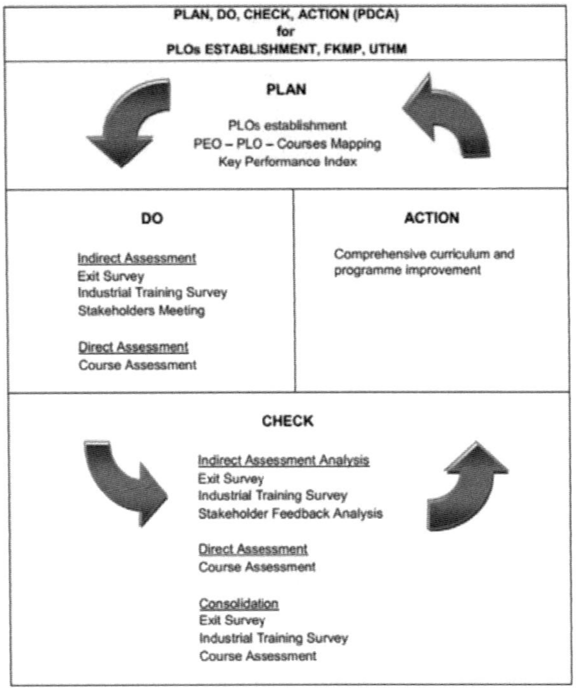

Figure 1. The Conceptual Process for PLOs Establishment and Review

Plan

The Planning Stage is mainly steered by Malaysian education system governing bodies such as Ministry of Higher Education and Board of Engineers Malaysia (Engineering Accreditation Department). Domains are standard and set by the Washington Accord. In Malaysia, the body in charge of coordinating all accredited engineering program is Board of Engineers Malaysia (BEM) and supported by Accreditation Engineer Department (EAD). There are 12 suggested POs outlined in the manual and disclosure for guidelines. EAC has three major domains of Affective, Cognitive and Psychomotor. Nonetheless, it is also the establishment of PLOs is based on the needs of program's constituencies/stakeholders' expectations and interest.

The PLOs are very specific to attain specified PEOs. This is translated through the PEOs to PLOs mapping which is further relates to Course Learning Outcomes (CLOs). In order to enable the measurement of the PLOs achievement, key performance index of the PLOs is determined by the attainment of score for related courses.

Do

The implementation of PLOs is divided into two types of assessments namely indirect and direct method. The indirect method comprises of exit survey, industrial training survey and stakeholders meeting. Stakeholders include the industries, external examiners, visiting professors, adjunct professors and students are involved in the assessment exercises to ensure their views are given due consideration when implementing improvements to the programs. The direct method mainly sourced from the achievement of the CLOs.

Check

The monitoring and review stage include analysis of indirect assessment, direct assessment and consolidation of indirect and direct assessment. Consolidation of indirect and direct assessment involves correlation of exit survey, industrial training survey and course assessment analysis. Input from various parties are analysed by the relevant committees in the Faculty to assess the achievement of the PLOs. Outputs from each assessment are utilised by the Faculty in identifying any issue that needs rectifications or improvement.

Action

The Faculty will then outline the necessary plan of actions and table them to the Faculty academic committee meeting for their endorsement. The endorseplan and actions to be implemented may be issued to the Senate meeting for the approval.

3. Evaluation of Programme Learning Outcomes

There are two methods of evaluating the PLOs, namely (i) indirect assessment and (ii) direct assessment. The tools used in theses assessments are summarised as in the following Table .

Table 3. Summary of Different Types of Assessments

Types of Assessment	Platform	Method/Tools	Output
Indirect	Industrial Advisory Committee (IAC)	Annual Meeting	Reports
	Adjunct Professor	Annual Meeting	Reports
	External Examiner	Annual Meeting	Reports
	Exit Survey	Survey	Reports
	Industrial Training Survey	Survey	Reports
Direct	Summative Evaluation of Students	CEPAT	Students' achievement
	Formative Evaluation of Students	CEPAT / SAS / OBESYS	

IOP Conf. Series: Materials Science and Engineering **165** (2017) 012031 doi:10.1088/1757-899X/165/1/012031

3.1. Indirect Assessment

The PLOs attainment is assessed continuously by the Faculty through indirect assessment. The types of indirect evaluations and their mechanisms are explained as follows:

(i) Industry Survey

The industry survey is a tool to identify the needs and gaps of skill in industry based on the employer's perception in terms of students' abilities to attain the PLOs. The industry survey was conducted yearly during the Semester 3 of the third year academic session. The respondents are the employers of whom the FKMP students are undergoing the industrial training, including the managers, senior engineers and practicing engineers.

(ii) Exit Survey

The Exit Survey refers to an indirect measurement method of students' self-evaluation or individual perception for the assessment of qualities as prescribed in PLOs, which was conducted upon graduation.

(iii) Assessment by External Examiners/Adjunct Professor/Visiting Professors/IAC

The review from the External Examiners, Adjunct Professor, Visiting Professors and IAC regarding the appropriateness of the Faculty PLOs are taken during periodic review through annual stakeholder meetings.

3.2. Direct Assessment

PLOs in FKMP were analysed by using online and offline systems. The established of management system (online and offline) used are divided into three systems as stated below:

(i) Offline system - Course Evaluation and Performance Analysis Tool (CEPAT)

(ii) Online system - Students Assessment System (SAS)

(iii) Online System – Outcome Based Education System (OBEsys)

The CEPAT assessment template, which is an offline excel-based programming developed by the Faculty member, is capable in analysing the relations of the assessments for each courses to the course learning outcomes (CLOs) and further to PLOs. The SAS and OBEsys are online database system developed by the University (Information Technology Centre), which offer similar function with CEPAT with extra features. The different functions and features of the systems are as summarised in Table Error! **No text of specified style in document.**.

Table Error! No text of specified style in document.: Summary of Functions and Features of Different Systems of Assessments

	System	Basic Function (Marks compilation)	Features
1.	CEPAT	Continuous assessment	• Offline System
		Final examination	• Analysis of CLO achievement • Mapping of CLO achievement to PLOs • Analysis of CLO and PLO KPI achievement
2.	SAS	Continuous assessment	• Online System linked to TCIS
		Final examination	• Systemised compilation of courses CQI
3.	OBEsys		

			• Online system linked to TCIS • Analysis of PLO achievement

CEPAT and SAS were used to store continuous assessment and final examination marks of all students by the respective lecturer. CEPAT is managed by the Faculty's OBE Committee meanwhile SAS is managed by the University. SAS is linked to the Total Campus Integrated System (TCIS). TCIS is an online system which integrates necessary information of personal, teaching and learning activities of each individual staff, regardless of academic or supporting staffs.

The course coordinators were given the authority to set up the mark weightage for continuous assessment and a final examination in SAS which is linked to the TCIS system. The continuous assessment includes quizzes, assignments, projects and tests. The total mark of continuous assessment is 50% and remaining 50% are allocated for final examination marks for the courses. Respective lecturers can fill up the marks after the weightage has been set up in TCIS by course coordinator.

In SAS, overall marks were analysed to represent the CLOs achievement percentages. The analysis results would show the overall report which consists of the continuous assessment, overall marks, analyses of CLO and PLO achievement. This system allows the student to check their grades on-line. The lecturers and the Faculty management team are able to monitor and evaluate the result of the examination for student accordingly.

Currently, FKMP with CAD (Centre for Academic Development and Training) and Information Technology Centre have developed a system which manages the PLOs assessment known as 'OBEsys'. OBEsys is managed by the Faculty's OBE committee and the Information Technology Centre. This system is capable to manage and measure the PLOs assessment.

4. Continuous Quality Improvement

Continuous CQI is essential, in improving the quality of the programme. **Error! Reference source not found.** summarises the action taken and further planned CQI to review the programme PLOs based on the direct and indirect assessment, respectively. Seven of aspect has been revised include both in direct and indirect assessments.

(i) Revision of PLO

The potential improvement are due to EAC outlined a new set of PLOs as in EAC Manual 2012. The Faculty structured a new set of PLOs, which are in accordance with the PLOs in the EAC manual 2012. The number of PLOs increases to 13, taking into consideration the requirement of EAC and MOHE. The PLOs were approved by the Senate and implemented from Semester 1, Session 2013/2014.

(ii) Addition of critical problem solving and complex engineering activities in PLOs

This potential improvement need implementation of CPS and CEA in PLOs. The Faculty has self-initiated addressing the implementation of complex problem solving in the curriculum. The courses mapped to PLO 1, PLO 4, PLO 10 and PLO 11 are identified for the embedment of complex engineering problems, while the courses with PLO 2 and PLO 3 are identified for the embedment of complex engineering activities. The full implementation of the complex problem solving is targeted to start from Semester 1, 2016/2017.

(iii) Review on PLOs assessment tool - exit survey of student

The potential improvement are identified as verification of exit survey by subject matter experts. As the action taken, the designed questionnaire could be reviewed by subject matter experts during the pilot test before the distributions gauge the graduates' satisfaction on their PLOs achievement upon graduation.

(iv) Review on PLOs assessment tool –industry survey

Potential improvement are highlighted as mechanism to increase the number of respondents. Currently the industry survey is conducted annually during the Semester 3. The respondents are the employers of whom the FKMP students are undergoing the industrial training. For CQI, the Faculty has initiated a facebook group in an effort to encourage information dissemination to the active students and graduates. In long run, it helps the faculty to reach out to the industry during the reviewing of the PLOs.

(v) Review on PLOs assessment tool - external examiner review

This aspect of reviewed has been raised by External Examiner: The PLOs should be in accordance to the EAC 2012 manual, issues related to the complex problem solving should be addressed and PLOs need to be arranged into technical section and generic skill section. Some actions have been taken by continuous consultation with the external examiners was maintained. Faculty has taken measures to the raised issues:

- The revision of the PLOs has been made in accordance to the manual EAC 2012 and MOHE requirement
- The integration of complex problem solving has been initiated since year 2013
- The PLOs are categorised into cognitive domain, psychomotor domain and affective domain.
- The PLOs related to the cognitive domain are PLO 1, PLO 4, PLO 10 and PLO 11, the PLOs related to the psychomotor domain are PLO 2 and PLO 3, while the PLOs related to the affective domain are PLO 5, PLO 6, PLO 7, PLO 8, PLO 9, PLO 12 and PLO 13.

(vi) PLOs assessment

This potential improvement focused on measurement of overall PLOs assessment. As an action taken, CEPAT software was introduced to the Faculty member which measure real achievement as required. Furthermore CEPAT has also come out with detailed course performance analysis that satisfies the OBE requirement.

(vii) Course improvement

This involved with improvement in CLOs assessment. As an action taken, improvement of CLO assessment in current and previous semester can be accessed through TCIS System by Course Coordinator. Meanwhile, a SAS system shows an example of CQI report proving CLO assessment for two subsequent semesters. These CQI report will be attached to the SAS report submitted to Departmental Examination Committee level. Other Improvement is on individual attainment of PLOs. The action has been taken on using OBEsys to show the attainment of 13 PLOs related to CLOs for each courses reflected for individual student.

5. Conclusion

PLOs evaluation method in direct and indirect assessments have contribute to continuous quality improvement for the faculty in several aspect includes CPS and CEA, indirect assessment tool, external examiner review, measurement of overall PLOs assessment, CLOs assessment and PLOs attainments

6. Reference

1. Engineering Programme Accreditation Manual 2012, Engineering Accreditation Council, Board of Engineers Malaysia.

Application of Modified Flower Pollination Algorithm on Mechanical Engineering Design Problem

Ong Kok Meng, Ong Pauline*, Sia Chee Kiong, Hanani Abdul Wahab, Noormaziah Jafferi

ongkokmeng93@hotmail.com, {ongp*, sia, nani, normazia}@uthm.edu.my

Abstract. The aim of the optimization is to obtain the best solution among other solutions in order to achieve the objective of the problem without evaluation on all possible solutions. In this study, an improved flower pollination algorithm, namely, the Modified Flower Pollination Algorithms (MFPA) is developed. Comprising of the elements of chaos theory, frog leaping local search and adaptive inertia weight, the performance of MFPA is evaluated in optimizing five benchmark mechanical engineering design problems - tubular column design, speed reducer, gear train, tension/compression spring design and pressure vessel. The obtained results are listed and compared with the results of the other state-of-art algorithms. Assessment shows that the MFPA gives promising result in finding the optimal design for all considered mechanical engineering problems.

1. Introduction

Optimization is numerical methods which search and identify the best candidate from a collection of alternatives without evaluation on all possible alternatives [2]. In engineering field, engineers are often required to build the smallest heat exchanger that can achieve the desired heat transfer or to build the lowest cost engine for a car. In this regard, engineer will design a product by finding the best combination of parameters using optimization technique that will satisfy human needs, and this process is known as engineering design optimization. Therefore, engineering design optimization is very important to solve real world problem by finding the best solution with the least work.

To initiate the optimization solution procedure, the engineering design problem is derived in mathematical form, consisting of the objective function, and probably with set of constraints. The objective function is then solved by using optimization algorithm in order to find its optimal variables, which either maximizes or minimizes the underlying problem. For instance, optimal variables obtained from the optimization algorithm will give the objective function the minimum value for the case study of designing lowest cost engine for a car.

Most of the optimization problems can be mathematically written as:

$$\begin{aligned} &Minimize \quad f_i(x), \qquad (i = 1,2,\dots,M),\\ &subject\ to \quad \emptyset_j(x) = 0, \quad (j = 1,2,\dots,J),\\ &\psi_k(x) \le 0, \quad (k = 1,2,\dots,K), \end{aligned} \tag{1}$$

Where $f_i(x)$, $\emptyset_j(x)$ and $\psi_k(x)$ are functions of the design variables, $x = (x_1, x_2, \dots, x_n)^T$. $f_i(x)$ is known as objective function and the design variables can be real continuous, discrete or the mixed of them. Based on optimization equations in Equation (1), if J=K=0, there is no constraint for that particular optimization problem. For optimization problems which are classified as equality-constrained problem, their K=0 and J≥1, meanwhile for those which are classified as inequality-constrained problem, their J=0 and K≥1 [3].

In order to solve the optimization problem, the development of robust algorithms in solving complex optimization problems initiated since 1960s, where the differential calculus methods form its foundation. The invention of high-speed digital computers in the middle of twentieth results in the arising of new areas in optimization theory - the bio-inspired algorithm, offers an important breakthrough in optimization. Since then, the bio-inspired optimization algorithm has been the in thing.

Nature has evolved more than millions of years and she has found her best solution in overcoming problems that she has met. The success of how nature finds its way in solving problem triggers the formulation of bio-inspired algorithms [3]. Exemplary, the genetic algorithm (GA), perhaps, one of the most widely used bio-inspired techniques, is a model of biological evolution based on the Charles Darwin's theory [4]. The particle swarm optimization (PSO) algorithm, on the other hand, imitates the behavior of social organisms. The manner of how each particle communicates and shares the information to others forms the basis of PSO [5].

The success of GA and PSO in solving complex optimization problems prompts the formulation of more bio-inspired optimization algorithms, including the flower pollination algorithm (FPA) [6]. In nature, flower pollination is a marvelous process where the evolutionary objective such as reproduction is maximized efficiently. There are two types of pollination - abiotic and biotic, in which the FPA is derived based on such fascinating characteristics. The pioneer work of FPA shows its superiority over GA and PSO in the considered benchmark functions [6], but with fewer control parameters. The comprehensive ability for optimization, simplicity, and yet rapid convergence of FPA have gained widespread implementation in diverse areas; including optimization of wireless sensor network life time [7], retinal vessel segmentation [8], designing compensator damping controller [9], and allocation of capacitors [10].

This paper presents a modified flower pollination algorithm (MFPA), attempting to optimize five benchmark Mechanical Engineering design problems.

The paper is organized as follows. The basic idea of flower pollination process which leads to the formulation of FPA is first introduced. Subsequently, the chaotic maps, frog leaping local search and inertia weight are introduced as well as the procedure of proposed MFPA are presented. The performance of the MFPA is then evaluated on five benchmark and the results are compared with the results of the other state-of-art algorithm.

2. Flower Pollination Algorithm

There are between 5% and 20% of the flowering plants around 105 million years ago, but it increases to 80% after 40 million years. Scientists believe that the flowering plants start to dominate the landscape 125 million years ago during the Cretaceous Period. The mystery and yet amazing evolution of flowering plant drives Xin-She Yang to deep in his studies and develop FPA for global optimization [6].

The main factor causes the evolution of flowering plant is reproduction via pollination. Pollination can be categorized into two types - abiotic and biotic. Abiotic pollination does not require any pollinators to transfer the pollen. It depends on wind and water to transfer the pollen from one flower to another, known as wind pollination and water pollination, respectively. Meanwhile, biotic pollination requires pollinators to transfer the pollen. The examples of pollinator are beetle, fly, bee, butterfly, moth, bird and bat [11]. Furthermore, there are two types of pollination mechanism, specifically, self-pollination and cross-pollination. Self-pollination occurs when pollen is transferred from one flower to another same species flower, whereas cross-pollination occurs when pollen is transferred from one flower to another flower of the different species [11].

The aforesaid characteristics of flower pollination process are translated into the FPA by Xin-She Yang [6]. In the FPA, the biotic and cross-pollination are considered as global searching, as it has high probability occurring at long distance. In addition, the pollinators such as bees, birds and bats might behave as Levy flight behavior which obey a Levy distribution [12]. Meanwhile, the abiotic and self-pollination are considered as local searching in the FPA, as it may occur at short distance only. For simplicity, the developed FPA is idealized based on the assumptions of [6]:

1) Pollen-carrying pollinators perform the pollination using Lévy flight, where such biotic and crosspollination are considered as global searching.
2) Abiotic and self-pollination are equivalent to local searching.
3) Flower constancy is equivalent to a reproduction probability. It is proportional to the similarity of two flowers involved.
4) The switch probability, p ϵ [0,1] controls both local and global pollination. Local pollination has significant fraction p and dominates overall pollination activities due to physical proximity and wind.

A set of location updating formula are required to implement these idealized rules. First, for global pollination process, the pollinators such as bee carried the flower pollens and travelled over a long distance (Rule 1). This pollination process can be expressed mathematically as:

$$x_i^{t+1} = x_i^t + L(x_i^t - g_*) \tag{2}$$

Where x_i^t is the pollen i or solution vector x_i at iteration t, and g_* is the current best solution found among all solutions at current iteration. The Lévy flight is used to imitate the flying pattern of the insect [13]. Therefore, L>0 is drawn from a Lévy distribution

$$L \sim \frac{\lambda \Gamma(\lambda) \sin(\frac{\pi\lambda}{2})}{\pi} \cdot \frac{1}{s^{1+\lambda}}, \qquad (s \gg s_o \gg 0) \tag{3}$$

Where $\Gamma(\lambda)$ is the standard gamma function. This distribution is valid for large steps s>0.

Considering the flower constancy, ϵ the local pollination process (Rule 2) can be represented mathematically as:

$$x_i^{t+1} = x_i^t + \epsilon(x_j^t - x_k^t) \tag{4}$$

Where x_j^t and x_k^t represent pollen from different flowers of the same species. The use of flower constancy ϵ in Rule 3 ensures that the same species of flowers (similar solutions) are chosen more frequently and thus encouraging the exploitation. According to Rule 4, switch probability is bias towards local pollination. In the pioneer work of the FPA, p=0.8 works well for most applications.

3. Chaotic Maps

Chaos is a random state found in the non-linear dynamical deterministic system, possesses non-period, non-converging and bounded properties [14]. The use of chaotic sequences is more beneficial than the random sequences due to its non-repetition and ergodicity properties. Borrowing the advantages of ergodicity, non-repetition and randomness of the chaotic sequences, the chaotic map is replacing the random sequences in generating the initial population in the FPA in this study. This is to ensure that the diversity of the initial population can be improved, where the distribution of the initial population is more uniform. Ten different chaotic maps are listed in Table 1 and circle map is selected for the integration with FPA.

Table-1: Ten different chaotic maps

Name	Formula
Chebyshev map	$x_{n+1} = \cos^{-1}(n \times \arccos(x_n))$
Circle map	$x_{n+1} = (x_n + 0.2 - (0.5/2\pi)\sin(2\pi x_n)) \times mod(1)$
Gauss map	$x_{n+1} = \begin{cases} 0, & x_n = 0 \\ (1/x_n) \times mod(1), & x_n \in (0,1) \end{cases}$
Iterative map	$x_{n+1} = \sin(0.7\pi/x_n)$
Logistic map	$x_{n+1} = 4x_n(1 - x_n)$
Piecewise map	$x_{n+1} = \begin{cases} 2.5x_n, & x_n \in (0,0.4) \\ (x_n - 0.4)/0.1, & x_n \in (0.4,0.5) \\ (0.6 - x_n)/0.1, & x_n \in (0.5,0.6) \\ (1 - x_n)/0.4, & x_n \in (0.6,1.0) \end{cases}$
Sine map	$x_{n+1} = \sin(x_n \times \pi)$
Singer map	$x_{n+1} = 1.07(7.86x_n - 23.31(x_n)^2 + 28.75(x_n)^3 - 13.302875(x_n)^4)$
Sinusoidal map	$x_{n+1} = 2.3(x_n)^2 \times \sin(x_n \times \pi)$
Tent map	$x_{n+1} = \begin{cases} x_n/0.7, & x_n < 0.7 \\ (10/3) \times (1 - x_n), & x_n \geq 0.7 \end{cases}$

4. Frog Leaping Local Search

The Shuffled Frog-Leaping Algorithm (SFLA) was proposed by Eusuff, Lansey and Pasha in 2003. SFLA is classified as memetic metaheuristic algorithm. The creation of SFLA is inspired by the foraging behaviours of frogs.

First, a population size of frog is determined and stayed in a wetland (search space). Each frog in the wetland hold the information about the amount of food (fitness) in that location (solution) within the wetland. It is then divided into a number of groups of frogs known as number of memeplexes.

In each group, the frog that holds the least amount of food (worst fitness) is moved to new location in wetland due to information sharing from the frog that holds the most amount of food (best fitness) in that particular group. If the frog that holds the least amount of food is moved to a location where the amount of food is lesser, means no improvement, the frog will then randomly move to another new location in wetland. This process is repeated in each group and it is known as the iteration within each memeplex.

When the process undergone by each group of frogs happened for a period of time, all groups of frogs are gathered and joined as one population size for information sharing among those frogs. If the largest amount of food (convergence criteria) found by the frog in that population size is unsatisfied, the population size of frog is then divided into a number of groups of frogs again and continually searching for food until they find the largest amount of food in that wetland. The flowchart of SFLA is illustrated in Figure 1

In SFLA, the position of the frog with the worst fitness in a group is changed due to equations as follows:

$$Change\ in\ frog\ position\ (D_i) = rand() \cdot C \cdot (X_b - X_w) \quad (5)$$

$$\text{New position } X_w = \text{current position } X_w + D_i, \ (D_{max} \gg D_i \gg -D_{max}) \quad (6)$$

where rand() a random number between 0 and 1, C is a search-acceleration factor, X_b is best fitness, X_w is worst fitness and D_{max} is the maximum allowed change in a frog's position.

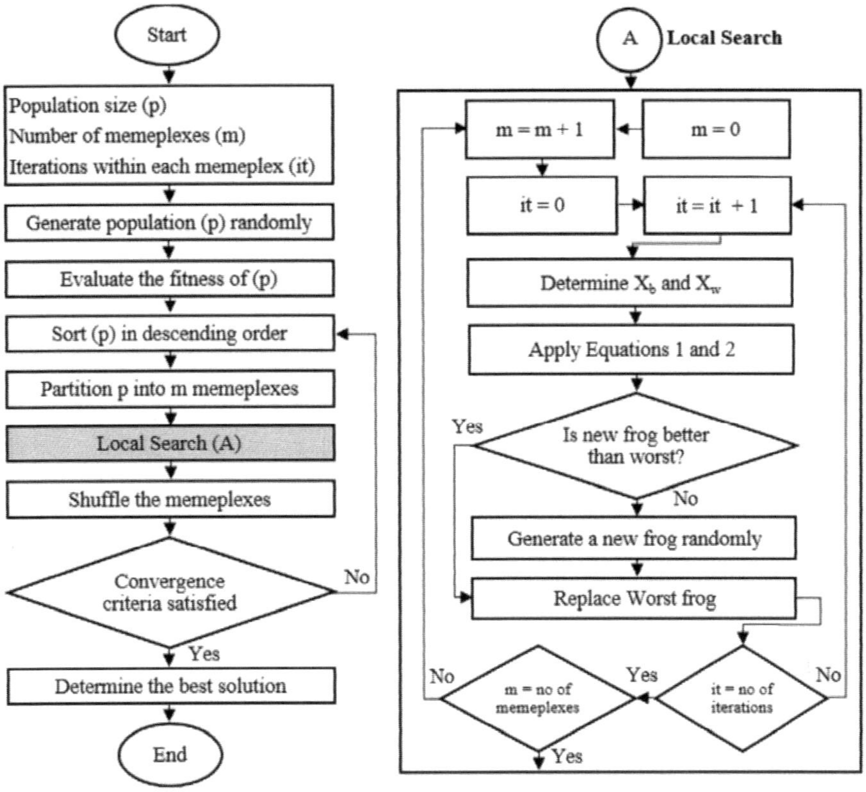

Figure-1: Flowchart of the SFLA

5. Inertia Weight

In basic FPA, the global searching of Levy flight is a random walk. Thus, there is no guaranteed on its fast convergence. To improve the performance of FPA in global searching, inertia weight is integrated into Levy flight. The formula of global searching of Levy flight in FPA is expressed in equation (2). The levy flight in equation (2) is used to imitate the fly pattern of the insect. To improve this formula, the Levy flight, L is multiplied by inertia weight, w as shown as follows:

$$x_i^{t+1} = x_i^t + wL(x_i^t - g_*) \quad (7)$$

$$w = w_{min}\left(1 + \frac{w_{max}}{\sqrt{t}} \times \tanh\left(\frac{g_*}{g_1}\right)\right) \quad (8)$$

Where g_1 is the best fitness value in first iteration, t denotes the current iteration number and g_* is the obtained best fitness value in current iteration.

The inertia weight, w determines how far the new flower is located from the previous flower. As the number of iteration, t increases, the best fitness value, g_* in current iteration decreases for minimization case. Wherefore, the value of w will be decreased from a larger value to a final value, w_{min} at last. This indicates that the larger the difference of the new fitness of the current best flower from the fitness of the best flower in first iteration, the nearer the new location of flower will be shifted from previous location. To start with identifying the optimal value of w_{min} and w_{max}, the value of w_{min} and w_{max} are set in the range of 1 to 10. Initially, w_{min} is set to 1 whereas w_{max} is set to 10 and then a parametric study is carried out to identify the most appropriate parameter. From the simulations, $w_{min} = 1$ and $w_{max} = 8$ works better for most applications. Thus, by integrating the inertia weight into Lévy flight, the search space exploration is improved more entirely and the solution can effectively move away from a local optima at the early phase and accelerate convergence at the later phase.

6. Modified Flower Pollination Algorithm

In the MFPA, the initial population is generated using the circle map, frog leaping local search is performed by each solution and when rand>p, modified Levy flight with integration of inertia weight in global pollination is performed on that particular solution. The steps involved in the MFPA are as follows:

Step 1: Parameter Initialization
Initialize the relevant parameters of population size, n, dimension of search space, d, maximum iteration, *max_iter*, switch probability, p, range of search space [*Lb*, *Ub*]], number of memeplexes, m and iterations within each memeplex, *it*.

Step 2: Generate Initial Population using the best chaotic map
The n initial population is generated randomly, where $Pop = (flower_1, flower_2, \ldots, flower_n)$, where $flower_i = (x_{i1}, x_{i2}, \ldots, x_{id})$, $i = 1, \ldots, n$ and d is dimension of the search variables. Then, each dimension of the search variable in first flower, x_{1j} is randomly generated between zero and one where $x_{1j} \in [0,1]$ and $j = 1, \ldots, d$. Subsequently, each dimension of the search variable in the rest of the flower, x_{kj} is updated using the circle map as in Table 1 where $k = 2, \ldots, n$. The generated chaotic sequences are further mapped within the solution space using $X_{ij} = Lb + (Ub - Lb)\, x_{ij}$.

Step 3: Find the Best Solution
The fitness value of each solution is calculated and the best solution g* is determined.

Step 4: Perform the Frog Leaping Local Search
For each solution, frog leaping local search is performed by following the flowchart in Figure 1.

Step 5: Perform the Global Search of Flower Pollination Algorithm
For each solution, an evenly distributed number, rand is generated from uniform distribution [0,1]. If rand>p, global pollination is performed by using Equation (7), otherwise, the solution remain in its position.

Step 6: Update the Solution
The fitness value of each new solution is evaluated. The historical position is updated through comparison with the new solution. Subsequently, the best solution g* is updated.

Step 7: Check Termination Condition
If the termination condition is satisfied, the optimal solution g* is displayed. Otherwise, Step 4 is repeated.

7. Numerical Simulations

In order to validate the feasibility of the MFPA, the algorithm is tested by means of optimizing five benchmark Mechanical Engineering design problems. For each benchmark Mechanical Engineering design problem, MFPA is simulated for 30 trials. For all simulations, the switch probability, p, is assigned as 0.8 in accordance with the pioneer work.

Each Mechanical Engineering design problem will be begun with it objective function. Afterward, the design variables are listed with their boundaries. The constraint of those particular problem will be listed too if available. The information above will be encoded into MFPA and the algorithm will then solve this particular problem. Lastly, the result from solving each Mechanical Engineering design problem by using MFPA will be listed and compared with the result of the other state-of-art algorithm.

7.1. Tubular Column Design

A tubular section as shown in Figure 2 is designed to carry a compressive load $P = 2500\ kg_f$ at minimum cost. Both end of the tubular column are hinge joints. The properties of the material that made up this column are listed in Table 2.

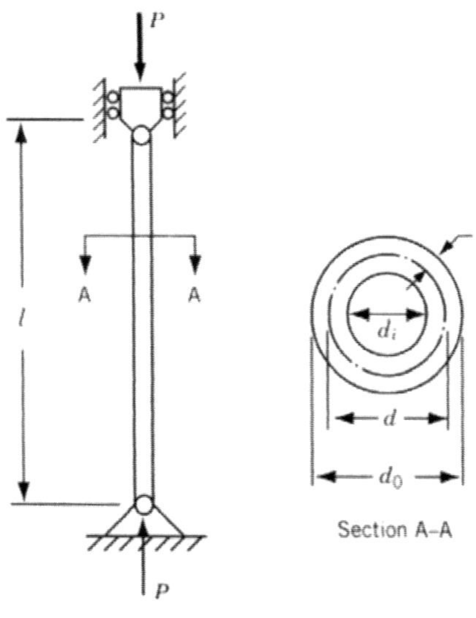

Figure-2: Tubular Column

Table-2: Properties of the Material

Yield Stress, σ_y (kg_f/cm^2)	500
Modulus of elasticity, E (kg_f/cm^2)	0.85×10^6
Weight density, ρ (kg_f/cm^3)	0.0025

The length (L) of the column is 250cm. The stress induced in the column should be less than the buckling stress (constraint g_1) as well as the yield stress (constraint g_2). The mean diameter of the column is restricted to lie between 2 and 14 cm (design variable 1), and columns with thicknesses outside the range 0.2 to 0.8 cm (design variable 2) are not available in the market. Hence, the design variables in this problem are mean diameter of the column in centimetres, d, and thickness of the column, t. The cost of the column includes material and construction costs can be taken as 5W + 2d, where W is the weight in kilograms force [15].

The objective function of this model is given as following:

$$\begin{aligned} Minimize\ f(d,t) &= 5W + 2d \\ &= 5pl\pi dt + 2d \\ &= 9.8dt + 2d \end{aligned} \qquad (9)$$

Subject to

$$g_1 = \frac{P}{\pi dt\sigma_y} - 1 \leq 0$$

$$g_2 = \frac{8PL^2}{\pi^3 Edt(d^2 + t^2)} - 1 \leq 0$$

Where

$$2 \ll d \ll 14$$
$$0.2 \ll t \ll 0.8$$

Table 3 shows the results provided by MFPA in optimizing tubular column design problem. The population size, n of MFPA in optimizing this problem is 30 and maximum number of iteration for MFPA to optimizing this problem is 100. Table 3 also shows the result comparison between MFPA and Cuckoo Search algorithm (CS) [16]. Table 4 illustrates the comparison of the results reported by Hsu and Liu [17] as well as Rao [15] and MFPA. Bold sets in Table 4 are violated sets which they do not fulfill the condition of constraint.

Table-3: Result Comparison between MFPA and CS

	CS	MFPA
Mean diameter, d	5.45139	5.4512
Thickness, t	0.29196	0.29197
f_{min}	26.50034	26.49995
Maximum number of iteration	15000	100
Average elapsed time of each run (s)	2.56919	0.365889

Table-4: Best Solutions for the Tabular Column Example

	Hsu and Liu	Rao	MFPA
d	5.4507	5.44	5.4512
t	0.292	0.293	0.29197
g_1	-3.4537×10^{-5}	-1.4873×10^{-3}	-2.3519×10^{-5}
g_2	$\mathbf{1.3171 \times 10^{-4}}$	$\mathbf{2.5802 \times 10^{-3}}$	-3.9619×10^{-5}
F_{min}	26.49912	26.50042	26.49995

In Table 4, MFPA has outperformed than CS in term of number of iteration and average elapsed time of each run. This is because MFPA can optimize the tubular column design problem almost seven times faster than CS in only 100 number of iteration. In this table, the method provided by Hsu and Liu as well as Rao are not feasible due to the violation of second constraint (g_2). Therefore, MFPA provides the best results among other algorithms in this particular problem.

7.2. Speed Reducer

A speed reducer [18] as illustrated in Figure 3 is a benchmark Mechanical Engineering design optimization problem with the face width (b), module of teeth (m), number of teeth on pinion (z), length of shaft 1 between bearings (l_1), length of shaft 2 between bearings (l_2), diameter of shaft 1 (d_1), and diameter of shaft 2 (d_2). Minimizing the total weight of the speed reducer is the objective of this problem. The constraints involve limitations on the bending stress of the gear teeth, surface stress, transverse deflections of shafts 1 and 2 due to transmitted force, and stresses in shafts 1 and 2.

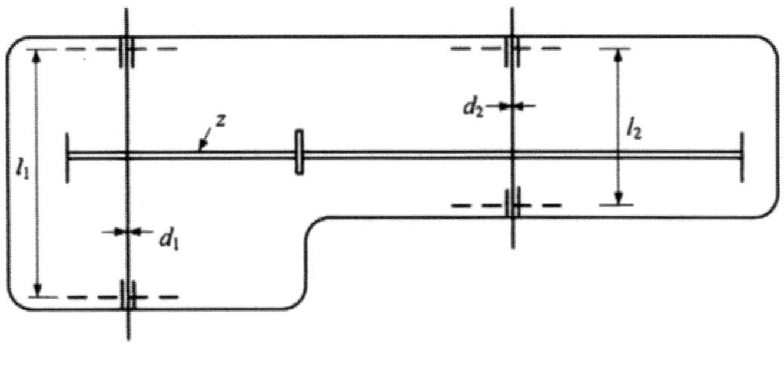

Figure-3: Speed Reducer

The objective function of speed reducer is given as following:

$$\begin{gathered} Minimize: f(b,m,z,l_1,l_2,d_1,d_2) = \\ 0.7854bm^2(3.3333z^2 + 14.9334z - 43.0934) \\ -1.508b(d_1^2 + d_2^2) + 7.477(d_1^3 + d_2^3) \\ +0.7854(l_1d_1^2 + l_2d_2^2) \end{gathered} \tag{10}$$

Subject to:

$$g_1 = \frac{27}{bm^2z} - 1 \ll 0$$

$$g_2 = \frac{397.5}{bm^2z^2} - 1 \ll 0$$

$$g_3 = \frac{1.93l_1^3}{mzd_1^4} - 1 \ll 0$$

$$g_4 = \frac{1.93l_2^3}{mzd_2^4} - 1 \ll 0$$

$$g_5 = \frac{1}{110d_1^3}\sqrt{\left(\frac{745l_1}{mz}\right)^2 + 16.9 \times 10^6 - 1 \ll 0}$$

$$g_6 = \frac{1}{85d_2^3}\sqrt{\left(\frac{745l_2}{mz}\right)^2 + 157.5 \times 10^6 - 1 \ll 0}$$

$$g_7 = \frac{mz}{40} - 1 \ll 0$$

$$g_8 = \frac{5m}{b} - 1 \ll 0$$

$$g_9 = \frac{b}{12m} - 1 \ll 0$$

$$g_{10} = \frac{1.5d_1 + 1.9}{l_1} - 1 \ll 0$$

$$g_{11} = \frac{1.1d_2 + 1.9}{l_2} - 1 \ll 0$$

The result provided by MFPA in optimizing speed reducer problem is shown in Table 5. The simple bounds of each design variable of the speed reducer problem are also shown in this table.

Table-5: Speed Reducer Problem Optimized by MFPA

	Bound	MFPA
b	2.6-3.6	3.5
m	0.7-0.8	0.7
z	17-28	17
l_1	7.3-8.3	7.3
l_2	7.8-8.3	7.8005
d_1	2.9-3.9	3.35021
d_2	5.0-5.5	5.28668
Objective function value		2996.2195
Popular size		90
Maximum Number of iteration		50
Time of each run (s)		1.325822

Table 6 shows a comparison of the results given by MFPA and other methods which are the result reported by Kuang, Rao, & Chen [19], Ray & Saini, [20], Akhtar, Tai, & Ray [21], Mezura-Montes, Coello, & Landa-Becerra [22] and Gandomi Yang, & Alavi. [16]. Bold sets in Table 6 are violated sets which they do not fulfill the condition of constraint.

In Table 6, the result obtain by MFPA is better than those results reported by Akhtar, Tai, & Ray, Mezura-Montes, Coello, & Landa-Becerra and Gandomi Yang, & Alavi. Although the minimum objective value reported by Kuang, Rao, & Chen and Ray & Saini are better than those of MFPA, the reported minimum objective values are not feasible. This is due to the violation of sixth constraints (g_6) in the result of Kuang, Rao, & Chen. Meanwhile, for the result of Ray & Saini, there are violation in fifth and sixth constraints (g_5 & g_6). Therefore, MFPA has outperformed than other opponents as it has obtained the least objective value for speed reducer problem.

Table-6: Results Obtained from Different Algorithm on Optimizing Speed Reducer Problem

	Kuang, Rao, & Chen (1998)	Ray & Saini, (2001)	Akhtar, Tai, & Ray (2002)	Mezura-Montes, Coello, & Landa-Becerra (2003)	Gandomi et al., (2013)	MFPA (2016)
F_{min}	2876.117623	2732.9006	3008.08	3025.005	3000.9810	2996.2195
b	3.6	3.514185	3.506122	3.506163	3.5015	3.5
m	0.7	0.700005	0.700006	0.700831	0.7000	0.7
z	17	17	17	17	17.0000	17
l_1	7.3	7.497343	7.549126	7.460181	7.6050	7.3
l_2	7.8	7.8346	7.85933	7.962143	7.8181	7.8005
d_1	3.4	2.9018	3.365576	3.3629	3.3520	3.35021
d_2	5	5.0022	5.289773	5.3090	5.2875	5.28668
g_1	-0.0996	-0.00777	-0.0755	-0.0777	-0.0743	-0.0739
g_2	-0.2203	-0.2012	-0.1994	-0.2013	-0.1983	-0.1980
g_3	-0.5279	-0.0360	-0.4562	-0.4741	-0.4349	-0.4992
g_4	-0.8769	-0.8754	-0.8994	-0.8971	-0.9008	-0.9015
g_5	-0.0433	**0.5395**	-0.0132	-0.0110	-0.0011	0
g_6	**0.1821**	**0.1805**	-0.0017	-0.0125	-0.0004	0
g_7	-0.7025	-0.7025	-0.7025	-0.7022	-0.7025	-0.7025
g_8	-0.0278	-0.0040	-0.0017	-0.0006	-0.0004	0
g_9	-0.5714	-0.5816	-0.5826	-0.5831	-0.5832	-0.5833
g_{10}	-0.0411	-0.1660	-0.0796	-0.0691	-0.0890	-0.0513
g_{11}	-0.0513	-0.0552	-0.0179	-0.0279	-0.0130	-0.0109

7.3. Gear Train

Sandgren has proposed a gear train mechanical design problem as shown in Figure 4 [23]. It is an unconstrained problem which has four integer variables. The objective of this problem is to produce a gear ratio as possible as 1/6.931. The objective function of this model is as follow:

$$\text{Minimize:} f(T_d, T_b, T_a, T_f) = \frac{1}{6.931} - \frac{T_d T_b}{T_a T_f} \quad (11)$$

Where T_i is the number of teeth of the gear i and they are all integers varying in the range of 12 to 60.

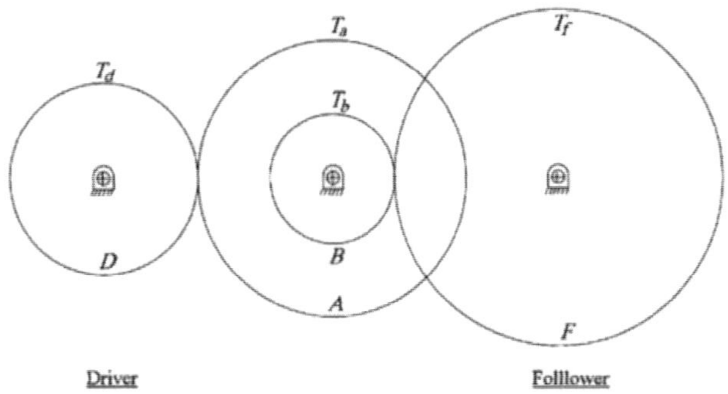

Figure-4: Gear Train

The result obtained by MFPA in optimizing gear train design problem is shown in Table 7. This table also shows the comparison result obtained by MFPA and CS [16]. In Table 8, it shows the comparison result obtained by MFPA and other methods which are the result reported by Sandgren [23], Kannan & Kramer [24] and Deb & Goyal [25].

Table-7: Result Obtained by CS and MFPA in Optimizing Gear Train Model

	CS	MFPA
Population size	25	30
Maximum Number of iteration	5000	100
T_d	19	17
T_b	16	28
T_a	43	60
T_f	49	55
Gear ratio	0.144281	0.144242
F_{min}	-1.643×10^{-6}	3.69×10^{-5}
Time of each run	3.25003	0.349064

In Table 7, MFPA has outperformed than CS in term of time of each run. This is because MFPA optimizes the gear train model almost ten times faster than CS although CS obtains better minimum objective value. In Table 8 shows that MFPA able to optimize gear train model and give a competitive solution as compare to other methods in a shorter time. Therefore, MFPA is a competitive algorithm in this gear train model as compare to other algorithms.

Table-8: Comparison Results of Different Optimization Algorithm in Optimizing Gear Train Model

	Sandgren (1990)	Kannan & Kramer (1994)	Deb & Goyal (1996)	MFPA (2016)
T_d	18	13	19	17
T_b	22	15	16	28
T_a	45	33	49	60
T_f	60	41	43	55
Gear ratio	0.146667	0.144124	0.144281	0.144242
F_{min}	-2.387×10^{-3}	1.552×10^{-4}	-1.643×10^{-6}	3.69×10^{-5}

7.4. Tension/Compression Spring Design

Designing a spring that can fulfill a mechanical task is a crucial part in a design process. A few design variables are needed to be determined that fulfil the constraints of the problem. Therefore, optimization is here to tackle these problems. Tension or compression spring design is one of the mechanical problem as shown in Figure 5. Minimizing the weight of a tension or compression spring is the objective of this problem. In the same time, it has to fulfill the constraints of this problem which are minimum deflection, shear stress, surge frequency, as well as limits on outside diameter and on design variables.

There are three variables in this problem which are wire diameter x_1, the mean coil diameter x_2 and the number of active coils x_3. The objective function of this problem is shown as follow:

$$Minimize: f(\vec{x}) = (x_3 + 2)x_2x_1^2 \tag{12}$$

Subject to:

$$g_1(\vec{x}) = 1 - \frac{x_2^3x_3}{71785x_1^4} \ll 0$$

$$g_2(\vec{x}) = \frac{4x_2^2 - x_1x_2}{12566(x_2x_1^3) - x_1^4} + \frac{1}{5108x_1^2} - 1 \ll 0$$

$$g_3(\vec{x}) = 1 - \frac{140.45x_1}{x_2^2x_3} \ll 0$$

$$g_4(\vec{x}) = \frac{x_2 + x_1}{1.5} - 1 \ll 0$$

Where $0.05 \ll x_1 \ll 2.0$, $0.25 \ll x_2 \ll 1.3$, and $2.0 \ll x_3 \ll 15.0$.

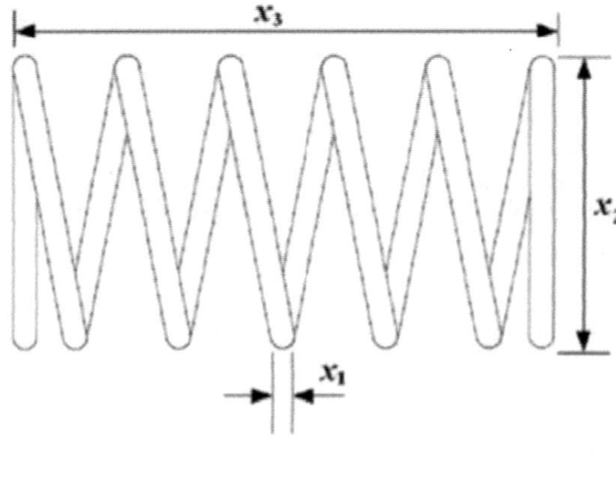

Figure-5: Tension/compression spring

Table 9 shows the best solution obtained from MFPA in optimizing tension or compression spring design problem. The best solution obtained from MFPA is then compared with the result obtained from other algorithms in Table 10.

Table-9: Best result obtained from MFPA

	MFPA
Population size	90
Maximum Number of iteration	100
x_1	0.0517621
x_2	0.358477
x_3	11.1866
F_{min}	0.012665
Time of each run (s)	1.619263

Table-10: Comparison results of different optimization algorithm in optimizing tension or compression spring design problem

	SiC-PSO	G-QPSO(2)	T-Cell	MFPA
x_1	0.051690	0.051515	0.051622	0.0517621
x_2	0.356750	0.352529	0.355105	0.358477
x_3	11.287126	11.538862	11.384534	11.1866
F_{min}	0.012665	0.012665	0.012665	0.012665

The result at the second column in Table 10 is obtained from Simple Constrained Particle Swarm Optimizer (SiC-PSO) which is proposed by Cagnina and Susana at 2008 [26]. Leandro dos Santos Coelho has proposed Quantum-behaved Particle Swarm Optimizer (G-QPSO) [27] whereas Victoria S. Aragón and Susana C. Esquivel have proposed T-Cell algorithm [28]. Their results on optimizing tension or compression spring design problem are listed in this table at third and fourth column respectively. As can be seen in this table, all of the algorithms can achieve the minimum objective value of 0.012665. Therefore, MFPA is a competitive algorithm as other optimization algorithms in this problem.

7.5. Pressure Vessel

A compressed air tank is designed to withstand a working pressure of 3000 psi and has a minimum volume of 750 ft^3. It is made of a cylindrical pressure vessel capped at both ends by hemispherical heads as illustrated in Figure 6. ASME boiler and pressure vessel code are followed to design this compressed air tank. The objective of this problem is to minimize the total cost which including material cost, forming cost and welding cost. The design variables of this problem are thickness x_1, thickness of the head x_2, the inner radius x_3 and the length of the cylindrical section of the vessel x_4. The thickness x_1 and thickness of the head x_2 are discrete values which are integer multiples of 0.0625 inch [23].

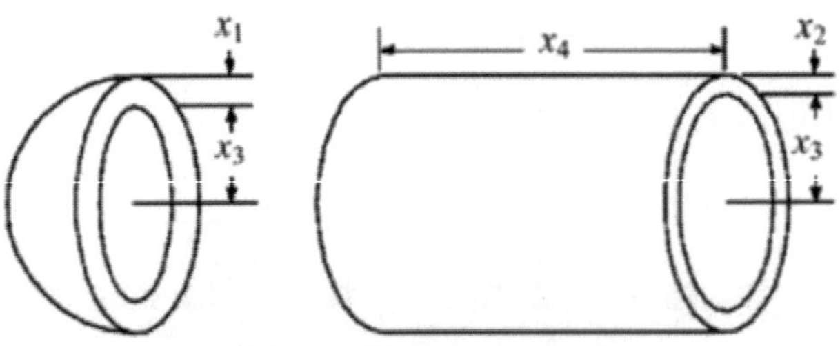

Figure-6: Pressure vessel

The objective function of this model is as follows:

$$Minimize: f(\vec{x}) = 0.6224x_1x_3x_4 + 1.7781x_2x_3^2 \\ +3.1661x_1^2x_4 + 19.84x_1^2x_3 \quad (13)$$

Subject to:

$$g_1(\vec{x}) = -x_1 + 0.0193x_3 \ll 0$$
$$g_2(\vec{x}) = -x_2 + 0.00954x_3 \ll 0$$
$$g_3(\vec{x}) = -\pi x_3^2 x_4 - \frac{4}{3}\pi x_3^3 + 1296000 \ll 0$$
$$g_4(\vec{x}) = x_4 - 240 \ll 0$$

where $1 \times 0.0625 \ll x_1, x_2 \ll 99 \times 0.0625$ and $10.0 \ll x_3, x_4 \ll 200.0$.

Table 11 shows the best solution obtained from MFPA in optimizing pressure vessel problem. It also shows the comparison result obtained by MFPA and CS in this table [16]. The best solution obtained from MFPA is then compared with the result obtained from other algorithms such as Stochastic Ranking (SR) which is reported by Runarsson and Yao [29], Artificial Immune System Genetic Algorithm (AIS-GA) which is proposed by Bernardino, Barbosa and Lemonge [30] as well as T-cell algorithm which is proposed by Victoria S. Aragón and Susana C. Esquivel [28] in Table 12.

Table-11: Best result obtained from CS and MFPA in optimizing pressure vessel problem

	CS	MFPA
Population size	25	90
Maximum Number of iteration	15000	100
x_1	0.8125	0.8125
x_2	0.4375	0.4375
x_3	42.0984456	42.09845
x_4	176.6365958	176.6366
F_{min}	6059.7143348	6059.7143
Time of each run (s)	2.50472	1.426891

Table-12: Comparison results of different optimization algorithm in optimizing pressure vessel problem

	SR	AIS-GA	T-Cell	MFPA
x_1	1.1250	0.8125	0.8125	0.8125
x_2	0.5625	0.4375	0.4375	0.4375
x_3	58.1267	42.0931	42.098429	42.09845

x_4	44.5941	176.7031	190.787695	176.6366
F_{min}	6832.583	6060.367	6390.554	6059.7143

In Table 11, MFPA has outperformed than CS in term of number of iteration and time of each run. This is because MFPA optimizes the pressure vessel problem almost one second faster than CS in only 100 number of iteration. In Table 12 shows that MFPA has produced outstanding result in optimizing pressure vessel problem as it gives the lowest minimum function value among other algorithms proposed by other researchers. Therefore, MFPA is a powerful algorithm in optimizing this pressure vessel problem.

8. Conclusion

In this study, MFPA is used to optimize five benchmarks Mechanical Engineering design problems. In tabular column design, MFPA has optimized the model almost seven times faster than CS and in the same time, enable to obtain the least magnitude of objective function. In speed reducer, MFPA has obtained the least magnitude of objective function compare to other algorithms proposed by other researchers in latest research. Furthermore, MFPA has optimized the gear train model ten times faster than CS. Apart from that, MFPA enable to optimize tension/compression spring design as other algorithms. Lastly, MFPA has optimized pressure vessel model one second faster than CS by only 100 number of iteration. Therefore, these results prove that MFPA is feasible in optimizing five benchmark Mechanical Engineering design problems as well as real world Mechanical Engineering design problem.

Acknowledgments

Financial support from the Malaysian government with cooperation of Universiti Tun Hussein Onn (UTHM) in the form of FRGS Vot 1490 is gratefully acknowledged.

References

[1] Gomez, A. G., & Gruender, J. L. (2011). Engineering Your Future: A Project-Based Introduction to Engineering. Wildwood: Great Lakes Press, Incorporated. Retrieved from https://books.google.com.my/books?id=kMOzNAEACAAJ

[2] Ravindran, A., Ragsdell, K. M., & Reklaitis, G. V. (2006). Engineering Optimization: Methods and Application. Journal of Chemical Information and Modeling (2nd ed., Vol. 53). Hoboken, New Jersey: John Wiley & Sons, Inc. http://doi.org/10.1017/CBO9781107415324.004

[3] Yang, X. S. (2010). Engineering Optimization: An Introduction with Metaheuristic Applications. Engineering Optimization: An Introduction with Metaheuristic Applications. Hoboken, New Jersey: John Wiley & Sons, Inc. http://doi.org/10.1002/9780470640425

[4] J. H. Holland, Adaptation in Natural and Artificial Systems. (University of Michigan Press, 1975).

[5] J. Kennedy and R. Eberhart, presented at the Proceedings of IEEE International Conference on Neural Networks, 1995 (unpublished).

[6] X. S. Yang, in Lecture Notes in Computer Science (including subseries Lecture Notes in Artificial Intelligence and Lecture Notes in Bioinformatics) (2012), Vol. 7445 LNCS, pp. 240-249.

[7] M. Sharawi, E. Emary, I. A. Saroit and H. El-Mahdy, Int J Soft Comput Eng 4, 54-59 (2014).

[8] E. Emary, H. M. Zawbaa, A. E. Hassanien, M. F. Tolba and V. Snášel, presented at the Proceedings of the Fifth International Conference on Innovations in Bio-Inspired Computing and Applications IBICA 2014, 2014 (unpublished).

[9] Y. Abdelaziz and E. S. Ali, Electr Pow Comp Syst 43, 1268-1277 (2015).

[10] Abdelaziz, E. Ali and S. A. Elazim, Int J Eng Sci Techno 19, 610-618 (2016).

[11] Glover, Understanding Flowers and Flowering: An Intergrated Approach (Oxford University Press, New York, 2007).
[12] Pavlyukevich, J Comput Phys 226, 1830-1844 (2007).
[13] A. M. Reynolds and M. A. Frye, Free-flight odor tracking in Drosophila is consistent with an optimal intermittent scale-free search 2, e354 (2007).
[14] 12. B. Alatas, Expert Syst Appl 37, 5682-5687 (2010).
[15] Rao, S. S. (2009). Engineering Optimization: Theory and Practice (4th ed.). Hoboken, New Jersey: John Wiley & Sons, Inc. http://doi.org/10.1080/03052150500066646
[16] Gandomi, A. H., Yang, X. S., & Alavi, A. H. (2013). Cuckoo search algorithm: A metaheuristic approach to solve structural optimization problems. Engineering with Computers, 29(1), 17–35. http://doi.org/10.1007/s00366-011-0241-y
[17] Hsu, Y.-L., & Liu, T.-C. (2007). Developing a fuzzy proportional-derivative controller optimization engine for engineering design optimization problems. Engineering Optimization, 39(6), 679–700. http://doi.org/10.1080/03052150701252664
[18] Gandomi, A. H., & Yang, X.-S. (2011). Benchmark Problems in Structural Optimization. In S. Koziel & X.-S. Yang (Eds.), Computational Optimization, Methods and Algorithms (pp. 259–281). Berlin, Heidelberg: Springer Berlin Heidelberg. http://doi.org/10.1007/978-3-642-20859-1_12
[19] KU, K. J., RAO, S. S., & CHEN, L. I. (1998). TAGUCHI-AIDED SEARCH METHOD FOR DESIGN OPTIMIZATION OF ENGINEERING SYSTEMS. Engineering Optimization, 30(1), 1–23. http://doi.org/10.1080/03052159808941235
[20] RAY, T., & SAINI, P. (2001). ENGINEERING DESIGN OPTIMIZATION USING A SWARM WITH AN INTELLIGENT INFORMATION SHARING AMONG INDIVIDUALS. Engineering Optimization, 33(6), 735–748. http://doi.org/10.1080/03052150108940941
[21] Akhtar, S., Tai, K., & Ray, T. (2002). A Socio-Behavioural Simulation Model for Engineering Design Optimization. Engineering Optimization, 34(4), 341–354. http://doi.org/10.1080/03052150212723
[22] Mezura-Montes, E., Coello, C. A. C., & Landa-Becerra, R. (2003). Engineering optimization using simple evolutionary algorithm. Proceedings 15th IEEE International Conference on Tools with Artificial Intelligence, (December 2003), 149–156. http://doi.org/10.1109/TAI.2003.1250183
[23] Sandgren, E. (1990). Nonlinear Integer and Discrete Programming in Mechanical Design Optimization. Journal of Mechanical Design, 112(2), 223–229. Retrieved from http://dx.doi.org/10.1115/1.2912596
[24] Kannan, B. K., & Kramer, S. N. (1994). An Augmented Lagrange Multiplier Based Method for Mixed Integer Discrete Continuous Optimization and Its Applications to Mechanical Design. Journal of Mechanical Design, 116(2), 405–411. Retrieved from http://dx.doi.org/10.1115/1.2919393
[25] Deb, K., & Goyal, M. (1996). A combined genetic adaptive search (GeneAS) for engineering design. Computer Science and Informatics, 26(1), 30–45. http://doi.org/citeulike-article-id:9625478
[26] Cagnina, L. C., Esquivel, S. C., Nacional, U., Luis, D. S., Luis, S., & Coello, C. A. C. (2008). Solving Engineering Optimization Problems with the Simple Constrained Particle Swarm Optimizer 1 Introduction 2 Literature review 3 Our proposed approach : SiC-PSO, 32, 319–326.
[27] Coelho, L. dos S. (2010). Gaussian quantum-behaved particle swarm optimization approaches for constrained engineering design problems. Expert Systems with Applications, 37(2), 1676–1683. http://doi.org/10.1016/j.eswa.2009.06.044

[28] Aragón, V. S. ., Esquivel, S. C. ., & Coello, C. A. C. . B. C. (2010). A modified version of a T-Cell Algorithm for constrained optimization problems. International Journal for Numerical Methods in Engineering, 84(3), 351–378. http://doi.org/10.1002/nme.2914
[29] Runarsson, T. P., & Yao, X. (2000). Stochastic ranking for constrained evolutionary optimization. IEEE Transactions on Evolutionary Computation. http://doi.org/10.1109/4235.873238
[30] Bernardino, H. S., Barbosa, H. J. C., & Lemonge, A. C. C. (2007). A hybrid genetic algorithm for constrained optimization problems in mechanical engineering. 2007 IEEE Congress on Evolutionary Computation. http://doi.org/10.1109/CEC.2007.4424532

IOP Conf. Series: Materials Science and Engineering **165** (2017) 012033 doi:10.1088/1757-899X/165/1/012033

Assessment and Evaluation for Programme Learning Outcomes in Faculty of Mechanical and Manufacturing Engineering, Universiti Tun Hussein Onn Malaysia

Ong Pauline*, Hariati Taib, Mohd AzwirAzlan, Nurul Fitriah Nasir, Mohd Azham Azmi, Saliha Md Salleh, Badrul Aisham Md Zain, Shahruddin Mahzan, Muhd Hafeez Zainulabidin, Sufizar Ahmad, Nasrull Abd Rahman, Zamani Ngali, Hamimah Abd Rahman

Faculty of Mechanical and Manufacturing Engineering, Universiti Tun Hussein Onn Malaysia (UTHM), 86400 Parit Raja, Batu Pahat, Johor.

{ongp, hariati, azwir, fitriah, azham, saliha, aisham, sharudin, hafeez, sufizar, mnasrull, zamani, hamimah}@uthm.edu.my

Abstract. Universiti Tun Hussein Onn Malaysia (UTHM) is the 15th of the 20 public universities established in Malaysia. UTHM consists of eight faculties. One of them is the Faculty of Mechanical and Manufacturing Engineering (FKMP). The programme offered – Bachelor of Mechanical Engineering with Honours (BDD), undergoes continuous auditing and accreditation by the Engineering Accreditation Council (EAC). To fulfill the requirement for accreditation, EAC requires the faculty to assess and evaluate the programme learning outcomes' (PLOs) attainment of the graduates, such that all graduates are targeted to achieve the performance indicator (PI) upon graduation. This paper discusses the methods developed by the FKMP for PLOs' measurement and evaluation. Two approaches are utilized in this regards, (i) the direct assessment based on students' performance in courses, and (ii) the indirect assessment based the industrial and graduates' perception. The PLOs' attainment for graduates of cohort 2010-2012 are analysed in this study. Consolidation data shows that the PI for majority of the PLOs are achieved, indicating that the programme offered complies with the standard expected by different stakeholders.

1. Introduction

All engineering programmes in Malaysia are monitored closely by the Engineering Accreditation Council (EAC) through auditing and accreditation practices such that their qualities are ensured. Each engineering program is expected to fulfill the eight components of the accreditation requirements, in which the Programme Learning Outcomes (PLOs) are one of the elements being assessed by the EAC.

PLOs are statements that describe what students are expected to know and be able to perform or attain by the time of graduation, that are linked to the domain of knowledge (cognitive), skill (psychomotor) and soft skill (affective) [1]. The programme of Bachelor of Mechanical Engineering with Honours (BDD) offered by the Faculty of Mechanical and Manufacturing Engineering (FKMP), Universiti Tun Hussein Onn Malaysia (UTHM) has 13 PLOs, which are established according to the Engineering Programme Accreditation Manual 2012 and the Ministry of Higher Education requirements. Table 1 summarises the PLOs of the BDD programme.

Table 1: The 13 PLOs of the BDD Programme offered by FKMP, UTHM

PLO	Attribute	Domain	Description
1	Engineering Knowledge	Cognitive	Apply knowledge of mathematics, science, engineering fundamentals and an engineering specialisation to the solution of complex engineering problems.
2	Modern Tool Usage	Psychomotor	Create, select and apply appropriate techniques, resources, and modern engineering and IT tools, including prediction and modelling, to complex engineering activities, with an understating of the limitation.
3	Communication	Psychomotor	Communicate effectively on complex engineering activities with the engineering community and with society at large, such as being able to comprehend and write effective reports and design documentation, make effective presentations, and give and receive clear instructions.
4	Investigation	Cognitive	Conduct investigation into complex problems using research based knowledge and research methods including design of experiments, analysis and interpretation of data, and synthesis of information to provide valid conclusion.
5	Individual & Teamwork	Affective	Function effectively as an individual, and as a number or leader in diverse teams and in multi-disciplinary setting.
6	Lifelong Learning	Affective	Recognise the need for, and have preparation and ability to engage in independent and life-long learning in the broadest context of technological change.
7	Entrepreneurship	Affective	Recognise the importance of entrepreneurship in mechanical engineering and its' related discipline.
8	Ethics	Affective	Apply ethical principles and commit to professional ethics and responsibilities and norms of engineering practice.
9	Project Management & Finance	Affective	Demonstrate knowledge and understanding of engineering and management principles and apply these to one's own work, as a member and leader in a team, to management projects and in multidisciplinary environments.
10	Design/ Development of Solution	Cognitive	Design solution for complex engineering problems and design systems, components or processes that meet specified need with appropriate consideration for public health and safety, cultural, societal, and environmental considerations.
11	Problem Analysis	Cognitive	Identify, formulate, research literature and analyse complex engineering problems reaching substantiated conclusions using first principles of mathematics, natural sciences and engineering sciences.
12	Environmental and Sustainability	Affective	Understand the impact of professional engineering solutions in societal and environmental contexts and demonstrate knowledge of and need for sustainable development.
13	The Engineer & Society	Affective	Apply reasoning informed by contextual knowledge to assess society, health, safety, legal and cultural issues and the consequent, responsibilities relevant to professional engineering practice.

Different methods, exemplary, formative and summative assessments, questionnaires, and interviews, can be used to assess the PLOs' achievement. These methods, in general, can be classified into two categories: direct and indirect approaches [2]. The direct assessment approach systematically evaluate the students' performance in the form of tests, final examinations, assignments, quizzes, projects and laboratory works, based on the established rubric. The indirect assessment approach, on the other hand, assesses the stakeholders' perception in terms of students' abilities to attain the PLOs. The consolidation of both direct and indirect assessment is useful to determine the quality of the programme and to provide analysis for continual quality improvement in the programme.

The PLOs are related to the course learning outcomes (CLOs). CLOs are statements that describe what students are expected to know and be able to perform or attain by the time of completing the courses, that are linked to the domain of knowledge (cognitive), skill (psychomotor) and soft skill (affective) Each course has its own course learning outcomes (CLOs), and has various delivery and assessment methods. Each CLO of the particular course is mapped to the particular PLO. Through the CLO-PLO mapping, the contributions of the CLOs towards the achievement of the PLOs can be evaluated. Figure 1 gives the CLO-PLO mapping of BDA24202 Computer Programming. There are 3 CLOs in this course, in which the CLO1, CLO2 and CLO3 are mapped to PLO1 Engineering Knowledge, PLO2 Modern Tool Usage and PLO6 Lifelong Learning, respectively. The performance indicator (PI) of each CLO is set as 100% students get 40% marks and above. In order to ensure that the CLOs are attained in the end results, the course lecturer develops the learning and teaching plans, selects the delivery and assessment method, and designs the assessment questions with proper rubric at the beginning of each semester. For instance, the lab sheets are used to assess whether the students are able to develop the program using C language (CLO2), which is related to the PLO2 Modern Tool Usage. Upon completing the course, the attainment of each CLO is measured and its contribution towards the particular PLO is analysed, which will be discussed further in the following sections.

Course Learning Outcome, Delivery and Assessment Template

Faculty: Faculty of Mechanical and Manufacturing Engineering
Programme: Bachelor of Mechanical Engineering with Honours
Course: Computer Programming
Code: BDA 24202

No	Course Learning Outcomes	Compliance to PLO													Delivery	Assessment	KPI
		PLO 1	PLO 2	PLO 3	PLO 4	PLO 5	PLO 6	PLO 7	PLO 8	PLO 9	PLO 10	PLO 11	PLO 12	PLO 13			
1	Apply the problem solving technique using algorithm and construct the structure of program in C language. (C3, LO1)	✓ C3													Lecture	Test, Final exam	100% of students must get 40% of marks and above
2	Develop the program to solve problem using C language. (P4, LO2)		✓ P4												Laboratory work	Lab evaluation	100% of students must get 40% of marks and above
3	Present the project findings in oral and written form for the individual or group assignment (A2, LO6)						✓ A2								Project	Presentation, report	100% of students must get 40% of marks and above

Figure 1: CLO-PLO mapping of BDA24202 Computer Programming

2. Methods Used to Assess the Attainment of Programme Learning Outcomes

There are two methods of assessing the PLOs, namely (i) indirect assessment and (ii) direct assessment. The tools used in theses assessments are summarised as in Table 2.

2.1. Indirect Assessment Approach

The PLOs attainment is assessed continuously by the FKMP through indirect assessment. The types of indirect evaluations and their mechanisms are explained as follows:

Table 2: Summary of Different Types of Assessments

Types of Assessment	Platform	Method/Tools	Output
Indirect	Industrial Advisory Committee (IAC)	Annual Meeting	Reports
	Adjunct Professor	Annual Meeting	Reports
	External Examiner	Annual Meeting	Reports
	Exit Survey	Survey	Reports
	Industrial Training Survey	Survey	Reports
Direct	Summative Evaluation of Students	CEPAT	Students' achievement
	Formative Evaluation of Students	CEPAT / SAS / OBESYS	

(i) Industry Survey

The industry survey is a tool to identify the needs and gaps of skill in industry based on the employer's perception in terms of students' abilities to attain the PLOs. The industry survey was conducted yearly during the Semester 3 of the third year academic session. The respondents are the employers of whom the FKMP students are undergoing the industrial training, including the managers, senior engineers and practicing engineers. The KPI set was 80% scored in both good and excellent index for each PLOs, as detailed out in the questionnaire.

(ii) Exit Survey

The Exit Survey refers to an indirect measurement method of students' self-evaluation or individual perception for the assessment of qualities as prescribed in PLOs, which was conducted upon graduation. The graduates were required to assess themselves based on the questions as detailed out in the questionnaire. The KPI set was 80% scored in both good and excellent index for each PLO, as detailed out in the questionnaire.

(iii) Assessment by External Examiners/Adjunct Professor/Visiting Professors/Industrial Advisory Community(IAC)

The review from the External Examiners, Adjunct Professor, Visiting Professors and IAC regarding the appropriateness of the Faculty PLOs are taken during periodic review through various meetings.

2.2. Direct Assessment Approach

PLOs in FKMP were analysed by using online and offline systems. The established of management system (online and offline) used are divided into three systems as stated below:

(i) Offline system - Course Evaluation and Performance Analysis Tool (CEPAT)
(ii) Online system - Students Assessment System (SAS)
(iii) Online System – Outcome Based Education System (OBEsys) evaluation

The CEPAT assessment template, which is an offline excel-based programming developed by the Faculty member, is capable in analysing the relations of the assessments for each courses to the CLOs and further to PLOs. The SAS and OBEsys are online database system developed by the University (Information Technology Centre), which offer similar function with CEPAT with extra features. The different functions and features of the systems are as summarised in Table 3.

Table 3: Summary of Functions and Features of Different Systems of Assessments

	System	Basic Function (Marks compilation)	Features
1.	**CEPAT**	Continuous assessment Final examination	• Offline System • Analysis of CLO achievement • Mapping of CLO achievement to PLOs • Analysis of CLO and PLO KPI achievement
2.	**SAS**	Continuous assessment Final examination	• Online System linked to TCIS • Systemised compilation of courses CQI
3.	**OBEsys**		• Online system linked to TCIS • Analysis of PLO achievement

CEPAT and SAS were used to store continuous assessment and final examination marks of all students by the respective lecturer. CEPAT is managed by the Faculty's OBE Committee meanwhile SAS is managed by the University. SAS is linked to the Total Campus Integrated System (TCIS). TCIS is an online system which integrates necessary information of personal, teaching and learning activities of each individual staff, regardless of academic or supporting staffs. The OBEsys system is managed by the Faculty's OBE committee and is developed by the Information Technology Centre of UTHM. This system is capable to manage and measure the individual PLOs assessment.

The CEPAT is the main document used by the FKMP to measure the PLOs attainment. At the beginning of the semester, the course coordinators were given the authority to set up the mark weightage for each CLO towards the particular PLO marks. The mapping between the CLOs and PLOs were fixed by the faculty [3], but the mark weightage for each CLO can be varied in every semester, depending on the learning and teaching plan. The course coordinator will set up the weightage in TCIS and CEPAT, and the respective lecturers can fill up the marks in SAS and CEPAT afterwards. Figure 2 shows an example of the weightage distribution for each CLO to the PLO for BDA24202 Computer Programming in CEPAT.

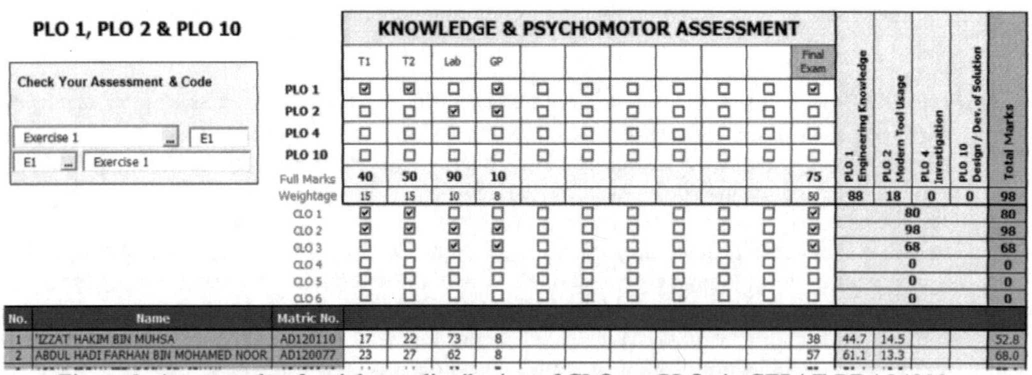

Figure 2: An example of weightage distribution of CLOs to PLOs in CEPAT BDA24202

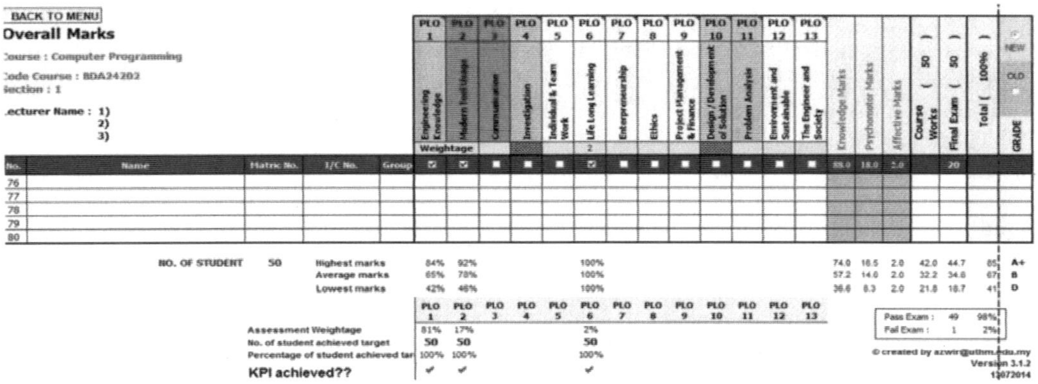

Figure 3: An example of attainment of each PLO in CEPAT BDA24202

Table 4: Cumulative PLO Attainments for Cohort 2010

				Emphasis To The PLO												
	No.	Code	Course	PLO 1	PLO 2	PLO 3	PLO 4	PLO 5	PLO 6	PLO 7	PLO 8	PLO 9	PLO 10	PLO 11	PLO 12	PLO 13
Engineering Courses	1	BDA 10102	Engineering Drawing	80%	80%	100%										
	2	BDA 10203	Statics	88%			99%	99%					99%	99%		
	3	BDA 10302	Material Science	82%	89%	99%										
	4	BDA 10402	Solid Mechanics I	78%			100%	100%					100%	100%		
	5	BDA 10502	Fluid Mechanics	76%		100%	97%	94%					97%	97%		
	6	BDA 10602	Creativity & Innovation	100%				90%		100%		100%	100%			
	7	BDA 17001	Engineering Laboratory I													
	8	BDA 18001	Mechanical Engineering Practice I	100%				100%	100%							
	9	BDA 18101	Mechanical Engineering Practice II	99%	99%		100%	100%		100%			100%	100%		
	10	BDA 20103	Dynamics	88%		100%	100%	100%						100%		
	11	BDA 20202	Thermodynamics I	59%			100%	100%						100%		
	12	BDA 20303	Mechanics of Machine	94%			100%	100%						100%		
	13	BDA 20402	Engineering Materials Selection	95%	99%	100%										
	14	BDA 20502	Computer Aided Design	100%	99%				100%				100%			
	15	BDA 27001	Engineering Laboratory II		97%	98%		99%								
	16	BDA 27101	Engineering Laboratory III		99%	99%	98%	100%					99%	99%		
	17	BDA 27201	Engineering Laboratory IV		99%	99%	92%	99%								
	18	BDA 28001	Mechanical Engineering Practice III	98%	100%			96%								
	19	BDA 28101	Mechanical Engineering Practice IV	97%	100%		99%	100%		99%			99%	99%		
	20	BDA 30203	Fluid Mechanics II	83%			100%	100%						100%		
	21	BDA 30303	Solid Mechanics II	90%			97%	100%						97%		
	22	BDA 30403	Thermodynamics II	81%			99%	100%						99%		
	23	BDA 30502	Manufacturing Technology	96%	98%	100%				97%						
	24	BDA 30603	Heat Transfer	94%			100%	99%						100%		
	25	BDA 30703	Control Engineering	85%			100%	99%						100%		
	26	BDA 30803	Mechanical Engineering Design	74%	95%				100%				100%			
	27	BDA 37001	Engineering Laboratory V		99%	99%	77%	99%								
	28	BDA 37101	Engineering Laboratory VI		99%	99%	75%	99%								
	29	BDA 38004	Industrial Training		100%	100%			100%		100%					100%
	30	BDA 40102	Industrial Engineering	93%		99%	100%							100%		
	31	BDA 40203	Engineering Design	100%	100%								100%		100%	
	32	BDA 40303	Finite Element Method	96%	99%		98%							97%		
	33	BDA 40603	Noise and Vibration	93%	96%		100%	98%								
	34	BDA 49002	Bachelor Degree Project I	99%		99%			99%						99%	
	35	BDA 49104	Bachelor Degree Project II	97%	97%	96%	96%					96%			97%	
	36	BD* 4**03	Elective I	99%		99%	100%							99%		
	37	BD* 4**03	Elective II	93%		100%	100%							100%		
	38	BD* 4**03	Elective III	93%		99%	100%							99%		
	39	BDA 40402	Management and Professional Ethics	99%							99%	99%				100%
	40	BDA 40502	Engineers & Society	97%								100%			100%	
	41	BPK 20802	Entrepreneurship	100%	100%					100%	100%					100%
	42	BPK 30902	Engineering Economy	98%	100%					98%						
	43	BTI 10202	Computer Programming	54%	100%				100%							
			Total Responder	36	22	18	24	22	6	6	3	4	10	19	4	3
			Responder Percentage (%)	87.8	244.4	94.7	150.0	100.0	46.2	150.0	100.0	80.0	250.0	126.7	100.0	150.0
			Average Score (Student achievement) Results	90%	97%	99%	97%	99%	100%	99%	100%	99%	99%	99%	99%	100%

The attainment of PLOs will be assessed in continuous assessment including quizzes, assignments, projects and tests. The total mark of continuous assessment is 50% and remaining 50% are allocated for final examination marks for the courses. The marks from continuous assessment and final examination will contribute to the CLOs marks, based on the weightage that has been determined at the beginning of the semester. The contribution of each CLO to the particular PLO, depending on the weight dissemination, thus can be analysed. The PI of each CLO is set as 100% students get 40% marks and above. At course level, the individual PLO attainment of each course is measured based on the total percentage of students who get at least 40% score and above, as shown in Figure 3.

Once the CLOs assessments are completed for all courses, the respective PLO mapped to the CLO will be assessed. An example of cumulative CLO attainment that contribute to PLOs achievement for cohort 2010 is presented in Table 4.

3. Analysis and Evaluation of Programme Learning Outcomes Attainment

This section discusses the PLOs attainment results from both indirect and direct measurement approaches.

3.1. Analysis of PLOs Achievement by Indirect Assessment

(i) Industrial Survey

Figure 4 presents the analysis of the PLOs' achievement through the industrial survey for cohort 2010.

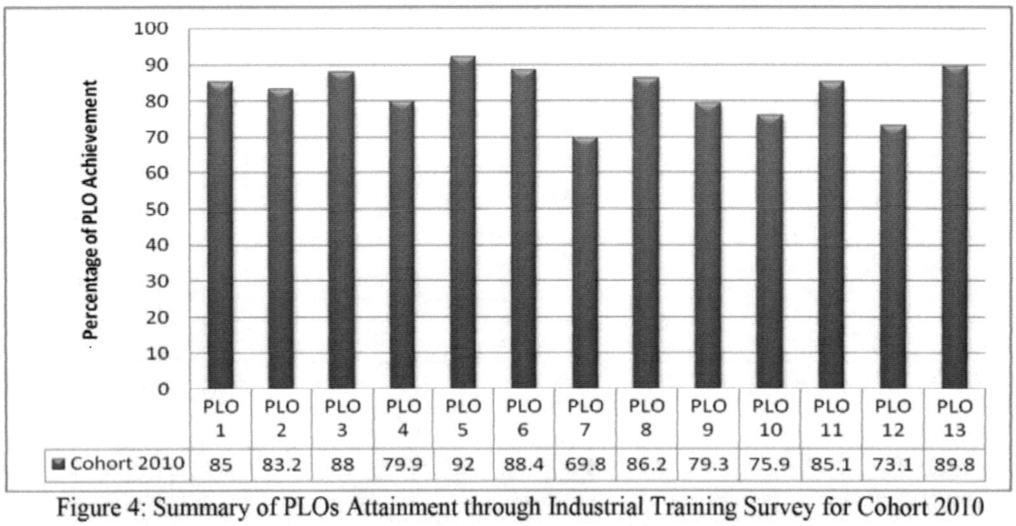

Figure 4: Summary of PLOs Attainment through Industrial Training Survey for Cohort 2010

The survey indicated that employers confirmed good attainment of FKMP students' in terms of different attributes, in relation to the different PLOs, as shown in the PLOs attainment of PLO 1 (Engineering Knowledge), PLO 2 (Modern Tool Usage), PLO 3 (Communication), PLO 5 (Individual & Team Work), PLO 6 (Lifelong Learning), PLO 8 (Ethics), PLO 11 (Problem Analysis), and PLO 13 (The Engineer and Society). It is thus indicating that more than 80% of the respondents agree that the students are noted as both good and excellent index for these PLOs.

(ii) Exit Survey

Error! Reference source not found.5 presents the alumni self-perception of competence at time of graduation, for graduates of cohort 2010. As shown in Figure 5, the PIs for PLO 3 (Communication), PLO 5 (Individual and Team Work), PLO 6 (Lifelong Learning), PLO 8 (Ethics), PLO 9 (Project

Management & Finance) and PLO 13 (The Engineer and Society) are achieved this cohort. Thus this implies that the alumni agree that the current PLOs are able to equip them with the skill needs at workplace.

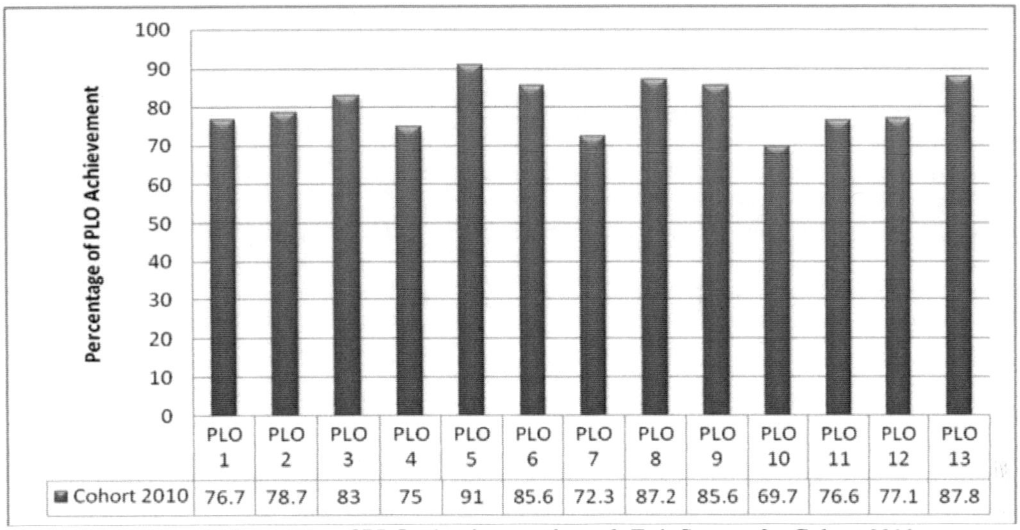

Figure 5: Summary of PLOs Attainment through Exit Survey for Cohort 2010

3.2. Analysis of PLOs Achievement by Direct Assessment

The results of PLOs achievement by direct measurement for cohort 2010 are shown in Figure 6. It is clear that the percentage of courses from this cohort at every PLO which attained minimum 80% score. It is clear that all cohorts have attained the targeted PI.

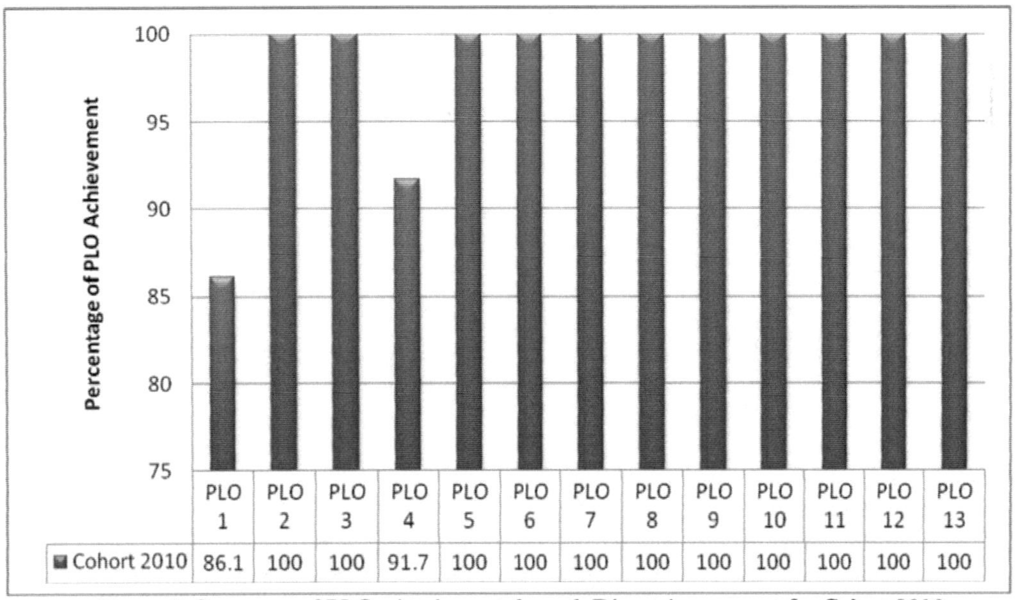

Figure 6: Summary of PLOs Attainment through Direct Assessment for Cohort 2010

3.3. Consolidation of PLOs Achievement from Indirect and Direct Assessment

Consolidation of PLOs achievement data is performed by integrating direct assessment data (represented by University) and indirect data (represented by Student Survey and Industry Survey). These data are plotted into sets of triangular diagrams based on the respective PLOs.

Figure 7 shows the triangulation diagrams for cohort 2010 of these aforementioned PLOs assessment methods. The red dot at the centre of each triangle is the reference point, which indicates balance agreement on the PLOs achievement between the university, industries and students. Meanwhile the blue dot indicates the result of consolidated data obtained from university assessment and, industry and student survey. If the blue dot coincides with the red dot, it shows that the PLOs achievement is satisfactory for the three parties. The consolidation analyses ratify that almost all PLOs are well attained by the students as both blue and red dots coincide well between them. For example, for PLO 5, 35.3%, 32.1% and 32.5% PLOs achievement from the perspective of university, student and industry are achieved, where the percentages are near to the ideal balance point, which is represented by 33.33%. Figure 7(g), (j) and (l) reveal that the perception of PLO7, PLO10 and PLO12 attainment by the industries and students is slightly low compared to the direct assessment data (university).

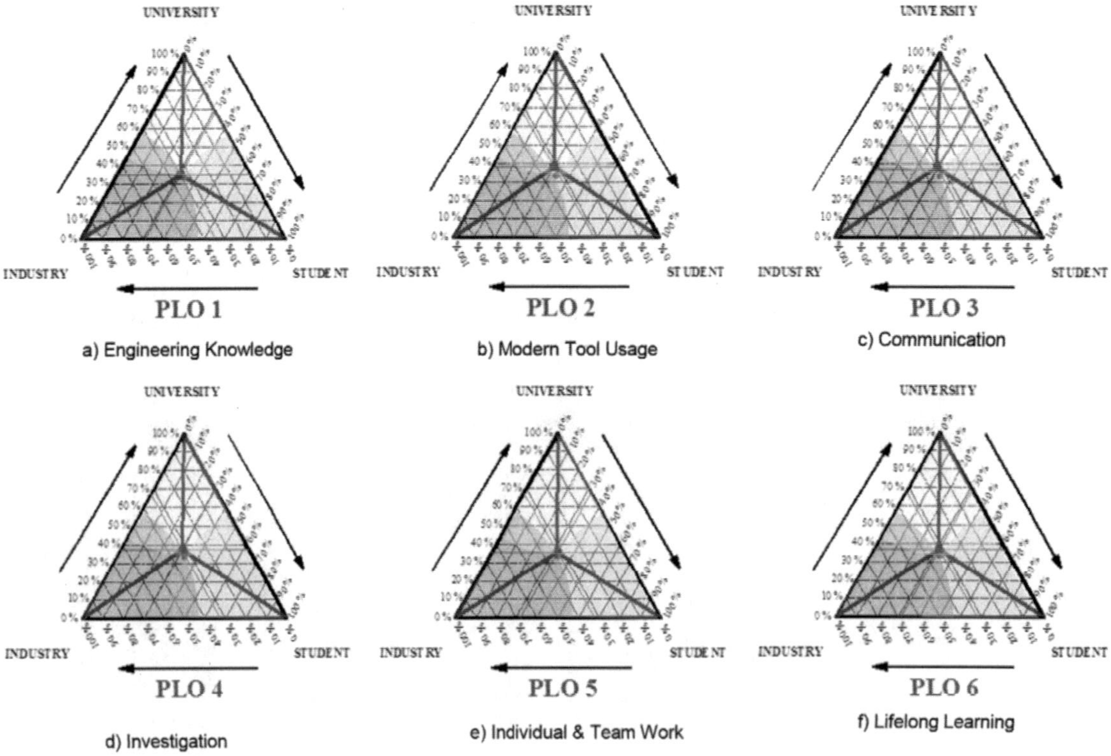

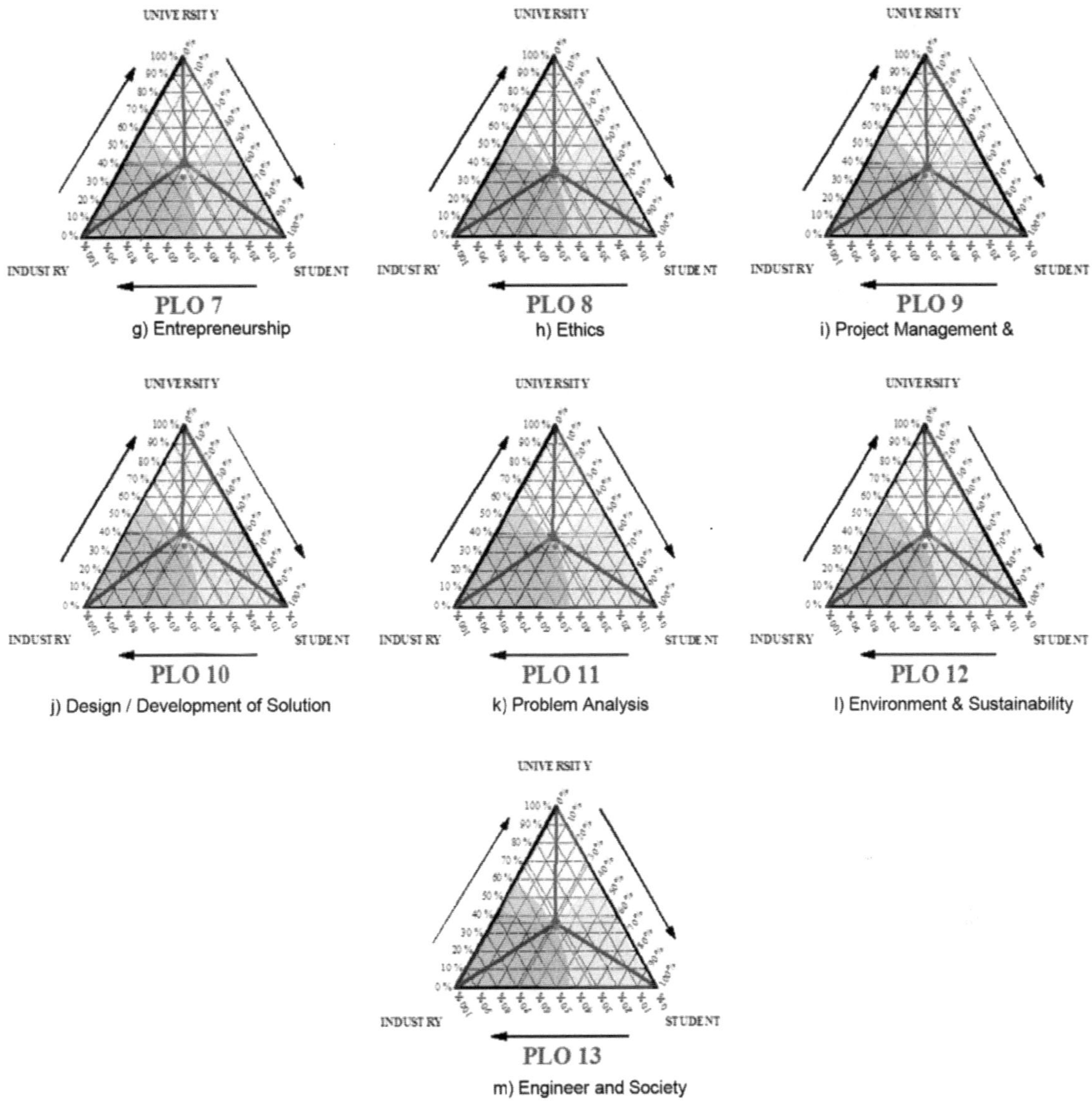

Figure 7: Triangulation Diagram for Cohort 2010

4. Conclusion

Through data consolidation, in whole, the analysis indicates that PLOs achievement of the BDD programme is agreeable between the three parties and indeed a rather balanced programme. However, at course level, the PLOs attainment of some courses can be further improved through some continual quality improvement (CQI) strategies. The Faculty monitors this concern through OBEsys and CEPAT. At the end of the semester, the respective lecturers have to identify the causes of underachievement and propose the potential action taken in OBEsys and CEPAT. The proposed CQI can be viewed by other lecturers in the coming semester, where necessary action can be taken.

References

[1] Engineering Programme Accreditation Manual 2012, Engineering Accreditation Council.
This reference has two entries but the second one is not numbered (it uses the 'Reference (no number)' style.
[2] Breslow, L. (2007). Methods of Measuring Learning Outcomes and Value Added. Teaching and Learning Laboratory, Massachussets Institute of Technology.
[3] Faculty of Mechanical and Manufacturing Engineering (2016). Universiti Tun Hussein Onn Malaysia, Course Profile Handbook V5.1.

IOP Conf. Series: Materials Science and Engineering **165** (2017) 012034 doi:10.1088/1757-899X/165/1/012034

Numerical Analysis on the Effect of Boom Sprayer Collecting Plate Angle to the Distribution of Granular Fertilizers

[a] **Eng Pei Ying** [b] **Zamani Ngali** [c] **Rosman Tukiman**

[a,b,c] University Tun Hussein Onn Malaysia. Faculty Mechanical and Manufacturing, 86400 Batu Pahat Johor , Malaysia
[a]gd140091@siswa.uthm.edu.my, [b]zamani@uthm.edu.my, [c]rosman@uthm.edu.my

Abstract. Optimization of boom sprayer collecting plate angle is a tedious procedure if it is done fully experimental. This paper demonstrates that the optimization process is more practical by simulation analysis validate through logical reflection of particles. This study is carrying out through simulating the distribution parts of the boom sprayer by using the commercial software, ANSYS. The multiphysics capabilities of ANSYS enable ANSYS to carrying out this simulation. The simulation is carrying out by manipulating the angle of the collecting plate, 32°, 60°,90° and 120° of the boom sprayer to find the optimum range of angle that will produce a good distribution for different sizes of the granular fertilizers and air velocity of the blower. The constant variables in this simulation are the atmospheric pressure of 1 atm and the particles size of Potassium K is 1mm. There are 60 per cent of the images produce by ANSYS, through observing the number of stream lines and the angle of distribution show that the optimum angle is between 32° to 60°. For further study, in order to increase the accuracy, the simulation is further validate through experiment. It is preferred to carry up the experiment through scaled down model without causing any changes to the current design and in order to be carrying out in the lab.

1. Introduction

1.1. Current practice in boom sprayer collecting plate angle optimization

Optimization in this study is defined as finding the best parameters to produce the best distribution in distributing the fertilizers [1]. Optimization can be done both in experimental and simulation. However, carrying out full experiment is not preferable. First, the actual size of the boom sprayer is too bulky for lab scales. It requires a lot of space. Second, manipulating the parameters might interfere the actual geometry of the boom sprayer. Thirdly, if the experimental is carrying outdoor, the results might affect by a lot of others factors such as wind factors and the condition of the road, which makes the study complicated [2]. Forth, it is very costly because the usage of materials such as petrol must be consider and the usage of fertilizer will be in large scales. Firth, It took a longer time to finish one experiment whereby simulation can generate a lot of results at the same time by using all the available computer in the lab.

In these papers, the main focus is how the readjust of the angle of collecting plate to help to improve the distribution [3]. Further study can be made by validate the results through scaled down experiment and by manipulating more parameters such as the distance of the each blow head, the length of the collecting plate and even the sizes and shape of the boom. It depends on your modelling. Either simulation or scaled down experimental can be done in the lab, making the work of design much more easier and comfortable compare to outdoor. The result of the simulation and conclusion produce in this paper help to contributing parts of ideas for researches to redesign boom sprayer for a better distribution under various condition.

Table 1.0 show the summary of previous study of researches which manipulating parameter of the boom sprayer. Some manipulated the air velocity of the blower and some the nozzles distance. So far manipulating the angle of the collecting plate still requires a lot of study. The function of the collecting plate is to direct the fertilizer from the boom to the field. Therefore, the distribution will change when the angle of the collecting plate is changed. This is a very tedious work which involved total of 40 experiments under two conditions. The first one is that the angle of the collecting plate from blow head 2 to 7 are the same. Another condition is the angle of the collecting plate of the blow head nearest to the inlet must not larger than the angle of the collecting plate further from the inlet. The reason is explained in Table 1.1. Conducting 40 experiments require a high cost and time consume. This can all be minimize by using simulation.

Table 1.0 : The summary of the previous study [4], [5], [6]

Author/year	Title	Findings
Dieter Foque, Pascal Braekman, Jan G. Pieters, David Nuyttens / 2012	A vertical spray boom application technique for conical bay laurel (Laurus nobilis) plants.	-The most efficient spray application technique was found to be an extended range flat fan, an air included flat fan or a hollow cone nozzle all at 4900 L ha^{-1} resulting in the highest relative spray deposits and the most uniform spray distribution in the canopy.
Dieter Foque, Jan G. Pieters, David Nuyttens / 2012	Spray deposition and distribution in a bay laurel crop as affected by nozzle type, air assistance and spray direction when using vertical spray booms	-Nozzle choice, spray boom setting, spray distance and the air speed of air assisted sprayers require careful consideration. -application made with an extended range standard flat fan nozzle without air support, directed straight toward the crop and used with a fixed spray distance of 30 cm to the stem generally produce the best spray results in the considered, conically pruned bay laurel crop.
D.Nuyttens, S.Windey, B. Sonck/2004	Optimisation of a Vertical Spray Boom	-Tests were conducted in greenhouses with two different spray systems both

	for Greenhouse Spray Application	equipped with vertical spray boom. - spray with 0.35 m nozzle spacing provided a much better spray distribution than that with 0.50 m nozzle spacing. -The optimal spray distance for 80° flat fan nozzles with the 0.35 m nozzle spacing was about 0.30 m.

1.2. Proposed procedure in boom sprayer collecting plate angle optimization

By using orthogonal array in arranging the parameters, there are total of 40 experiments. As mentioned earlier, the angle of all the blow head from 2 to 7 is the same and that the angle of the collecting plate of the blow head nearer to the inlet must not more than those that further away from the inlet.

Table 1.1 shows an explanation for the second consideration. The black circles of the four images had a fix position which is at the end of the left hand side of the dotted line in each of the images. It represent the area of particles flow before touching the collecting plate. The bigger the circle, the larger the possibility of the particles exit the blow head. Conclusion is the larger the angle of collecting plate, bigger the distribution area. The blow head nearer to the inlet require a smaller area of distribution because most of the particles will drop at the blow head which is nearer to the inlet.

The commercial simulation software used is ANSYS. The ANSYS software is use as the analysis software computing fluid dynamics to simulate the flow in the boom pipe and collecting plate [7]. ANSYS license provided by the University of Tun Hussein Onn, CFD lab able to mesh up to 510000 and above number of nodes. CFX is one of the simulation tool used in ANSYS to provides facility to simulate particles flow in fluid through computer modeling. The result is produce by the software through the displays of images and data to predict the performance of the design.

Table 1.1: The area of distribution versus the angle of collecting plate

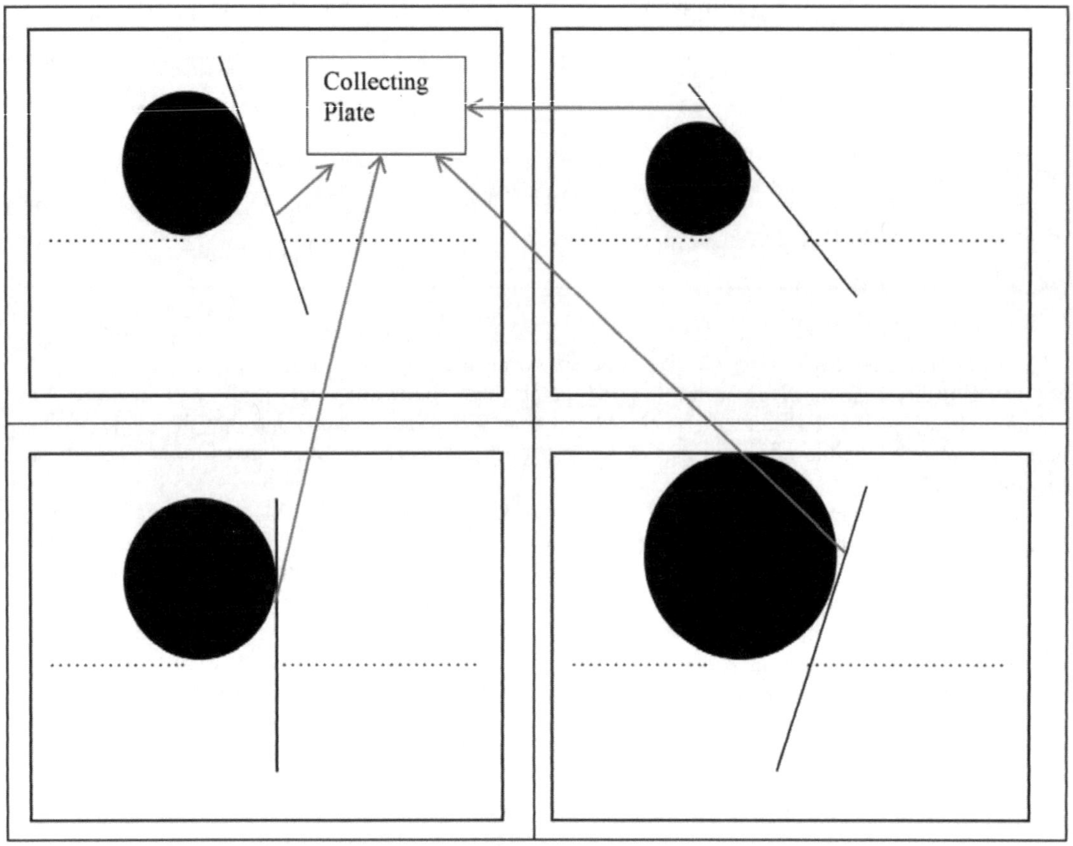

2. Methodology via Engineering software manipulations

2.1. Boom sprayer and the ambient modelling and the preliminary setup

Figure 2.0 show the overall process of the methodology. ANSYS software procedure is necessary to make the mesh before carrying out an analysis. Meshing can help to get the results on the surface or in geometry. Mesh size is very important because it plays an important role for obtaining a more accurate and effective. Detail of solid works that have been stored in IGES format (*.igs) will be imported into ANSYS software [8]. Figure 2.1 show the model of the boom sprayer by using solid work.

After meshing, the domain is established for the geometry. The domain of air and fertilizer. parts of the input and output [9]. Fluid type and its properties have been added in the domain. Parameter set to air as a continuous fluid while the boom has been designated as a solid spread as fertilizer to be discharged from the hopper. Because this study is a multi-phase fluid, homogeneous model should be selected. Figure 2.2 show the pre-processor. This is how it display after setting up all boundaries.

After all properties completed inserted domain, boundary conditions are added to the domain. Four different boundary conditions created in that part of the air intake, part of the product and the wall. At the exit pressure is only relatively fixed. In addition, the characteristics of air and fertilizer in certain areas [10]. Table 2.0 show the modeling properties of the boom sprayer and its boundaries

Turbulence model is used to increase the accuracy of the result by combining with Navier–Stokes equations. In this simulation, model K-Epsilon is been used. K-Epsilon is the most common turbulence model and it involved on two equations. The first transported variable is turbulent kinetic energy. The second transported variable in this case is the turbulent dissipation. Good result is provided that the pressure gradient is small [11]. In this case study, it has a low pressure gradient because there is only one pump that provide constant flow rate. The involvement equation is lesser, therefore the running time of the simulation is been reduce.

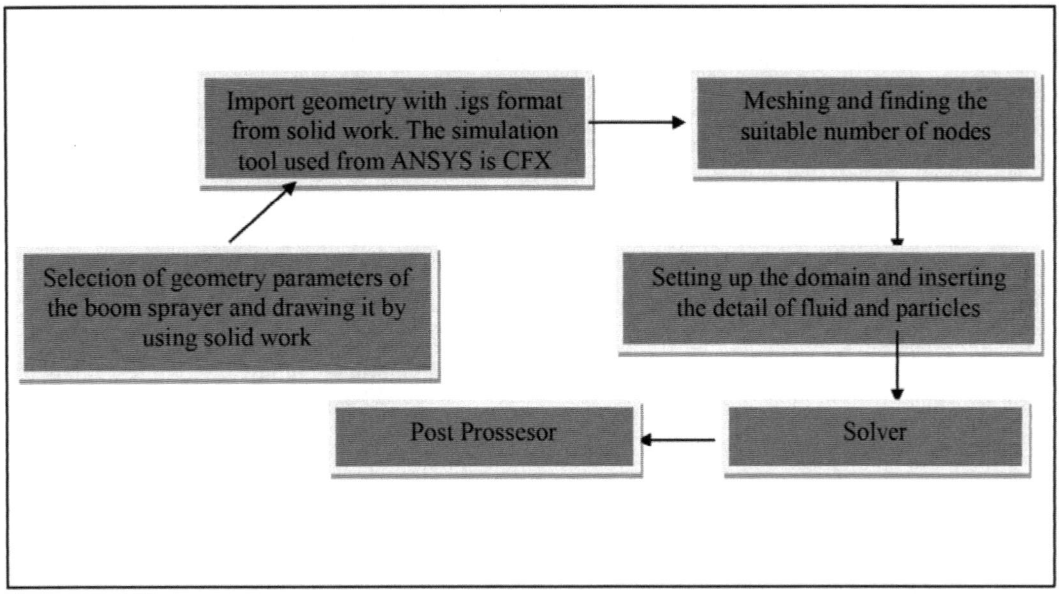

Figure 2.0: the overall process of the methodology [12]

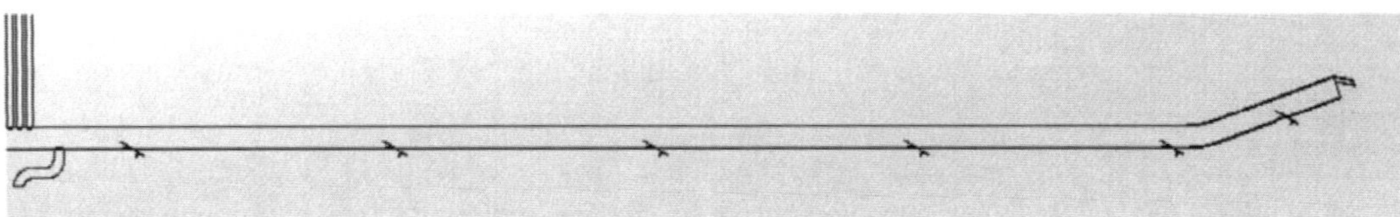

Figure 2.1: The model of the boom sprayer

Table 2.0: Modeling properties of the boom sprayer and its boundaries.

Fluid and particle definition	Air at 25°C	Continuous fluid
	Particle	Particle transport solid
Domain model	Reference pressure	1 atm
Fluid specific models	Air	Non buoyant
	Particle	Density difference
Urea	Specific diameter	2 mm, 4mm
	Gravity X and Z Dirn.	0 m/s^2
Buoyancy model	Gravity Y Dirn.	- 9.81 m/s^2
	Bouy. Ref. Density	1.226 kg/m^3
Boundary condition		
Air inlet	Flow regime	Subsonic
	Normal speed	8 ms^{-1}, 24 ms^{-1}
	Flow regime	Subsonic
Particle inlet	Normal speed	0.2 ms^{-1}
	Fluid value	0.2 ms^{-1}
	Number of position	10000
	Flow regime	Subsonic
Outlet (opening)	Mass & momentum	Normal speed
	Pressure	0 Pa

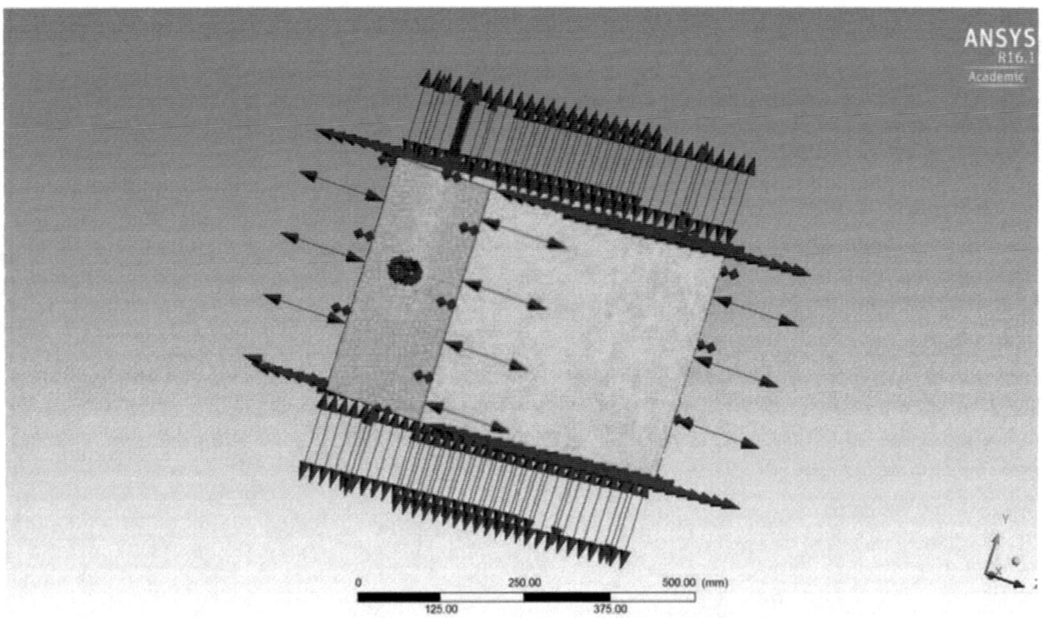

Figure 2.2: Pre-Processor

2.2 Solver, Post Processing procedure and Parameter selections

CFX solver will pre solutions and calculate the parameters set out above. Accuracy and time taken by CFX solver to calculate the mixture was largely dependent on the geometry of the mesh. Element and node value higher will result in a more accurate and also will take a longer time to complete [13].

The results of flow simulations obtained from CFX post. To see the flow behaviour, flow streamlines for air flow and the flow of steel. Airflow and steel will be incorporated into the flow streamlines different. This is because the flow streamlines have a different starting point. Although both fluid flows coupled with flow line is different but the behaviour of the mixture can be seen after the two current lines intersect each other. Contour plus in his decision to see the effect of these two current lines and contours. Distribution of velocity and pressure can be seen in the current line and contour.

The results of flow simulations obtained The results of flow simulations obtained from CFX post manipulating the parameters that will affect the distribution in order to find the optimum parameters to produce a good distribution. The parameters are the sizes of the granular fertilizers, air velocity of the blower and the angle of the collecting plate. Reference pressure is at atmospheric pressure of 1 atm and the particles size of Potassium K is 1mm. Table 2.1 show the parameters for the overall simulation.

The manipulation variables is angle of the collecting plate, the air velocity of the blower, granular fertilizers size for each Nitrogen, N and granular fertilizers size for Phosphorus, P. 2mm and 4mm are both the maximum and the minimum sizes of Nitrogen and Phosphorus. This is to study the effect of the granular size toward the flow. Same goes to air velocity of the blower whereby the minimum and maximum values had been selected. Each angle of the collecting plate is 30° different from one and another. This is because we want to find the range of the optimum angle that will affect the distribution. 32° is chosen instead of 30° because 32° is the minimum of the bending of the collecting plate which touches the other parts of the blow head as show in Figure 2.3. The blue line represent the collecting plate when it is at 30°.

Table 2.1: Parameters for overall simulation.

Level Factor	1	2	3	4
Granular fertilizers size for each Nitrogen, N (mm)	2	4	-	-
Granular fertilizer size for each Phosphorus, P (mm)	2	4	-	-
Air velocity of the blower	8m/s	24m/s	-	-
Angel of the collecting plate 1 (°)	32	60	90	120
Angel of the collecting plate 2 (°)	32	60	90	120
Angel of the collecting plate 3 (°)	32	60	90	120
Angel of the collecting plate 4 (°)	32	60	90	120
Angel of the collecting plate 5 (°)	32	60	90	120
Angel of the collecting plate 6 (°)	32	60	90	120

2.5.1 Orthogonal Array

Based on the parameters that have been made, the Taguchi method was chosen to simplify and expedite the experimental method to be executed. From Table 2.3, there are 2-4 levels and 9 factors controlled during the experiment. The Table 2.4 shows the collector array by each blow head with the combination of 4 different angle of collecting plate that has been introduced by Taguchi without the combination of the sizes of granular fertilizers and air velocity of the blower. A represents 32°, B represents 60°, C represents 90° and D represents 120°. This paper only cover having the angle of the collecting plate for each blow head the same. In order to see the effect of the each angle toward it distribution. The total number of experiment is 40.

The Table 2.3 : The number of experience taken by inserting the 3 types of fertilizers.

Experiment	Granular sizes of each Nitrogen (mm)	Granular sizes of each Phosphorus (mm)	Granular sizes of each Potassium (mm)	Air velocity (m/s)	Blow head 1	Blow head 2	Blow head 3	Blow head 4	Blow head 5	Blow head 6
1	2	-	-	8	A	A	A	A	A	A
2	2	-	-	24	A	A	A	A	A	A
3	2	-	-	8	B	B	B	B	B	B
4	2	-	-	24	B	B	B	B	B	B
5	2	-	-	8	C	C	C	C	C	C
6	2	-	-	24	C	C	C	C	C	C
7	2	-	-	8	D	D	D	D	D	D
8	2	-	-	24	D	D	D	D	D	D
9	4	-	-	8	A	A	A	A	A	A
10	4	-	-	24	A	A	A	A	A	A
11	4	-	-	8	B	B	B	B	B	B
12	4	-	-	24	B	B	B	B	B	B
13	4	-	-	8	C	C	C	C	C	C
14	4	-	-	24	C	C	C	C	C	C
15	4	-	-	8	D	D	D	D	D	D
16	4	-	-	24	D	D	B	D	D	D
17	-	2	-	8	A	A	A	A	A	A
18	-	2	-	24	A	A	A	A	A	A
19	-	2	-	8	B	B	B	B	B	B
20	-	2	-	24	B	B	B	B	B	B
21	-	2	-	8	C	C	C	C	C	C
22	-	2	-	24	C	C	C	C	C	C
23	-	2	-	8	D	D	D	D	D	D
24	-	2	-	24	D	D	D	D	D	D
25	-	4	-	8	A	A	A	A	A	A
26	-	4	-	24	A	A	A	A	A	A
27	-	4	-	8	B	B	B	B	B	B
28	-	4	-	24	B	B	B	B	B	B
29	-	4	-	8	C	C	C	C	C	C
30	-	4	-	24	C	C	C	C	C	C
31	-	4	-	8	D	D	D	D	D	D
32	-	4	-	24	D	D	D	D	D	D
33	-	-	1	8	A	A	A	A	A	A
34	-	-	1	24	A	A	A	A	A	A
35	-	-	1	8	B	B	B	B	B	B
36	-	-	1	24	B	B	B	B	B	B
37	-	-	1	8	C	C	C	C	C	C
38	-	-	1	24	C	C	C	C	C	C
39	-	-	1	8	D	D	D	D	D	D
40	-	-	1	24	D	D	D	D	D	D

Figure 2.3: The angle of collecting plate of the actual blow head

3. Result and Discussion of the simulation

Table 3.1 show the good samples results of the various angle of collecting plates, particles sizes and air blower velocity which refer to Table 2.3. Whereby, Table 3.2 show the bad samples results of the various angle of collecting plates, particles sizes and air blower velocity which refer to Table 2.3. The nitrogen, N fertilizer is green color, the phosphorus, P is grey color and the potassium, K is orange color. These samples are the best result among the 40 experiments that is taken. Through analysis of the images, there is a big different between both Table 3.1 and Table 3.2. Table 3.2 shows the different between the result produce by Table 3.0 and Table 3.1.

Table 3.0: The good samples result of various angle of collecting plates, particles sizes and air blower velocity which refer to Table 2.3

Experiment	Good result
1	
18	
34	

IOP Conf. Series: Materials Science and Engineering **165** (2017) 012034 doi:10.1088/1757-899X/165/1/012034

Table 3.1 : The bad samples result of various angle of collecting plates, particles sizes and air blower velocity which refer to Table 2.3

Experiment	Bad result
15	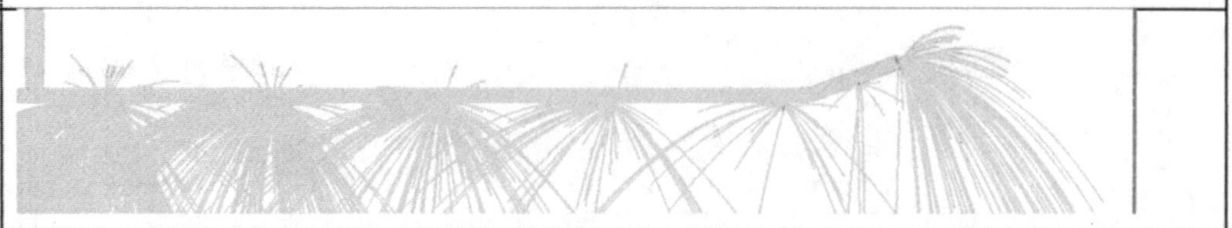
23	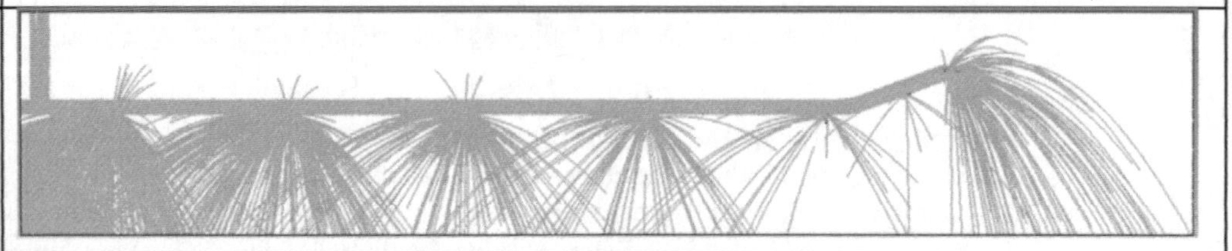
37	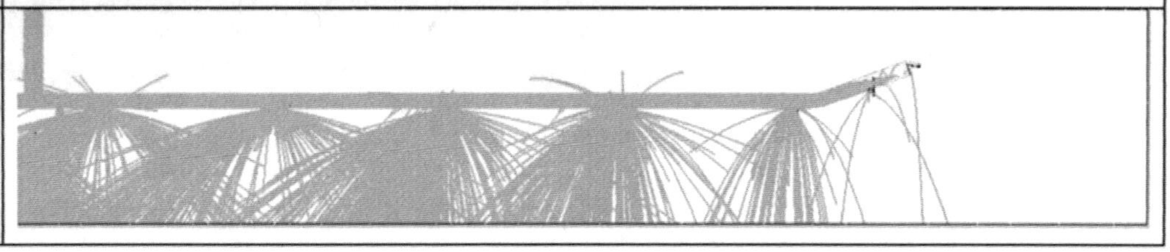

Table 3.2: The different between the result produce by Table 3.0 and Table 3.1

No	Table 3.1	Table 3.2
1.	The streamline of every blow head is nearly the same	The streamline at the first half of the boom and the second half of the boom is very much different
2.	The streamline of every blow head is a lot	The streamline at the last blow head is very little
3.	The angle of distribution is very even	The angle of distribution is not even

Table 3.3 show the logical explanation of having the range angle of collecting plate of 32° to 60° that will produce a better distribution. Orange color line represents the direction of flow of fertilizers from the inlet to it touches the collecting plate. Red color lines represent the possibilities of the direction of the fertilizer after it bound from the collecting plate to the nozzle. By comparing the 4 images, the angle which is less than 90° has a better shape of distribution. The image (a) and (b) show that the red lines producing a shape of a triangle with the left and right spread in much event compare to image (c) and (d). Therefore, it is logical to say that the angle of range 32° to 60° will produce a better distribution compare to angle 90° and above.

Table 3.2: The different between the result produce by Table 3.0 and Table 3.1

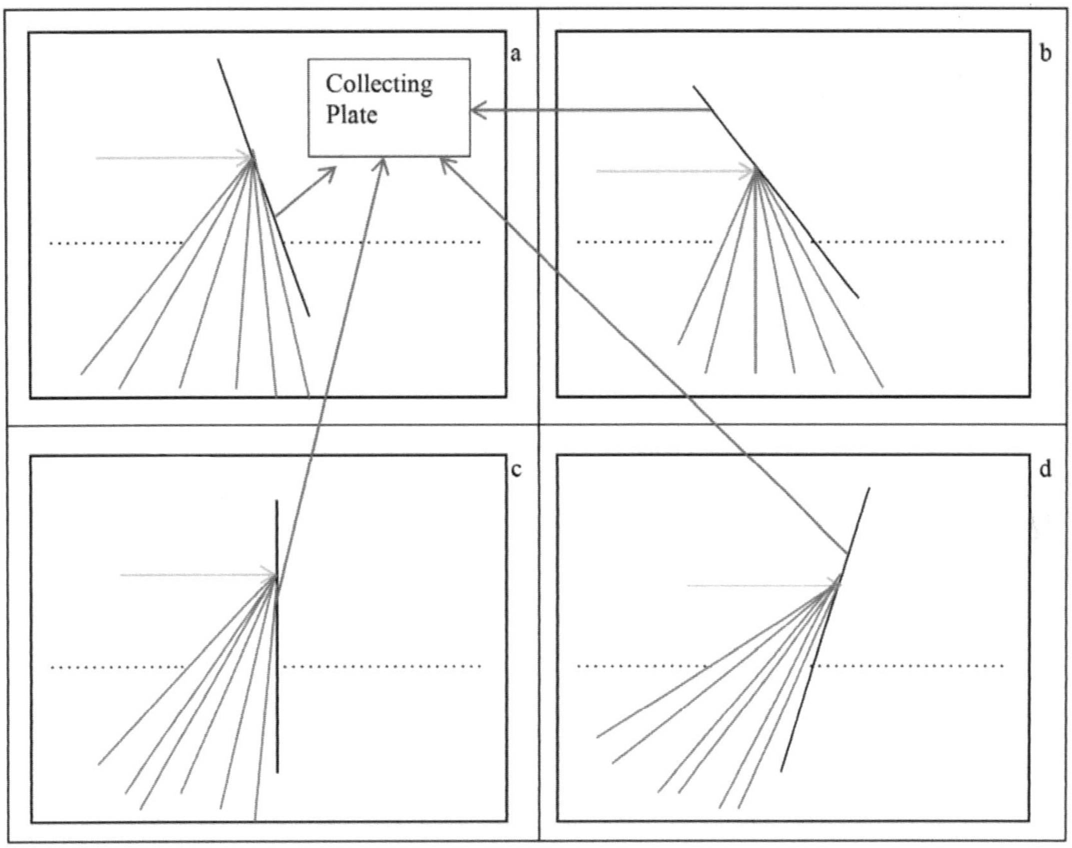

4. Conclusion

4.1. With the invention of commercial simulation software, optimization of boom sprayer collecting plate angle is much easier compare to fully be done by experimental. Simulation is more practical for this paper because the number of experiment that needed to be carried out is a lot, 40 experiments. Through using the simulation, one time can generate a lot of results by utilized all the computer that is available in the lab.

4.2. The commercial software as mentioned is actually ANSYS. ANSYS has multiphysics capabilities and it is user friendly. The manipulate variables for the simulation are the angle of the collecting plate, 32°, 60°, 90° and 120° of the boom sprayer, different sizes of the granular fertilizers and air velocity of the blower. Whereby, the constant variables in this simulation are the atmospheric pressure of 1 atm and the particles size of Potassium K is 1mm. 60 per cent of the images produce by ANSYS, show that the optimum angle is between 32° to 60°. The way of analyse is through observing the number of stream lines and the angle of distribution. By using simple reflection of particles as explained in Table 3.2 prove that the optimum range of angle is logical.

Further study can be done by validating the result through experiment. It is preferred to carry up the experiment through scaled down model without causing any changes to the current design and in order to be carrying out in the lab.

References

[1] Fletcher, R. (1987). Practical Methods of Optimization. JOHN WILEY SONS INC ONE WILEY DR (Vol. 2nd). https://doi.org/10.2307/2008742

[2] Abubakar, M. S. A., Ahmad, D., Jamarei, O., Samsuddin, S., & Norhisam, M. (2011). Evaluation of a dual-purpose chemical applicator for paddy fields. American Journal of Applied Sciences, 8(May), 362–367. doi:10.1109/ICOM.2011.5937188

[3] At, I. A. (2007). Operation Manual.

[4] Foqué, D., Braekman, P., Pieters, J. G., & Nuyttens, D. (2012). A vertical spray boom application technique for conical bay laurel (Laurus nobilis) plants. *Crop Protection*, *41*, 113–121. https://doi.org/10.1016/j.cropro.2012.05.011

[5] Foqué, D., Pieters, J. G., & Nuyttens, D. (2012). Spray deposition and distribution in a bay laurel crop as affected by nozzle type, air assistance and spray direction when using vertical spray booms. *Crop Protection*, *41*, 77–87. https://doi.org/10.1016/j.cropro.2012.05.020

[6] Nuyttens, D., Windey, S., & Sonck, B. (2004). Optimisation of a vertical spray boom for greenhouse spray applications. *Biosystems Engineering*, *89*, 417–423. https://doi.org/10.1016/j.biosystemseng.2004.08.016

[7] Bibuli, M., Caccia, M., & Lapierre, L. (2007). Path-following algorithms and experiments for an autonomous surface vehicle. IFAC Proceedings Volumes (IFAC-PapersOnline), 7, 81–86. doi:10.1002/rob

[8] Butler Ellis, M. C., Lane, a. G., O'Sullivan, C. M., Miller, P. C. H., & Glass, C. R. (2010). Bystander exposure to pesticide spray drift: New data for model development and validation. Biosystems Engineering, 107(3), 162–168. doi:10.1016/j.biosystemseng.2010.05.017

[9] Corradini, E., de Moura, M. R., & Mattoso, L. H. C. (2010). A preliminary study of the incorparation of NPK fertilizer into chitosan nanoparticles. Express Polymer Letters, 4(8), 509–515. doi:10.3144/expresspolymlett.2010.64

[10] Duarah, I., Deka, M., Saikia, N., & Deka Boruah, H. P. (2011). Phosphate solubilizers enhance NPK fertilizer use efficiency in rice and legume cultivation. 3 Biotech. doi:10.1007/s13205-011-0028-2

[11] Periasamy, M., Sivakumar, D. B., & Kumar, T. S. (2012). Dampening Flow Induced Vibration Due To Branching Of Duct at Elbow, *2*(4), 1001–1004.

[12] Sun, J., & Miao, Y. (2011a).Modeling and simulation of the agricultural sprayer boom leveling system, 613–618. doi:10.1109/ICMTMA.2011.439

[13] Sun, J., & Miao, Y. (2011b). Simulation and controller design for an agricultural sprayer boom leveling system. Proceedings - 3rd International Conference on Measuring Technology and Mechatronics Automation, ICMTMA 2011, 3, 245–248. doi:10.1109/ICMTMA.2011.632

Acknowledgement

I acknowledge Universities Tun Hussein Onn Malaysia for providing me facilities for me to write this paper. This research was supported financially by the ministry of higher education Malaysia (FRGS/2/2013/STWN03/UTHM/03/1) also known as Grant FRGS 1420.

Optimization of the EMI shielding effectiveness of fine and ultrafine POFA powder mix with OPC powder using Flower Pollination Algorithm

O L C Narong[1], C K Sia [1], S K Yee[3], P Ong [2], A Zainudin[1], N H M Nor[1] and N A Kasim[3]

[1.] Department of Engineering Design and Materials, University of Tun Hussein Onn Malaysia, Parit Raja 86400, Batu Pahat, Johor, Malaysia

[2.] Department of Mechanic Engineering, University of Tun Hussein Onn Malaysia, Parit Raja 86400, Batu Pahat, Johor, Malaysia

[3.] Research Centre for Applied Electromagnetic, Faculty of Electrical and Electronic Engineering, University Tun Hussein Onn Malaysia, Parit Raja 86400, Batu Pahat, Johor, Malaysia

E-mail: gd150077@siswa.uthm.edu.my

Abstract. In order to solve the electromagnetic interference (EMI) issue and provide a new application for palm oil fuel ash (POFA), POFA was used as the cement filler for enhancing the EMI absorption of cement-based composites. POFA was refined by using water precipitation for 24 hours to remove the filthiness and distinguish the layer 1 (floated) and layer 2 (sink) of POFA. Both layers POFA were dried for 24 hours at 100 ± 5 °C and grind separately for sieve at 140 μm (Fine) and 45 μm sizes (Ultrafine). The microstructure and element content of the both layers POFA were characterized by scanning electron microscope (SEM) and energy-dispersive X-ray spectroscopy (EDS) respectively. The results showed layer 1 POFA has potentialities for EMI shielding effectiveness (SE) due to its higher carbon content and porous structure. The study reveals that EMI SE also influenced by the particle size of POFA, where smaller particle size can increase 5 % to 13 % of EMI SE. When the specimen consists of 50% POFA with passing through 45 μm sieve, the EMI was shield -13.08 dB in between 50 MHz to 2 GHz range. Flower Pollination Algorithm (FPA) proves that POFA passing 45 μm sieve with 50% mixed to OPC is optimal parameter. The error between experimental and FPA simulation data is below 1.2 for both layers POFA.

1. Introduction

Most of the advance technologies in electrical and electronic devices, especially the wireless and communication systems are polluted by electromagnetic wave (EW) pollution, which can be harmful to health and may cause of death [1], [2]. Electromagnetic interference (EMI) is the disruption of operation of an electronic device when it is in the vicinity of an electromagnetic field (EMF) in the radio frequency (RF) spectrum that is caused by another electronic device. Electromagnetic pollution has become part of people daily life due to the proliferation of the high speed electrical and electronic systems such as a mobile phone based station, mobile phone, radar, television and radio transmitters and so on that ease our daily life. The studies on electromagnetic fields (EMF) have been motivated primarily by public for health considerations [3]. A high frequency range of electromagnetic wave able to increase human body temperature approximately 1–5 °C. Researchers have reported the severity of the physiological effects produced by the small temperature surge. Some of the effects include, organ malformations, temporary infertility in males, brain lesions, and blood effect [4], [5]. The human brain is sensitive to EMI field and it promotes the brain tumor risk [1], [6]–[8]. Besides that, the device such as cordless telephones, home entertainment systems, computers, and certain medical devices can fail to operate properly in the presence of strong RF fields [8], [9]. Hence, it is important to reduce the EMI radiation.

The current study stated that EMI shielding in building have been shielded by carbon filling cement materials and metal filling cement composite [10]. The carbon filling materials consist of graphite, carbon black, coke and carbon fiber. These carbon materials have comparatively high conductivity and EMI shielding [11]. The metal based fillers consist of metal powder, metal fiber and metal alloys, which are used in cement matrix composites [10]. The metals such as silver, copper, and nickel was used for long time as conductive components in polymer due to their high conductivity [10].

In recent years, Palm oil fuel ash (POFA) is attractive alternative filler commonly found as agricultural waste that is uncontrollably dumped in landfills, especially in South East Asia [12]. As the second largest global palm oil producer, Malaysia has the highest daily wastage [12]. The issue of environmental pollution resulting from disposal of palm oil fuel ash (POFA) has attracted the attention of researchers to explore its potential as an alternative fuel [13] or other field. Some researcher also used the POFA as cement additive or cement replacement as the POFA have similar characteristic with Ordinary Portland Cement (OPC) [12], [14]–[26]. Some researchers have reported that the elements in POFA contribute the most to the microwave absorption is carbon (C) and iron III oxide (Fe_2O_3) [3], [10], [27]–[32].

There are also many other high conductive fillers like silver, copper, and nickel can be use. However, these metal fillers are not only high in cost but very complicated in processing so they are not commonly used in cement matrix composite [10], therefore the POFA was choose to use as filler to mitigate against the EMI. The motivation of using POFA in EMI shielding is to reduce the agricultural waste that is uncontrollably dumped in landfills. Besides that, POFA is containing high carbon and iron III oxide which can be used to prevent the EMF radiation.

The objective of this study is to measure the EMI SE of layer 1 and layer 2 POFA by mixed with OPC at 5%, 20%, 35%, and 50% mixing percentage. Furthermore, this study also to compare the POFA with passing through 140 μm sieve and 45 μm sieve in EMI SE performance after mixed OPC for both layers. The frequency range of the EMI SE measurement is 50 MHz to 2 GHz. At the end of the experiment, the data are optimized by Flower Pollination Algorithm (FPA) method to get the optimum mixing percentage. Lastly, the experimental data were compared to the simulation data for calculating the error.

2. Experimental

2.1 Materials
The main raw material that used in this research is Palm Oil Fuel Ash (POFA) collected from Genting Plantations Oil Mill Ayer Hitam branch in Johor, Malaysia. The palm fibre, palm shell, empty palm fruit bunches burn as fuel at 800 °C ± 50 °C in the boiler at the palm mill. Ordinary Portland Cement (OPC) type I powder is collected from Civil Engineering Advance Material Laboratory, Faculty of Civil and Environmental Engineering (FKAAS), UTHM. The OPC was used for all sample mixtures.

2.2 POFA Refining Process
One kilogram of POFA is filled into a graduated cylinder. The water from the laboratory is then added into the same cylinder until 80% full. The mixture of POFA and water is stirred for 30 minutes by

using a portable stirrer. After the stirring process, the first-water has been used was removed due to the filthiness in the POFA. Again, the water from the laboratory is then added into the same cylinder until 80% full and left the cylinder for 24-hours. After 24-hour of precipitation, two main layers of POFA appeared, namely layer 1 and layer 2. The layer 1 POFA was withdrawn using a spatula as it is floated. Then layers 2 was removed by using a syringe withdrawal method as given in Figure 1 [33].

After separating the layers of POFA, POFA was placed in different containers. Then, the POFA layer 1 and layer 2 was drying in oven for 24 hours at temperature of 100°C. Due to the high humidity of POFA, the drying time must be 24 hours and above. This is to make sure the POFA is completely dry and to avoid the stickiness during grinding process.

Figure 1: Schematic diagram of syringe withdrawal method, (1) Layer 1 and (2) Layer 2

2.3 Physical properties and chemical compositions of POFA

All layers were examined by Energy Dispersive X-ray Spectroscopy (EDS) to identify the element's content of three random point in the POFA. Then the microstructure characteristic of each layer was explored using Scanning Electron Microscope (SEM).

2.4 Sample preparation

After drying process, both layers were grind using planetary micro mill Pulverisette 7 Classic Line machine. The grinding speed is 250 RPM with 10 grinding balls. After grinding process, the POFA was sieved by using Vibratory Sieve Shake Analysette 3 Pro. The sieve sizes are 140 μm and 45 μm. Both layers were sieved separately with the selected sieve size. The POFA passing through 140 μm sieve was declared as Fine POFA (FP) and POFA passing through 45 μm sieve was declared as Ultrafine POFA (UP). After sieving process, the OPC is partially replaced by POFA at the ratio of 5%, 20%, 35% and 50%. The detail of all samples is given in Table 1 to Table 4. Detail in Table 1 and Table 2 are for POFA layer 1 and detail in Table 3 and Table 4 for POFA layer 2. The 100% OPC was represented as CP where the CP is used as reference for other samples. CPF is representing OPC mixed with Fine POFA and CPU is representing OPC mixed with Ultrafine POFA.

Table 1: The mixing ratio of CPF and OPC for layer 1

Name of the samples	% Weight of cement (OPC)	%Weight of layer 1 POFA based on particle size (FP)
CP 100%	100	0
1-CPF 5%	95	5
1-CPF 20%	80	20
1-CPF 35%	65	35
1-CPF 50%	50	50

Table 2: The mixing ratio of CPU and OPC for layer 1

Name of the samples	% Weight of cement (OPC)	%Weight of layer 1 POFA based on particle size (UP)
CP 100%	100	0
1-CPU 5%	95	5
1-CPU 20%	80	20
1-CPU 35%	65	35
1-CPU 50%	50	50

Table 3: The mixing ratio of CPF and OPC for layer 2

Name of the samples	% Weight of cement (OPC)	%Weight of layer 2 POFA based on particle size (FP)
CP 100%	100	0
2-CPF 5%	95	5
2-CPF 20%	80	20
2-CPF 35%	65	35
2-CPF 50%	50	50

Table 4: The mixing ratio of CPU and OPC for layer 2

Name of the samples	% Weight of cement (OPC)	%Weight of layer 1 POFA based on particle size (UP)
CP 100%	100	0
2-CPU 5%	95	5
2-CPU 20%	80	20
2-CPU 35%	65	35
2-CPU 50%	50	50

2.5 Transmission Measurement

Transmission measurement (S_{21}) is the experimental measurement method of the SE. SE can separate into the product of three terms each represents one of the phenomena of reflection loss (R_{dB}), absorption loss (A_{dB}) and multiple reflection loss (M_{dB}) [10]. The SE of a shield is defined in decibels (dB) and its magnitude can be writes as $SE_{dB} = R_{dB} + A_{dB} + M_{dB}$ [10]. The signal is in the radio frequency (RF) form. The RF was supplied from port 1 to port 2 where the sample is placed in between the port 1 and port 2 transmission lines. Hundreds of data were obtained by Network Analyser (NA) and calculated in MATLAB by using equation (1). APC-7 connectors are used due to the samples are in powder form. This measurement carried out by using APC 7 connectors as sample holder. During this measurement, both APC 7 connectors are connected to a NA. NA is a device used to measure the ratio of transmission and reflection of electromagnetic waves through a device under test. In this work the device under test is the APC 7 connectors. The APC 7 connectors are shown in Figure 2. Figure 3 shows how the powder-formed-samples are placed in the APC 7 connector.

Figure 2: APC 7 connector tail (1) and head (2)

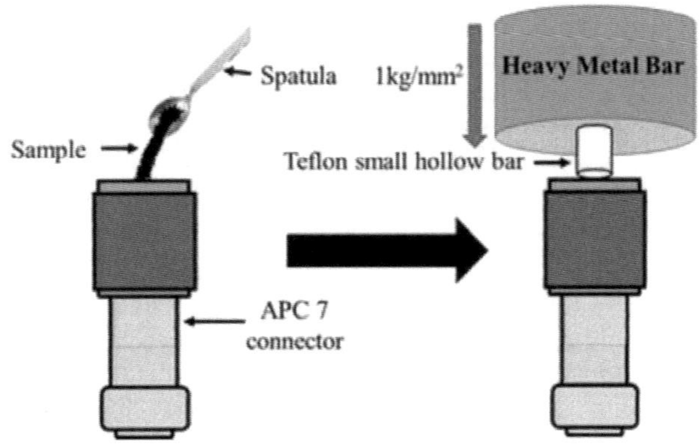

Figure 3: The method of insert sample into the APC 7 connector

The inner diameter of the APC 7 connector 3 mm and outer diameter is 7 mm. The sample is toroidal form when it is inside the APC 7 connector. The sample in the APC 7 connectors is compressed by using a heavy metal bar, where the pressing load approximately 1kg/mm^2. This is to ensure the sample is completely occupied in the APC-7 connector space. After the measurement is completed, the APC-7 connector was clean by using high pressure air. The blow process was needed to remove the remaining sample in the APC 7 connector before the next measurement. Figure 4 shown the schematic diagram of dielectric measurement method. In this measurement the SE is obtained based on equation (1). S_{21} represents the power reserved at antenna 2 relative to the power input to antenna 1. Two types of S_{21} here, which are $S_{21\text{-}without}$ and $S_{21\text{-}with}$. $S_{21\text{-}without}$ representing the S_{21} when the APC 7 connector does not contain any sample, while $S_{21\text{-}with}$ represent the S_{21} when the APC 7 connector is filled with the sample which is in the powder form.

$$SE(dB) = \text{Forward gain} = 20\log\left[\frac{S21-with}{S21-without}\right] \tag{1}$$

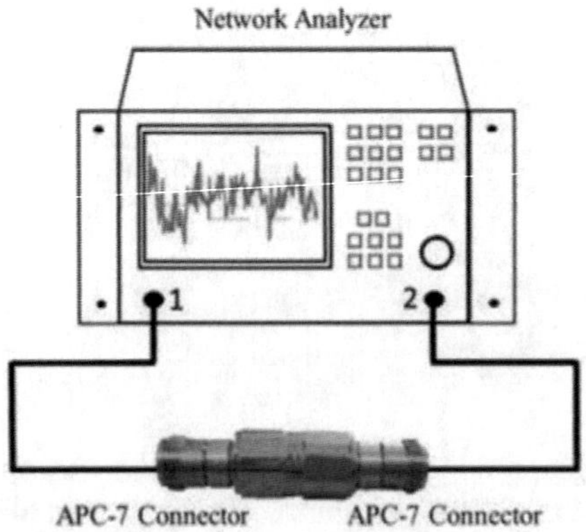

Figure 4: Schematic diagram of transmission measurement method

2.6 Flower pollination algorithm (FPA)
FPA was developed by Xin-She Yang in 2012 [34]. This algorithm method is a new nature-inspired algorithm, based on the characteristics of flowering plants [34], [35]. Flower pollination algorithm (FPA) uses the following four rules in optimization [34]:

- Biotic and cross-pollination can be considered as a process of global pollination process, and pollen-carrying pollinators move in a way which obeys Lévy flights (Rule 1).
- For local pollination, abiotic and self-pollination are used (Rule 2).
- Pollinators such as insects can develop flower constancy, which is equivalent to a reproduction probability that is proportional to the similarity of two flowers involved (Rule 3).
- The interaction or switching of local pollination and global pollination can be controlled by a switch probability $p \in [0, 1]$, with a slight bias towards local pollination (Rule 4).

To run the FPA, a set of objective function equation must develop before running the optimization in MATLAB software. The objective function equation is as equation (2) shown. This objective function equation was developed in Microsoft Excel.

$$Y = b_0 + b_1x_1 + b_2x_2 + b_3x_1x_2 \quad (2)$$

where Y is the dependent variable or estimation for the response variables of the designed experiment (EMI shielding effectiveness), and b is the estimation for the regression coefficient computed by least squares approach. The pseudo code of the proposed Flower Pollination Algorithm (FPA) can represent as follows [34]:

*Objective min or max f (**x**), **x** = (x$_1$, x$_2$, ..., x$_d$)*
Initialize a population of n flowers/pollen gametes with random solutions
Find the best solution g∗ in the initial population
Define a switch probability p $\in$ [0, 1]
Define a stopping criterion (either a fixed number of generations/iterations or accuracy)
while *(t <MaxGeneration)*
  for *i = 1: n (all n flowers in the population)*
    if *rand < p,*
      Draw a (d-dimensional) step vector L which obeys a Lévy distribution

Global pollination via $x_i^{t+1} = x_i^t + L(g_* - x_i^t)$
else
Draw ϵ *from a uniform distribution in [0,1]*
Do local pollination via $x_i^{t+1} = x_i^t + \epsilon(x_j^t - x_k^t)$
end if
Evaluate new solutions
If new solutions are better, update them in the population
end for
Find the current best solution g_*
end while
Output the best solution found

3 Results and discussions

3.1 Physical properties and chemical compositions of POFA

The EDS and SEM result were discussed in this subsection. From the precipitation process, the POFA is split into two main layers as in Figure 5. Based on the EDS result (Table 5), the carbon content of layer 1 is higher than layer 2, which is 85.89%. Figure 6 shows the SEM morphology images of both layers. It can be seen that the layer 1 (Figure 6 (1)) shows that the particle is fully porous like light foam. This explains the reason why layer 1 is floating on the surface. The SEM morphology image of layer 2 (Figure 6 (2)) shows less porous compared to layer 1. It is observed that layer 2 still contains the unburn palm shell and this is the reason layer 2 was submerged in the water.

Figure 5: The layer of POFA, (1) Layer 1 and (2) Layer 2

Table 5: Compound in each layer of POFA

Compound	Layer 1 *(%)*	Layer 2 *(%)*
C	85.89	81.68
MgO	1.27	1.12
Al_2O_3	-	1.72
SiO_2	5.60	8.22
P_2O_5	-	-
K_2O	3.25	2.52
CaO	3.98	2.01
FeO	-	2.73
Total	**100**	**100**

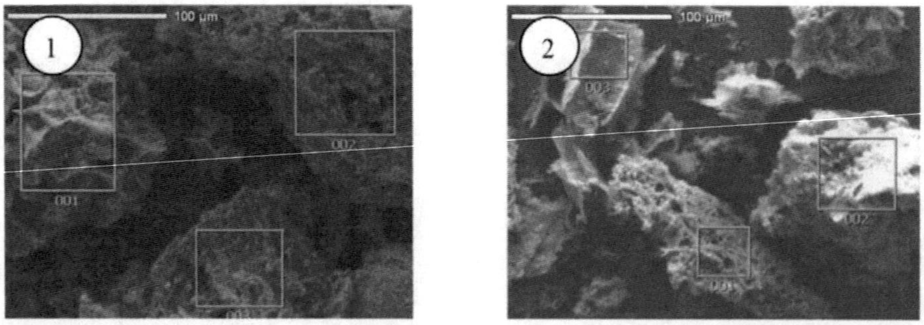

Figure 6: SEM morphology image of (1) layer 1 and (2) layer 2 POFA

3.2 Experimental SE measurement

The SE of POFA mix with OPC for layer 1 (CPU vs CPF) was shown in Figure 7. The results show that all ultrafine POFA samples (CPU) were achieved better SE than fine POFA (CPF). For POFA layer 1, the ultrafine POFA (CPU) can increase the SE approximately at 13%. The results show that the different particle size of POFA can improve the SE. The highest SE was achieved by ultrafine POFA (CPU) at 50% of mixing ratio, about -13.53 dB SE at 1.52 GHz. The fine POFA (CPF) layer 1 mix OPC with 50% of the ratio can achieve only -12.11 dB at 2 GHz.

The SE of POFA mix with OPC for layer 2 (CPU vs CPF) was shown in Figure 8. The results also show that all ultrafine POFA samples (CPU) were achieved better SE than fine POFA (CPF). This result was the same pattern as layer 1 because ultrafine POFA (CPU) has better SE than fine POFA (CPF). The ultrafine POFA (CPU) can increase the SE approximately at 5 %. The highest SE for layer 2 was achieved by ultrafine POFA (CPU) at 50% of mixing ratio, about -12.82 dB SE at 2 GHz. The fine POFA (CPF) layer 1 mix OPC with 50% of the ratio can achieve -12.46 dB at 1.5 GHz.

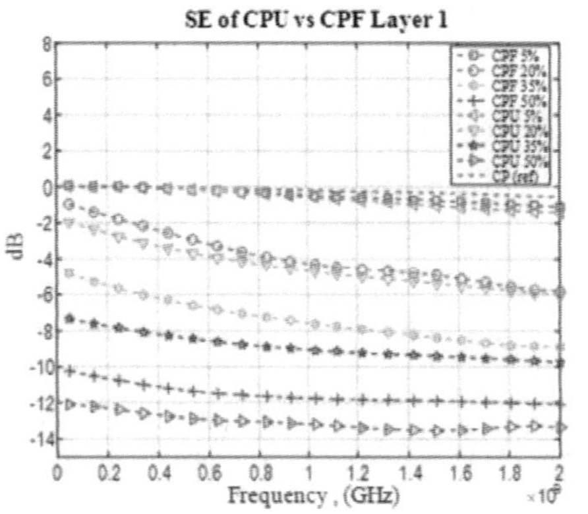

Figure 7: Shielding effectiveness of layer 1 (CPU vs CPF)

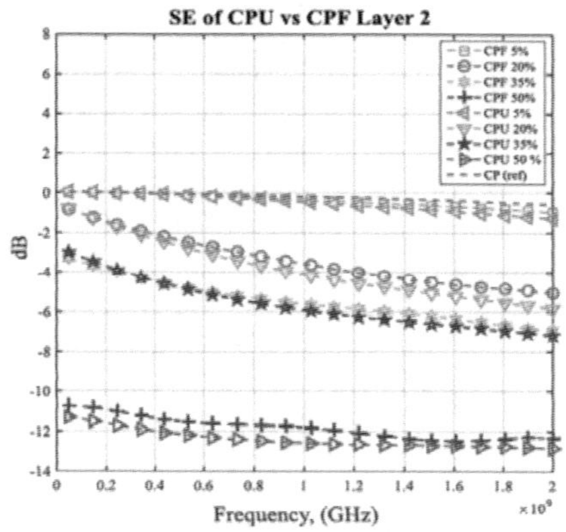

Figure 8: Shielding effectiveness of layer 2 (CPU vs CPF)

In Figure 9 and Figure 10, POFA layer 1 shows better SE results because of the layer 1 POFA has the high carbon content. With the existence of carbon in POFA, this helps to transform the EM field into heat [21], [27], as the carbon has good conductivity [36], [37]. The carbon was found in both layers of POFA. The carbon/carbon black has been used as floor heating elements, electronic equipment related material in various display components and magnetic recording materials [36], and electromagnetic interference (EMI) shielding [37]. Overall results were proving that the POFA replacement in OPC can improve the SE compared to the 100% OPC (CP). The smaller particle size

POFA can improve with a certain percentage of SE because of the small particle can fulfill the space and giving a good shielding especially in complex space. Overall, the layer 1 POFA show better SE compare layer 2 POFA.

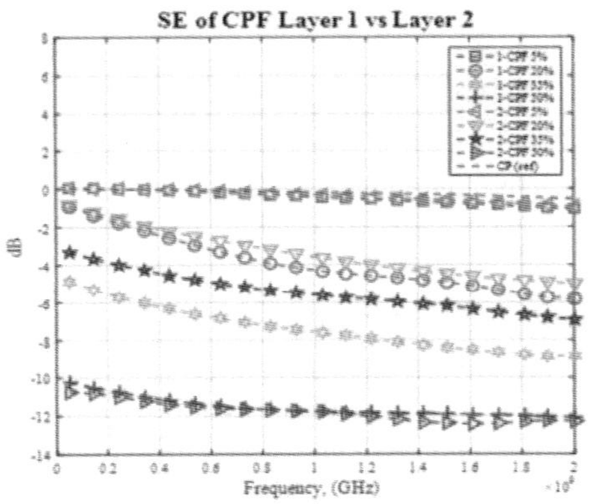

Figure 9: Shielding effectiveness of CPF (Layer 1 vs Layer 2)

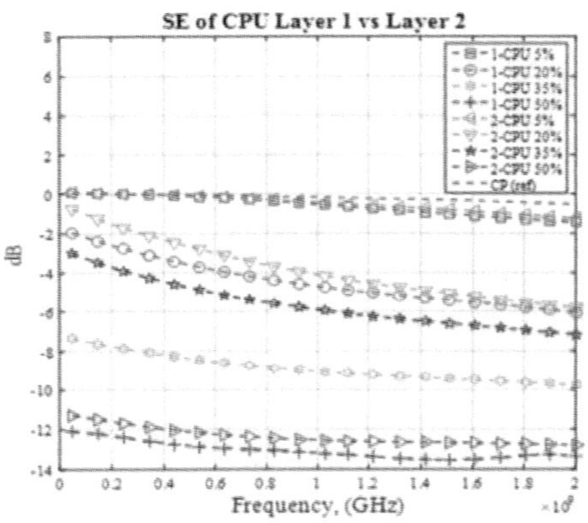

Figure 10: Shielding effectiveness of CPU (Layer 1 vs Layer 2)

3.3 FPA optimization and simulation

In this subsection was discussing about the FPA optimization results. This optimization method was used to validate the experimental and FPA simulation data. In addition, the FPA can obtain the optimal sieve size and optimal mixing ratio POFA in OPC for EMI SE. Table 6 shows the regression statistics result for POFA layer 1, where the R square is equal to 0.998733, which is a perfect fit. The biggest SE value (with ignore the positive and negative sign), the better regression line fit the data. From the analysis of variance (ANOVA) at Table 7 shown, the result is statistically significant due to the significant F value less than 0.05. In addition, all variables are significant to the EMI SE due to three over four P value less than 0.05. Base on the Table 8, the sign and magnitude of regression coefficient signify the relative influence of each parameter of the output response (EMI SE) with a negative sign indicating an antagonistic effect whilst a positive one signifies a synergistic effect. The mathematical interaction model of EMI SE of POFA layer 1 was developed based on the sign and magnitude of regression coefficient as shown in equation (3).

$$Y = b_0 + b_1x_1 + b_2x_2 + b_3x_1x_2$$
$$Y = -6.311 + 0.458x_1 + -5.901x_2 + 0.371x_1x_2 \quad (3)$$

Table 6: Regression Statistics for layer 1

Regression Statistics	
Multiple R	0.999367
R Square	0.998733
Adjusted R Square	0.997784
Standard Error	0.223131
Observations	8

Table 7: ANOVA for layer 1

	df	SS	MS	F	Significance F
Regression	3	157.0416	52.34719	1051.418	3.01E-06
Residual	4	0.199149	0.049787		
Total	7	157.2407			

Table 8: Objective function information for layer 1

Variable	Coefficients	Standard Error	t Stat	P-value	Lower 95%	Upper 95%	Lower 95.0%	Upper 95.0%
b0	-6.311	0.079	-79.993	0.000	-6.530	-6.092	-6.530	-6.092
b1	0.458	0.079	5.806	0.004	0.239	0.677	0.239	0.677
b2	-5.901	0.106	-55.752	0.000	-6.195	-5.607	-6.195	-5.607
b3	0.371	0.106	3.507	0.025	0.077	0.665	0.077	0.665

For layer 1, the prediction result suggested that 45 µm sieve size and 50% of mixing ratio is the optimum value in SE for layer 1. This is the same with experimental result where experimental result also shows better SE at 45 µm sieve size and 50% of mixing ratio. The SE from FPA optimization method is -13.04 dB where it has just obtained only 0.04 error compare to the experimental result. The target has set to achieve -14 dB, but the optimization result only obtains -13.04 dB.

Table 9 shows the regression statistics result for POFA layer 2, where the R square is equal to 0. 937431098, which is a practicable fit. From the analysis of variance (ANOVA) at Table 10 shown, the POFA layer 2 also statistically significant due to the significant F value less than 0.05. In addition, all variables are considered significant to the EMI SE due to the two over four P value less than 0.05 but well less than POFA layer 1. Base on the Table 11, the sign and magnitude of regression coefficient signify the relative influence of each parameter of the output response (EMI SE) with a negative sign indicating an antagonistic effect whilst a positive one signifies a synergistic effect which is the same trend with POFA layer 1. The mathematical interaction model of EMI SE for POFA layer 2 was developed as shown in equation (4).

$$Y = b_0 + b_1x_1 + b_2x_2 + b_3x_1x_2$$
$$Y = -5.436 + 0.171x_1 + -5.542x_2 + 0.080x_1x_2 \quad (4)$$

Table 9: Regression Statistics for layer 2

Regression Statistics	
Multiple R	0.968210255
R Square	0.937431098
Adjusted R Square	0.890504422
Standard Error	1.510563977
Observations	8

Table 10: ANOVA for layer 2

	df	*SS*	*MS*	*F*	*Significance F*
Regression	3	136.7473959	45.58247	19.97651	0.007185
Residual	4	9.127214114	2.281804		
Total	7	145.87461			

Table 11: Objective function information for layer 2

Variable	Coefficients	Standard Error	t Stat	P-value	Lower 95%	Upper 95%	Lower 95.0%	Upper 95.0%
b0	-5.436	0.534	-10.178	0.001	-6.918	-3.953	-6.918	-3.953
b1	0.171	0.534	0.320	0.765	-1.312	1.654	-1.312	1.654
b2	-5.542	0.717	-7.734	0.002	-7.531	-3.552	-7.531	-3.552
b3	0.080	0.717	0.111	0.917	-1.910	2.069	-1.910	2.069

For POFA layer 2, the prediction result also suggested that 45 μm sieve size and 50% of mixing ratio is the optimum value in SE. The SE from FPA optimization method is -11.23 dB where the 1.16 error is compared to the experimental result. In the FPA software, the target has set to achieve -14 dB, but the optimization result only obtains -11.23 dB. Table 12 shows the comparison of SE between FPA simulation and experimental data.

Table 12: Comparison of SE between FPA simulation and experimental data

Samples name	Layers	Sieve sizes (μm)	Mixing percentages (%)	EMI SE Average (dB)		Error
				FPA Simulation	Experimental	
1-CPU 50%	1	45	50%	-13.04	-13.08	0.04
2-CPU 50%	2	45	50%	-11.23	-12.39	1.16

4. Conclusion

In conclusion, this study has proven that the POFA has the potential to be used as the additive for cement based material to enhance the EMI SE in the building. The element that contributed the most to EMI SE in POFA is carbon as the carbon has good conductivity. The high porosity of particle in POFA also gives a good absorbing to the EMF. The highest SE was achieved by layer 1 ultrafine POFA (CPU) with passing 45 μm sieve at 50% of mixing ratio, about -13.53 dB at 1.52 GHz. While the fine POFA (CPF) layer 1 mix OPC with 50% of the ratio can achieve only -12.11 dB at 2 GHz. The layer 1 POFA shows better EMI SE compare to layer 2 POFA because the layer 1 POFA contain higher carbon content. This study has found that the smaller particle size of POFA can improve the SE in 5% to 13% due to the smaller particle. The smaller particle POFA can cover up the complex space in the cement structure to avoid the EMF leaked. Based on FPA simulation data, the error is below 1.2 for both layers. FPA optimization prediction has come out comparable result with experimental data. By using FPA optimization method can help researcher obtain the optimum sieve size and mixing percentage. This is a good sign to improve the next study.

Acknowledgment
This research was financed by Fundamental Research Grant Scheme (FRGS) Vot No. 1490 and Office for Research, Innovation, Commercialization and Consultancy Management (ORICC).

References

[1] C. Beall, E. Delzell, P. Cole, and I. Brill, "Brain tumors among electronics industry workers.," *Epidemiology*, vol. 7, no. 2, pp. 125–130, 1996.
[2] S. Xie, Y. Yang, G. Hou, J. Wang, and Z. Ji, "Development of layer structured wave absorbing mineral wool boards for indoor electromagnetic radiation protection," *J. Build. Eng.*, vol. 5, pp. 79–85, 2016.
[3] G. Bantsis, C. Sikalidis, M. Betsiou, T. Yioultsis, and T. Xenos, "Electromagnetic absorption,

reflection and interference shielding in X-band frequency range of low cost ceramic building bricks and sandwich type ceramic tiles using mill scale waste as an admixture.," *Ceram. Int.*, vol. 37, no. 8, pp. 3535–3545, 2011.

[4] T. Wessapan, S. Srisawatdhisukul, and P. Rattanadecho, "Numerical analysis of specific absorption rate and heat transfer in the human body exposed to leakage electromagnetic field at 915 MHz and 2450 MHz," *J. Heat Transfer*, vol. 133, no. 5, p. 51101, 2011.

[5] P. Keangin, K. Vafai, and P. Rattanadecho, "Electromagnetic field effects on biological materials," *Int. J. Heat Mass Transf.*, vol. 65, pp. 389–399, 2013.

[6] D. P. Loomis and D. A. Savitz, "Mortality from brain cancer and leukaemia among electrical workers.," *Br. J. Ind. Med.*, vol. 47, no. 9, pp. 633–8, 1990.

[7] B. B. Levitt and H. Lai, "Biological effects from exposure to electromagnetic radiation emitted by cell tower base stations and other antenna arrays," *Environ. Rev.*, vol. 18, no. NA, pp. 369–395, 2010.

[8] E. E. Fesenko, V. R. Makar, E. G. Novoselova, and V. B. Sadovnikov, "Microwaves and cellular immunity. I. Effect of whole body microwave irradiation on tumor necrosis factor production in mouse cells.," *Bioelectrochem. Bioenerg.*, vol. 49, no. 1, pp. 29–35, 1999.

[9] A. Miszczyk and K. Darowicki, "Study of anticorrosion and microwave absorption properties of NiZn ferrite pigments," *Anti-Corrosion Methods Mater.*, vol. 58, no. 1, pp. 13–21, 2011.

[10] H. Guan, S. Liu, Y. Duan, and J. Cheng, "Cement based electromagnetic shielding and absorbing building materials," *Cem. Concr. Compos.*, vol. 28, no. 5, pp. 468–474, 2006.

[11] D. D. L. Chung, "Materials for electromagnetic interference shielding," *J. Mater. Eng. Perform.*, vol. 9, no. 3, pp. 350–354, 2000.

[12] N. H. A. Khalid, M. W. Hussin, J. Mirza, N. F. Ariffin, M. A. Ismail, H.-S. Lee, A. Mohamed, and R. P. Jaya, "Palm oil fuel ash as potential green micro-filler in polymer concrete," *Constr. Build. Mater.*, vol. 102, pp. 950–960, 2016.

[13] W. A. N. W. Mohamed and N. R. Abdullah, "Combustion Characteristics of Palm Shells and Palm Fibers Using an Inclined Grate Combustor," *Carbon N. Y.*, vol. 57, pp. 48–94.

[14] A. S. M. A. Awal, I. A. Shehu, and M. Ismail, "Effect of cooling regime on the residual performance of high-volume palm oil fuel ash concrete exposed to high temperatures," *Constr. Build. Mater.*, vol. 98, pp. 875–883, 2015.

[15] S. K. Nguong and U. Teknologi, "A Short-Term Investigation On High Volume Palm Oil Fuel Ash (POFA) Concrete," *35th Conf. OUR WORLD Concr. Struct. 25 - 27 August 2010, Singapore*, 2010.

[16] W. Tangchirapat, T. Saeting, C. Jaturapitakkul, K. Kiattikomol, and A. Siripanichgorn, "Use of waste ash from palm oil industry in concrete," *Waste Manag.*, vol. 27, no. 1, pp. 81–88, 2007.

[17] V. Sata, C. Jaturapitakkul, and K. Kiattikomol, "Utilization of palm oil fuel ash in high-strength concrete," *J. Mater. Civ. Eng.*, vol. 16, no. 6, pp. 623–628, 2004.

[18] M. A. Salih, A. A. Abang Ali, and N. Farzadnia, "Characterization of mechanical and microstructural properties of palm oil fuel ash geopolymer cement paste," *Constr. Build. Mater.*, vol. 65, pp. 592–603, 2014.

[19] N. Ranjbar, A. Behnia, B. Alsubari, P. Moradi Birgani, and M. Z. Jumaat, "Durability and mechanical properties of self-compacting concrete incorporating palm oil fuel ash," *J. Clean. Prod.*, vol. 112, pp. 723–730, 2015.

[20] M. A. A. Rajak, Z. A. Majid, and M. Ismail, "Morphological Characteristics of Hardened Cement Pastes Incorporating Nano-palm Oil Fuel Ash," *Procedia Manuf.*, vol. 2, no. February, pp. 512–518, 2015.

[21] H. Noorvand, A. A. A. Ali, R. Demirboga, H. Noorvand, and N. Farzadnia, "Physical and chemical characteristics of unground palm oil fuel ash cement mortars with nanosilica," *Constr. Build. Mater.*, vol. 48, pp. 1104–1113, 2013.

[22] A. Awal and S. K. Nguong, "A short-term investigation on high volume palm oil fuel ash (POFA) concrete," *Proc. 35th Conf. our World Concr. Struct.*, pp. 185–192, 2010.

[23] K. Muthusamy and Z. N. Azzimah, "Exploratory Study of Palm Oil Fuel Ash as Partial Cement Replacement in Oil Palm Shell Lightweight Aggregate Concrete," vol. 8, no. 2, pp. 150–152, 2014.
[24] M. R. Karim, M. F. M. Zain, M. Jamil, and F. C. Lai, "Fabrication of a non-cement binder using slag, palm oil fuel ash and rice husk ash with sodium hydroxide," *Constr. Build. Mater.*, vol. 49, pp. 894–902, 2013.
[25] P. Chindaprasirt, S. Rukzon, and V. Sirivivatnanon, "Resistance to chloride penetration of blended Portland cement mortar containing palm oil fuel ash, rice husk ash and fly ash," *Constr. Build. Mater.*, vol. 22, no. 5, pp. 932–938, 2008.
[26] S. Bamaga, M. A. Ismail, and M. W. Hussin, "Chloride Resistance of Concrete Containing Palm Oil Fuel Ash," vol. 1, no. December, pp. 158–166, 2010.
[27] H. Abdullah, A. Zanal, M. N. Taib, I. R. Mohamed Noordin, W. K. Wan Ali, R. Ariffin, S. Abdullah, R. Baharudin, and A. T. Abdullah, "Microwave Absorber Coating Material Using Oil Palm Ash," *Adv. Mater. Res.*, vol. 512–515, pp. 1941–1944, 2012.
[28] L. Baoyi, D. Yuping, and L. Shunhua, "The electromagnetic characteristics of fly ash and absorbing properties of cement-based composites using fly ash as cement replacement," *Constr. Build. Mater.*, vol. 27, no. 1, pp. 184–188, 2012.
[29] R. Che, C. Wang, Y. Ni, and B. Yu, "Preparation and microwave absorbing properties of the core-nanoshell composite absorbers with the magnetic fly-ash hollow cenosphere as core," *J. Environ. Sci.*, vol. 23, pp. S74–S77, 2011.
[30] Z. Dou, G. Wu, X. Huang, D. Sun, and L. Jiang, "Electromagnetic shielding effectiveness of aluminum alloy–fly ash composites," *Compos. Part A Appl. Sci. Manuf.*, vol. 38, no. 1, pp. 186–191, 2007.
[31] M. Mishra, A. P. Singh, and S. K. Dhawan, "Expanded graphite–nanoferrite–fly ash composites for shielding of electromagnetic pollution," *J. Alloys Compd.*, vol. 557, pp. 244–251, 2013.
[32] J. Y. Cao and D. D. L. Chung, "Use of fly ash as an admixture for electromagnetic interference shielding," *Cem. Concr. Res.*, vol. 34, no. February, pp. 1889–1892, 2004.
[33] M. B. Melwanki and M. R. Fuh, "Dispersive liquid-liquid microextraction combined with semi-automated in-syringe back extraction as a new approach for the sample preparation of ionizable organic compounds prior to liquid chromatography," *J. Chromatogr. A*, vol. 1198–1199, no. 1–2, pp. 1–6, 2008.
[34] X. S. Yang, M. Karamanoglu, and X. He, "Multi-objective flower algorithm for optimization," *Procedia Comput. Sci.*, vol. 18, pp. 861–868, 2013.
[35] X.-S. Yang, "Flower pollination algorithm for global optimization," in *International Conference on Unconventional Computing and Natural Computation*, 2012, pp. 240–249.
[36] W. Zhang, A. A. Dehghani-Sanij, and R. S. Blackburn, "Carbon based conductive polymer composites," *J. Mater. Sci.*, vol. 42, no. 10, pp. 3408–3418, 2007.
[37] S. K. De and J. R. White, *Short fibre-polymer composites*. Elsevier, 1996.

IOP Conf. Series: Materials Science and Engineering **165** (2017) 012036 doi:10.1088/1757-899X/165/1/012036

Taguchi design and flower pollination algorithm application to optimize the shrinkage of triaxial porcelain containing palm oil fuel ash

A Zainudin[1*], C K Sia[1], P Ong[2], O L C Narong[1] and N H M Nor[1]

[1]Department of Material and Design Engineering, Faculty of Mechanical and Manufacturing Engineering, Universiti Tun Hussein Onn Malaysia, 86400 Parit Raja, Batu Pahat, Johor, Malaysia

[2]Department of Mechanics Engineering, Faculty of Mechanical and Manufacturing Engineering, Universiti Tun Hussein Onn Malaysia, 86400 Parit Raja, Batu Pahat, Johor, Malaysia

Email: *gd150136@siswa.uthm.edu.my

Abstract. In the preparation of triaxial porcelain from Palm Oil Fuel Ash (POFA), a new parameter variable must be determined. The parameters involved are the particle size of POFA, percentage of POFA in triaxial porcelain composition, moulding pressure, sintering temperature and soaking time. Meanwhile, the shrinkage is the dependent variable. The optimization process was investigated using a hybrid Taguchi design and flower pollination algorithm (FPA). The interaction model of shrinkage was derived from regression analysis and found that the shrinkage is highly dependent on the sintering temperature followed by POFA composition, moulding pressure, POFA particle size and soaking time. The interaction between sintering temperature and soaking time highly affects the shrinkage. From the FPA process, targeted shrinkage approaching zero values were predicted for 142 μm particle sizes of POFA, 22.5 wt% of POFA, 3.4 tonne moulding pressure, 948.5 °C sintering temperature and 264 minutes soaking time.

Keywords: Triaxial porcelain; Palm Oil Fuel Ash; Shrinkage; Taguchi Design; Flower Pollination Algorithm

1. Introduction

Conventional triaxial porcelain body mixed contains three main components; plastic material, fluxing agent and filler. Usually the plastic material is kaolinite, the fluxing agent is feldspar and the filler is quartz. Formulations of triaxial porcelain (SiO_2– Al_2O_3–KNaO) usually involve 25 wt.% of plastic component, 25 wt.% silica and 50 wt.% feldspar (generally sodium feldspar) for soft porcelain and 50 wt.% of clay, 25 wt.% silica and 25 wt.% feldspar (generally potassium feldspar) for hard porcelain[1,2]. Clay provides plasticity and green strength to the body and acts as a binder. Feldspar acts as a fluxing material which does mean that generates low melting phase and helps vitrification. Quartz is a filler material which remains stable in the normal firing range and reduces distortions and shrinkage[3]. Many studies on triaxial porcelains report that quartz significantly affects triaxial porcelain properties [4–6] . Recently, researchers were tried to replace those natural raw materials with a variety of industrial waste materials. Furthermore, the fabrication of products of waste is an advantage that may give the manufacturer a highly competitive position in the market due to the economic issues involved. Blast furnace slag and coal fly ash are examples of waste material which has been converted into some raw material added in recent years [7–15].

Currently, Malaysia had large palm oil industry. Furthermore, Malaysia is the second largest global palm oil producer and exporter after Indonesia [16]. Palm oil industry had daily wastage of burning of empty fruit bunch, fiber and palm oil shell in the boiler. This wastage was called as POFA. Currently, POFA usage is very limited and uncontrollable, and most of it is disposed of towards landfills. Consequently, it has caused numerous environmental harms [17]. Earlier, POFA was applied for variety of application and typically used in concrete application. In concrete, POFA was used as filler to increase the strength of concrete [18]. Besides, POFA was used as pozzolanic material which can improve the durability of the concrete [19]. Previously, the potential of POFA in ceramic application was applied. POFA was heated at 600 °C at 1.5 hours to produce treated POFA (TPOFA). Their study consist of volume shrinkage, porosity, bulk density, bending strength and compressive strength [20]. The properties accomplished still track the standard requirement of porcelain products.

Shrinkage is the rate at which the body shrinks during drying and firing. In manufacturing process, the shrinkage is a great importance issues. It is broadly used in the control of the end product size. During the sintering process, the success of firing depends in control of shrinkage consequently. Currently, minimum shrinkage achieved is 18 % with 2 wt% of TPOFA additive [20]. However, the shrinkage from blast furnace slag at present achieves value less than 1% for 6-13 wt % additive [10]. Sometimes, negative shrinkage was attained means that the specimen expanded upon sintering process. The negative shrinkage was observed after 1120 °C indicating initiating of over firing [21]. Overall, the approaching zero shrinkage must be targeted by researchers [22].

Various styles design of experiment (DOE) was applied to optimize the optimal parameter of triaxial porcelain process. The familiar DOE applied is the mixture design where it optimizes the best composition of three raw materials in form. Mixture design functional need fixed other processing parameters like sintering temperature and soaking time [23–27]. Earlier, Taguchi design was applied in investigation of high temperature co-fired ceramics sintering conditions. The investigation only ran nine experiments (L_9 orthogonal array table) to discover the optimal peak temperature, dwell time and heating rate [28]. Recently, Taguchi design application constant the composition of all raw materials [28]. The Taguchi design also was compared with other DOE or statistical tools for validation purpose [29].

Taguchi design usually involve signal to noise ratio (S/N) analysis. The S/N ratio limited to recognize the optimum parameter directly from the list of variable in Taguchi design table. So, the researchers tried a lot of alternatives to develop the result precisely such as cuckoo search algorithm, FPA, swarm algorithm, genetic algorithm and others. FPA is proposed to this paper to solve the problem of optimal triaxial porcelain parameter. It has only one key parameter p (switch probability) which makes the algorithm easier to implement and faster to reach optimal solution. FPA originally created by Yang [30]. Nowadays, FPA is well-known for researchers to optimize several application problems.

In this investigation, shrinkage was studied on hard triaxial porcelain with mixed POFA as filler material. The Taguchi design through orthogonal array was applied upon study to design an experiment with minimum number of test. FPA was applied to determine the optimal variable of parameter to accomplish targeted shrinkage.

2. Experimental

2.1. Methods

Four main raw materials were prepared including kaolin, potash feldspar, quartz and POFA. The kaolin, potash feldspar and quartz were already comes in the form of fine powder. Kaolin was supplied by Kaolin (M) Sdn Bhd, Malaysia with grade KM40. Besides, potash feldspar and quartz was supplied from BG Oil Sdn Bhd, Malaysia. Meanwhile, POFA was supplied from Genting Ayer Hitam Oil Mill, Johor, Malaysia. It is originally supplied in form of large particles and impurities. POFA was sieve by

160 μm and 63 μm sieves to create the unground POFA (UPOFA) and ground POFA (GPOFA) respectively. Six different porcelain samples were prepared as per the batch composition provided in Table 1. The standard composition of hard porcelain denoted are 50 wt. % kaolin, 25 wt. % quartz and 25 wt. % potash feldspar [1,2].

Table 1: Batch compositions

Batch	Percentage composition of Raw Material (wt %)				
	Kaolin	Potash Feldspar	Quartz	UPOFA	GPOFA
QU5	50	25	20	5	0
QU15	50	25	10	15	0
QU25	50	25	0	25	0
QG5	50	25	20	0	5
QG15	50	25	10	0	15
QG25	50	25	0	0	25

All raw materials were mixed based on the percentage composition of variable POFA composition, 5-25 wt% as shown in Table 1. Six mixtures were prepared by using a FRITSCH's planetary mono mill. Each mixing process in the machine was milled at 250RPM by 30 minutes. The dried mixture was compacted manually by using Carver Press. The compaction with variable pressure, 2-4 tonne based on Taguchi design in Table 2. The specimens are in the form of cylinder pellet of 13 mm in diameter with variable thickness, 3-4 mm. At least five compacted specimens were prepared for each similar experiment which need $5 \times 18 = 90$ specimens overall [31]. The compacted specimen was sintered in an electric furnace with variable temperature, 800-1200°C and variable soaking time, 60-300 minutes as shown in Table 2. Meanwhile, heating and cooling rate was fixed at 5 °C/min. The thickness and diameter of specimen before and after the sintering process was measured by using the Vernier caliper. The volume of specimens for both before and after the sintering process was calculated to determine the shrinkage of the specimen based on Eq. (1):

$$Shrinkage = \frac{V_p - V_s}{V_p} \times 100 \quad (1)$$

where $V_p and\ V_s$ are the specimen's volume before and after sintered, respectively.

2.2. Taguchi design

The Taguchi method, developing the orthogonal arrays from experimental design theory to study a large number of variables with minor number of experiments, was chosen to optimize the preparation condition of triaxial porcelain from POFA. POFA particle size, POFA composition, moulding pressure, sintering temperature and soaking time were chosen as factors and an appropriate L18 orthogonal array with five factors at mixed level of two and three levels was selected to conduct the experiment as shown in Table 2.

Normally, a full factorial design would require $2^1 \times 3^4 = 162$ experimental runs. Apparently, it is impossible to do these experimental runs and accompanies with very high cost, while with using Taguchi method simply 18 experiments are needed based on the L18 orthogonal array, as shown in Table 2. The parameter variables were normalized between (-1, 1) using Eq. (2) in order to ascertain the direct comparison of the observations from different scales, and thus the individual influence of each parameter in characterizing the output response can be explored:

$$X_n = 2\frac{X - \bar{X}}{X_{max} - X_{min}} \quad (2)$$

where X_n is the normalized value of the parameter variables POFA particle size, POFA composition, moulding pressure, sintering temperature and soaking time, X is the absolute value of the actual response, $\bar{X}$ is the mean value of X_{max} and X_{min} whereas X_{max} and X_{min} is the maximum and minimum value for a particular parameter variable, respectively. The normalized result was shown in Table 2.

Table 2: Taguchi L18 and Experimental Results

No.	Parameters										Experimental values for shrinkage (%)
	Actual value					Coded value					
	POFA Particle Size (μm)	POFA composition (Wt %)	Molding Pressure (Tonne)	Sintering Temperature (°C)	Soaking Time (Minutes)	X_A	X_B	X_C	X_D	X_E	
1	160	5	2	800	60	-1	-1	-1	-1	-1	-2.59
2	160	5	3	1000	180	-1	-1	0	0	1	15.82
3	160	5	4	1200	300	-1	-1	1	1	0	17.88
4	160	15	2	800	180	-1	0	-1	-1	1	-6.25
5	160	15	3	1000	300	-1	0	0	0	0	3.79
6	160	15	4	1200	60	-1	0	1	1	-1	16.51
7	160	25	2	1000	60	-1	1	-1	0	-1	2.05
8	160	25	3	1200	180	-1	1	0	1	1	9.59
9	160	25	4	800	300	-1	1	1	-1	0	-3.32
10	63	5	2	1200	300	1	-1	-1	1	0	22.27
11	63	5	3	800	60	1	-1	0	-1	-1	-1.32
12	63	5	4	1000	180	1	-1	1	0	1	14.80
13	63	15	2	1000	300	1	0	-1	0	0	2.31
14	63	15	3	1200	60	1	0	0	1	-1	17.62
15	63	15	4	800	180	1	0	1	-1	1	-8.76
16	63	25	2	1200	180	1	1	-1	1	1	24.08
17	63	25	3	800	300	1	1	0	-1	0	-2.97
18	63	25	4	1000	60	1	1	1	0	-1	4.76

2.3. Flower pollination algorithm

The coded values of the experimental results were then modeled with the following interaction model equation (Eq. (3)).

$$Y = \beta_0 + \sum_{i=1}^{k} \beta_i X_i + \sum_{i<j}^{k} \beta_{ij} X_i X_j \tag{3}$$

where Y is the dependent variable or estimation for the response variables of the designed experiment (shrinkage), k is the number of normalized parameter variables (where $k = 5$ in this study), and β is the estimation for the regression coefficient computed by least squares approach. The $X_i X_j$ term in this equation account for two variable interaction effects of the parameter variables with the response variables, shrinkage.

Interpolation on the relationship of the parameter variables and the dependent response variables will be examined by using the FPA [30]. The pseudo code of the FPA applied in Matlab software was shown in Figure 1.

Flower Pollination Algorithm (or simply Flower Algorithm)

Objective min or max $f(\boldsymbol{x}), \boldsymbol{x} = (x_1, x_2, \dots, x_d)$
Initialize a population of n flowers/pollen gametes with random solutions
Find the best solution $\boldsymbol{g}_*$, in the initial solution
Define a switch probability p Є [0,1]

```
Define a stopping criterion (either a fixed number of generations/iterations or accuracy)
while (t <MaxGeneration)
  for i=1: n (all n flowers in the population)
    if rand < p,
      Draw a (d-dimensional step vector L which obeys a Levy distribution
      Global pollination via x_i^{t+1} = x_i^t + L(g_* - x_i^t)
    else
      Draw Є from a uniform distribution in [0,1]
      Do local pollination via x_i^{t+1} = x_i^t + Є(x_j^t - x_k^t)
    end if
    Evaluate new solutions
    If new solutions are better, update them in the population
  end for
    Find the current best solution g_*
end while
Output the best solution found
```

Figure 1: Pseudo code of the proposed Flower Pollination Algorithm (FPA) [30]

By setting the population size 25 and total number of iteration 5000, the coding was run to develop the best result of targeted shrinkage value.

3. Result and discussion

3.1. Interaction modelling

Table 3 presents the regression statistics result, which R square equal to 0.98, which is a very good fit. The 98 % of the variation in shrinkage is explained by all independent variables investigated. The closer to 1, the better the regression line fits the data. Meanwhile, from the analysis of variance (ANOVA) (Table 4) the result are statistically insignificant due to significant F value greater than 0.05. In addition, all variables are statistically insignificant to the shrinkage due to all P value greater than 0.05. But, the investigations still consider all variables in modelling equation. The sign and magnitude of regression coefficient signify the relative influence of each parameter on the output response (shrinkage) with a negative sign indicating an antagonistic effect whilst a positive one signifies a synergistic effect (Table 5). The mathematical interaction model of shrinkage was developed based on the sign and magnitude of regression coefficient as shown in Eq. (4).

Table 3: Regression Statistics

Regression Statistics	
Multiple R (Correlation coefficient)	0.976345669
R Square (Coefficient of determination)	0.953250865
Adjusted R Square	0.602632353
Standard Error	6.566039377
Observation	18

Table 4: ANOVA

	df	SS	MS	F	Significance F
Regression	15	1758.209371	117.2139581	2.718769352	0.301680675
Residual	2	86.22574619	43.1128731		
Total	17	1844.435117			

Table 5: Coefficient of mathematical interaction model

	Coefficients	Standard Error	t Stat	P-value	Lower 95%	Upper 95%
Intercept	7.01	1.54763	4.53264	0.04539	0.35594	13.67377
X_A	0.44	2.26297	0.19454	0.86372	-9.29652	10.17700
X_B	-3.86	2.74290	-1.40841	0.29435	-15.66486	7.93863

X_C	1.37	2.57094	0.53305	0.64730	-9.69144	12.43231
X_D	9.73	2.60124	3.73976	0.06465	-1.46423	20.92025
X_E	0.03	2.58811	0.01252	0.99115	-11.10334	11.16815
X_AX_B	0.88	2.25534	0.39089	0.73359	-8.82236	10.58556
X_AX_C	-0.37	4.16723	-0.08871	0.93739	-18.29984	17.56046
X_AX_D	2.40	2.70300	0.88817	0.46815	-9.22935	14.03082
X_AX_E	0.71	2.41077	0.29287	0.79721	-9.66667	11.07876
X_BX_C	-2.06	3.47227	-0.59439	0.61253	-17.00388	12.87608
X_BX_D	-0.44	4.28342	-0.10250	0.92771	-18.86912	17.99103
X_BX_E	-2.00	3.92068	-0.50886	0.66143	-18.86444	14.87425
X_CX_D	-1.89	4.95305	-0.38214	0.73914	-23.20406	19.41849
X_CX_E	-2.67	4.34327	-0.61474	0.60135	-21.35756	16.01762
X_DX_E	3.11	3.23923	0.96071	0.43807	-10.82533	17.04925

$$\begin{aligned} Shrinkage = {} & 7.01 + 0.44X_A - 3.86X_B + 1.37X_C + 9.73X_D + 0.03X_E + 0.88X_AX_B \\ & - 0.37X_AX_C + 2.4X_AX_D + 0.71X_AX_E - 2.06X_BX_C - 0.44X_BX_D \\ & - 2X_BX_E - 1.89X_CX_D - 2.67X_CX_E + 3.11X_DX_E \end{aligned} \quad (4)$$

3.2. Shrinkage

Shrinkage is highly dependent on the sintering temperature (X_D), as shown in Eq. (4), followed by POFA composition (X_B), moulding pressure (X_C), POFA particle size (X_A) and soaking time (X_E). In terms of parameter variable interaction, shrinkage is highly dependent on the interaction X_{DE} (sintering temperature and soaking time) followed by X_{CE} (moulding pressure and soaking time), X_{AD} (POFA particle size and sintering temperature), X_{BC} (POFA composition and moulding pressure), X_{BE} (POFA composition and soaking time), X_{CD} (moulding pressure and sintering temperature), X_{AE} (POFA particle size and soaking time), X_{BD} (POFA composition and sintering temperature) and X_{AC} (POFA particle size and moulding pressure).

The FPA gives the best solution for approaching zero shrinkage as follow: 142 µm POFA particle size; 22.5 wt% of POFA composition; 3.4 tonne Carver moulding pressure; 948.5 °C sintering temperature and 264 minutes soaking time. Besides, the FPA also can generate other results as shown in Table 6.

Table 6: List of optimal result of shrinkage

No	Parameter					Shrinkage (%)
	POFA size (µm)	POFA Composition (wt %)	Molding Pressure (Tonne)	Sintering Temperature (°C)	Soaking Time (min)	
1	90	21.3	2.0	925.5	298	0.031
2	134	9.2	2.0	946.8	60	0.005
3	146	19.9	2.3	911.8	122	0.015
4	159	24.4	2.7	932.6	162	0.047
5	68	17.1	2.0	912.4	63	0.002
6	63	7.6	2.0	876.3	109	0.023
7	143	12.5	2.1	932.8	60	0.039
8	116	15.8	2.0	932.8	60	0.054
9	83	24.8	2.1	922.8	84	0.065
10	79	25.0	4.0	1072.1	300	0.030

4. Conclusions

The present study deals with experimental investigation in triaxial porcelain preparation containing waste material POFA. The experiment has been carried out based on Taguchi L18 orthogonal array and optimized the process parameters for targeted shrinkage. Mathematical interaction model of shrinkage in terms of all parameter variables have been developed using regression analysis and checked for its adequacy. The model developed was inserted into FPA command in Matlab software to develop the best optimal parameter variable for targeted shrinkage. The following conclusions may be drawn from the results of this investigation.

- From mathematical intersection model, the shrinkage is highly dependent on the sintering temperature followed by POFA composition, moulding pressure, POFA particle size and soaking time.
- In terms of interaction between two parameters, the shrinkage is highly dependent on interaction X_{DE} (sintering temperature and soaking time) followed by X_{CE} (moulding pressure and soaking time), X_{AD} (POFA particle size and sintering temperature), X_{BC} (POFA composition and moulding pressure), X_{BE} (POFA composition and soaking time), X_{CD} (moulding pressure and sintering temperature), X_{AE} (POFA particle size and soaking time), X_{BD} (POFA composition and sintering temperature) and X_{AC} (POFA particle size and moulding pressure).
- To achieve approaching zero shrinkage is as follow: 142 μm POFA particle size; 22.5 wt% of POFA composition; 3.4 tonne Carver moulding pressure; 948.5 °C sintering temperature and 264 minutes soaking time.
- The optimal parameter variable listed limited for the single objective only: approaching zero shrinkage.

Acknowledgement
The research was financed by Fundamental Research Grant Scheme (FRGS) Vot No. 1490

References

[1] E. Kamseu, C. Leonelli, D.N. Boccaccini, P. Veronesi, P. Miselli, G. Pellacani, U. Chinje Melo., Characterisation of porcelain compositions using two china clays from Cameroon, Ceramics International. 33 (2007) 851–857.
[2] H. Boussak, H. Chemani, A. Serier, Characterization of porcelain tableware formulation containing bentonite clay, International Journal of Physical Sciences. 10 (2015) 38–45.
[3] W.M. Cam, U. Senapati, Porcelain-Raw Materials, Processing, Phase Evolution, and Mechanical Behavior, Jornal of the American Ceramic Sociaty. 81 (1998) 3–20.
[4] A.D.N. Junior, D. Hotza, V.C. Soler, E.S. Vilches, Effect of quartz particle size on the mechanical behaviour of porcelain tile subjected to different cooling rates, Journal of the European Ceramic Society. 29 (2009) 1039–1046.
[5] G. Stathis, A. Ekonomakou, C.J. Stournaras, C. Ftikos, Effect of firing conditions, filler grain size and quartz content on bending strength and physical properties of sanitaryware porcelain, Journal of the European Ceramic Society. 24 (2004) 2357–2366.
[6] S.I. Warshaw, R. Seider, Comparison of Strength of Triaxial Porcelains Containing Alumina and Silica, Journal of the American Ceramic Society. 50 (1967) 337–343.
[7] K. Dana, S.K. Das, Partial substitution of feldspar by B.F. slag in triaxial porcelain: Phase and microstructural evolution, Journal of the European Ceramic Society. 24 (2004) 3833–3839.
[8] E. Karamanova, G. Avdeev, A. Karamanov, Ceramics from blast furnace slag, kaolin and quartz, Journal of the European Ceramic Society. 31 (2011) 989–998.
[9] I. Ozdemir, S. Yilmaz, Processing of unglazed ceramic tiles from blast furnace slag, Journal of Materials Processing Technology. 183 (2007) 13–17.
[10] Z. Bayer Ozturk, E. Eren Gultekin, Preparation of ceramic wall tiling derived from blast furnace slag, Ceramics International. 41 (2015) 12020–12026.
[11] S. Ghosh, M. Das, S. Chakrabarti, S. Ghatak, Development of ceramic tiles from common clay and blast furnace slag, Ceramics International. 28 (2002) 393–400.
[12] A. Zimmer, C.P. Bergmann, Fly ash of mineral coal as ceramic tiles raw material., Waste Management (New York, N.Y.). 27 (2007) 59–68.

[13] K. Dana, S. Das, S.K. Das, Effect of substitution of fly ash for quartz in triaxial kaolin–quartz–feldspar system, Journal of the European Ceramic Society. 24 (2004) 3169–3175.

[14] A. Olgun, Y. Erdogan, Y. Ayhan, B. Zeybek, Development of ceramic tiles from coal fly ash and tincal ore waste, Ceramics International. 31 (2005) 153–158.

[15] M. Aineto, A. Acosta, I. Iglesias, The role of a coal gasification fly ash as clay additive in building ceramic, Journal of the European Ceramic Society. 26 (2006) 3783–3787.

[16] MPOB, Overview of the Malaysian Oil Palm Industry 2015, Malaysia Palm Oil Board. (2016). http://www.mpob.gov.my/ (accessed September 20, 2016).

[17] E. Khankhaje, M.W. Hussin, J. Mirza, M. Rafieizonooz, M.R. Salim, H.C. Siong, M. Naqiuddin M. Warid, On blended cement and geopolymer concretes containing palm oil fuel ash, Materials & Design. 89 (2015) 385–398.

[18] S.K. Lim, C.S. Tan, O.Y. Lim, Y.L. Lee, Fresh and hardened properties of lightweight foamed concrete with palm oil fuel ash as filler, Construction and Building Materials. 46 (2013) 39–47.

[19] W. Kroehong, T. Sinsiri, C. Jaturapitakkul, Effect of Palm Oil Fuel Ash Fineness on Packing Effect and Pozzolanic Reaction of Blended Cement Paste, Procedia Engineering. 14 (2011) 361–369.

[20] J. Usman, M. Zaky, Z. Arifin, Effects of palm oil fuel ash composition on the properties and morphology of porcelain-palm oil fuel ash composite, Jurnal Teknologi (Sciences & Engineering). 70 (2014) 5–10.

[21] E. Rambaldi, W.M. Carty, A. Tucci, L. Esposito, Using waste glass as a partial flux substitution and pyroplastic deformation of a porcelain stoneware tile body, Ceramics International. 33 (2007) 727–733.

[22] S. Salem, A. Salem, Shrinkage prediction during non-isothermal sintering in the presence liquid phase: New kinetic model, Part I, Thermochimica Acta. 575 (2014) 322–330. doi:10.1016/j.tca.2013.11.004.

[23] B.K. Ngun, H. Mohamad, K. Katsumata, K. Okada, Z.A. Ahmad, Using design of mixture experiments to optimize triaxial ceramic tile compositions incorporating Cambodian clays, Applied Clay Science. 87 (2014) 97–107.

[24] S.. Correia, D. Hotza, A.. Segadães, Simultaneous optimization of linear firing shrinkage and water absorption of triaxial ceramic bodies using experiments design, Ceramics International. 30 (2004) 917–922.

[25] S.L. Correia, K.A.S. Curto, D. Hotza, A.M. Segadães, Using statistical techniques to model the flexural strength of dried triaxial ceramic bodies, Journal of the European Ceramic Society. 24 (2004) 2813–2818.

[26] S.L. Correia, G. Dienstmann, M. V Folgueras, A.M. Segadaes, Effect of quartz sand replacement by agate rejects in triaxial porcelain., Journal of Hazardous Materials. 163 (2009) 315–22.

[27] M. Dal Bó, A.M. Bernardin, D. Hotza, Formulation of ceramic engobes with recycled glass using mixture design, Journal of Cleaner Production. 69 (2014) 243–249.

[28] D. Jurków, J. Stiernstedt, Investigation of High Temperature Co-fired Ceramics sintering conditions using Taguchi Design of the experiment, Ceramics International. 40 (2014) 10447–10455.

[29] M.-L. Huang, Y.-H. Hung, Z.-S. Yang, Validation of a method using Taguchi, response surface, neural network, and genetic algorithm, Measurement. 94 (2016) 284–294.

[30] X. Yang, Flower Pollination Algorithm for Global Optimization, Unconventional Computation and Natural Computation 2012, Lecture Notes in Computer Science. 7445 (2012) 240–249.

[31] ASTM C326, Standard Test Method for Drying and Firing Shrinkages of Ceramic Whiteware Clays, ASTM International. (1982).